Pollution and Fish Health
in
Tropical Ecosystems

Pollution and Fish Health
in
Tropical Ecosystems

Editors

Eduardo Alves de Almeida
Depto. de Química e Ciências Ambientais
IBILCE–UNESP
Sao Jose do Rio Preto
Brazil

Ciro Alberto de Oliveira Ribeiro
Federal University of Paraná
Department of Cellular Biology
Curitiba, Paraná
Brazil

CRC Press
Taylor & Francis Group
Boca Raton London New York

CRC Press is an imprint of the
Taylor & Francis Group, an **informa** business
A SCIENCE PUBLISHERS BOOK

CRC Press
Taylor & Francis Group
6000 Broken Sound Parkway NW, Suite 300
Boca Raton, FL 33487-2742

© 2014 Copyright reserved
CRC Press is an imprint of Taylor & Francis Group, an Informa business

Library of Congress Cataloging-in-Publication Data

Pollution and fish health in tropical ecosystems / editors, Eduardo Alves de Almeida, Ciro Alberto de Oliveira Ribeiro.
       pages cm
  "A CRC title."
  Includes bibliographical references and index.
   ISBN 978-1-4822-1287-7 (hardcover : alk. paper) 1. Marine ecosystem health--Tropics. 2. Marine ecosystem health--Brazil. 3. Ecosystem health--Tropics. 4. Ecosystem health--Brazil. 5. Aquatic ecology--Tropics. 6. Aquatic ecology--Brazil. 7. Water--Pollution--Environmental aspects--Tropics. 8. Water--Pollution--Environmental aspects--Brazil. 9. Fishes--Effect of water pollution on--Tropics. 10. Fishes--Effect of water pollution on--Brazil. I. Almeida, Eduardo Alves de, editor of compilation. II. Ribeiro, Ciro Alberto de Oliveira, 1960- editor of compilation.
   QH541.5.W3P65 2014
   577.0913--dc23
                                                      2013038262

Visit the Taylor & Francis Web site at
http://www.taylorandfrancis.com

CRC Press Web site at
http://www.crcpress.com

Science Publishers Web site at
http://www.scipub.net

Foreword

Because a large number of chemicals directly used in our daily lives or generated by our cars and industries still have their final destination in the aquatic compartment of our planetary ecosystem, the health of small and large organisms living in lakes, rivers, estuaries, oceans and even below the Antarctic ice sheet should stay among our top priorities in protection and conservation measures to be developed by national agencies and international conventions. Healthy molluscs, crabs and fish mean high quality food for human consumption but also, and above all, healthy aquatic ecosystems where organisms in complex interactions with each other can reproduce and grow, and hence contribute to maintain the biodiversity of life on our small blue planet. Determining the health status of aquatic invertebrates and fish is a quite complex task currently addressed by a large world community of scientists sharing their technical skills and disciplinary knowledge to develop and apply indicators, markers and models to understand the behaviour of chemicals in water and sediment and to elucidate the responses of living cells and tissues to the aggression of xenobiotics.

This book is the best example ever of such an effort of dedicated scientists to share their expertises and recent results in an attempt to describe the effects of pollutants on fish health, with a special emphasis on tropical species. Fish physiology and water temperature are two governing factors that must be taken under consideration when modelling fish health in presence of toxicants. For a number of fish biomarkers using biochemical reactions such as enzyme kinetics or oxidative stress responses, temperature can be considered as a confounding factor when comparing toxic responses in fish living under tropical conditions to those experiencing high seasonal variations in northern countries. The use of sub-tropical and tropical fish species in experimental protocols is still an important challenge for South American fish toxicologists because most models have been developed for northern species and differences in fish physiology are raising important problems in the interpretation of toxicological data. What the authors are proposing in this book is a synthesis of the most important and challenging research topics related to fish toxicology, and more generally, to aquatic ecotoxicology where fish species are considered as a pivotal component of tropical ecosystems.

Emilien Pelletier,
Professor of Marine Ecotoxicology
Université du Québec à Rimouski, Qc, Canada

Preface

This book contains information obtained from authentic and highly regarded sources, with a wide variety of references listed.

Brazilian and other tropical aquatic ecosystems house the highest diversity and the largest number of aquatic organisms species in the world, but unfortunately these regions also tends to be responsible for the poorest investments in environmental research. Significant investment in research is necessary for understanding the ecological complexity in these ecosystems, and even more financial and scientific resources are required to estimate the real impact of human activities. The high degree of ecological complexity present in tropical aquatic ecosystems also means that there is a comparable level of fragility, making it more difficult to study the interactions between ecological factors and the increasing presence of chemicals. This book is focused on studies about species that are native to Brazil. The experiences of the individual authors have been brought together to make a collection within a chapter with a common goal. Thus, this volume is designed to be both a textbook and a general reference book. It includes the integration and synthesis of ecotoxicological research with tropical fish species. Topics from population to molecular effects of chemicals are covered, but the many complex pathways and mechanisms of chemically-induced disorders could only be briefly mentioned.

The effort of authors is not enough to reveal and explain the problems resulting from the chemicals present in the tropical and aquatic ecosystem. The authors recognize this challenge, and they have done their best to emphasize the importance of ecotoxicological studies as a tool for political statements in these regions.

Additionally, the authors are grateful to state and federal agencies for their financial support. Importantly, the authors would like to express a warmest thanks to the students who, through their academic studies, contributed to the basis of this book.

The editors

Contents

Foreword   v

Preface   vii

1. **Introduction—Pollution and Fish Health in Tropical**   1
   Ecosystems: A Brief Summary on Current Challenges
   and Perspectives
   Eduardo Alves de Almeida, Aline Cristina Ferreira Rodrigues and
   Ciro Alberto de Oliveira Ribeiro

2. **Genes and Proteins Related with Biotransformation in**   15
   Tropical Fishes: Perspectives in Aquatic Toxicology
   Juliano Zanette

3. **Nuclear Receptors in Fish and Pollutant Interactions**   35
   Afonso C.D. Bainy, Jacó J. Mattos and *Marília N. Siebert*

4. **Fish Neurotoxic Pollutants**   51
   Helena Cristina Silva de Assis and *Maritana Mela*

5. **Pollutants and Oxidative Stress**   84
   Francisco Filipak Neto

6. **Genotoxicity and Mutagenicity**   132
   Marta Margarete Cestari

7. **The Use of Fish Biomarkers in the Evaluation of Water**   164
   Pollution
   Thiago E.M. Parente and *Rachel Ann Hauser-Davis*

8. **Blood Parameters of Estuarine and Marine Fish as**   182
   Non-Destructive Pollution Biomarkers
   R. Seriani, D.M.S. Abessa, C.D.S. Pereira, A.A. Kirschbaum,
   L.D. Abujamara, L.M. Buruaem, C. Félix, G.C.R. Turatti,
   L.R.G.B. Prado, E.C.P.M. Sousa and *M.J.T. Ranzani-Paiva*

9. **Histopathological Markers in Fish Health Assessment**   206
   Ciro Alberto de Oliveira Ribeiro and *Marisa Fernandes Narciso*

10. **Emerging Contaminants and Endocrine System Dysfunction**   243
    Daniele Dietrich Moura Costa

11. **Nanoecotoxicology in Fish Species**   312
    Juliane Ventura-Lima, Alessandra Martins da Rocha,
    Marlize Ferreira-Cravo, André Luís da Rosa Seixas,
    Carmen Luiza de Azevedo Costa, Isabel Soares Chaves,
    Josencler Luis Ribas Ferreira, Rafaela Elias Letts,
    Lucas Freitas Cordeiro, Glauce R. Gouveia and *José María Monserrat*

12. **Effect of Pollutants on Condition Index**   338
    Larissa Paola Rodrigues Venancio and
    Claudia Regina Bonini Domingos

13. **Behavioral Biomarkers and Pollution Risks to Fish Health and Biodiversity**   350
    Paulo Sérgio Martins de Carvalho

Index   379

Color Plate Section   389

Introduction—Pollution and Fish Health in Tropical Ecosystems

A Brief Summary on Current Challenges and Perspectives

Eduardo Alves de Almeida,[1],* Aline Cristina Ferreira Rodrigues[1] and Ciro Alberto de Oliveira Ribeiro[2]

Pollutants and Fish Health—A Complex Relationship

Aquatic contaminants include classic chemicals such as pesticides, metals, polycyclic aromatic hydrocarbons, polychlorinated biphenyls, organic compounds from sewage, pulp mill effluents, and what is known as emerging contaminants or "old compounds with a new face" (these include pharmaceuticals, nanotechnology residues, anti-flame compounds and cyanotoxins, amongst others). These contaminants have been a cause for significant concern in recent years. In most cases, the impact caused by

[1]Department of Chemistry and Environmental Sciences, São Paulo State University, Rua Cristovao Colombo 2265, Sao Jose do Rio Preto SP, BRASIL CEP 15054-000.
[2]Cellular Toxicology Laboratory, Cellular Biology Department, C. Postal 19031, Federal University of Paraná, CEP: 81.531-980 Curitiba–PR Brazil.
Email: ciro@ufpr.br
*Corresponding author: ealmeida@ibilce.unesp.br

environmental contaminants depends on the contaminant's toxic effects on the exposed biota, leading to alterations in physiology that compromises the life cycles of organisms. The American Chemical Society reported the existence of an estimated 10 million substances; approximately 70,000 are used daily, but only 2,000 have toxic effects that are well characterized in the literature (Mozeto and Zagato 2006). Because the final destination of most anthropogenic pollutants is the aquatic environment, fish are largely affected by aquatic pollutants. They represent one of the most diverse class of aquatic vertebrates, with about 32,500 described species (Fish base 2013) inhabiting the most diverse aquatic ecosystems.

It has been substantially proven that the health status of fish is inherently dependent on the quality of the aquatic environment. Additionally, exposure to toxic compounds may lead to physiological disorders that are responsible for increases in their susceptibility to opportunistic pathogenic organisms such as parasites, leading to diseases that can accelerate mortality. Sub-lethal concentrations of pesticides (Cossarini-Dunier 1987), metals (O'Neill 1981), and sewage components (Secombes et al. 1991, 1992) have been documented to affect the immune system of fish, increasing their susceptibility to infections. Moreover, the adverse water quality resulting from the presence of pollutants is often reflected in oxygen depletion, changes in pH values, larger microbial populations, and higher than usual amounts of organic material in water, all of which can have potentially negative effects on fish health (Austin 1999). For this reason, the maintenance of appropriate aquatic environmental conditions is a multidisciplinary issue of primordial importance for the adequate preservation of fish species.

Fish represent a class of organisms with a great diversity of habits, and studies have shown that the impact of human activities on aquatic ecosystems is responsible for relevant shifts in species distribution and abundance (Pörtner et al. 2010). The effects of pollutants on fish depend on the physico-chemical characteristics of the environment, as well as on the trophic position, age, sex and physiological status of the affected fish. For this reason, studies on different fish species exposed to different classes of contaminants are fundamental for recognizing the susceptibility of different species to chemical exposure. All these aspects are essential for establishing policies that limit pollutant concentrations in aquatic ecosystems to levels that better represent the water quality requirements of different fish species. The inherent complexity of this topic represents an enormous challenge for aquatic toxicologists, especially if one considers the high variety of biotic and abiotic factors that account for optimal environmental conditions for each fish species (i.e., temperature, salinity, dissolved oxygen, food availability, pH, hardness, suspended organic matter, as well as the reproductive cycles, body composition and physiology of the organisms). The problem becomes even more complex when one considers the fact that pollutants are often present in the environment at very complex mixtures that make it difficult to investigate exposure risks. The responses can vary significantly from one fish to another or even from one tissue to another in the same fish (Jordan et al. 2012). This variation makes it difficult to interpret the data and establish typical or expected responses of different species to specific contaminants.

Fish toxicologists have been able to elucidate the mechanisms of toxicity of many toxic compounds, especially after the development of OMICs technologies, which currently allows for the monitoring of thousands of genes and proteins that have been altered by pollutant exposure. Nevertheless, a series of new challenges has emerged in recent years, especially with respect to complex mixtures, emerging contaminants, and climate changes.

The challenge of complex mixtures

Many studies attempt to understand the effects of isolated contaminants on aquatic organisms, and they propose sophisticated mechanisms of action for different chemicals. However, it is known that many of the isolated effects that a contaminant has on fish are quite different when this contaminant is present in combination with other compounds. In other words, the interactions between different classes of contaminants can generate responses or effects in fish that are completely distinct from those observed when the contaminant is present in isolation under natural or experimental conditions. One compound can make the effect of another more powerful. A compound can also generate additive or subtractive effects of another compound. It should be considered that the toxic action of any compound is completely dependent on all of the factors that govern the toxicokinetics and toxicodynamics of xenobiotics in the organism; the toxicity depends on an equation that includes the mechanisms of absorption, distribution, disposition, biotransformation and excretion of the compound. Moreover, all these factors affect the bioaccumulation factor and the compound's residence time in the organism, which is also related to the composition of tissues and fluids in the organism (blood volume, fat and protein contents, etc.) as well as to the organism's physiology (metabolic and cardiac rate, blood pressure, reproductive cycles, etc.). These relationships clearly show that the alterations made to target cells or tissues by specific compounds significantly influence the mechanisms of action of other chemicals that may also be present in a contaminated aquatic environment.

An example to illustrate these interactions was recently published by Trídico et al. (2010). In this study, fish were exposed to mixtures of the polycyclic aromatic hydrocarbon benzo[*a*]pyrene (BaP) and the organophosphate pesticide diazinon. BaP is a well-known inducer of cytochrome P450 1A (CYP1A) in fish. In most studies in which fish are exposed to BaP, CYP1A induction consistently occurs. However, thionic organophosphate pesticides can have a contrary effect, due to the binding of the sulfur atom from the pesticide to the catalytic center of the enzyme, a reaction which significantly inhibits enzyme activity. Thus, though increases in CYP1A are expected when fish are exposed to BaP, this response is less dramatic if organophosphate pesticides are present in combination with BaP in the environment. According to Teuschler (2007), the assessment of the risks that pollutant mixtures pose to exposed organisms often involves only simple notions of additivity and is often based on determinations of toxicological similarity or dissimilarity among the components of the mixture. Nevertheless, the chemical variability in the composition of a complex mixture can be responsible for many atypical or unexpected responses from the exposed organism due to chemical interactions between the different compounds present in the mixture, including the diversity of synergistic, antagonistic or even additive effects. Thus, knowledge about the toxic effects of the components of mixtures does not always make it easier to interpret the observed effects. New methods are necessary if researchers want to better estimate the effects of complex mixtures on fish. Indeed, the development of such new methods requires careful design and conduct of toxicological research to prove or disprove their utility and establish the boundaries for their credible application.

The challenge of emerging contaminants

For many years, environmental toxicologists have been engaged in studies on the effects of relevant or more significant aquatic contaminants in an attempt to identify substances that deserve immediate regulation, such as metals, pesticides, polycyclic aromatic hydrocarbons, and polychlorinated biphenyls. The identification of these substances would establish concentration limits that could be considered suitable for preserving the status of the aquatic environment. In theory, these studies would provide support for the regulatory agencies to create effective legislation that would regulate the concentrations of contaminants in different types of aquatic ecosystems.

More recently, new concerns have arisen from the finding that much of the compounds that were previously considered to be non-priority can also pose risks to the aquatic biota, generally after long term exposures and at very low concentrations. These chemicals, which lack regulations in most countries and whose effects on the environment are still unknown, have been termed "Emerging Contaminants" ECs (Deblonde et al. 2011). They include the phthalates, bisfenol A, pharmaceuticals, nanoengineered materials, flame retardants and personal care products, among others (see Box 1.1).

Box 1.1. Descriptions and main characteristics of some emergent contaminants.	
Phtalates	Often called "plasticizers," phtalates are a group of chemicals used to make plastics more flexible and harder to break. They are used in hundreds of products, such as vinyl flooring, adhesives, detergents, lubricating oils, automotive plastics, plastic clothes (raincoats), and personal-care products (soaps, shampoos, hair sprays, and nail polishes). According to Deblonde et al. (2011), phthalates have been used for 50 years and 3 million tons are produced per year around the world.
Bisfenol A	Bisfenol A (BPA) is also used to make certain plastics and epoxy resins; it has been in commercial use since 1957 and around 3.6 million tons (8 billion pounds) of BPA are used by manufacturers annually.
Pharmaceuticals	Pharmaceuticals include synthetic hormones, anti inflammatory, anti epileptic medications, statins, antidepressants, beta blockers, and antibiotics (Miège et al. 2009). After use by humans, they are excreted unaltered or metabolized through urine or feces into sewage (Kim et al. 2007; Roberts and Thomas 2006). The world per capita consumption of pharmaceuticals is 15 g, and this value is 50g–150 g in developed countries (Zhang et al. 2008), a data that highlights the relevance of the environmental problems caused by these compounds.
Nanomaterials	Nanomaterials (NM) are generally described as materials that have at least one dimension on the scale of 1–100 nm (National Nanotechnology Initiative 2000). NMs are used commercially in products such as electronic components, scratch-free paint, sports equipment, cosmetics, food color additives, and surface coatings (Mahmoudi et al. 2011). The number of NM-based publications has increased significantly over the years; however, the majority of publications are focused on the synthesis and development of novel nanomaterials, and less than one percent have focused on the biological impact of NMs (Sharifi et al. 2011).
Flame retardants	Flame retardants are substances used in plastics, textiles, electronic circuits and other materials in order to prevent fires, and are generally composed of brominates organic compounds. The total world production of all brominates flame retardants in 1992 was estimated to be approximately 150,000 metric tons/year (Wit 2002).
Personal care products	Personal care products (PCPs) are a diverse group of compounds used in soaps, lotions, toothpaste, fragrances, sunscreens, and many other products used daily. The primary classes of PCPs include disinfectants, fragrances, insect repellants, preservatives, and UV filters, and are among the most commonly detected compounds in surface waters throughout the world (Peck 2006; Brausch and Rand 2011).

Around the world, emerging contaminants have been produced and consumed by humans on a large scale in recent years. As a consequence, the concentration of these chemicals in aquatic environments is rapidly increasing. Many of these chemicals can have negative effects on fish, even at very low concentrations, and especially on their endocrine systems.

The endocrine system of organisms is prepared to respond to biological signals that are mediated by neurotransmitters and hormones at very low concentrations, so if a chemical has properties that allow it to interact with receptors so that the endocrine system can be activated or blocked, an endocrine disruption effect occurs; this will be discussed at length in Chapter 10.

Again, the issues regarding contaminant mixtures also include emerging contaminants, since the main sources of these compounds in the aquatic environment are often present in a very complex mixture. In accordance with this, Yan et al. (2013) showed that goldfish exposed to 17β-estradiol (E2), a natural hormone, for 10 days presented significant increases in aromatase activity, an effect that was suppressed when fish were exposed to E2 in combination with ketoconazole. In addition, ketoconazole alone significantly inhibited CYP1A activity, but it had an opposite effect when the fish were exposed to it and E2 at the same time. In another study, Yan et al. (2012) evaluated the effect of E2 and BaP exposure on goldfish, and observed that E2 increased the expression of vitellogenin while BaP increased the expression of CYP1A. When BaP at low concentration and E2 were combined, a significant reduction in vitellogenin and CYP1A was observed. These examples are consistent with the findings on emerging contaminants (E2), and they illustrate diverse effects when other chemicals (diazinon, ketoconazole and BaP) are presented in a mixture.

The challenge of climate changes

Recent literature has shown that world temperatures have risen over the last 20 years, both in air (Tett et al. 1999) and water (Barnet et al. 2005), driven by human activities, and that these changes may rapidly alter global ecosystems. The ability of aquatic organisms to adapt to these increases is limited, and there is recent evidence that global warming has caused some species to exceed their adaptation capacities, leading to population declines (Schiedel et al. 2007). Additionally, studies on a variety of freshwater fish species have shown that the upper temperature tolerance limits are lowered in the presence of certain organic chemicals (Cossins and Bowler 1987; Patra et al. 2007). The inverse is also true; that is to say, it has been shown that temperature increase can also increase the toxicity of many chemical compounds (Schiedek et al. 2007). Rising temperatures often cause increases in the rate of uptake of contaminants due to increases in ventilation, in response to an increased metabolic rate and decreased oxygen concentration in the water (Kennedy and Walsh 1997). Moreover, Sollid et al. (2003) have shown that, under hypoxic conditions, the surface areas of the gills increases 7.5-fold with a concomitant decrease in cell mass in order increase oxygen uptake due to a higher area for oxygen diffusion and a thinner barrier of cells to support oxygen diffusion. In accordance with all of these physiological changes, Terzi and Verep (2012), for example, showed that the toxicity of mercury chloride in rainbow trout increases as the water temperature increases. As for the effects of variations on oxygen concentration, Fleming and Di Giulio (2012) showed a severe increase in the toxicity of mixtures of different polycyclic aromatic hydrocarbons in zebrafish when the fish were forced into a state of hypoxia, compared to fish exposed to the same pollutants, but under a state of normoxia.

Global warming also affects climate variables that affect the dynamics of many chemicals in the environment. Changes in winds, precipitation and ocean currents, for example, alter the transport, transfer and deposition of chemicals (Macdonald et al. 2005), which, in turn, alters the dispersion of the pollutants in the environment. Increases in water temperature can also modify the partitioning of chemicals between sediment and water, or between suspended particles and water, a situation which alters the bioavailability of chemicals to fish. Another consequence linked to global warming is oceanic acidification due to increasing atmospheric CO_2 levels. Increases in atmospheric carbon dioxide levels shift the carbon dioxide-carbonate equilibrium in water toward acidic direction, causing the acidification of oceans (Nikinmaa 2013). It seems obvious that the acidification of oceans represents a serious problem for fish, which must adjust their physiology to cope with this new situation. Species that are less tolerant to changes in water pH may not be able to adapt to changes imposed by oceanic acidification, which would make them more susceptible to environmental contaminants. In addition, ocean acidification can affect the speciation of metals (Millero et al. 2009) by changing their toxicity. Though still of recent origin, this situation is quite worrying, and much study is required on this subject. It currently represents a very relevant challenge for aquatic toxicologists, since climate change is taking place very rapidly.

The Lack of More Studies on Tropical Fish Species

Despite the numerous challenges to studies on fish toxicology, scientists have made significant efforts to identify and propose model species to predict the pollution risks for fish. The studies on species from temperate areas have been more frequent, despite the lower diversity of species found in this region of the planet. The main reason for this tendency is likely the better development of environmental toxicology science in developed countries. Some examples of fish that are frequently used as models in toxicology studies are the common carp (*Cyprinus carpio*), the rainbow trout (*Oncorhynchus mykiss*), the Japanese killfish (*Oryzias latipes*), the European sea bass (*Dicentrarchus labrax*), and the goldfish (*Carassius auratus*). Nevertheless, important tropical fish species have been also adopted as relevant model organisms in toxicological studies, including the zebrafish (*Danio rerio*) and the Nile tilapia (*Oreochromis niloticus*). When choosing a model fish species for a toxicological study, several aspects should be considered, especially "the question to be answered." A very interesting discussion on this subject was published in 2007 in the journal *Zebrafish* (DOI: 10.1089/zeb.2006.9998). The paper, entitled "Fish Models in Toxicology," presents a roundtable discussion among very relevant fish toxicologists on important aspects to be considered when choosing fish species to use in toxicology experiments. The paper starts with the following six questions:

1. *What species should we consider to be fish models for toxicology, and why? Consider both laboratory and extra-laboratory fish species.*
2. *Are the current fish models for assessing the impact of toxic chemicals adequate to represent the vast majority of ecologically important fish species, given potential differences in feeding, reproductive strategies, or other life history parameters?*
3. *How is the development of -omic technologies going to advance our understanding of fish ecotoxicology? What further developments are going to promote this advancement?*
4. *What is needed to make research involving fish models for human environmental health and toxicology accepted, when the frequently heard comment is: "There are perfectly good rodent models" for many human diseases?*
5. *In what ways has research using fish models enhanced our understanding of human environmental health and toxicology, and what are the areas in which fish models are going to have the biggest impact in the future?*
6. *What tools do we need to ensure that research using fish models will make an impact on human environmental health and toxicology? Alternatively, what tools do we already have that are poised to support this?*

All of these questions are undoubtedly important when we are searching for a *model* organism; that is, an organism that should be extensively studied in order to understand particular biological effects of contaminants, with the expectation that discoveries made in this model organism will provide insight on the effects of the same contaminant on other organisms, even when the organisms are not from the same species. In most toxicological studies on fish, this is an important aspect to be considered when scientists are trying to establish the typical responses of one representative fish species to specific contaminants or chemical mixtures that could be extrapolated to other fish species. Although this approach has the advantage of eliciting important conserved responses from fish after their exposure to pollutants, it fails when the expected effect varies as a result of changes in the ecosystem or due to physiological particularities of different species. Even defending the use of fish models in toxicology,

Michael J. Carvan in the above mentioned roundtable discussion pointed that most model fish species cannot represent the great diversity of fishes: "there can be dramatic differences in physiology that would complicate the extrapolation mechanism to many of the world's fishes as it relates to the validity of biomarker or the assessment of susceptibility to specific toxicants. This is where comparative studies become so important". If the high diversification of niches and habits among the numerous fish species is considered, some species will be more or less tolerant to fluctuations in many physical-chemical parameters, including dissolved oxygen, salinity, temperature, pH and ammonia levels. Additionally, other species will present different metabolic rates, cardiac frequencies or even better molecular and biochemistry machinery to deal with the toxicants.

An interesting review by Hall and Anderson (1995) on the effect of salinity on the toxicity of chemicals present in aquatic ecosystems in numerous aquatic species, including fish, showed a large variability in pollutant toxicity depending on the environmental conditions. In general, most of the tested pollutants increased in toxicity as the salinity decreased in most of the tested species. A decrease in salinity is associated with decreases in pollutant solubility, which alters the pollutant's bioavailability. Thus, it is expected that freshwater and estuarine fish will experience higher toxicity from pollutants compared to marine fish exposed to the same pollutant at the same concentration.

Another example involves anoxia-tolerant fish species. Recently, experiments were performed using two species of fish with distinct habits: the nektonic Nile tilapia (*Oreochromis niloticus*) and the benthonic armored suckermouth catfish (*Pterygoplichthys anisitsi*) (Fig. 1.1) (unpublished data). The fish were exposed to 2 mg/L of the organophosphate pesticide diazinon for two and seven days. While the catfish survived for seven days to this exposure, the tilapias died after 24 hours, results which clearly demonstrate the greater susceptibility of tilapias to diazinon at this concentration. This data confirms that different species tend to present different responses to chemicals dissolved in water, and this difference is certainly due to the numerous biological factors inherent to each species.

Fish absorb lipophilic pollutants through epithelial surfaces of the skin, gills and intestine (Wood 2001). Thus, the morphological characteristics of these two species may explain the greater tolerance that the armored suckermouth catfish has to water contamination. There is a relationship between the morphometric parameters of the gills and fish habits. In general, fast-swimming fish have a larger gas exchange surface, while slow-moving fish and air-breathing fish have reduced gas exchange surfaces (Wilson and Laurent 2002). More active fish require more energy and, consequently, a greater supply of molecular oxygen. When the lamellar surface of the gills of the two species, *P. anisitsi* and *O. niloticus*, are compared, it can be

Figure 1.1. The armored suckermouth catfish *Pterygoplichthys anisitsi* (A) and the Nile tilapia *Oreochromis niloticus* (B).

noted that the lamellar surface of *P. anisitsi* is reduced, which means that this species has a smaller area for absorption of oxygen and therefore for contaminants as well. Tilapias swim actively, while the armored catfish stays at the bottom of the aquarium and only occasionally rises to the water's surface to get air. Cruz reported that, under normoxia, most catfish species do not breathe air during the day, and they absorb oxygen directly from the water; however, they usually swim to the surface to capture air during night or during cases of hypoxia. *P. anisitsi* also come up for air during the day. Oliveira et al. (2001) and Cruz et al. (2009, 2012) showed that this species comes up for air even when experiencing normoxia, and showed that the stomach of this species is anatomically and histologically modified to capture air and absorb oxygen, which contributes to the breathing process of the species. According to Val (1995), Mattias et al. (1996), and Graham (1997), several species of Loricariidae were found to have anatomical modifications in the digestive tract for accessorial oxygenation. In the case of *O. niloticus*, breathing is exclusive to the gills, ensuring an efficient use of the oxygen dissolved in the water (Fernandes 1996), and thus justifying their large epithelial surface.

A comparative picture of the gills of the tilapia and the catfish is shown in Fig. 1.2. The picture shows that the tilapia has a significantly larger gill surface when compared to the catfish. This difference implies a lower surface area for pollutant absorption in the catfish, and, when combined

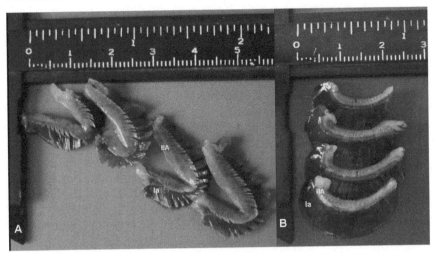

Figure 1.2. Gills of the fish (A) *Pterygoplichthy sanisitsi.* (B) *Oreochromis niloticus.* BA: branchial arch. la: lamella.

with the lower rate of ventilation in the catfish's gills due to accessory respiration in the stomach, even less pollutant intake will occur in this species through the gill. Moreover, it should be also considered that the body of *P. anisitsi* is covered by a longitudinal series of large bony plates (Nico et al. 2009), instead of the squame and thick skin that are typical of the tilapia. These factors—lower surface area, less ventilation through the gills, and the presence of bony plates in the body—may all work together to minimize the absorption of the contaminants in the environment, and they may explain the ability of the armored suckermouth catfish to tolerate more environmental disturbances and contamination.

These were only two examples to illustrate how variable the responses of different fish to pollutants can be, and how changes to the environment can also account for changes to pollutant toxicity. Discovering how all the environmental variables imaginable can influence the effects of pollutants on different fish species is another key challenge to overcome, especially in tropical zones, which contain the greatest diversity and richness of fish species in the world. Considering only freshwater environments, neotropical ichthyofauna are represented by approximately 4,500 described species (Reis 2003). Therefore, any attempt to elucidate the influences that all of these factors can have on tropical fish responses to pollutants is valid. It would contribute to a better understanding of the dynamics of pollutant effects in these organisms, and it would generate important information for regulatory policies, such as specific legislation for the protection of tropical aquatic ecosystems.

There is a variety of published books that focus on fish responses to environmental pollutants, such as *The Toxicology of Fishes*, edited by Di Giulio and Hinton (2008). Nevertheless, articles about the effects of pollutants on tropical fish are still limited in the literature, and this fact encouraged us to edit this book. Thus, the main goal of the current book is to bring together relevant South American fish toxicologists to present their research and studies on tropical fish species. The book was planned in order to compile both general aspects and the state of the art regarding the effects of pollutants on fish health, with a special emphasis on tropical species. The 13 chapters that make up this book were organized following a crescent biological level of organization that begins with the first line of molecular and cellular responses to xenobiotics in fish and ends with the effects at higher structural levels, such as tissue morphology and fish behavior. Current and relevant topics regarding the toxicity of emerging contaminants (Chapter 10) and nanomaterials (Chapter 11) are also discussed.

Acknowledgements

E.A. Almeida and C.A. Oliveira Ribeiro would like to thanks for the Brazilian funding agencies CAPES, FAPESP and CNPq.

Keywords: Fish health, pollution, tropical ecosystems, climate changes, contaminant mixtures, emerging contaminants

References

Austin, B. 1999. The effects of pollution on fish health. J. Appl. Microbiol. 85: 234S–242S.

Barnett, T.P., D.W. Pierce, K.M.A. Rao, P.L. Gleckler, B.D. Santer, J.M. Gregory and W.M. Washington. 2005. Penetration of human-induced warming into the world's oceans. Science 309: 284–287.

Brausch, J.M. and G.M. Rand. 2011. A review of personal care products in the aquatic environment: environmental concentrations and toxicity. Chemosphere 82: 1518–1532.

Carvan, M.J., E.P. Gallagher, A. Goskoyr, M.E. Hahn and D.G. Joakim Larssom. 2007. Fish models in toxicology. Zebrafish 4(1): 9–20.

Cossarini-Dunier, M. 1987. Effect of the pesticides atrazine and lindane and of manganese ions on cellular immunity of carp, *Cyprinus carpio*. J. Fish Biol. 31: 67–73.

Cossins, A.R. and K. Bowler. 1987. Temperature biology of animals. Chapman and Hall, London.

Cruz, A.L. 2007. O comportamento respiratório e a cascata de O_2 no cascudo de respiração bimodal *Pterygoplichthys anisitsi* Eigenmann e Kennedy, 1903 (Teleostei, Loricariidae). PhD Thesis. São Carlos: Universidade Federal de São Carlos. 161 p.

Cruz, A.L., A.C. Pedretti and M.N. Fernandes. 2009. Stereological estimation of the surface area and oxygen diffusing capacity of the respiratory stomach of the air-breathing armored catfish *Pterygoplichthys anisitsi* (Teleostei: Loricariidae). J. Morphol. 270: 601–614.

Deblonde, T., C. Cossu-Leguille and P. Hartemann. 2011. Emerging pollutants in wastewater: a review of the literature. Int. J. Hyg. Environ. Health 214: 442–448.

Di Giulio, R.T. and D.E. Hinton. 2008. The toxicology of fishes. 1st ed. CRC Press, Boca Raton, USA, 1071 p.

Fishbase. 2013. http://www.fishbase.org/search.php.

Fleming, C.R. and R.T. Di Giulio. 2011. The role of CYP1A inhibition in the embryotoxic interactions between hypoxia and polycyclic aromatic hydrocarbons (PAHs) and PAH mixtures in zebrafish (*Danio rerio*). Ecotoxicology 20(6): 1300–1314.

Hall, L.W. and R.D. Anderson. 1995. The influence of salinity on the toxicity of various classes of chemicals to aquatic biota. Crit. Rev. Toxicol. 25(4): 281–346.

Jordan, J., A. Zare, L.J. Jackson, H.R. Habibi and A.M. Weljie. 2012. Environmental contaminant mixtures at ambient concentrations invoke a metabolic stress response in goldfish not predicted from exposure to individual compounds alone. J. Proteome Res. 11(2): 1133–1143.

Kennedy, C.J. and P.J. Walsh. 1997. Effects of temperature on xenobiotic metabolism. *In*: C.M. Wood and D.G. McDonald (eds.). Global Warming—Implications for Freshwater and Marine Fish, Cambridge University Press, Ontario, Canada, pp. 303–324.

Kim, S.D., J. Cho, I.S. Kim, B.J. Vanderford and S.A. Snyder. 2007. Occurrence and removal of pharmaceuticals and endocrine disruptors in South Korean surface, drinking and waste waters. Water Res. 41: 1013–1021.

Macdonald, R.W., T. Harner and J. Fyfe. 2005. Recent climate change in the Artic and its impact on contaminant pathways and interpretation of temporal trend data. Sci. Tot. Environ. 342: 5–86.

Mahmoudi, M., K. Azadmanesh, M.A. Shokrgozar and S. Laurent. 2011. Effect of nanoparticles on the cell life cycle. Chem. Rev. 111: 3407–3432.

Mattias, A.T., S.E. Moron and M.N. Fernandes. 1996. Aquatic respiration during hypoxia of the facultative air-breathing *Hoplerythrinus unitaeniatus*: a comparison with the water-breathing *Hoplias malabaricus*. *In*: A.L. Val, V.M. Almeida-Val and D.J. Randall (eds.). Physiology and Biochemistry of the Fishes of the Amazon. INPA, Manaus 402 p.

Miège, C., J.-M. Choubert, L. Ribeiro, M. Eusèbe and M. Coquery. 2009. Le devenir des résidus pharmaceutiques dans les stations d'épuration d'eaux usées. Une synthèse de la littérature. Tech. Sci. Méth. 11 : 75–94.

Millero, F.J., R. Woosley, B. Ditrolio and J. Waters. 2009. Effect of ocean acidification on the speciation of metals in seawater. Oceanography 22: 72–85.

Mozeto, A.A. and P.A. Zagatto. 2006. Introdução de agentes químicos no ambiente. *In*: P.A. Zagatto and E. Bertoletti (eds.). Ecotoxicologia Aquática. Rima Editora, São Carlos, Brasil 464 p.

National Science and Technology Council. 2000. National nanotechnology initiative: the initiative and its implementation plan. USA.

Nico, L.G., H.L. Jelks and T. Tuten. 2009. Non-native suckermouth armored catfishes in florida: description of nest burrows and burrow colonies with assessment of shoreline conditions. ANSRP Bull. 9: 1–30.

Nikinmaa, M. 2013. Climate change and ocean acidification—Interactions with aquatic toxicology. Aquat. Toxicol. 126: 365–372.

Oliveira, C., S.R. Taboga, A.L. Smarra and G.O. Bonilla-Rodriguez. 2001. Microscopical aspects of accessory air breathing through a modified stomach in the armoured catfish *Liposarcus anisitsi* (Siluriformes, Loricariidae). Cytobios 105: 153–162.

O'Neill, J.G. 1981. The humoral immune response of *Salmo trutta* L. and *Cyprinus carpio* L. exposed to heavy metals. J. Fish Biol. 19: 297–306.

Patra, R.W., J.C. Chapman, E.P. Lim and P.C. Gehrke. 2007. The effects of three organic chemicals on the upper thermal tolerances of four freshwater fishes. Environ. Toxicol. Chem. 26: 1454–1459.

Peck, A.M. 2006. Analytical methods for the determination of persistent ingredients of personal care products in environmental matrices. Anal. Bioanal. Chem. 386: 907–939.

Pörtner, H.O., P.M. Schulte, C.M. Wood and F. Schiemer. 2010. Niche dimensions in fishes: an integrative view. Physiol. Biochem. Zool. 83(5): 808–826.

Reis, R.E., S.O. Kullander and C.J. Ferraris. 2003. Check list of the freshwater fishes of South and Central América. 1 ed. Porto Alegre: EDIPUCRS. 729 p.

Roberts, P.H. and K.V. Thomas. 2006. The occurrence of selected pharmaceuticals in wastewater effluent and surface waters of the lower Tyne catchment. Sci. Total Environ. 356: 143–153.

Schiedek, D., B. Sundelin, J.W. Readman and R.W. Macdonald. 2007. Interactions between climate change and contaminants. Mar. Pollut. Bull. 54(12): 1845–156.

Secombes, C.J., T.C. Fletcher, J.A. O'Flynn, M.J. Costello, R. Stagg and D.F. Houlihan. 1991. Immunocompetence as a measure of the biological effects of sewage sludge pollution in fish. Comp. Biochem. Physiol. 100C: 133–136.

Secombes, C.J., T.C. Fletcher, A. White, M.J. Costello, R. Stagg and D.F. Houlihan. 1992. Effects of sewage sludge on immune response in the dab, *Limanda limanda* (L.). Aquat. Toxicol. 23: 217–230.

Sharifi, S., S. Behzadi, S. Laurent, M.L. Forrest, P. Stroeve and M. Mahmoudi. 2012. Toxicity of nanomaterials. Chem. Soc. Rev. 41: 2323–2343.

Sollid, J., P. De Angelis, K. Gundersen and G.E. Nilsson. 2003. Hypoxia induces adaptive and reversible gross morphological changes in crucian carp gills. J. Exp. Biol. 206(20): 3667–3673.

Terzi, E. and B. Verep. 2012. Effects of water hardness and temperature on the acute toxicity of mercuric chloride on rainbow trout (*Oncorhynchus mykiss*). Toxicol. Ind. Health 28(6): 499–504.

Tett, S.F.B., P.A. Stott, M.R. Allen, W.J. Ingram and J.F.B. Mitchell. 1999. Causes of twentieth-century temperature change near the Earth's surface. Nature 339: 569–572.

Teuschler, L.K. 2007. Deciding which chemical mixtures risk assessment methods work best for what mixtures. Toxicol. Appl. Pharmacol. 223: 139–147.

Trídico, C.P., A.C. Ferreira Rodrigues, L. Nogueira, D.C. da Silva, A. Benedito Moreira and E.A. Almeida. 2010. Biochemical biomarkers in *Oreochromis niloticus* exposed to mixtures of benzo[a]pyrene and diazinon. Ecotoxicol. Environ. Saf. 73(5): 858–863.

Val, A.L. 1995. Oxygen transfer in fish: morphological and molecular adjustments. Braz. J. Med. Biol. Res. 28: 1119–1127.

Wilson, J.M. and P. Laurent. 2002. Fish gill morphology: inside. J. Exp. Zool. 293: 192–213.

Wit, C.A. 2002. An overview of brominated flame retardants in the environment. Chemosphere 46: 583–624.

Wood, C.M. 2001. Toxic responses of the gill. *In*: D. Schlenk and W.H. Benson (eds.). Target Organ Toxicity in Marine and Freshwater Teleosts. Taylor & Francis, New York 416 p.

Yan, Z., G. Lu and J. He. 2012. Reciprocal inhibiting interactive mechanism between the estrogen receptor and aryl hydrocarbon receptor signaling pathways in goldfish (*Carassius auratus*) exposed to 17β-estradiol and benzo[a]pyrene. Comp. Biochem. Physiol. C Toxicol. Pharmacol. 156(1): 17–23.

Yan, Z., G. Lu, D. Wu, Q. Ye and Z. Xie. 2013. Interaction of 17β-estradiol and ketoconazole on endocrine function in goldfish (*Carassius auratus*). Aquat. Toxicol. 132–133C: 19–25.

Zhang, Y., S.U. Geissen and C. Gal. 2008. Carbamazepine and diclofenac: removal in wastewater treatment plants and occurrence in water bodies. Chemosphere 73(8): 1151–1161.

CHAPTER 2

Genes and Proteins Related with Biotransformation in Tropical Fishes

Perspectives in Aquatic Toxicology

Juliano Zanette

Introduction

The recent advances in molecular biology and biochemistry have led to important discoveries and new insights in the field of aquatic toxicology. Fishes that are used in laboratory (e.g., zebrafish *Danio rerio*) and environmental studies (e.g., North American killifish *Fundulus heteroclitus*) are considered model organisms for such purposes. This chapter aims to review the current literature regarding some genes and proteins involved in the biotransformation of organic molecules in well-known model fishes, and exploring how the new discoveries could be important and applicable to a wide range of South American tropical fish, with a major focus on the study of cytochrome P450 of family 1 (CYP1s) gene/proteins and cyprinodontiforme fishes.

Instituto de Ciências Biológicas (ICB), Universidade Federal do Rio Grande (FURG), Av. Itália Km 8, Rio Grande, RS, 96208-060, Brazil.
Email: julianozanette@furg.br; biozanette@hotmail.com

The Diversity of Biotransformation Proteins and Related Genes in Zebrafish

An important issue to be considered in the study of contaminant biotransformation in fishes is the diversity and the current nomenclature employed for proteins and genes involved in those biochemical processes. As one genome after another becomes sequenced, it has been imperative to consider in the gene/protein annotations the complexity of genes, genetic architecture, gene expression and evolutionary relatedness across species (Nebert and Wain 2003). It is also important to consider that a huge diversity of protein isoforms are involved in the biotransformation of xenobiotics, and those proteins are grouped into super families and/or families composing a unique "set" of isoforms in a given animal species. In humans, for example, 57 known Cytochrome P450s (CYP) genes and 33 pseudogenes are arranged into 18 families and 42 subfamilies, and a fraction of those CYPs are key enzymes involved in the oxidative or reductive biotransformation of xenobiotics (Nebert and Russell 2002). Other important redox enzymes are the Flavin-containing monooxygenases (FMO), aldo-keto reductases (AKR), epoxide hydrolase (EPHX), NAD(P)H-quinone oxidoreductase (NOQ1) and aldehyde dehydrogenases (ALDH). In contrast to the CYPs, much less is known about many of the substrates of those mentioned redox enzymes, even in humans (Goldstone 2008). The list of biotransformation proteins also includes conjugative enzymes such glutathione-*S*-transferases (GST), microsomal glutathione *S*-transferases (MGST or MAPEG), sulfotransferases (SULT), UDP-glucuronosyl transferases (UGT), *N*-acetyl transferases (NAT) and the membrane ATP binding cassette (ABC) efflux transporters (Goldstone 2008).

The Table 2.1 lists some of those gene/proteins and the proposed nomenclature used for superfamily, family, subfamily and classes, depending on the group considered (Human and Zebrafish). Zebrafish belongs to the order Cypriniformes that also includes economically important species, such as the carp *Cyprinus carpio*. Unfortunately, there are no native species of Cypriniformes inhabiting the South and Central America tropical environment according to the Brazilian National Museum (Buckup and Menezes 2003) and the Fish Base (Froese and Pauly 2012). Therefore, some differences could be expected in the gene/proteins of tropical fishes that inhabit these regions, compared with the model zebrafish, due to the huge phylogenetic distances between these species. The superorders Cypriniformes and Cyprinodontiformes were separated about 290 million years ago; the Cypriniformes and Siluriformes about 290 million years ago; while Cypriniformes and Perciformes were separated more recently, around 144 million years ago (Steinke et al. 2006). Future studies in model species from other orders that are better represented in the tropical environment

Table 2.1. Expected number of genes involved in the biotransformation of organic contaminants in *Homo sapiens* (Hs) and zebrafish *Danio rerio* (Dr). The numbers were estimated by other studies using bioinformatic tools and the current genomic and transcriptomic public data available for Hs and Dr (e.g., Genbank). Pseudo genes were not included in the number count. * and + denotes families and subfamilies (classes in the case of GSTs) that exist only in zebrafish but not in humans or that exist only in humans but not in zebrafish, respectively. Nf = not found.

Function	Super family	Family	Sub Family or Class	Hs	Dr	Reference
Oxidative	CYP	1	a,b, c*,d*	3	5	(Goldstone 2008; Goldstone et al. 2010)
		2	u, r, a+, b+, c+, d+, e+, f+, j+, s+, w+, y*, k*, n*, p*, v*, x*, aa*, ad*, ae*	16	48	(Goldstone et al. 2010)
		3	a, c*	4	5	(Goldstone et al. 2010)
		4	v, f, b+, z+, x+, a+,t*	12	4	(Goldstone et al. 2010)
	FMO	1-5	Nf	6	4	(Goldstone 2008; Rodriguez-Fuentes et al. 2008)
	ALDH	1-9, 16, 18	Nf	19	21	(Sophos and Vasiliou 2003; Goldstone 2008; Zhou et al. 2012)
Conjugative	GST	GST	Alpha, Kappa, Mu, Pi, Theta, Omega, Zeta, Sigma, Rho*	21	Nf	(Goldstone 2008; Li et al. 2010)
		MGST	1,2,3	6	7	(Frova 2006; Goldstone 2008)
	SULT	1, 2, 3*, 4, 6	Nf	13	16	(Goldstone 2008; Liu et al. 2010)
	UGT	1, 2, 5*	Nf	19	40	(Goldstone 2008; Huang and Wu 2010)
	NAT	1, 2	1a-d*, 2a-b*	2	6	(Sim et al. 2008)
Reductive	AKR	Nf	Nf	8	12	(Goldstone 2008)
	EPHX	Nf	Nf	2	Nf	(Goldstone 2008)
	NOQ1	Nf	Nf	2	Nf	(Goldstone 2008)
Efflux transport	ABC	B	Nf	11	9	(Dean and Annilo 2005)
		C	Nf	12	11	(Dean and Annilo 2005)
		G	Nf	5	5	(Dean and Annilo 2005)

of South America (e.g., Characiformes, Siluriformes, Perciformes and Cyprinodontiformes) could lead to a more accurate picture of the complete set of gene/proteins in South American tropical species. Even considering the lack of species closely related to zebrafish in the tropical environments of South America, the set of genes related to biotransformation that have been revealed—based on zebrafish—is still probably the best starting point to understand the nomenclature that could be used for new gene/proteins identified in alternative fish species.

The criterion used for gene/protein nomenclature has varied historically. Originally, enzyme annotation was made based on the catalytic function (e.g., urease—enzyme that hydrolyzes urea), and more recently by considering the analysis of amino acid sequences, gene structure and other criteria by means of bioinformatic tools. A root symbol is very much encouraged by the Human Gene Nomenclature Committee (HGNC) as the basis for a hierarchical series of genes (e.g., for the ABC family, subfamily A, isoforms ABCA1, ABCA2, ABCA3, ABCA4) (Nebert and Vasiliou 2004). To date, more than 130 gene superfamilies and large gene families have followed this same format (Nebert and Wain 2003). The nomenclature used for genes in fishes generally follows the same roots used in human nomenclature, although some differences in the classes and subfamilies could exist. For example, the CYP family's 1–4 are present in both fish and human, but although there are subfamilies that are common, some of those are exclusive for fish and other subfamilies are exclusive for human (see Table 2.1).

One of the earliest examples of the nomenclature approach for homologous genes was the cytochrome P450 (CYP) gene super family, in which it was agreed that approximately 40% or more amino acid identity allows two proteins to be placed in the same family and about 55% or greater identity allows two proteins to be assigned to the same subfamily (Nelson et al. 2004). But differing from the ABC super family, that have each family annotated by letters (e.g., ABCA, ABCB and ABCC), each CYP enzyme family is currently annotated by Arabic numbers (e.g., CYP1, CYP2 and CYP3), CYP subfamilies by letters (e.g., CYP1A and CYP1B), and the Arabic number which follows the letter denotes the individual protein isoform within the subfamily (e.g., CYP1A1, CYP1A2 and CYP1B1) (Nelson et al. 2004). The gene nomenclature follows the same idea as the protein nomenclature, but is italicized (e.g., *CYP1A1, CYP1A2* and *CYP1B1*). The zebrafish have a total of 94 CYP genes, distributed among 18 gene families, and the same 18 families are also found in humans. The Table 2.1 shows only the families 1–4, that includes the major enzymes recognized to be involved in drug and xenobiotic metabolism in human and zebrafish (Goldstone et al. 2010). The CYP2 is the largest CYP family in zebrafish (47 genes), and only two of those genes (CYP2R and CYP2U1) are orthologs to humans, based on sequence identity and synteny analysis (Goldstone et al. 2010). The function of most of those fish CYP2s are unknown and probably could be important for xenobiotic biotransformation. In mammals and fish, CYP3 proteins are very abundant in the liver and gastrointestinal tract. In humans, the CYP3s are probably the major enzymes involved in the drug metabolism. Human CYP3s metabolizes more than 120 pharmacological compounds, and endogenous substrates like sterols, arachidonic acid, bile salts, and vitamin D (Nebert and Russell 2002; Verslycke et al. 2006) and it is estimated that 40%–60% of all pharmacological compounds consumed by humans are metabolized by the CYP3A4 isoform (Guengerich 1999). The zebrafish isoform CYP3A65 is 54% identical in the amino acid sequence to human CYP3A4 (Goldstone et al. 2010) and it is similarly inducible in the transcriptional level by PXR (Pregane X Receptor; see Chapter 3) agonist compounds (Bresolin et al. 2005).

The current nomenclature system for aldehyde dehydrogenases (ALDHs) functions similarly to CYPs; thus, an ALDH protein from one gene family is defined as having less than 40% amino-acid identity comparing to genes from another ALDH families. Two members of the same ALDH subfamily exhibit approximately >60% amino-acid identity (Vasiliou et al. 1999). The ALDH of the family 2 (ALDH2) is especially important to oxidize acetaldehyde, which is responsible for some of the deleterious effects of ethanol. Recombinant zebrafish ALDH2 protein expressed using the baculovirus expression system shows similar kinetic properties compared with the recombinant human ALDH2 protein. Thus, it indicates that zebrafish, and possibly other fishes, are suitable models for studying ethanol metabolism and, therefore, ethanol toxicity (Lassen et al. 2005). It was also shown that the side effect of nifurtimox, a 5-nitrofuran anti-trypanosome prodrug used in humans, could be dependent of ALDH2 using the zebrafish model (Zhou et al. 2012). The full complement of the ALDHs isoforms in zebrafish, compared with human ALDH complement, have been shown elsewhere (Zhou et al. 2012).

In mammals, five families of distinct flavin-containing monooxygenases (FMO) genes have been identified and classified based on the amino acid sequence and named FMO 1–5. These isoforms differ in tissue distribution, regulation, and substrate specificity. In zebrafish, four FMO genes were found in the chromosome 20 (Rodriguez-Fuentes et al. 2008). FMO proteins are found in the smooth endoplasmic reticulum of multiple tissues that catalyze the oxidation of soft nucleophilic heteroatom substances to their respective oxides. FMO substrates include alkaloids, pesticides, and pharmaceutical compounds (Rodriguez-Fuentes et al. 2008).

The glutathione *S*-transferases (GSTs) are divided in two families in animals, the MAPEG (membrane associated proteins involved in eicosanoid and glutathione metabolism) and the soluble or cytosolic GSTs (also termed canonical by some authors) (Frova 2006). The cytosolic GSTs are grouped in distinct classes designated by Greek letters (e.g., alpha, pi and theta). Sequence mean identity within class is typically >40%. However, pair wise comparisons often indicate a much broader range. Interclass identities are significantly lower, usually <25% in mammals (Frova 2006). It is important to note that fishes possess the class "rho" that does not exist in humans and other mammals (Carletti et al. 2008).

The members of sulfotransferases (SULT) gene family share at least 40%–45% amino acid sequence identity, while members of the subfamilies further divided within each SULT gene family are >60% identical in their amino acid sequences (Liu et al. 2010). SULT proteins generally catalyze the transfer of a sulfonate group from the active sulfate, 3'-phosphoadenosine 5'-phosphosulfate (PAPS), to low-molecular weight substrate compounds containing hydroxyl or amino group(s). Fifteen zebrafish SULT proteins have been heterologously expressed and characterized for substrate specificity, and similar to many human SULT1 enzymes, zebrafish SULT1s and SULT3 (this last one is present in fish but not in humans), but not SULT 2 and 6, appears to play a crucial role in the metabolism and detoxification of xenobiotics (Liu et al. 2010).

The UDP-glucuronosyl transferases (UGT) enzymes can catalyze the conjugation of a vast number of lipophilic xenobiotics and endobiotics with the UDP-glucuronic acid. This glucuronidation reaction converts hydrophobic aglycones to water-soluble glucuronides and enhances their excretion from the body (Huang and Wu 2010). Comparative analyses showed that both "a" and "b" clusters of the zebrafish UGT1 and UGT2 genes have orthologs in other teleosts, suggesting that they may be resulted from the "fish-specific" whole-genome duplication event. The UGT5 genes are a novel family of UGT genes, that have the highest number of isoforms in zebrafish (40% of the total UGTs) and exist in teleosts and amphibians (Huang and Wu 2010).

Arylamine N-acetyltransferases (NATs) catalyze the N-acetylation of arylamines, arylhydroxylamines and arylhydrazines with the acetyl group being transferred from acetyl-Coenzyme A and are also likely to be important for future toxicological investigations. Six NAT genes were identified in the zebrafish genome, compared with two in humans (Sim et al. 2008). Human NAT genes have been arbitrarily assigned as NAT1 or NAT2, and no characterization of the proteins has yet been carried out.

The aldo-keto reductases (AKR), epoxide hydrolase (EPHX) and NAD(P)H-quinone oxidoreductase (NOQ1) are poorly characterized in terms of substrate specificity, even in humans. There are many homologues for those genes in zebrafish, that possibly are involved in xenobiotic metabolism and physiological normal functions (Stegeman et al. 2010).

The ATP-binding cassette (ABC) transporters bind and hydrolyze ATP and can use this energy to pump compounds across the membrane (Dean and Annilo 2005). The chordates have eight ABC subfamilies designated ABC A through H, but only a subset of these families includes proteins known to export toxicants: the ABCB, ABCC, and ABCG (Goldstone 2008); thus, the other families were omitted in Table 1. The isoform ABCB1 is also known as Multidrug Resistance transporter or P-glycoprotein (MDR or PGP); ABCC1-6 and 10 are the Multidrug Resistance Proteins MRP1-6 and 7, respectively; and the ABCG2 is known as Multidrug Transporter (MXR). Most MRPs are involved in the cellular export of toxic compounds either complexed with glutathione or co-transported with glutathione (Dean and Annilo 2005). The above mentioned isoforms of families ABCB1, ABCC and ABCG2 are probably the major ABC transporter for drugs and other environmental xenobiotics in both humans and zebrafish.

The CYP1s in Fish: Implications for Environmental Monitoring and Toxicology

Compared with humans, who possess three CYP1 genes, named *CYP1A1*, *CYP1A2* and *CYP1B1* (Nebert and Russell 2002), the zebrafish *Danio rerio* (Cypriniforme) possess five genes: *CYP1A, CYP1B1, CYP1C1, CYP1C2* and *CYP1D1* (Goldstone et al. 2009; Goldstone et al. 2010). The same set of five *CYP1* members apparently is present in other model fishes, such the killifish *Fundulus heteroclitus* (Cyprinodontiformes), the medaka *Oryzias latipes* (Beloniformes) and the stickleback *Gasterosteus aculeatus* (Gasterosteiformes) (Zanette et al. 2009). Notably, Salmoniformes, such as the rainbow trout *Oncorhynchus mykiss* and the *Salmo salar* have additional CYP1As and/ or CYP1Cs members (see Table 2.2), likely reflecting more recent genome duplication in the Salmoniformes. The common molecular phylogeny for the CYP1 genes in several species thus supports the hypothesis that the CYP1As and CYP1Ds diverged from a common CYP1A/CYP1D ancestor, and the CYP1Bs and CYP1Cs from a common CYP1B/CYP1C ancestor (Goldstone et al. 2007; Goldstone et al. 2009).

Table 2.2. Summary of the CYP1 members in humans and fishes.

Organism	CYP1s	Reference
Human/mouse	1A1, 1A2, 1B1	(Nebert and Russell 2002)
Model fishes (zebrafish, medaka, stickleback, fundulus)	1A, 1B1, 1C1, 1C2, 1D1	(Zanette et al. 2009)
Fish (rainbow trout)	1A1, 1A3, 1B1, 1C1, 1C2, 1C3, 1D1*	(Jonsson et al. 2010)
Tropical fish (*Poecilia vivipara*)	1A, 1B1, 1C1, 1C2*, 1D1*	(Dorrington et al. 2012)

* nucleotide sequences unknown so far.

The five fish CYP1s present a markedly distinct organ-specific distribution in the gene expression level, suggesting that different physiological roles could exist (Jonsson et al. 2007; Goldstone et al. 2009). CYP1A is highly expressed in liver, as well as in many extrahepatic organs, although the other four CYP1 genes are more highly expressed than CYP1A in some extrahepatic organs (Jonsson et al. 2007; Zanette et al. 2009).

Environmental monitoring studies have extensively used the CYP1A isoform as a biomarker in fish since the induction of CYP enzymes by organic contaminants was first suggested in the 1970s (e.g., (Payne 1976)). CYP1A in fish organs is often measured by enzymatic activity assay, i.e., EROD, and protein detection, i.e., by Western blot (Bucheli and Fent 1995) and immunohistochemistry (VanVeld et al. 1997). Recently, the analysis of *CYP1A* transcript levels in fish organs, using reverse transcriptions followed by real time PCR (RT-qPCR), have been compared with EROD and considered a sensitive biomarker for organic contaminant exposure as well (Pina et al. 2007).

The well-known mechanism associated with the CYP1A induction involves the binding of chemicals with the cytosolic aryl hydrocarbon receptor (AHR; see Chapter 3) protein, the transport to the nucleous, dimerization of AHR with the AHR nuclear translocator (ARNT) and the transcriptional activation of *CYP1A* gene expression. The transcriptional activation depends on the recognition of xenobiotic responsive elements (XRE) (also known as dioxin responsive elements, DRE) in the *CYP1A* promoter region (Hahn 2002).

Compounds that have high affinity to bind to AHR, and potential to induce CYP1A, include organic planar molecules with aromatic rings, some polycyclic aromatic hydrocarbons (PAHs), polychlorinated biphenyl (PCB), dibenzo-p-dioxin (PCDD), and dibenzofuran (PCDF) congeners, that represent risk for humans and wildlife (Bucheli and Fent 1995). The PCDD 2,3,7,8-Tetrachlorodibenzo-p-dioxin (TCDD) seems to be the strongest inducer, and among the 209 existing PCB congeners, the non-*ortho* substituted PCB 77, PCB 81, PCB 126 and PCB 169, seem to be the most potent inducers. In addition, complex environmental mixtures such as cigarette smoke, diesel emissions, urban air, motorcycle exhaust, carbon black, jet fuel, and metal ore and fumes, that occur through media such as air, soil, water and food and are considered toxic, up-regulate gene expression of *CYP1A* in a more consistent way than other genes, as shown on microarray study analyses after exposure of mammalian cells or living organisms (Sen et al. 2007). The exposure of the fish Atlantic cod (*Gadus morhua*), using an *ex vivo* gill EROD assay shows that environmental mixtures that came from petroleum platforms, such as crude oil and produced water resulted in a concentration-dependent induction of gill EROD too (Abrahamson et al. 2008).

In addition to *CYP1As*, other *CYP1s*, notably the newly identified *CYP1B1* and *CYP1Cs*, have the potential to be responsive biomarkers of exposure to AHR agonist contaminants in fishes, as it has been shown using RT-qPCR (Wang et al. 2006; Zanette et al. 2009). Smaller responsiveness by *CYP1C2* and the lack of *CYP1D1* induction by potent AHR agonists in two different fish species suggests that levels in the expression of these two genes would not be suitable markers for exposure to such chemicals (Zanette et al. 2009). There was no induction of *CYP1D1* in zebrafish embryos or adults treated with PCB126 or TCDD (Goldstone et al. 2009). The lack of *CYP1D1* induction could be associated with a lack of functional AHR response elements (AHREs, also called XREs) (Goldstone et al. 2009). In zebrafish, there are only two putative AHREs (of unproven function) in the promoter of zebrafish *CYP1D1*, in contrast to the 22 putative AHREs and three proven functional AHREs for *CYP1A* (ZeRuth and Pollenz 2007; Goldstone et al. 2009). A variety of other response elements were identified in the promoter region of zebrafish *CYP1D1* (Goldstone et al. 2009) and thus the possibility of induction via other receptors and agonists still remains to be elucidated, both for the fishes *CYP1Cs* and *CYP1D1*. Considering the variation in the magnitude of induction of the five *CYP1s*, it can be concluded that the choice of the target organ in combination with the *CYP1* isoform could be useful to optimize the sensitivity in the use of *CYP1s* as biomarkers (Zanette et al. 2009). Based on those studies using zebrafish and killifish, we could expect that the *CYP1A*, *CYP1B1* and *CYP1C1* could be explored as biomarkers in tropical fishes for monitoring organic contamination.

Regarding the catalytic activity, the five zebrafish CYP1s seem to have distinct substrate specificities, as measured by *in vitro* assays with the five recombinant expressed proteins (Scornaienchi et al. 2010). In humans, the members of family 1 are very important in the metabolism of endogenous/exogenous organic molecules. CYP1A1 and CYP1B1 are efficient in the PAH metabolism; CYP1A1 inactivates prostaglandins; CYP1A2 and CYP1B1 hydroxylate estrogens; and CYP1A2 also metabolizes arylamines, N-heterocyclic compounds and a large number of therapeutic drugs (Nebert and Russell 2002). The induction of CYP1A by planar compounds is often correlated with the generation of toxic effects, the so-called dioxin-like toxicity. In this sense, mammalian CYP1A1 and CYP1A2 induction has been used to estimate toxic equivalent factor (TEF) to dioxin-like compounds, helping in the evaluation of risk for isolated chemicals and mixtures (Toyoshiba et al. 2004). Those calculations were used, for example, to rank the toxic potential for the existing 209 PCB congeners (Pereira 2004).

Studies with Biotransformation Genes and Proteins in Tropical Fishes

A higher diversity of fish species is present in tropical regions of South and Central America, compared with North America, and some of these tropical fishes are potential model candidates. Table 2.3 shows some of the native tropical fish species generally used for aquatic toxicology studies in Brazil and other countries in South and Central America. It could be noted that very few genes identified in those fishes codify for biotransformation enzymes and the majority of those genes are especially valuable for ecological, population and phylogenetic studies, but not so much for aquatic toxicology. For example, mitochondrial genes such the control region (COI), ATP synthase, cytochrome b, cytochrome c oxidase (COX), ribosomal genes and microsatellites, have been identified in most of those species. Opportunely, some *CYP1A* sequences were recently identified in the live bearers (Cyprinodontiformes) *Poecilia vivipara* (Dorrington et al. 2012; Stacke Ferreira et al. 2012) and *Jenynsia multidentata* (Stacke Ferreira et al. 2012) and the cyclid (Perciformes) *Astronotus ocellatus* (dos Anjos et al. 2011). Additional genes related to biotransformation in *P. vivipara*, including the CYP1s, CYP1B1 and CYP1C1 (Dorrington et al. 2012), CYP2, GST, UDPGT and ABC members have also been identified (Mattos et al. 2009) (waiting for GenBank submission).

Following the example of the Atlantic killifish *Fundulus heteroclitus* (Cyprinodontiforme, Fundilidae), extensively studied in North America (Burnett et al. 2007), Cyprinodontiforme fishes that inhabit South and Central America, e.g., *J. multidentata* (Anablepidae) (from Argentina to Brazil) and *P. vivipara* (Poecilidade) (from Argentina to Central America) (Froese and Pauly 2012) (Table 2.3), have been suggested as "novel" model species in environmental toxicology (Cazenave et al. 2008; Amado et al. 2009; Ame et al. 2009; Mattos et al. 2009; Dorrington et al. 2012; Stacke Ferreira et al. 2012). The use of model Cyprinodontiforme fishes in toxicology, relies on the study of the fate of contaminants in the aquatic environment (Burnett et al. 2007; Matson et al. 2008) and also in the understanding of the genetic/evolutionary aspects involved in adaptation to survive in extreme polluted conditions (Williams and Oleksiak 2008; Wirgin et al. 2011) that helps to elucidate fundamental biochemical and molecular mechanisms in toxicology (Hahn et al. 2004).

Although poorly studied so far, *J. multidentata* and *P. vivipara*, as well as other Cyprinodontiforme live bearer fishes from South America, possess one or more of the following features that support their use as model species in toxicology: (1) geographically widespread (Fig. 2.1) and adapted to live in a wide range of stressful conditions (e.g., pollution, salinity, oxygen and temperature stresses) (Gomes and Monteiro 2008);

Table 2.3. Some of the major tropical fish species used in environmental toxicology in Brazil and other countries from South and Central America. The name of the specie is followed by the reference that suggests its uses for aquatic toxicology. The geographical distribution of fishes was assessed based on the *Fish Base* (Froese and Pauly 2012). The common names used in Brazil and USA for the tropical fishes are also shown. NCBI webpage searching (http://www.ncbi.nlm.nih.gov) was used to estimate the number of biomedical literature citations and abstracts available (PUB), the number of nucleotide sequences submitted in the NCBI (Nuc), the number of nucleotide sequences available that codify biotransformation enzymes (BT), and the number of nucleotide sequences available that codify CYP1 proteins (CYP1), for each one of the fish species.

Order	Tropical Species and related reference	Common	Distribution*	PUB	Nuc	BT	CYP1
Characiformes (Characins)	Astyanax bimaculatus (Gomes et al. 2012)	Lambari, Tetra	~11 countries	20	44	-	-
	Hoplias malabaricus (Ramsdorf et al. 2009)	Traíra, Trahira	~14 countries	109	103	-	-
	Geophagus brasiliensis (Linde-Arias et al. 2008)	Cará, Pearl eartheater	BRA, URY	28	62	-	-
	Stellifer brasiliensis (Dantas et al. 2012)	Boca-de-rato	BRA	2	9	-	-
	Prochilodus lineatus (Pereira et al. 2012)	Curimbatá, Curimbata	ARG, BRA, PRY, URY	88	409	-	-
Siluriformes (Catfish)	Cathorops spixii (Azevedo et al. 2009)	Bagre	BRA, COL, GUF, GUY, SUR, VEN	11	14	-	-
	Netuma barba (Rodriguez-Cea et al. 2006)	Bagre	ARG,BRA,URY	3	20	-	-
	Rhamdia quelen (Cericato et al. 2009)	Jundiá, Jundia bagre	~21 countries	69	39	-	-
	Sciades herzbergii (Possatto et al. 2011; Carvalho-Neta et al. 2012)	bagre	~9 countries	7	8	-	-
Perciformes (Perch-likes)	Centropomus parallelus (Kirschbaum et al. 2009)	Robalo, Fat snook	~28 countries	10	62	-	-
	Trichiurus lepturus (Barletta et al. 2012)	Peixe-espada, Atlantic cutlassfish	Cosmopolitan	17	281	-	-
	Astronotus ocellatus (dos Anjos et al. 2011)	Acará-Açú, Oscar	ARG, BRA, COL, GUF, PER	63	182	1	1
Cyprinodontiformes (Rivulines, killifishes and live bearers)	Jenynsia multidentata (Stacke Ferreira et al. 2012)	Barrigudinho, Onesided livebearer	ARG,BRA,URY	21	13	2	1
	Poecilia vivipara (Dorrington et al. 2012; Stacke Ferreira et al. 2012)	Guarú, Guppy	ARG, BRA, GUF, GUY, MTQ, PRI, SUR, TTO, URY, VEN	13	22	3	3

*South and Central America

Figure 2.1. Distribution of *Fundulus heteroclitus* (+), *Poecilia vivipara* (•), *Jenynsia multidentata* (Δ) and *Phalloceros caudimaculatus* (*) Cyprinodontiforme fishes (adapted from (Froese and Pauly 2012)).

(2) most are ovoviviparous, being potential models to understand mother-embryo toxicological interaction during the embryonic development; (3) a huge diversity of species exists, and possibly a diversity of mechanisms of resistance/sensitivity to chemical stress; (4) some species are rare, thus possibly endangered; (5) Similarly to *Poecilia reticulata*, that have been used for decades as models to study reproduction behavior (Schroder and Peters 1988), those fishes have similar courtship behavior (sigmoid display and gonopodial thrust) and could be useful for behavioral testing against xenobiotics (e.g., endocrine disruptors).

Additional Poecilidae species distributed in South America, such as the genus *Phalloceros*, *Cnesterodon* and *Phallotorynus*, represented by many species (~22, ~9 and ~5, respectively) (Lucinda 2005; Lucinda et al. 2005; Lucinda 2008) could serve as surrogate models for toxicological studies. While *P. vivipara* and *J. multidentata* are limited, being distributed in mixohaline environments next to the coast (such mangroves and estuaries), the last three genus are also widespread in freshwater environments in South America.

The use of RT-qPCR allowed the evaluation of gene transcription levels in the Cyprinodontiforme live bearer fishes. The *P. vivipara CYP1A, CYP1B1* and *CYP1C1* appears to have similar basal expression compared with *F. heteroclitus* and zebrafish; *CYP1A* is most strongly expressed in the liver, while *CYP1B1*, and *CYP1C1* were most strongly expressed in the gill and intestine respectively (Dorrington et al. 2012). *P. vivipara* injected with 30 µg of 3-methylcholanthrene (3-MC) per g of fish, induced *CYP1A, CYP1B1*, and *CYP1C1* significantly (20–120-fold) in the liver, intestine and gill after 24 hours. In addition to the laboratory exposure, the significantly high levels of *CYP1A* and *CYP1C1* in gills (10–15-fold) and *CYP1B1* in liver (23-fold) in fish from a contaminated urban mangrove, relative to fish from a reference site, indicates the potential of *CYP1* gene as a biomarker for environmental monitoring using *P. vivipara* (Dorrington et al. 2012).

CYP1A was also strongly induced in the liver and gills of *J. multidentata* (~185-fold and ~20-fold, respectively) and *P. vivipara* (188-fold and 739-fold, respectively) after 24-hr exposure to 1 µM of the synthetic CYP1A inducer β-naphthoflavone (BNF) (Stacke Ferreira et al. 2012). The difference in the levels of induction between both fishes suggests that different patterns for *CYP1A* induction could be present and possibly will reflect differences in the detoxification/bioactivational processes that are linked to CYP1A activity in different fishes. It is also possible that the observed differences between species rely on the characteristic interval of time that each organ/species shows the "peak" of *CYP1A* induction. Small differences in the exposure time (a few hours) can strongly influence the levels of *CYP1A* induction in different organs in the putterfish (Kim et al. 2008). Interestingly, depending on the species analyzed, different responses were observed in the oxidative stress associated parameters (see Chapter 5 for more information about oxidative stress) which suggest distinct resistance/susceptibility to PAHs (Stacke Ferreira et al. 2012). Population studies can be subsequently carried out in South American Cyprinodontiformes in order to identify mechanisms in detoxification associated with genes and proteins that enable the fish to adapt to highly contaminated sites, similar to what has been described in North American fishes (Williams and Oleksiak 2008; Wirgin et al. 2011). In summary, the responsiveness of *CYP1* genes indicates that *P. vivipara, J. multidentata* and *P. caudimaculatus* are suitable as models for environmental toxicology studies and environmental assessment of pollution in tropical ecosystems (Dorrington et al. 2012; Stacke Ferreira et al. 2012).

In addition to the Cyprinodontiforme live bearers, *CYP1A* gene have been also identified recently in an abundant Amazonian fish species that has economic importance as an ornamental and food fish, the Perciform Acará-Açú, also known as Oscar, *Astronotus* ocellatus (dos Anjos et al. 2011). Transcript expression of CYP1A has been shown to be a sensitive marker for oil exposure in the Amazonian native fish *A. ocellatus,* and that those responses are similar to the zebrafish model. *A. ocellatus* is very tolerant to hypoxia and can be found in lakes, the marginal zone of rivers and among floating stands of herbaceous plants, with special preference for lentic habitats (dos Anjos et al. 2011). In contrast to tambaqui (*Colossoma macropomum*), another cyclid that has been also used in aquatic toxicological studies earlier (Matsuo et al. 2006), *A. oscellatus* does not undergo seasonal migrations and possibly could be more suitable for biomonitoring studies.

Some examples of methodological strategies used for the identification of new genes in tropical fishes such the perciformes cyclids and the cyprinodontiforme live bearers were the RT-PCR using degenerate primers (dos Anjos et al. 2011; Dorrington et al. 2012; Stacke Ferreira et al. 2012) and Subtractive Suppressive Hybridization (SSH) (Mattos et al. 2009) followed by cloning and sequencing. Our research group has met with success in following this strategy, using the same set of degenerate primers, for cloning the CYP1A transcript from the poecilid *Phalloceros caudimaculatus* (manuscript in preparation), *J. multidentata* and *P. vivipara* (Stacke Ferreira et al. 2012), thus showing that those degenerate primers could be used in alternative "surrogate" fish species. Alternatively, the recent advances in the next-generation sequencing methods (Loman et al. 2012) have the potential to accelerate the identification of genes in these tropical fishes, similar to what has been done recently for the Cyprinodontiforme fish *Poecilia reticulata* (Fraser et al. 2011) and *Fundulus heteroclitus* (Oleksiak et al. 2011). The comparisons between the high-throughput sequencing platforms, such the 454 GS Junior (Roche), MiSeq (Illumina) and Ion Torrent PGM (Life Technologies), that could be used for such purposes, have been discussed elsewhere (Loman et al. 2012). For example, the recent *de novo* assembly of the guppy (*Poecilia reticulata*) transcriptome using 454 sequence reads, resulted in 1,162,670 reads assembled into 54,921 contigs, creating a reference transcriptome for the guppy (Fraser et al. 2011) that potentially will serve to identify key genes involved in xenobiotic biotransformation in the guppy.

Conclusion

The current genomic and transcriptomic information available for the zebrafish model allowed the recent uncovering of a number of genes that are possibly involved in xenobiotic biotransformation and toxicity. We could expect that the recent emergence of the next-generation sequencing platforms such the 454 GS Junior (Roche), MiSeq (Illumina) and Ion Torrent PGM (Life Technologies) will accelerate the uncovering of genomes and transcriptomes of many non-model fish species in the next few years, including some tropical fishes. These sequencing platforms also allow the identification of transcripts that are differentially expressed for different conditions (e.g., exposure to organic contaminants), thus serving to find key genes involved in the biotransformation process in model, and non-model fishes. Trying to find the functional roles of those genes in the toxicological mechanisms will require additional studies on the protein level. In this chapter, we have presented some potential fish species that could be useful in aquatic toxicology for monitoring purposes and comparative toxicology. Following the example of the killifish *F. heteroclitus*, extensively used in North America, recent studies focused on the identification and use of CYP1s in the South American Cyprinodontiforme live bearers (e.g., *P. vivipara*, *J. multidentata* and *P. caudimaculatus*) indicate that these fishes could serve as surrogate and complementary models in aquatic toxicology in tropical environments.

Acknowledgements

The author would like to thank the Brazilian agencies CNPq (482768/2010-0), INCT-TA (573949/2008-5) and FAPERGs (12/1328-5).

Keywords: Monitoring, zebrafish, cyprinodontiformes, guppy, biotransformation, genes, Cytochrome P450, CYP, GST, transcriptome

References

Abrahamson, A., I. Brandt, B. Brunstrom, R.C. Sundt and E.H. Jorgensen. 2008. Monitoring contaminants from oil production at sea by measuring gill EROD activity in Atlantic cod (*Gadus morhua*). Environ. Pollut. 153: 169–175.

Amado, L.L., M.L. Garcia, P.B. Ramos, R.F. Freitas, B. Zafalon, J.L. Ferreira, J.S. Yunes and J.M. Monserrat. 2009. A method to measure total antioxidant capacity against peroxyl radicals in aquatic organisms: application to evaluate microcystins toxicity. Sci. Total Environ. 407: 2115–23.

Ame, M.V., M.V. Baroni, L.N. Galanti, J.L. Bocco and D.A. Wunderlin. 2009. Effects of microcystin-LR on the expression of P-glycoprotein in *Jenynsia multidentata*. Chemosphere 74: 1179–1186.

Azevedo, J.S., W.S. Fernandez, L.A. Farias, D.T.I. Favaro and E.S. Braga. 2009. Use of *Cathorops spixii* as bioindicator of pollution of trace metals in the Santos Bay, Brazil. Ecotoxicology 18: 577–586.

Barletta, M., L.R. Lucena, M.F. Costa, S.C. Barbosa-Cintra and F.J. Cysneiros. 2012. The interaction rainfall vs. weight as determinant of total mercury concentration in fish from a tropical estuary. Environ. Pollut. 167: 1–6.

Bresolin, T., M. de Freitas Rebelo and A. Celso Dias Bainy. 2005. Expression of PXR, CYP3A and MDR1 genes in liver of zebrafish. Comp. Biochem. Physiol. C Toxicol. Pharmacol. 140: 403–7.

Bucheli, T.D. and K. Fent. 1995. Induction of Cytochrome-P450 as a biomarker for environmental contamination in aquatic ecosystems. Crit. Rev. Environ. Sci. Technol. 25: 201–268.

Buckup, P.A. and N.A. Menezes. 2003. "Catálogo dos Peixes Marinhos e de Água Doce do Brasil." 2.ed., from http://www.mnrj.ufrj.br/catalogo/

Burnett, K.G., L.J. Bain, W.S. Baldwin, G.V. Callard, S. Cohen, R.T. Di Giulio, D.H. Evans, M. Gomez-Chiarri, M.E. Hahn, C.A. Hoover, S.I. Karchner, F. Katoh, D.L. MacLatchy, W.S. Marshall, J.N. Meyer, D.E. Nacci, M.F. Oleksiak, B.B. Rees, T.D. Singer, J.J. Stegeman, D.W. Towle, P.A. Van Veld, W.K. Vogelbein, A. Whitehead, R.N. Winn and D.L. Crawford. 2007. Fundulus as the premier teleost model in environmental biology: opportunities for new insights using genomics. Comp. Biochem. Physiol. D-Gen. Proteom. 2: 257–286.

Carletti, E., M. Sulpizio, T. Bucciarelli, P. Del Boccio, L. Federici and C. Di Ilio. 2008. Glutathione transferases from Anguilla anguilla liver: identification, cloning and functional characterization. Aquat. Toxicol. 90: 48–57.

Carvalho-Neta, R.N., A.R. Torres, Jr. and A.L. Abreu-Silva. 2012. Biomarkers in catfish *Sciades herzbergii* (Teleostei: Ariidae) from polluted and non-polluted areas (Sao Marcos' Bay, Northeastern Brazil). Appl. Biochem. Biotechnol. 166: 1314–27.

Cazenave, J., M.L. Nores, M. Miceli, M.P. Diaz, D.A. Wunderlin and M.A. Bistoni. 2008. Changes in the swimming activity and the glutathione S-transferase activity of *Jenynsia multidentata* fed with microcystin-RR. Water Res. 42: 1299–307.

Cericato, L., J.G. Neto, L.C. Kreutz, R.M. Quevedo, J.G. da Rosa, G. Koakoski, L. Centenaro, E. Pottker, A. Marqueze and L.J. Barcellos. 2009. Responsiveness of the interrenal tissue of Jundia (*Rhamdia quelen*) to an *in vivo* ACTH test following acute exposure to sublethal concentrations of agrichemicals. Comp. Biochem. Physiol. C Toxicol. Pharmacol. 149: 363–7.

Dantas, D.V., M. Barletta and M.F. da Costa. 2012. The seasonal and spatial patterns of ingestion of polyfilament nylon fragments by estuarine drums (Sciaenidae). Environ. Sci. Pollut. Res. Int. 19: 600–6.

Dean, M. and T. Annilo. 2005. Evolution of the ATP-binding cassette (ABC) transporter superfamily in vertebrates. An. Rev. Gen. Human Genetics 6: 123–142.

Dorrington, T., J. Zanette, F.L. Zacchi, J.J. Stegeman and A.C. Bainy. 2012. Basal and 3-methylcholanthrene-induced expression of cytochrome P450 1A, 1B and 1C genes in the Brazilian guppy, *Poecilia vivipara*. Aquat. Toxicol. 124–125C: 106–113.

Dos Anjos, N.A., T. Schulze, W. Brack, A.L. Val, K. Schirmer and S. Scholz. 2011. Identification and evaluation of cyp1a transcript expression in fish as molecular biomarker for petroleum contamination in tropical fresh water ecosystems. Aquat. Toxicol. 103: 46–52.

Fraser, B.A., C.J. Weadick, I. Janowitz, F.H. Rodd and K.A. Hughes. 2011. Sequencing and characterization of the guppy (*Poecilia reticulata*) transcriptome. BMC Genomics 12: 202.

Froese, R. and D. Pauly. 2012. FishBase. World Wide Web electronic publication, version (10/2012). 2012, from www.fishbase.org.

Frova, C. 2006. Glutathione transferases in the genomics era: New insights and perspectives. Biomol. Eng.. 23: 149–169.

Goldstone, J.V. 2008. Environmental sensing and response genes in cnidaria: the chemical defensome in the sea anemone *Nematostella vectensis*. Cell Biol. Toxicol. 24: 483–502.

Goldstone, J.V., H.M.H. Goldstone, A.M. Morrison, A. Tarrant, S.E. Kern, B.R. Woodin and J.J. Stegeman. 2007. Cytochrome p450 1 genes in early deuterostomes (tunicates and sea urchins) and vertebrates (chicken and frog): Origin and diversification of the CYP1 gene family. Mol. Biol. Evol. 24: 2619–2631.

Goldstone, J.V., M.E. Jonsson, L. Behrendt, B.R. Woodin, M.J. Jenny, D.R. Nelson and J.J. Stegeman. 2009. Cytochrome P450 1D1: A novel CYP1A-related gene that is not transcriptionally activated by PCB126 or TCDD. Arch. Biochem. Bioph. 482: 7–16.

Goldstone, J.V., A.G. McArthur, A. Kubota, J. Zanette, T. Parente, M.E. Jonsson, D.R. Nelson and J.J. Stegeman. 2010. Identification and developmental expression of the full complement of Cytochrome P450 genes in Zebrafish. BMC Genomics 11: 643.

Gomes, I.D., A.A. Nascimento, A. Sales and F.G. Araujo. 2012. Can fish gill anomalies be used to assess water quality in freshwater neotropical systems? Environ. Monit. Assess. 184: 5523–31.

Gomes, J.L. and L.R. Monteiro. 2008. Morphological divergence patterns among populations of *Poecilia vivipara* (Teleostei Poeciliidae): test of an ecomorphological paradigm. Biol. J. Linnean Soc. 93: 799–812.

Guengerich, F.P. 1999. Cytochrome P-450 3A4: Regulation and role in drug metabolism. An. Rev. Pharmacol. Toxicol. 39: 1–17.

Hahn, M.E. 2002. Aryl hydrocarbon receptors: diversity and evolution. Chem. Biol. Interac. 141: 131–160.

Hahn, M.E., S.I. Karchner, D.G. Franks and R.R. Merson. 2004. Aryl hydrocarbon receptor polymorphisms and dioxin resistance in Atlantic killifish (*Fundulus heteroclitus*). Pharmacogenetics. 14: 131–143.

Huang, H. and Q. Wu. 2010. Cloning and comparative analyses of the zebrafish Ugt repertoire reveal its evolutionary diversity. PLoS One. 5: e9144.

Jonsson, M.E., K. Gao, J.A. Olsson, J.V. Goldstone and I. Brandt. 2010. Induction patterns of new CYP1 genes in environmentally exposed rainbow trout. Aquat. Toxicol. 98: 311–21.

Jonsson, M.E., M.J. Jenny, B.R. Woodin, M.E. Hahn and J.J. Stegeman. 2007. Role of AHR2 in the expression of novel cytochrome p450 1 family genes, cell cycle genes, and morphological defects in developing zebra fish exposed to 3,3 ',4,4 ',5-pentachlorobiphenyl or 2,3,7,8-tetrachlorodibenzo-p-dioxin. Toxicol. Sci. 100: 180–193.

Jonsson, M.E., R. Orrego, B.R. Woodin, J.V. Goldstone and J.J. Stegeman. 2007. Basal and 3,3',4,4',5-pentachlorobiphenyl-induced expression of cytochrome P450 1A, 1B and 1C genes in zebrafish. Toxicol. Ap. Pharmacol. 221: 29–41.

Kim, J.H., S. Raisuddin, J.S. Ki, J.S. Lee and K.N. Han. 2008. Molecular cloning and beta-naphthoflavone-induced expression of a cytochrome P450 1A (CYP1A) gene from an anadromous river pufferfish, *Takifugu obscurus*. Mar. Pollut. Bull. 57: 433–440.

Kirschbaum, A.A., R. Seriani, C.D. Pereira, A. Assuncao, D.M. de Souza Abessa, M.M. Rotundo and M.J. Ranzani-Paiva. 2009. Cytogenotoxicity biomarkers in fat snook *Centropomus parallelus* from Cananeia and Sao Vicente estuaries, SP, Brazil. Genet. Mol. Biol. 32: 151–4.

Lassen, N., T. Estey, R.L. Tanguay, A. Pappa, M.J. Reimers and V. Vasiliou. 2005. Molecular cloning, baculovirus expression, and tissue distribution of the zebrafish aldehyde dehydrogenase 2. Drug Metab. Dispos. 33: 649–56.

Li, G., P. Xie, H. Li, J. Chen, L. Hao and Q. Xiong. 2010. Quantitative profiling of mRNA expression of glutathione S-transferase superfamily genes in various tissues of bighead carp (*Aristichthys nobilis*). J. Biochem. Mol. Toxicol. 24: 250–259.

Linde-Arias, A.R., A.F. Inacio, L.A. Novo, C. de Alburquerque and J.C. Moreira. 2008. Multibiomarker approach in fish to assess the impact of pollution in a large Brazilian river, Paraiba do Sul. Environ. Pollut. 156: 974–9.

Liu, T.A., S. Bhuiyan, M.Y. Liu, T. Sugahara, Y. Sakakibara, M. Suiko, S. Yasuda, Y. Kakuta, M. Kimura, F.E. Williams and M.C. Liu. 2010. Zebrafish as a model for the study of the phase II cytosolic sulfotransferases. Curr. Drug. Metab. 11: 538–46.

Loman, N.J., R.V. Misra, T.J. Dallman, C. Constantinidou, S.E. Gharbia, J. Wain and M.J. Pallen. 2012. Performance comparison of benchtop high-throughput sequencing platforms (vol 30, pg 434, 2012). Nat. Biotechnol. 30: 562–562.

Lucinda, P.H.F. 2005. Systematics of the genus Cnesterodon Garman, 1895 (Cyprinodontiformes: Poeciliidae: Poeciliinae). Neotrop. Ichthyol. 3: 259–270.

Lucinda, P.H.F. 2008. Systematics and biogeography of the genus Phalloceros Eigenmann, 1907 (Cyprinodontiformes : Poeciliidae : Poeciliinae), with the description of twenty-one new species. Neotrop. Ichthyol. 6: 113–158.

Lucinda, P.H.F., R.D. Rosa and R.E. Reis. 2005. Systematics and biogeography of the genus Phallotorynus Henn, 1916 (Cyprinodontiformes: Poeciliidae: Poeciliinae), with description of three new species. Copeia: 609–631.

Matson, C.W., B.W. Clark, M.J. Jenny, C.R. Fleming, M.E. Hahn and R.T. Di Giulio. 2008. Development of the morpholino gene knockdown technique in *Fundulus heteroclitus*: A tool for studying molecular mechanisms in an established environmental model. Aquatic Toxicol. 87: 289–295.

Matsuo, A.Y., B.R. Woodin, C.M. Reddy, A.L. Val and J.J. Stegeman. 2006. Humic substances and crude oil induce cytochrome P450 1A expression in the Amazonian fish species *Colossoma macropomum* (Tambaqui). Environ. Sci. Technol. 40: 2851–8.

Mattos, J.J., M.N. Siebert, K.H. Luchmann, N. Granucci, T. Dorrington, P.H. Stoco, E.C. Grisard and A.C. Bainy. 2009. Differential gene expression in *Poecilia vivipara* exposed to diesel oil water accommodated fraction. Mar. Environ. Res.

Nebert, D.W. and D.W. Russell. 2002. Clinical importance of the cytochromes P450. Lancet. 360: 1155–1162.

Nebert, D.W. and V. Vasiliou. 2004. Analysis of the glutathione S-transferase (GST) gene family. Hum. Genomics. 1: 460–4.

Nebert, D.W. and H.M. Wain. 2003. Update on human genome completion and annotations: gene nomenclature. Hum. Genomics. 1: 66–71.

Nelson, D.R., D.C. Zeldin, S.M.G. Hoffman, L.J. Maltais, H.M. Wain and D.W. Nebert. 2004. Comparison of cytochrome P450 (CYP) genes from the mouse and human genomes, including nomenclature recommendations for genes, pseudogenes and alternative-splice variants. Pharmacogenetics. 14: 1–18.

Oleksiak, M.F., S.I. Karchner, M.J. Jenny, D.G. Franks, D.B.M. Welch and M.E. Hahn. 2011. Transcriptomic assessment of resistance to effects of an aryl hydrocarbon receptor (AHR) agonist in embryos of Atlantic killifish (*Fundulus heteroclitus*) from a marine Superfund site. BMC Genomics. 12.

Payne, J.F. 1976. Field evaluation of benzopyrene hydroxylase induction as a monitor for marine petroleum pollution. Science 191: 945–946.

Pereira, B.F., R.M. Da Silva Alves, D.L. Pitol, J.A. Senhorini, R. De Cassia Gimenes De Alcantara Rocha and F.H. Caetano. 2012. Effects of exposition to polluted environments on blood cells of the fish *Prochilodus lineatus*. Microsc. Res. Tech. 75: 571–5.

Pereira, M.D. 2004. Polychlorinated dibenzo-P-dioxins (PCDD), dibenzofurans (PCDF) and polychlorinated biphenyls (PCB): Main sources, environmental behaviour and risk to man and biota. Quimica Nova. 27: 934–943.

Pina, B., M. Casado and L. Quiros. 2007. Analysis of gene expression as a new tool in ecotoxicology and environmental monitoring. Trac-Trends Anal. Chem. 26: 1145–1154.

Possatto, F.E., M. Barletta, M.F. Costa, J.A. do Sul and D.V. Dantas. 2011. Plastic debris ingestion by marine catfish: an unexpected fisheries impact. Mar. Pollut. Bull. 62: 1098–102.

Ramsdorf, W.A., M.V. Ferraro, C.A. Oliveira-Ribeiro, J.R. Costa and M.M. Cestari. 2009. Genotoxic evaluation of different doses of inorganic lead (PbII) in *Hoplias malabaricus*. Environ. Monit. Assess. 158: 77–85.

Rodriguez-Cea, A., A.R. Linde Arias, M.R. Fernandez de la Campa, J. Costa Moreira and A. Sanz-Medel. 2006. Metal speciation of metallothionein in white sea catfish, *Netuma barba*, and pearl cichlid, *Geophagus brasiliensis*, by orthogonal liquid chromatography coupled to ICP-MS detection. Talanta. 69: 963–9.

Rodriguez-Fuentes, G., R. Aparicio-Fabre, Q. Li and D. Schlenk. 2008. Osmotic regulation of a novel flavin-containing monooxygenase in primary cultured cells from rainbow trout (*Oncorhynchus mykiss*). Drug Metab. Dispos. 36: 1212–7.

Schroder, J.H. and K. Peters. 1988. Differential courtship activity of competing guppy males (*Poecilia reticulata* Peters; Pisces: Poeciliidae) as an indicator for low concentrations of aquatic pollutants. Bull. Environ. Contam. Toxicol. 40: 396–404.

Scornaienchi, M.L., C. Thornton, K.L. Willett and J.Y. Wilson. 2010. Functional differences in the cytochrome P450 1 family enzymes from Zebrafish (*Danio rerio*) using heterologously expressed proteins. Arch. Biochem. Biophys. 502: 17–22.

Sen, B., B. Mahadevan and D.M. DeMarini. 2007. Transcriptional responses to complex mixtures —A review. Mutation Research-Reviews in Mutat. Res. 636: 144–177.

Sim, E., N. Lack, C.J. Wang, H. Long, I. Westwood, E. Fullam and A. Kawamura. 2008. Arylamine N-acetyltransferases: structural and functional implications of polymorphisms. Toxicology 254: 170–83.

Sophos, N.A. and V. Vasiliou. 2003. Aldehyde dehydrogenase gene superfamily: the 2002 update. Chem. Biol. Interact. 143–144: 5–22.

Stacke Ferreira, R., J.M. Monserrat, J.L. Ribas Ferreira, A.C. Kalb, J. Stegeman, A.C. Dias Bainy and J. Zanette. 2012. Biomarkers of organic contamination in the South American fish *Poecilia vivipara* and *Jenynsia multidentata*. J. Toxicol. Environ. Health A. 75: 1023–34.

Stegeman, J.J., J.V. Goldstone and M.E. Hahn. 2010. Perspectives on zebrafish as a model in environmental toxicology. *In*: S.F. Perry, M. Ekker, A.P. Farrel and C.J. Brauner (eds.). Fish Physiology, Zebrafish. Elsevier, London, UK. 468 p.

Steinke, D., W. Salzburger and A. Meyer. 2006. Novel relationships among ten fish model species revealed based on a phylogenomic analysis using ESTs. J. Mol. Evol. 62: 772–784.

Toyoshiba, H., N.J. Walker, A.J. Bailer and C.J. Portier. 2004. Evaluation of toxic equivalency factors for induction of cytochromes P450 CYP1A1 and CYP1A2 enzyme activity by dioxin-like compounds. Toxicol. Ap. Pharmacol. 194: 156–168.

VanVeld, P.A., W.K. Vogelbein, M.K. Cochran, A. Goksoyr and J.J. Stegeman. 1997. Route-specific cellular expression of cytochrome P4501A (CYP1A) in fish (*Fundulus heteroclitus*) following exposure to aqueous and dietary benzo[a]pyrene. Toxicol. Ap. Pharmacol. 142: 348–359.

Vasiliou, V., A. Bairoch, K.F. Tipton and D.W. Nebert. 1999. Eukaryotic aldehyde dehydrogenase (ALDH) genes: human polymorphisms, and recommended nomenclature based on divergent evolution and chromosomal mapping. Pharmacogenetics. 9: 421–34.

Verslycke, T., J.V. Goldstone and J.J. Stegeman. 2006. Isolation and phylogeny of novel cytochrome P450 genes from tunicates (*Ciona* spp.): A CYP3 line in early deuterostomes? Mol. Phylogenet. Evol. 40: 760–771.

Wang, L., B.E. Scheffler and K.L. Willett. 2006. CYP1C1 messenger RNA expression is inducible by benzo[a]pyrene in *Fundulus heteroclitus* embryos and adults. Toxicol. Sci. 93: 331–340.

Williams, L.M. and M.F. Oleksiak. 2008. Signatures of selection in natural populations adapted to chronic pollution. BMC Evol. Biol. 8: 282.

Wirgin, I., N.K. Roy, M. Loftus, R.C. Chambers, D.G. Franks and M.E. Hahn. 2011. Mechanistic basis of resistance to PCBs in Atlantic tomcod from the Hudson River. Science 331: 1322–5.

Zanette, J., M.J. Jenny, J.V. Goldstone, B.R. Woodin, L.A. Watka, A.C.D. Bainy and J.J. Stegeman. 2009. New cytochrome P450 1B1, 1C2 and 1D1 genes in the killifish *Fundulus heteroclitus*: Basal expression and response of five killifish CYP1s to the AHR agonist PCB126. Aquatic Toxicol. 93: 234–243.

ZeRuth, G. and R.S. Pollenz. 2007. Functional analysis of cis-regulatory regions within the dioxin-inducible CYP1A promoter/enhancer region from zebrafish (*Danio rerio*). Chem. Biol. Interact. 170: 100–113.

Zhou, L., H. Ishizaki, M. Spitzer, K.L. Taylor, N.D. Temperley, S.L. Johnson, P. Brear, P. Gautier, Z. Zeng, A. Mitchell, V. Narayan, E.M. McNeil, D.W. Melton, T.K. Smith, M. Tyers, N.J. Westwood and E.E. Patton. 2012. ALDH2 mediates 5-nitrofuran activity in multiple species. Chem. Biol. 19: 883–92.

CHAPTER 3

Nuclear Receptors in Fish and Pollutant Interactions

Afonso C.D. Bainy, * *Jacó J. Mattos* and *Marília N. Siebert*

Introduction

The increasing list of xenobiotics discharged in the aquatic environment is a great risk to the conservation of biodiversity of aquatic organisms. Complex mixtures of organic compounds, inorganic metals, nanoparticles and other contaminants can cause metabolic disorders to the exposed organisms through different mechanisms of action; in some cases these disorders can cause mortality at very low levels of exposure.

The toxic effects of different compounds upon aquatic biota is directly associated with their chemical structure, which will determine the extra and intracellular bioavailability and reactivity. Some chemicals can globally affect the cell metabolism; for example, inhibiting enzymes responsible for ATP production or electron transfer blocking, which will lead to high levels of mortality. On the other hand, sub-lethal exposure to toxicants chemically similar to endobiotic compounds can elicit a multitude of biological responses, including signal transduction disruption, genetic damage and epigenetic disruption. These responses can be evaluated and quantified in aquatic organisms and used for risk assessment of susceptibility to chemical exposure in aquatic ecosystems.

Laboratory of Biomarkers of Aquatic Contamination and Immunochemistry, Department of Biochemistry, CCB, Universidade Federal de Santa Catarina.
*Corresponding author: afonso.bainy@.ufsc.br

There is a growing number of studies showing the role of Nuclear Receptors (NRs) as the main regulators of important pathways involved in the synthesis of Phase I, II and III Biotransformation enzymes in mammals and aquatic organisms.

The NRs superfamily is the largest group of eukaryotic transcription factors that participates as regulators of metabolism, differentiation, apoptosis, biotransformation and cell cycle mechanisms (Huang et al. 2011). The transcriptional activities of NRs are regulated by small, lipophilic molecules (Gronemeyer et al. 2004), including xenobiotics like pharmaceuticals and other chemicals. Altered function in NRs caused by xenobiotics activation has been related to pathologies in mammals (Casals-Casas and Desvergne 2011) and interaction among pesticides with the estrogen and androgen NRs has been associated to health disorders in fish (Soverchia et al. 2005; Sabo-Attwood et al. 2007).

NRs Structure

Nuclear receptors share a common structure divided into six parts, named A/B, C, D, E and F (Fig. 3.1). The C and E parts form two important domains in NRs structure: C region is a highly conserved and centrally located DNA-binding domain (DBD) and E region is a less conserved C-terminal ligand binding domain (LBD).

A/B domain is located in N-terminal region and is highly variable in length and amino acids sequence. This region contains transcriptional activation function termed AF-1, that functions as a ligand-independent transcriptional activator, and several autonomous transactivation domains (AD) (Robinson-Rechavi et al. 2003).

The nuclear receptor DBD is the most conserved region in NRs and one of the most prevalent DNA-interacting regions known. This domain is composed by a highly conserved 66 amino acid region of two zinc-binding loops and a pair of α helices. One of these helices interacts in a sequence-specific way with AGGTCA response element (Rastinejad 2001). Domain D is a less conserved region that acts as a flexible hinge between the C and E domains and also contains the nuclear localization signal (NLS) (Robinson-Rechavi et al. 2003). The E domain is the largest NRs domain and is responsible for many functions, especially, as suggested by the name, ligation to the ligand. In this domain, the 12 α-helixes secondary structure is better conserved than the primary amino acids sequence. This moderately

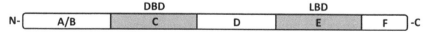

Figure 3.1. Nuclear receptors general structure.

conserved domain contains a binding pocket that interacts with its related hormone or ligand.

In addition to the ligand binding, LBD mediates homo-and heterodimerization, interaction with heat-shock proteins, ligand-dependent transcriptional activity, and in some cases, hormone reversible transcriptional repression. The LBDs contain two well-conserved regions: a "signature motif" or Ti and the COOH-terminal AF-2 motif responsible for ligand-dependent transcriptional activation that acts in recruitment of coactivator proteins (Aranda and Pascual 2001; Bain et al. 2007). Nuclear receptors may or may not contain a final domain in the C-terminus of the E domain, the F domain, whose sequence is extremely variable and whose structure and function are unknown.

Nuclear receptors classification

Nuclear receptors form a superfamily of phylogenetic related proteins. This phylogenetic relationship is the basis of NRs classification and nomenclature system. NR name is given by the form NRxyz where x is the sub-family, y is the group and z the gene in NR that have both DNA-binding and ligand-binding domains. In addition to these nuclear receptors, sub-family NR0 contains nuclear receptors that lack either of these domains, and are not represented in the phylogenetic tree (NRNC 1999).

The complete genome of the fly *Drosophila melanogaster* has 21 genes of this superfamily (King-Jones and Thummel 2005), humans have 48 (Robinson-Rechavi et al. 2003) and mice have 47 (Karpen and Trauner 2010). Compared to other vertebrates, fish have a higher number of nuclear receptor genes. The puffer fish genome (*Fugu rubripes*), for example, contains 68 NRs genes (Maglich et al. 2003) and 72 NR genes are present in *Tetraodon nigroviridis* genome (Metpally et al. 2007) (Table 3.1).

Table 3.1 compares NR genes found in human with two fish species: *F. rubripes* and *T. nigroviridis*. Fish NRs present significant similarity to human NR and all six families described in mammals are also present in fish; however, some new family members are present in fish, like COUP transcription factor (NR2F) family, ERRγ (NR3B3), RAR-related receptor (NR1F) family, present in *F. rubripes* and *T. nigroviridis* genome and RXRβ (NR2B2) present only in *T. nigroviridis* genome. The higher number of nuclear receptors found in fish, compared to mammals, is possibly related to a whole genome duplication event that took place before the divergence of fish lineages that occurred 320 million years ago; this duplication hypothesis coincided with the vast radiation of most modern ray-finned fishes (Christoffels et al. 2004; Vandepoele et al. 2004). It is likely that,

Table 3.1. Comparative number of nuclear receptor genes among *Homo sapiens*, *Tetraeodon nigroviridis* and *Fugu rubripes* with description.

	H. sapiens	*T. nigroviridis*	*F. rubripes*	Description
NR1	20	28	25	Comprises Thyroid Hormones Receptors (TR), Retinoic Acid Receptors (RAR) , Vitamin D Receptor (VDR), Pregnane X Receptor (PXR), Constitutive Androstane Receptor (CAR), Peroxisome Proliferator-Activated Receptor (PPAR) and others
NR2	11	18	16	Contains Retinoic Acid Receptor (RXR), important in heterodimerization and others orphan nuclear receptor as Hepatocyte Nuclear Factor 4 (HNF4) and COUP-TFI
NR3	9	16	15	Includes hormones receptors: Estradiol Receptors (ER), Glucocorticoid Receptor (GR), Mineralocorticoid Receptor (MR), Progesterone Receptor (PR) and Androgen Receptor (AR)
NR4	3	4	4	Contains orphan nuclear receptors NGFI-B, NURR1 and NOR1
NR5	2	3	3	Contains orphan nuclear receptors SF1, LRH-1 and FF1C
NR6	1	1	1	Includes the orphan nuclear receptor GCNF
NR0	2	2	4	Contains orphan nuclear receptor DAX-1 and SHP
TOTAL	48	72	68	

during the course of teleost evolution, after genome duplication, paralogue genes developed different functions in transcriptional regulation, obtaining different gene targets and tissue specific expression patterns.

In the following items, the biological roles of some of the most studied nuclear receptors in fish are addressed.

Estrogen receptor (ER)

The 17β–estradiol (E_2) is involved in several physiological processes like growth, development, behavior and reproduction in fishes and others vertebrates (Matthews et al. 2000). To exert its estrogenic effects, estradiol must interact with Estrogen Receptors (ERs) (Edwards 2005). It is well established that ERs play an important role in two different transduction pathways involving distinct receptor families. In the first pathway, ERs are members of nuclear receptor superfamily (NR3A1; NR3A2) acting as ligand-activated transcription factors to regulate gene transcription. In the second pathway, ERs are G protein-coupled receptors that initiate a rapid non-genomic signal transduction (Edwards 2005). Briefly, the genomic ER mechanism of gene regulation involves: 1) ER activation by 17β–estradiol, 2) ER homodimer formation, 3) ER binding in specific palindromic DNA sequence called estrogen responsive elements (ERE) at the promoter site of estrogen responsive genes, 4) recruitment of cofactors necessary for gene modulation, and 5) transcriptional activation of estrogen responsive genes (Boyce-Derricott et al. 2010). Among the estrogen-responsive genes are *zona radiata* proteins (*Zrps*) and vitellogenin (*Vtg*) (Arukwe et al. 2002; Arukwe and Goksoyr 2003). In fish, Zrps are involved in the egg shell hardening process and polyspermy prevention after fertilization (Arukwe et al. 2002). Vitellogenin is a precursor to egg yolk protein in oviparous vertebrates (Arukwe and Goksoyr 2003).

Similar to other animals, fishes have two *ER* subtypes, *ERα* (NR3A1) and *ERβ* (NR3A2) (Nagler et al. 2007; Boyce-Derricott et al. 2009). However, the *ERβ* is found as two different isoforms (*ERβ1; ERβ2*) in several fish species because of the whole genome duplication event that occurred in ray-finned fishes after they diverged from the lobe-finned fishes (Sarcopterygii) (Amores et al. 1998). Studies with E_2 exposure in zebrafish (*Danio rerio*) showed that ERα can up regulate its own mRNA and be up regulated by ERβ2 (Menuet et al. 2004). However, ERβ1 has no effect on *ERα* gene transcription (Menuet et al. 2004). This fact reveals how complex is gene expression network in fish when exposed to estrogenic compounds.

In addition to natural hormones, several classes of environmental contaminants, also known as endocrine disrupting compounds (EDCs; for more details, see chapter 10), could interact with ERs and regulate estrogen responsive genes (Navas and Segner 1998; Navas and Segner 2000; Yan et al. 2012). EDCs can be defined as an exogenous agent that interferes with the production, release, transport, metabolism, binding, action, or elimination of natural hormones in the body responsible for the maintenance of homeostasis and the regulation of developmental processes (Kavlock et al. 1996).

The EDCs are of high relevance for wildlife health, since endocrine signaling controls many essential physiological processes such as growth and development, stress response, and ultimately reproduction and population development. Some EDCs are estrogenic and anti-estrogenic contaminants that are continuously released in aquatic environment and cause negative effects on growth, development and reproduction in aquatic organisms (Arcand-Hoy and Benson 1998; Scholz and Mayer 2008). Among the contaminants that can interact with rainbow trout (*Oncorhynchus mykiss*) ER are: the pharmaceuticals Tamoxifen and its metabolite 4-hydroxitamoxifen, 17α-Ethynylestradiol, Diethylstilbestrol, Estradiol Benzoate; the pesticides methoxichlor and its bis-hydroxylated metabolite HPTE, dichlorodiphenyltrichloroethane (DDT) and its degradation product dichlorodiphenyldichloroethylene (DDE); the alkyl phenolic compound 4-*t*-octylphenol and the industrial product Bisphenol A (Matthews et al. 2000). These substances are ERs ligands, acting as potential agonists or antagonists to elicit estrogenic or anti-estrogenic effects (i.e., endocrine disruption), respectively (Scholz and Mayer 2008).

Some substances do not interact directly with ERs, but can elicit (anti) estrogenic effects. The aryl hydrocarbon receptor (AhR), a member of the helix–loop–helix–PAS family of gene regulatory protein, is activated by ligands such as 2,3,7,8-tetrachlorodibenzo-p-dioxin (TCDD) and polychlorinated biphenyls (PCBs) and polycyclic aromatic hydrocarbons (PAHs) (Navas and Segner 1998). These chemicals and the AhR agonists 3-methylcholanthrene (3MC) and β-naphthoflavone (βNF) were able to cause anti-estrogenic effects such as reduced Vtg synthesis or impaired gonad development in fish (Wannemacher et al. 1992; Anderson et al. 1996) or fish cells (Anderson et al. 1996; Navas and Segner 2000).

Several mechanisms of AhR-mediated anti-estrogenicity have been proposed (Navas and Segner 1998). One possibility is an enhanced metabolism of E_2 due to the AhR-mediated induction of CYP1A activity. However, some *in vitro* experiments with rainbow trout hepatocytes cell culture demonstrated that AhR agonists did not enhance the E_2 catabolism (Navas and Segner 2000). The other possible mechanisms are: (a) decreasing E_2 binding to the ER and ER binding to the ERE; (b) down-regulating ER protein by blockage of ER gene transcription and (c) blocking of estrogen responsive gene transcription through binding a repressor site within the promoter. However, rainbow trout treated simultaneously with E_2 and βNF did not present a decline of ER-ERE binding activity, although they showed a reduction of the ER-E_2 binding capacity (Anderson et al. 1996). Studies with fish hepatocytes cell culture demonstrated that βNF-induced anti-estrogenic effects may be dependent of AhR interaction with a xenobiotic response element (XRE) present in the regulatory regions of the VTG gene; this result supports the mechanism "c" above (Navas and Segner 2000). Although AhR ligands are able to elicit anti-estrogenic effects, βNF-E_2 co-exposure experiments also showed an increased estrogenic effect (i.e., augmented Vtg expression) depending on the dose of βNF and E_2 injected in rainbow trout juveniles (Anderson et al. 1996). One explanation for the βNF-E_2 synergism may involve ER activation by protein kinase-dependent phosphorylation, through an AhR-mediated mechanism, but more studies are necessary to test this hypothesis (Anderson et al. 1996). Another interesting issue is whether cross-talk between AhR and ER is bidirectional. A bidirectional inhibitory mechanism could arise from competition between the two receptors for a common co-activator (Bemanian et al. 2004). However, this would not primarily account for cross-talk between the AhR and ERα signaling pathways, since E2 was not able to promote an inhibitory effect on *CYP1A* gene transcription in rainbow trout cells treated with TCDD (Bemanian et al. 2004). One study reported a unidirectional inhibiting AhR-ER cross-talk, resulting in decreased *Vtg* mRNA induction in rainbow trout hepatocytes when EE2 was co-administrated with BNF (Grans et al. 2010). On the other hand, other authors proposed a bidirectional inhibitory cross-talk in Atlantic salmon co-exposed to nonylphenol and 3,3',4,4'-tetrachlorobiphenyl (PCB-77) (Mortensen and Arukwe 2007). The reason for the discrepancy between studies is not clear and requires further studies to be elucidated.

Given that EDC compounds are found in estuaries and rivers, it is reasonable to suggest that these compounds may represent risks as environmental (anti)estrogens, a risk that may be greater in areas close to the outlets of sanitary and industrial effluents. Consequently, it is necessary develop sensitive biomarkers to monitor the presence of these chemicals and related compounds in the environment. In addition, studies about critical population parameters such as offspring survival and recruitment are necessary to link molecular and biochemical alterations in endocrine system with higher levels of biological organization.

Peroxisome proliferator-activated receptors (PPARs)

Peroxisome proliferator-activated receptors (PPARs) are ligand-activated transcription factor members of the nuclear hormone receptors superfamily which play vital roles in a variety of cell process like regulation of energy homeostasis (especially glucose and lipid), cell proliferation, differentiation and survival (Fang et al. 2012). Three PPAR subtypes, *PPARα, PPARβ*, and *PPARγ*, have been identified in a wide range of vertebrates, including fish (Ibabe et al. 2002; Ibabe et al. 2004; Ibabe et al. 2005)).

In mammals, the three subtypes differ in tissue distribution and response to concentrations of their natural ligands. *PPARα* is expressed in many tissues, principally those with high fatty acid oxidation capacity. PPARα plays a central role in maintenance of lipid homeostasis, increasing the cellular capacity to mobilize and catabolize fatty acids. This function seems to be particularly active in the liver during starvation, where the oxidation of fatty acids is essential for energy production. Under these conditions, PPARα is probably activated by endogenous fatty acids and fatty acid derivatives. Beside natural ligands, PPARα can be activated by fibrates, widely used drugs that reduce serum lipids by increasing lipid oxidation and industrial phthalate-monoester plasticizers. The number of direct PPARα-target genes is large and includes many that encode enzymes that are involved in glucose, lipid and amino acid metabolism and also biotransformation enzymes such Cytochrome P450 superfamily members, UDP-glucuronosyl-transferase 1A9 and Multidrug resistance 2 (Pyper et al. 2010).

PPARγ acts in adipocyte differentiation and energy storage by adipocytes through regulation of the expression of adipocyte-specific genes involved in lipid storage and control of metabolism. PPARγ ligands are fatty acids, such as linoleic acid, fatty acid derivatives, such as prostaglandin J2, and the synthetic antidiabetic drugs, thiazolidinediones (Miard and Fajas 2005).

Of all three PPAR isoforms, PPARβ/δ is less studied, mostly because its near ubiquitous tissue expression in mammals suggests that it may have a more general housekeeping function (Kliewer et al. 1994). As others PPARs, PPARβ/δ can be activated by fatty acids and fatty acid derivatives besides synthetic and naturally occurring eicosanoids, such as prostaglandin A1, iloprost, 15d-J2, and carbaprostacyclin. This isoform regulates gene expression in diverse tissues including fat, skeletal muscle and heart. Its transcriptional regulatory activity enhances fatty acid catabolism and energy uncoupling, resulting in decreased triglyceride stores in fat tissue, improved endurance performance in skeletal muscle and enhanced cardiac contractility. PPARβ/δ also might play a protective role in the liver, down-regulating inflammatory signals during liver damage. Moreover, PPARβ/δ activation in liver suppresses hepatic glucose output, contributing to improved glucose homeostasis (Barish et al. 2006).

In fish a single ortholog of the human *PPARγ* and *PPARβ/δ* subtypes and two orthologs of human *PPARα* have been identified (Maglich et al. 2003; Metpally et al. 2007). In *Fugu rubripes*, all four *PPAR* genes were expressed in the gill, brain, gut, heart, liver and ovary of adults (Maglich et al. 2003). *PPAR* orthologs to mammalian α, β and γ subtypes have also been cloned in sea bass (*Dicentrarchus labrax*). In this species, *PPARα* is mainly expressed in the liver, *PPARγ* in adipose tissue, and *PPARβ* in all tissues, but mostly in the liver. *PPARs* expression pattern suggest that *PPARs* isoforms may have conserved function between fish and mammals (Boukouvala et al. 2004).

PPAR activation mechanism involves: a) PPAR ligand binding; b) heterodimer formation with another nuclear receptor, the retinoid X receptor (RXR); c) heterodimer binding at specific DNA regulatory element (peroxisome proliferator response element, PPRE) located in the promoter region of target genes d) transactivation of PPAR target genes such as Acyl-CoA synthetase , Acyl-CoA oxidase 1 and Malic Enzyme (Ibabe et al. 2002). However, target genes containing PPREs are not known in fish, except for genes of glutathione-S-transferase in the plaice (Leaver et al. 1997). It has been shown that a cross-talk between PPAR/RXR and ER may also occur. PPARs/RXRs were able to inhibit transactivation by the ER through competition for ERE binding (Keller et al. 1995).

Beside endogenous fatty acid and fatty acid derivatives ligands, PPARs can be also activated by xenobiotics like hypolipidemic drugs, phthalate ester, plasticizers and herbicides (Schoonjans et al. 1996). Natural fatty acids and synthetic hypolipidemic compounds can act as ligands of the sea bass PPARα and β, suggesting that PPAR in this fish has similar roles in mechanisms associated with lipid metabolism functions, as for mammalian PPAR (Boukouvala et al. 2004). However, it remains unclear whether piscine PPAR isotypes act through similar mechanisms and perform the same critical functions in lipid metabolism as they do in mammals (Leaver et al. 2005).

The increase in the volume and number of peroxisomes, concomitantly with peroxisomal β-oxidation enzymes induction, is characteristic of peroxisome proliferation and has been observed in several species after treatment with fibrate hypolipidemic drugs and a diverse array of environmental organic pollutants like phthalate ester plasticizers, chlorophenoxyacetate herbicides and other pesticides, steroids, solvents and diverse industrial chemicals, food derivatives, polycyclic aromatic hydrocarbons (PAHs), and polychlorinated biphenyls (PCBs) (Ibabe et al. 2002; Cajaraville et al. 2003; Ibabe et al. 2004; Ortiz-Zarragoitia et al. 2006). In zebrafish (*Danio rerio*), exposure to the pesticide methoxychlor, the phthalate dibutylphthalate, the alkylphenol 4-tert-octylphenol, and estrogens causes an increase in liver peroxisomal volume, surface and numerical densities together with a significant induction of acyl-CoA oxidase (AOX) activity (Ibabe et al. 2004). Induction of peroxisomal β-oxidation enzymes like AOX occurs through transcriptional activation of corresponding genes by PPARs (Issemann and Green 1990). In recent years, peroxisome proliferation has been proposed as a novel early warning biomarker that may be useful in marine pollution monitoring because of this significant response to pollutant exposure.

Pregnane X Receptor (PXR)

Pregnane X receptor (PXR) is a member of the nuclear hormone receptor (NHR) superfamily, commonly described as a promiscuous xenophore or a pharmacophore, since it can be activated by a wide array of structurally different endo and xenobiotics, including numerous pharmaceuticals and other environmental pollutants (Krasowski et al. 2005; Ekins et al. 2008). PXR regulates the expression of numerous genes that function in biotransformation and the disposition of xenobiotics and endobiotics upon binding as a heterodimer with RXR to an AG(G/T)TCA DNA motif in target promoter regions, such as phase 1 *cytochrome P4503A* (*CYP3A*) (Kliewer et al. 1998), phase II *glutathione S-transferases* (*GST*) (Gong et al. 2006), *UDP-glucuronosyltransferases* (*UGT*) (Zhou et al. 2005) and *sulfotransferases*

(*SULT*) (Runge-Morris and Kocarek 2005), phase III *multidrug resistance protein 1* (*MDR1*, also known as *ABCB1*) (Masuyama et al. 2005), *multidrug resistance-associated protein 2* (*MRP2*) (Maher et al. 2005), and *organic anion transporter polypeptide 2* (Guo et al. 2002), among others. Thus, this nuclear receptor shows a master role in the regulation of levels of toxic compounds in different tissues, particularly in liver.

PXR genes have been cloned from a variety of mammalian species, such as human, monkey, pig, dog, mouse and rat (Bertilsson et al. 1998; Moore and Kliewer 2000). In fish, the full sequence of *PXR* has been identified in zebrafish (Danio rerio) (Bainy and Stegeman 2002), fugu (*Takifugu rubripes*) (Genbank XP_003961757.1), rainbow trout (*Oncorhyncus mykiss*) (Wassmur et al. 2010), spotted green pufferfish (*Tetraodon nigroviridis*) (Genbank CAG05861.1) and Nile tilapia (*Oreochromis niloticus*) (Genbank BAM 36700.1). In adult zebrafish, *PXR* is expressed in the liver, kidney, ovary, gut, testes, gill, spleen and heart, and also in the early life stages (24, 48 and 96 h post-fertilization), suggesting that this receptor plays an important role during development (Bainy and Stegeman 2002).

Some studies have shown cross-species differences in PXR ligand specificity, including the selectivity for bile salts (Krasowski et al. 2011). Compared to mammalian *PXR*, zebrafish *PXR* is activated by a structurally narrow range of bile salts. Different authors have proposed that PXR has evolved due to the adaptation to evolutionary changes to bile salts structure or other endogenous molecules and function as xenobiotic sensor (Schuetz et al. 2001; Moore et al. 2002; Krasowski et al. 2005; Reschly et al. 2008).

Some authors suggest that *PXR* emerged together with *CAR* (*Constitutive Androstane Receptor*) and *Vitamin D receptor* (*VDR*) from an ancestral NR1I gene in early vertebrates, as a result of whole-genome duplications (Mathas et al. 2012). It has been proposed that PXR activity is possibly regulated by phosphorylation at several key amino acid residues (Lichti-Kaiser et al. 2009).

Bresolin et al. (2005) showed that fish injected with the mammalian CYP3A agonist, pregnenolone 16α-carboninitrile (PCN), induced the expression of hepatic *PXR*, *CYP3A* and *ABCB5*, suggesting that the intrinsic modulatory effect of PXR upon these genes is conserved in zebrafish. Incubation of carp (*Cyprinus carpio*) primary hepatocytes with the human PXR agonist rifampicin (RIF) showed an induction of all PXR target genes involved in phase I (*CYP2K*, *CYP3A*), phase II (*GSTα*, *GSTπ*) drug metabolism and drug transporters *MDR1* and *MRP2* (Corcoran et al. 2012). Likewise the PXR antagonist ketoconazole (KET) inhibited responses of *CYP2K* and *CYP3A*. However, exposure of these cells to the pharmaceuticals ibuprofen (IBU), clotrimazole (CTZ), clofibric acid (CFA) and propranolol

(PRP), found responses to IBU and CFA, but not CTZ or PRP. This is in contrast with mammals, where CTZ is a potent PXR-agonist, suggesting some divergence in the regulation pathways with those in mammals (Corcoran et al. 2012).

A recent study showed a significant induction of both *PXR* and *CYP3A4* in unfed female fathead minnows treated with CTZ, or pregnene-16alpha-olone, compared to fed females and males, which points to the need to understand the relationship between gender and diet with adaptation to contaminated habitats (Crago and Klaper 2011). Studies carried out in zebrafish showed evidence of PXR-mediated induction of enzyme expression, with increases in testosterone 6-beta-hydroxylation activity (a measure of cytochrome P450 3A activity in other species) and flurbiprofen 4-hydroxylation activity (measure of cytochrome P450 2C activity) following exposure to known PXR activators (Reschly et al. 2007). Likewise *in vivo* assays using zebrafish demonstrated increased hepatic transcription of another PXR target, multidrug resistance gene (*ABCB55*), following injection of the major zebrafish bile salt, 5alpha-cyprinol 27-sulfate (Reschly et al. 2007). The persistent DDT metabolite, 1,1-dichloro-2,2-bis(p-chlorophenyl) ethylene induced the transcription of *CYP3A* and *PXR* mRNA in liver of salmon parr, suggesting possible physiological and endocrine consequences of exposure to endocrine-disrupting chemicals for this species during smoltification (Mortensen and Arukwe 2006).

In recent years, there has been growing interest in the elucidation of comparative aspects of the evolutionary responses of PXR related with the toxic responses of different chemicals in the environment. It is quite plausible that this NR has a master role in the adaptation mechanisms of aquatic organisms chronically exposed to complex mixtures of contaminants.

Conclusion

Despite the growing interest in elucidating the role of NRs in the metabolism of endo and xenobiotics in fish, many questions are still unsolved. What are the environmental pressures that have promoted the acquisition and retention of transcriptional factors critical to the survival of species in contaminated environments? Why are some species or even individuals able to adapt more quickly to chemically stressed sites? Why are some endangered species disappearing from the environment with no clear justification? Why are some species which show more polymorphism in different genes better able to survive in contaminated sites? Answering these questions would help us to predict the magnitude and risk from contamination in aquatic systems.

Acknowledgements

Afonso C.D. Bainy is the recipient of CNPq productivity fellowship.

Keywords: Nuclear receptor (NR), Estrogen receptor (ER), Peroxisome proliferator activated receptor (PPAR), Pregnane X receptor (PXR), Retinoid X receptor (RXR), Aryl hydrocarbon receptor (AhR)

References Cited

Amores, A., A. Force, Y.L. Yan, L. Joly, C. Amemiya, A. Fritz, R.K. Ho, J. Langeland, V. Prince, Y.L. Wang, M. Westerfield, M. Ekker and J.H. Postlethwait. 1998. Zebrafish hox clusters and vertebrate genome evolution. Science 282: 1711–1714.

Anderson, M.J., H. Olsen, F. Matsumura and D.E. Hinton. 1996. *In vivo* modulation of 17 beta-estradiol-induced vitellogenin synthesis and estrogen receptor in rainbow trout (Oncorhynchus mykiss) liver cells by beta-naphthoflavone. Toxicol. App. Pharmacol. 137: 210–218.

Aranda, A. and A. Pascual. 2001. Nuclear hormone receptors and gene expression. Physiol. Rev. 81: 1269–1304.

Arcand-Hoy, L.D. and W.H. Benson. 1998. Fish reproduction: An ecologically relevant indicator of endocrine disruption. Environ. Toxicol. Chem. 17: 49–57.

Arukwe, A. and A. Goksoyr. 2003. Eggshell and egg yolk proteins in fish: hepatic proteins for the next generation: oogenetic, population, and evolutionary implications of endocrine disruption. Comp. Hepatol. 2: 4.

Arukwe, A., S.W. Kullman, K. Berg, A. Goksoyr and D.E. Hinton. 2002. Molecular cloning of rainbow trout (Oncorhynchus mykiss) eggshell zona radiata protein complementary DNA: mRNA expression in 17beta-estradiol- and nonylphenol-treated fish. Comp. Biochem. Physiol. Part B, Biochem. Mol. Biol. 132: 315–326.

Bain, D.L., A.F. Heneghan, K.D. Connaghan-Jones and M.T. Miura. 2007. Nuclear receptor structure: implications for function. Annual Rev. Physiol. 69: 201–220.

Bainy, A.C.D. and J.J. Stegeman. 2002. Cloning of a pregnane X receptor (PXR) in Zebrafish (Danio rerio). *In*: Abstract Book. SETAC 23rd Annual Meeting, 2002. Pensacola, Florida, EUA: SETAC Press, 301 p.

Barish, G.D., V.A. Narkar and R.M. Evans. 2006. PPAR delta: a dagger in the heart of the metabolic syndrome. J. Clin. Invest. 116: 590–597.

Bemanian, V., R. Male and A. Goksoyr. 2004. The aryl hydrocarbon receptor-mediated disruption of vitellogenin synthesis in the fish liver: Cross-talk between AHR- and ERalpha-signalling pathways. Comp. Hepatol. 3: 2.

Boukouvala, E., E. Antonopoulou, L. Favre-Krey, A. Diez, J.M. Bautista, M.J. Leaver, D.R. Tocher and G. Krey. 2004. Molecular characterization of three peroxisome proliferator-activated receptors from the sea bass (Dicentrarchus labrax). Lipids 39: 1085–1092.

Boyce-Derricott, J., J.J. Nagler and J.G. Cloud. 2009. Regulation of hepatic estrogen receptor isoform mRNA expression in rainbow trout (Oncorhynchus mykiss). Gen. Comp. Endocrinol. 161: 73–78.

Boyce-Derricott, J., J.J. Nagler and J.G. Cloud. 2010. Variation among rainbow trout (Oncorhynchus mykiss) estrogen receptor isoform 3' untranslated regions and the effect of 17beta-estradiol on mRNA stability in hepatocyte culture. DNA and Cell Biol. 29: 229–234.

Cajaraville, M.P., I. Cancio, A. Ibabe and A. Orbea. 2003. Peroxisome proliferation as a biomarker in environmental pollution assessment. Micr. Res. Techn. 61: 191–202.

Casals-Casas, C. and B. Desvergne. 2011. Endocrine disruptors: from endocrine to metabolic disruption. Ann. Rev. Physiol. 73: 135–162.

Christoffels, A., E.G. Koh, J.M. Chia, S. Brenner, S. Aparicio and B. Venkatesh. 2004. Fugu genome analysis provides evidence for a whole-genome duplication early during the evolution of ray-finned fishes. Mol. Biol. Evol. 21: 1146–1151.

Corcoran, J., A. Lange, M.J. Winter and C.R. Tyler. 2012. Effects of pharmaceuticals on the expression of genes involved in detoxification in a carp primary hepatocyte model. Environ. Sci. Technol. 46: 6306–6314.

Crago, J. and R.D. Klaper. 2011. Influence of gender, feeding regimen, and exposure duration on gene expression associated with xenobiotic metabolism in fathead minnows (Pimephales promelas). Comp. Biochem. Physiol. Toxicol. Pharmacol. 154: 208–212.

Edwards, D.P. 2005. Regulation of signal transduction pathways by estrogen and progesterone. Ann. Rev. Physiol. 67: 335–376.

Ekins, S., E.J. Reschly, L.R. Hagey and M.D. Krasowski. 2008. Evolution of pharmacologic specificity in the pregnane X receptor. BMC Evol. Biol. 8: 103.

Fang, C., X. Wu, Q. Huang, Y. Liao, L. Liu, L. Qiu, H. Shen and S. Dong. 2012. PFOS elicits transcriptional responses of the ER, AHR and PPAR pathways in Oryzias melastigma in a stage-specific manner. Aquat. Toxicol. 106–107: 9–19.

Gong, H., S.V. Singh, S.P. Singh, Y. Mu, J.H. Lee, S.P. Saini, D. Toma, S. Ren, V.E. Kagan, B.W. Day, P. Zimniak and W. Xie. 2006. Orphan nuclear receptor pregnane X receptor sensitizes oxidative stress responses in transgenic mice and cancerous cells. Mol. Endocrinol. 20: 279–290.

Grans, J., B. Wassmur and M.C. Celander. 2010. One-way inhibiting cross-talk between arylhydrocarbon receptor (AhR) and estrogen receptor (ER) signaling in primary cultures of rainbow trout hepatocytes. Aquat. Toxicol. 100: 263–270.

Gronemeyer, H., J.A. Gustafsson and V. Laudet. 2004. Principles for modulation of the nuclear receptor superfamily. Nature Rev. Drug Disc. 3: 950–964.

Guo, G.L., J. Staudinger, K. Ogura and C.D. Klaassen. 2002. Induction of rat organic anion transporting polypeptide 2 by pregnenolone-16alpha-carbonitrile is via interaction with pregnane X receptor. Mol. Pharmacol. 61: 832–839.

Huang, R., M. Xia, M.H. Cho, S. Sakamuru, P. Shinn, K.A. Houck, D.J. Dix, R.S. Judson, K.L. Witt, R.J. Kavlock, R.R. Tice and C.P. Austin. 2011. Chemical genomics profiling of environmental chemical modulation of human nuclear receptors. Environ. Health Persp. 119: 1142–1148.

Ibabe, A., E. Bilbao and M.P. Cajaraville. 2005. Expression of peroxisome proliferator-activated receptors in zebrafish (Danio rerio) depending on gender and developmental stage. Histochem. Cell Biol. 123: 75–87.

Ibabe, A., M. Grabenbauer, E. Baumgart, H.D. Fahimi and M.P. Cajaraville. 2002. Expression of peroxisome proliferator-activated receptors in zebrafish (Danio rerio). Histochem. Cell Biol. 118: 231–239.

Ibabe, A., M. Grabenbauer, E. Baumgart, A. Volkl, H.D. Fahimi and M.P. Cajaraville. 2004. Expression of peroxisome proliferator-activated receptors in the liver of gray mullet (Mugil cephalus). Acta Histochem. 106: 11–19.

Issemann, I. and S. Green. 1990. Activation of a member of the steroid hormone receptor superfamily by peroxisome proliferators. Nature 347: 645–650.

Karpen, S.J. and M. Trauner. 2010. The new therapeutic frontier—nuclear receptors and the liver. J. Hepatol. 52: 455–462.

Kavlock, R.J., G.P. Daston, C. DeRosa, P. Fenner-Crisp, L.E. Gray, S. Kaattari, G. Lucier, M. Luster, M.J. Mac, C. Maczka, R. Miller, J. Moore, R. Rolland, G. Scott, D.M. Sheehan, T. Sinks and H.A. Tilson. 1996. Research needs for the risk assessment of health and environmental effects of endocrine disruptors: a report of the U.S. EPA-sponsored workshop. Environ. Health Perspect. 104 Suppl. 4: 715–740.

Keller, H., F. Givel, M. Perroud and W. Wahli. 1995. Signaling cross-talk between peroxisome proliferator-activated receptor/retinoid X receptor and estrogen receptor through estrogen response elements. Mol. Endocrinol. 9: 794–804.

King-Jones, K. and C.S. Thummel. 2005. Nuclear receptors—a perspective from Drosophila. Nature Rev. Gen. 6: 311–323.

Kliewer, S.A., B.M. Forman, B. Blumberg, E.S. Ong, U. Borgmeyer, D.J. Mangelsdorf, K. Umesono and R.M. Evans. 1994. Differential expression and activation of a family of murine peroxisome proliferator-activated receptors. Proc. Natl. Acad. Sci. USA 91: 7355–7359.

Kliewer, S.A., J.T. Moore, L. Wade, J.L. Staudinger, M.A. Watson, S.A. Jones, D.D. McKee, B.B. Oliver, T.M. Willson, R.H. Zetterstrom, T. Perlmann and J.M. Lehmann. 1998. An orphan nuclear receptor activated by pregnanes defines a novel steroid signaling pathway. Cell 92: 73–82.

Krasowski, M.D., N. Ai, L.R. Hagey, E.M. Kollitz, S.W. Kullman, E.J. Reschly and S. Ekins. 2011. The evolution of farnesoid X, vitamin D, and pregnane X receptors: insights from the green-spotted pufferfish (*Tetraodon nigriviridis*) and other non-mammalian species. BMC Biochem. 12: 5.

Krasowski, M.D., K. Yasuda, L.R. Hagey and E.G. Schuetz. 2005. Evolution of the pregnane x receptor: adaptation to cross-species differences in biliary bile salts. Mol. Endocrinol. 19: 1720–1739.

Leaver, M.J., E. Boukouvala, E. Antonopoulou, A. Diez, L. Favre-Krey, M.T. Ezaz, J.M. Bautista, D.R. Tocher and G. Krey. 2005. Three peroxisome proliferator-activated receptor isotypes from each of two species of marine fish. Endocrinol. 146: 3150–3162.

Leaver, M.J., J. Wright and S.G. George. 1997. Structure and expression of a cluster of glutathione S-transferase genes from a marine fish, the plaice (Pleuronectes platessa). Biochem. J. 2: 405–412.

Lichti-Kaiser, K., C. Xu and J.L. Staudinger. 2009. Cyclic AMP-dependent protein kinase signaling modulates pregnane x receptor activity in a species-specific manner. J. Biol. Chem. 284: 6639–6649.

Maglich, J.M., J.A. Caravella, M.H. Lambert, T.M. Willson, J.T. Moore and L. Ramamurthy. 2003. The first completed genome sequence from a teleost fish (Fugu rubripes) adds significant diversity to the nuclear receptor superfamily. Nucleic Acids Res. 31: 4051–4058.

Maher, J.M., X. Cheng, A.L. Slitt, M.Z. Dieter and C.D. Klaassen. 2005. Induction of the multidrug resistance-associated protein family of transporters by chemical activators of receptor-mediated pathways in mouse liver. Drug Metb. Disp. 33: 956–962.

Masuyama, H., N. Suwaki, Y. Tateishi, H. Nakatsukasa, T. Segawa and Y. Hiramatsu. 2005. The pregnane X receptor regulates gene expression in a ligand- and promoter-selective fashion. Mol. Endocrinol. 19: 1170–1180.

Mathas, M., O. Burk, H. Qiu, C. Nusshag, U. Godtel-Armbrust, D. Baranyai, S. Deng, K. Romer, D. Nem, B. Windshugel and L. Wojnowski. 2012. Evolutionary history and functional characterization of the amphibian xenosensor CAR. Mol. Endocrinol. 26: 14–26.

Matthews, J., T. Celius, R. Halgren and T. Zacharewski. 2000. Differential estrogen receptor binding of estrogenic substances: a species comparison. J. Steroid Biochem. Mol. Biol. 74: 223–234.

Menuet, A., Y. Le Page, O. Torres, L. Kern, O. Kah and F. Pakdel. 2004. Analysis of the estrogen regulation of the zebrafish estrogen receptor (ER) reveals distinct effects of ERalpha, ERbeta1 and ERbeta2. J. Mol. Endocrinol. 32: 975–986.

Metpally, R.P., R. Vigneshwar and R. Sowdhamini. 2007. Genome inventory and analysis of nuclear hormone receptors in *Tetraodon nigroviridis*. J. Biosci. 32: 43–50.

Miard, S. and L. Fajas. 2005. Atypical transcriptional regulators and cofactors of PPARgamma. Int. J. Obes. 29 Suppl 1: S10–12.

Moore, L.B., J.M. Maglich, D.D. McKee, B. Wisely, T.M. Willson, S.A. Kliewer, M.H. Lambert and J.T. Moore. 2002. Pregnane X receptor (PXR), constitutive androstane receptor (CAR), and benzoate X receptor (BXR) define three pharmacologically distinct classes of nuclear receptors. Mol. Endocrinol. 16: 977–986.

Mortensen, A.S. and A. Arukwe. 2006. The persistent DDT metabolite, 1,1-dichloro-2,2-bis(p-chlorophenyl)ethylene, alters thyroid hormone-dependent genes, hepatic cytochrome

P4503A, and pregnane X receptor gene expressions in Atlantic salmon (Salmo salar) Parr. Environ. Toxicol. Chem. 25: 1607–1615.

Mortensen, A.S. and A. Arukwe. 2007. Targeted salmon gene array (SalArray): a toxicogenomic tool for gene expression profiling of interactions between estrogen and aryl hydrocarbon receptor signalling pathways. Chem. Res. Toxicol. 20: 474–488.

Nagler, J.J., T. Cavileer, J. Sullivan, D.G. Cyr and C. Rexroad, 3rd. 2007. The complete nuclear estrogen receptor family in the rainbow trout: discovery of the novel ERalpha2 and both ERbeta isoforms. Gene 392: 164–173.

Navas, J.M. and H. Segner. 1998. Antiestrogenic activity of anthropogenic and natural chemicals. Environ. Sci. Poll. Res. Int. 5: 75–82.

Navas, J.M. and H. Segner. 2000. Antiestrogenicity of beta-naphthoflavone and PAHs in cultured rainbow trout hepatocytes: evidence for a role of the arylhydrocarbon receptor. Aquat. Toxicol. 51: 79–92.

NRNC. 1999. A unified nomenclature system for the nuclear receptor superfamily. Cell 97: 161–163.

Ortiz-Zarragoitia, M., J.M. Trant and M.P. Cajaravillet. 2006. Effects of dibutylphthalate and ethynylestradiol on liver peroxisomes, reproduction, and development of zebrafish (Danio rerio). Environ. Toxicol. Chem. 25: 2394–2404.

Pyper, S.R., N. Viswakarma, S. Yu and J.K. Reddy. 2010. PPARalpha: energy combustion, hypolipidemia, inflammation and cancer. Nuclear. Rec. Signal. 8: e002.

Rastinejad, F. 2001. Retinoid X receptor and its partners in the nuclear receptor family. Curr. Op. Struct. Biol. 11: 33–38.

Reschly, E.J., N. Ai, S. Ekins, W.J. Welsh, L.R. Hagey, A.F. Hofmann and M.D. Krasowski. 2008. Evolution of the bile salt nuclear receptor FXR in vertebrates. J. Lipid Res. 49: 1577–1587.

Reschly, E.J., A.C. Bainy, J.J. Mattos, L.R. Hagey, N. Bahary, S.R. Mada, J. Ou, R. Venkataramanan and M.D. Krasowski. 2007. Functional evolution of the vitamin D and pregnane X receptors. BMC Evol. Biol. 7: 222.

Robinson-Rechavi, M., H. Escriva Garcia and V. Laudet. 2003. The nuclear receptor superfamily. J. Cell Sci. 116: 585–586.

Runge-Morris, M. and T.A. Kocarek. 2005. Regulation of sulfotransferases by xenobiotic receptors. Curr. Drug Metab. 6: 299–307.

Sabo-Attwood, T., J.L. Blum, K.J. Kroll, V. Patel, D. Birkholz, N.J. Szabo, S.Z. Fisher, R. McKenna, M. Campbell-Thompson and N.D. Denslow. 2007. Distinct expression and activity profiles of largemouth bass (Micropterus salmoides) estrogen receptors in response to estradiol and nonylphenol. J. Mol. Endocrinol. 39: 223–237.

Scholz, S. and I. Mayer. 2008. Molecular biomarkers of endocrine disruption in small model fish. Mol. Cell. Endocr. 293: 57–70.

Schoonjans, K., B. Staels and J. Auwerx. 1996. Role of the peroxisome proliferator-activated receptor (PPAR) in mediating the effects of fibrates and fatty acids on gene expression. J. Lipid Res. 37: 907–925.

Schuetz, E.G., S. Strom, K. Yasuda, V. Lecureur, M. Assem, C. Brimer, J. Lamba, R.B. Kim, V. Ramachandran, B.J. Komoroski, R. Venkataramanan, H. Cai, C.J. Sinal, F.J. Gonzalez and J.D. Schuetz. 2001. Disrupted bile acid homeostasis reveals an unexpected interaction among nuclear hormone receptors, transporters, and cytochrome P450. J. Biol. Chem. 276: 39411–39418.

Soverchia, L., B. Ruggeri, F. Palermo, G. Mosconi, G. Cardinaletti, G. Scortichini, G. Gatti and A.M. Polzonetti-Magni. 2005. Modulation of vitellogenin synthesis through estrogen receptor beta-1 in goldfish (Carassius auratus) juveniles exposed to 17-beta estradiol and nonylphenol. Toxicol. Appl. Pharmacol. 209: 236–243.

Vandepoele, K., W. De Vos, J.S. Taylor, A. Meyer and Y. Van de Peer. 2004. Major events in the genome evolution of vertebrates: paranome age and size differ considerably between ray-finned fishes and land vertebrates. Proc. Natl. Acad. Sci. USA 101: 1638–1643.

Wannemacher, R., A. Rebstock, E. Kulzer, D. Schrenk and K.W. Bock. 1992. Effects of 2,3,7,8-tetrachlorodibenzo-p-dioxin on reproduction and oogenesis in zebrafish (Brachydanio rerio). Chemosphere 24(9): 1361–1368.

Wassmur, B., J. Grans, P. Kling and M.C. Celander. 2010. Interactions of pharmaceuticals and other xenobiotics on hepatic pregnane X receptor and cytochrome P450 3A signaling pathway in rainbow trout (Oncorhynchus mykiss). Aquat. Toxicol. 100: 91–100.

Yan, Z., G. Lu and J. He. 2012. Reciprocal inhibiting interactive mechanism between the estrogen receptor and aryl hydrocarbon receptor signaling pathways in goldfish (Carassius auratus) exposed to 17beta-estradiol and benzo[a]pyrene. Comp. Biochem. Physiol. Toxicol. Pharmacol. 156: 17–23.

Zhou, J., J. Zhang and W. Xie. 2005. Xenobiotic nuclear receptor-mediated regulation of UDP-glucuronosyl-transferases. Curr. Drug Metab. 6: 289–298.

CHAPTER **4**

Fish Neurotoxic Pollutants

Helena Cristina Silva de Assis[1],* and Maritana Mela[2]

Introduction

Many toxic chemicals are released by industrial facilities and by agriculture practices; some of them end up in water and are known or suspected to be neurotoxicants. Pesticides are of particular concern because they are designed to be neurotoxic to insects and have toxic effects on non-target animals. Neurotoxicities may be neuropathologic or may alter neurochemical, electrophysiological or behavioral functions. The developing brain is more sensitive to chemical toxicants than the adult brain, and exposure during development has been implicated in neurological diseases and mental retardation (Anderson et al. 2000).

The central nervous system (CNS) integrates stimuli from sensory organs, and directs reactions to them. In fish, the spinal cord may control locomotion independently of the brain. It is structurally less complex than the brain, which commonly is subdivided by embryonic regions: myelencephalon, metencephalon, mesencephalon, diencephalon and telencephalon. Besides, the sensory systems of fish can be divided into vision, auditory, mechanosensory, electrosensory and chemoreception. In the assessment of neurotoxicity, these sensory organ systems are viewed as readouts of a functional nervous system (Groman 1982).

[1]Laboratory of Environmental Toxicology, Department of Pharmacology, Federal University of Paraná, PO. BOX 19031, 81.531-990 Curitiba, Paraná, Brazil.
Email: helassis@ufpr.br
[2]Federal University of Paraná, PO.BOX 19031, 81.531-990, Curitiba, Paraná, Brazil.
Email: maritana.mela@gmail.com
*Corresponding author

The basic structure of the teleost nervous system includes neurons, glial cells, ependymal-lining cells, astrocytes, phagocytic cells (macrophages), vascular elements, epithelial elements such as those comprising the primitive meninx and tela choroidea, and non-neural ectodermal and dural supporting tissues (Gibbins et al. 1995). These tissues are organized into central and peripheral components. The CNS includes the brain and the spinal cord. The peripheral nervous system (PNC) consists of mixed nerve-fiber bundles and ganglia (sympathetic and parasympathetic), which are most apparent along the spinal nerves (Zohar et al. 2010). The mixed nerve fibers have both myelinated and unmyelinated portions; the ganglia are composed of large multinucleated nerve cells and nerve fibers (Irie et al. 2011). The sympathetic ganglia are adjacent to the spinal cord whereas the parasympathetic ganglia are located close to or inside effector organs, the myenteric plexus, for example (Nilsson 2011).

There are a significant number of environmental pollutants that are neurotoxic to aquatic organisms such as pesticides, metals and cyanobacteria toxins. Many pesticides are designed to be neuroactive and antagonize synaptic neurotransmission in the CNS. Several neurotoxic compounds, such as some insecticides, are known to target the cholinergic nervous system. By far the largest classes are the organophosphate and carbamate compounds. The organochlorine insecticides also cause neurotoxicity and are used worldwide for public health (e.g., mosquito control) and agricultural production. As with synthetic pyrethroid insecticides, the neurotoxicity mechanism of oganochlorine compounds is the disruption of voltage-sensitive sodium channels. The neurotoxicological effects of metal exposures must be considered in context of the chemical form of the metal, route and duration of exposure, and toxicokinetics. The metals may also be linked covalently to the carbon atoms in an organic group such as methyl group. These compounds called organometallic are liposoluble and facilitate the action on CNS. Among the elements that are found methylated in the environment, we have lead and mercury. Neurotoxins derived from aquatic organisms produce toxicity through a variety of mechanisms. Toxins derived from cyanobacteria, for example, act on membrane sodium channels directly or on membranes to create ion permeabilities. Saxitoxin is one of the most potent natural neurotoxins known. It acts on the voltage-gated sodium channels of nerve cells, preventing normal cellular function and leading to paralysis.

The present chapter focuses on providing an overview of neurotoxic mechanism action, followed by neurotoxic effects for several classes of compounds. Examples of the neurotoxic effects on Brazilian fish species will be highlighted.

Neurotoxicity of Pesticides

The history of pesticide use and the development of modern pesticides has been well documented and discussed (Edwards 1970). The two decades following World War II witnessed extensive use of organochlorine compounds (OC) especially dichloro-diphenyltrichloroethane (DDT) in North America, cyclodienes (aldrin and dieldrin in particular) and hexachlorocyclohexane (HCH) in Great Britain and in Japan (Ishikura 1972; Matsumura 1972). The indiscriminate use of DDT and its environmental implications were appreciated only after Rachel Carson highlighted the environmental hazards of pesticides in her book *Silent Spring*. From 1950 onward, another group of pesticides, organophosphates compounds, gained in importance. In 1950s and 1960s, compounds like parathion and malathion partly replaced DDT and cyclodienes. Since the 1970s the pyrethroids have been widely used to control insect pests in agriculture and public health. By the mid-1990s, pyrethroid use had grown in the world insecticide market, ranking second only to organophosphates compounds among insecticide classes (Casida and Quistad 1998). One of the effects of pesticides is seen on the non target organisms, especially those inhabiting the aquatic environment.

Organochlorine Compounds

The organochlorine compounds are among the largest category of insecticides and are used worldwide for public health (e.g., mosquito control) and agricultural production. The organochlorine insecticides are a diverse group of agents belonging to distinct chemical classes including the dichlorodiphenylethane, the chlorinated cyclodien, and the chlorinated benzene and cychlohexane related structures. The properties (low volatility, chemical stability, lipid solubility, slow rate of biotransformation and degradation) that made these chemicals such effective insecticides also brought about their demise because of their persistence in the environment, bioconcentration, and biomagnification within various food chains (Ecobichon 1995).

Currently, the use of OC is prohibited or restricted worldwide, including in Brazil since 1981 (Penteado and Vaz 2001). However, they are still found in high concentration in organisms such as marine mammals (Yordy et al. 2010; Lailson-Brito et al. 2012). Additionally, OC can generate chronic and acute toxic effects and they have been related to hepatomegaly, thymic atrophy, immunosuppression, neurotoxicity and cancer in several species (ATSDR 2002).

The neurotoxicity mechanism of chlorinated ethanes is the voltage-sensitive sodium channels (VSSCs) disruption. Studies with invertebrate and vertebrate preparations established that DDT prolongs the falling phase of the action potential, which typically produces repetitive firing. This repetitive neuron firing leads to the hyperexcitability and tremors noted in intoxicated insects, mammals, birds, and fish. Studies by Narahashi (1994) demonstrated that DDT caused voltage-sensitive channels to remain in the open state longer than normal and close slowly resulting in an increased overall open time. Hyperexcitability is also associated with increased releases of neurotransmitters throughout the mammalian nervous system which likely reflect secondary responses but could represent an additional neurotoxic mechanism. Furthermore, these compounds inhibit gamma-aminobutyric acid (GABA) stimulated flows of chloride across membrane vesicles (see reviews by Coats 1990; Woolley 1995).

The cyclohexanes such as lindane, the gamma isomer of hexachlorohexane, are approximately 10 times more acutely toxic to the rat than the alpha isomer and roughly 100 times more potent than the beta and delta isomers. Although disruption of VSSCs and GABA receptors are well characterized mechanisms of acute organochlorine neurotoxicity, there is little evidence about/concerning the role of these mechanisms in developmental neurotoxicity. Lindane is known to inhibit mammalian embryonic development from the eight-cell stage up to the blastocyst stage (Scascitelli and Pacchierotti 2003).

Organochlorine compounds—neurotoxicity in fish

Organochlorine exposure affects behavior in many fish species. In unrestrained fish, cyclodiene exposure caused hyperactivity in response to stimuli, followed by recurrent tremors, rapid pectoral fin movement, and convulsions (Carlson et al. 1998). In rainbow trout (*Oncorhynchus mykiss*), endosulfan and endrin intoxication induced branchial tremors, increased cough rate, and increased pectoral fin movement, with eventual tetany and convulsions (Bradbury et al. 1991). Endosulfan (6,7,8,9,10,10-hexachloro-1,5,5a,6,9,9a-hexahydro-6,9-methano-2,4,3-benzo-dioxathiepin-3-oxide), an organochlorine insecticide and acaricide is no longer produced in the United States, but is still used in manufacturing processes and has been found in Brazilian soybean oil, which is a major energy source in commercial tilapia feeds and in fish muscle from Brazilian fish farms (Botaro et al. 2011). Effects associated with CNS were demonstrated in fish. At sublethal concentrations of endosulfan, medaka (*Oryzias latipes*) were less susceptible to predation, presumably due to hyperactivity; however, at endosulfan concentrations approximating the LC20 level, medaka were more susceptible to predation than control fish (Carlson et al. 1998). *In vivo* electrophysiological studies of sublethal endosulfan exposures in medaka demonstrated increased motorneuron amplitude peaks and significantly increased stimulus-response ratio on the Mauthner cell startle response. The hyper-responsiveness of the Mauthner cell to stimuli is consistent with cyclodiene acting at the picrotoxin site in the GABA receptor–chloride complex (Carlson et al. 1998). These compounds are extremely lipid-soluble and tend to concentrate in highest amounts in adipose tissue and also brain and liver.

Organophosphate and Carbamate Compounds

Organophosphates (OPs) and carbamates (CAs) compounds are commonly used as pesticides in agriculture, industry, and also for household purposes throughout the world. In addition, these chemicals are used as parasiticides in veterinary medicine.

One of the earliest organophosphate insecticides synthesized by Gerhard Schrader (a chemist at the I.G. Farben industrie) was parathion, which is still used worldwide. Prior to World War II, the German Ministry of Defense developed highly toxic OP compounds such as tabun, sarin, soman and diisopropyl phosphorofluoridate. After the War, thousands of OPs have been synthesized in the search for compounds with species selectivity, i.e., more toxic to insects and less toxic to mammals. Malathion is an example. This compound has been used for about half a century as the most popular insecticide (Gupta 2006). Agricultural use of insecticidal organophosphorus and carbamates increased in the 1960s as the use of organochlorine compounds declined.

The chemical group of carbamates is derived from carbamic acid and includes more than 25 compounds that are used as insecticides, fungicides and nematicides. As with OPs, the carbamates are easily hydrolyzed in alkaline solutions. The first pesticide carbamic acid esters were synthesized in the 1930s and were marketed as fungicides. As these aliphatic esters had presented poor insecticidal activity, enthusiasm for the compounds lay dormant until the mid-1950s, when renewed interest in insecticides having anticholinesterase activity, but reduced mammalian toxicity, lead to the synthesis of several potent aryl esters of methylcarbamic acid. The insecticide carbamates were synthesized on purely chemical grounds as analogues of the drug physostigmine, a toxic anticholinesterase alkaloid extracted from the seeds of the plant *Physostigma venenosum*, the Calabar bean (Ecobichon and Joy 1974).

OPs and CAs insecticides are primarily known to inhibit the acetylcholinesterase (AChE) enzyme, which plays an important role in neurotransmission, at cholinergic synapses, by the rapid hydrolysis of neurotransmitter acetylcholine to choline and acetate (Kavitha and Rao

2007). Although these compounds are neurotoxicants, they produce a variety of cholinergic and non-cholinergic effects. Latest evidence suggests that while cholinergic mechanisms play a critical role in the initial stage of toxicity, neuronal damage/death appears to occur through noncholinergic mechanisms. OPs and CAs are discussed here together because they both produce similar toxic effects in poisoned animals.

The toxic effects involve the parasympathetic system, sympathetic, motor and central nervous systems. The decreasing of AChE activity can affect swimming performance and compromise the ability of fish species to capture their prey. Alterations in the AChE activity have been demonstrated in various diseases and poisonings, suggesting that this enzyme could be an important physiological and pathological parameter (Gonçalves et al. 2010).

Over 35 years ago, tri-o-cresyl phosphate (TOCP) was known to produce delayed neurotoxic effects in human and chicken, characterized by ataxia and weakness of the limbs, developing 10–14 days after exposure. This syndrome is called OP-induced delayed polyneuropathy (OPIDP). TOCP and certain other compounds have minimal or no anti-AChE property; however they cause phosphorylation and aging (dealkylation) of a protein in neurons called neuropathy target esterase (NTE), and subsequently lead to OPIDP. Many compounds, such as DFP, N,N'-diisopropyl phosphorodiamidic fluoride (mipafox), tetraethyl pyrophosphate (TEPP), paraoxon, parathion, o-cresyl saligenin phosphate, and haloxon, are known to produce this syndrome (Moretto and Lotti 2006).

Brain tissue is particularly susceptible to oxidative damage, as it is rich in polyunsaturated fatty acids which easily undergo peroxidation. Moreover, the brain uses a relatively large amount of oxygen at rather low activities of antioxidative enzymes (Dringer 2000; Droge 2002).

Studies on the effect of exposure to organophosphate pesticides and changes in the antioxidative parameters in the brain are particularly important, since the lowered glutathione level appears to be the first indicator of oxidative stress in Parkinson's disease progression and because of evidence that oxidative stress is an important pathomechanism in other neurodegenerative diseases (Dringer 2000; Abdollahi et al. 2004).

Subacute intoxication with chlorfenvinphos showed increased activities of antioxidative enzymes and a decreased level of reduced glutathione (GSH) in the brain as well as increased level of malondialdehyde (Lukaszewicz-Hussain et al. 2007). In acute intoxication with malathion, some researchers have found intensification of lipid peroxidation in the cerebral cortex and hippocampus, decreased activity of glutathione reductase (GR), as well as lack of changes in glutathione peroxidase (GPx) activity and total glutathione level (Brocardo et al. 2005).

Neurological and behavioral activities of animals can be extremely sensitive to environmental contamination (Bretaud et al. 2000).

Organophosphate and carbamate compounds—neurotoxicity in Fish

Despite the wealth of acute toxicity data, relatively little is known about the developmental neurotoxicity of cholinergic agonists in fish. Existing data demonstrate that exposure either *in ovo* or as juvenile fish has detrimental consequences on learning and motorneuron development. Chlorpyrifos exposure produced hypoactivity in zebrafish (*Danio rerio*) hatchling swimming behavior (Levin et al. 2004). In addition, developmental chlorpyrifos exposure of zebrafish embryos has long-term effects on learning.

Adult zebrafish exposed to chlorpyrifos during development showed reduced choice accuracy and spatial discrimination (Levin et al. 2003). Behavioral effects have been seen in juveniles of other fish species as well. At concentrations of carbaryl or chlorpyrifos up to 10 times lower than 48-hour LC50 values, larval medaka (*Oryzias latipes*) were more susceptible to predation, although a consistent dose–response relationship between carbaryl exposure and susceptibility to predation was not observed (Carlson et al. 1998). *In vivo* electrophysiological studies of sublethal chlorpyrifos and carbaryl effects on the Mauthner cell startle response in larval medaka (Carlson et al. 1998) demonstrated an effect on neuromuscular junctions, as evidenced by a dose-related increase in the ratio of startle response to stimuli.

Studies showed that methyl parathion (MP) and chlorpyrifos (CPF) caused significant inhibition of brain AChE activity compared to control fish. There was dose-dependent decrease in the activity of AChE and the degree of inhibition was slightly higher in brain of CPF exposed fish than the MP exposure (Sharbidre et al. 2011). In the guppy *Poecilia reticulata* exposed to a sublethal concentration of chlorpyrifos (3.25µg/L) for six days, a 90% inhibition of the acetylcholinesterase activity was observed (Almeida et al. 2005).

Bhattacharyya (1985) reported carbofuran-induced acetylcholinesterase inhibition in the fish (*Channa punctatus* and *Anabas testudineus*). Brain AChE activity decreased significantly when these fish were exposed to carbofuran in laboratory conditions. Carbofuran inhibits AChE significantly in *Cyprinus carpio* exposed to a sublethal dose (Dembele et al. 2000).

Roex et al. (2003) also demonstrated that the inhibition of AChE activity is associated with chronic exposure to parathion, where *Danio rerio* increases the feed but decreases the AChE activity. The higher consumption is probably due to the increased activity also confirmed by Kumar and Chapman (1998) in fish. In addition, Sancho et al. (2000) reported that the effects of carbamate on AChE activity on the eyes of *Anguilla anguilla* did not return after a recovery treatment. According to the same authors, the consequences for swimming, feeding, locomotion and balance in exposed organisms are very important (Oliveira Ribeiro and Silva de Assis 2005).

Pyrethroid Compounds

Pyrethroids are synthetic derivatives of pyrethrins, which are toxic components contained in the flowers of *Chrysanthemum cinerariaefolium*. The synthetic pyrethroids are less persistent and less toxic to mammals and birds (Sayeed et al. 2003). The active ingredients are a group of compounds, commonly known as pyrethroid or pyrethrins, which are mixed esters of pyrethrolone and cinerolone with chrysanthemic and pyrethric acid. They affect the sodium current during depolarization and, as such, the rapid opening of the activation gate and the slow closing of the inactivation gate proceeds normally. Pyrethroid with an α-cyano group (deltamethrin, cypermethrin) cause an intensive repetitive activity in the form of long lasting trains of impulses (Perger and Szadkowski 1994).

Physiological and neurochemical studies of pyrethroid-intoxicated animals confirm that acute pyrethroid intoxication is associated with altered nerve function, principally involving neuroexcitatory effects, in the brain, spinal cord, and elements of the PNS. Although differences in sensitivity between brain regions are evident, there is not a single region of the nervous system that is the locus of pyrethroid intoxication and identifies a mechanism of toxic action.

Unfortunately, pyrethroids are highly toxic to a number of non-target organisms such as bees, freshwater fish and other aquatic organisms even at very low concentrations (Oudou et al. 2004). Non-target organisms such as aquatic invertebrates and fish are extremely sensitive to the neurotoxic effects of these insecticides when they enter surface water-courses (Bradbury and Coats 1989; Mittal et al. 1994) (Table 4.1).

In both mammals and insects, the primary mechanism of acute synthetic pyrethroid neurotoxicity is the disruption of voltage-sensitive sodium channels (VSSCs) (Soderlund et al. 2002). Type I pyrethroid insecticides prolong VSSC opening, allowing more sodium to cross the membrane and leading to repetitive firing of action potentials. Conversely, type II pyrethroids delay VSSCs inactivation, resulting in a depolarization-dependent block that prevents action potential generation (Soderlund et al. 2002). Although the primary mechanism of acute pyrethroid neurotoxicity is the disruption of VSSCs, evidence suggests that numerous secondary sites and mechanisms of action are possible (Soderlund et al. 2002). Synthetic pyrethroids have also been shown to affect voltage-gated potassium and chloride channels, as well as ligand-operated channels such as the GABA

Table 4.1. Rat lethal dose 50% (LD 50) and fish lethal concentration 50% (LC50) of pyrethroids.

Insecticide pyrethroid	LD 50 in rats (mg/kg)	LC 50 in fish (µg/L)
Cialothrin	114	0.54
Cipermethrin	250–4150	0.82
Deltamethrin	135–5000	0.91
Permethrin	430–4000	2.10

Reference: Solomon et al. 2010

receptor–ionophore complex, the nicotinic acetylcholine receptor, and the peripheral-type benzodiazepine receptor (Soderlund et al. 2002). These effects, however, are usually associated with physiologically unrealistic pyrethroid exposures or nonspecific interactions.

There is little consensus on which, if any, of the acute neurotoxicity mechanisms are applicable to pyrethroid developmental neurotoxicity (Shafer et al. 2004). Some pyrethroids insecticides show significant age differences in acute toxicity, with younger animals usually being more sensitive. Several studies also demonstrate persistent changes in motor activity, learning, and sexual activity following developmental pyrethroid exposure. None of these effects has been associated with a putative neurotoxic mechanism.

Pyrethroid compounds—neurotoxicity in fish

As reviewed by Bradbury and Coats (1989), acute pyrethroid intoxication in small aquarium fish typically causes loss of schooling behavior, followed by hyperactivity, erratic swimming, whole-body seizures, and loss of buoyancy. The stages of behavioral changes in fish are generally consistent with those observed in mammals (Bradbury and Coats 1989); however, an insufficient number of compounds have been studied in fish to differentiate pyrethroid intoxication syndromes as has been done with mammals (Soderlund et al. 2002). Because seizures are typically stimulus dependent in hypersensitized fish, it seems reasonable to assume that pyrethroid-induced coughs could trigger convulsions. The cough response itself could be a CNS-mediated component in the seizure syndrome, a side effect due to interactions with sensory receptors in the pharynx and gill arches, or direct irritation of gill tissue (Kumaraguru et al. 1982).

Unlike studies of acute toxicity, few studies have addressed the developmental neurotoxicity of synthetic pyrethroids in fish. Exposure to sublethal levels of permethrin *in ovo* caused a hatching delay in medaka (*Oryzias latipes*). Medaka hatchlings demonstrated hyperactivity, uncoordinated movement, and an inability to respond to stimuli. These hatchlings also failed to inflate their swimming bladder and had spinal curvatures (González-Doncel et al. 2003). Unfortunately, these effects still have to be correlated with a putative neurotoxicity mechanism, so it is unclear how pyrethroids may be inducing developmental neurotoxicity.

Bálint et al. (1995) observed 20% decrease in AChE activity of brain, heart, blood, liver and skeletal muscle of carp after three days exposure to deltamethrin. The ATPase had been demonstrated to be one of the targets of DDT and pyrethroids. Many researches showed that ATPase including cell membrane-associated Na^+-K^+-ATPase, and Ca^{2+}-Mg^{2+}-ATPase, mitochondrial Mg^{2+}-ATPase and Ca^{2+}-ATPase could be inhibited by DDT and pyrethroids (Luo and Bodnaryk 1988). Cipermethrin, also a pyrethroid, caused ATPase inhibition in brain, liver and kidney of carps exposed during 45 days. It can inhibit the ions sodium transport through the membrane, disrupting the cell ionic balance, which can cause irreversible damage in the tissues (Silva de Assis et al. 2009).

Neurotoxicity of Metals

Another consequence of increased industrialization, contamination of natural freshwaters by heavy metals, such as iron, zinc, lead, mercury, cadmium, nickel and manganese, has become a global problem (Abdullah et al. 2007). Certain metals have been known to be toxic in different ways, and some of them with specificity to the nervous system (e.g., neurotoxic).

In case of essential trace elements, e.g., copper, zinc and nickel, the optimal concentration ranges for fish growth and reproduction are narrow and both excess and deficiency are harmful to the fish (Brunelli et al. 2011). Some non-essential trace metals such as mercury, lead and cadmium are neurotoxic at concentrations observed in natural waters (Javed and Mahmood 2001; Mela et al. 2007). The toxicity of these metals is in part due to the fact that they accumulate in biological tissues, a process known as bioaccumulation (Méndez-Armenta and Ríos 2007). This occurs because the metal, once taken up into the body, is stored in particular organs, for example the liver or the kidney, and is excreted at a slow rate compared with its uptake (Santovito et al. 2012). This process of bioaccumulation of metals occurs in all animals, including food animals such as fish and cattle as well as humans (Mela et al. 2007; Meyer et al. 2008).

Mercury, lead and cadmium

Much of what we know about mercury toxicity in humans stems from several mass poisoning events that occurred in Japan during the 1950s and 1960s, and Iraq during the 1970s (Yorifuji et al. 2011). In Japan, a chemical factory discharged vast quantities of mercury (Hg) into several bays near fishing villages. Many people who consumed large amounts of fish from these bays became seriously ill or died over a period of several years (Syversen and Kaur 2012). In Iraq, thousands of people were poisoned by eating contaminated bread that was mistakenly made from seed grain treated with methylmercury (MeHg) (Syversen and Kaur 2012). From studying these cases, researchers have determined that the main target of MeHg toxicity is the CNS (Ventura et al. 2005). At the highest exposure levels experienced in these poisonings, MeHg toxicity symptoms included nervous system effects such as loss of coordination, blurred vision or blindness, and hearing and speech impairment (Warfvinge and Bruun 2000; Tanan et al. 2006; Mela et al. 2010, 2012). Scientists also discovered that the developing nervous systems of fetuses are particularly sensitive to the toxic effects of MeHg (Nitschke et al. 2000). Once mercury gets into water, much of it settles to the bottom where bacteria in the mud or sand convert it to the organic form of methylmercury (Grotto et al. 2010). Larger and older fish absorb more methylmercury as they eat other fish (Berzas Nevado et al. 2010). In this way, the amount of MeHg builds up as it passes through the food chain (Malm et al. 1995). Fish eliminate MeHg slowly, and so it builds up in fish in much greater concentrations than in the surrounding water (Mela et al. 2007). Eating fish is the main way that people get exposed to MeHg (Grotto et al. 2010). Each person's exposure depends on the amount of MeHg in the fish that they eat and how much and how often they eat fish (Vahter et al. 2007). Besides, we may be exposed to inorganic forms of mercury through dental amalgams or accidental spills, such as from a broken thermometer (Syversen and Kaur 2012).

The developing nervous system is especially susceptible to damage by MeHg (Ek et al. 2012) and this pollutant produces focal damage to specific areas in the brain (Ventura et al. 2004, 2005; Syversen and Kaur 2012). One hypothesis proposes that certain cells are susceptible because they cannot repair the initial damage to the protein synthesis machinery (Mela et al. 2012). The initial damage is nonselective, consistent with the chemical properties of MeHg. It has been discovered that microtubules are destroyed by this form of mercury and this effect may explain the inhibition of cell division and cell migration, processes that occur only in the developmental stages (Clarkson 1987). It has also been observed that MeHg inhibits DNA, RNA and protein synthesis, interacts with the cytoskeleton, alters the properties of biomembranes and disrupts axoplasmic transport. In the proliferating cell, MeHg inhibits mitosis and/or decreases the rate of the cell cycle (Nielsen and Andersen 1991).

In the 1960s, lead (Pb) emerged as a major public health problem (Ganjavi et al. 2010). The toxic effects of lead, like those of mercury, have been principally established in studies on people exposed to lead in the course of their work (Vieira et al. 2011). Lead is an extremely common metal that can be found almost anywhere. Some common sources of Pb include: water, paint, electric storage batteries, insecticides, auto body shops, gasoline, etc. (Monteiro et al. 2011). Once lead enters the body it interferes with normal cell function and physiological processes (Monteiro et al. 2011). Some of the physiological effects of Pb include harm done to the PNS and CNS, blood cells, metabolism of vitamin D and calcium, and reproductive toxicity (Luo et al. 2012). The nervous system seems to be the most sensitive to lead poisoning (Luo et al. 2012). Like mercury, lead crosses the placental barrier and accumulates in the foetus (Ganjavi et al. 2010). Consumption of food containing Pb is the major source of exposure for the general population (Atta et al. 1997).

A number of interesting hypotheses have been proposed for the mechanism of Pb toxicity on the nervous system (Clarkson 1987; Zheng et al. 2003). Lead is known to be a potent inhibitor of heme synthesis. The loss of heme containing enzymes should affect mitochondrial function, producing adverse effects on energy metabolism. The cells of the blood-brain barrier are especially rich in mitochondria, making the blood-brain barrier susceptible to lead poisoning (Zheng et al. 2003). Lead may affect brain function by interference with neurotransmitters such as GABA (Monteiro et al. 2011). There is mounting evidence that Pb interferes with membrane transport and binding of calcium ions. High levels of lead decrease transport of calcium and vice versa, therefore these two metals function as competitive inhibitors. Lead can enter through the same ion channels as calcium and regulate the activity of those channels to uptake more lead into the cell. Calcium enters the presynaptic terminal via the calcium channel in response to the arrival of a wave of depolarization. Inside the cell, calcium activates calmodulin, leading to the fusion of acetylcholine vesicles with the plasma membrane and the release of acetylcholine (Quintanar-Escorza et al. 2010). There is evidence that Pb competitively inhibits calcium entry. The findings of Kostial and Vouk (1957) can be explained by the action of Pb on this channel. They measured the magnitude of the endplate potential that results from calcium entry into the cell and were able to show competitive inhibition by lead and calculate a Lineweaver- Burke inhibitory constant in the micromolar range (Quintanar-Escorza et al. 2010).

Cadmium (Cd) was discovered as an element in 1817 and its industrial use was quite minor until 50 years ago (Méndez-Armenta and Ríos 2007). Of the hazards associated with exposure to cadmium, CNS disorders have been reported in the case of Itai-itai disease and in various clinical studies on children and exposed workers (Shukla et al. 1996). Friberg (1950) reported that 37% of workers in a Cd battery factory showed olfactory dysfunction after exposure to cadmium for about 20 years. Vorobjeva (1975) observed CNS symptoms such as headache and vertigo in cadmium-exposed workers. In animal studies, abnormal behavior patterns were observed in fish species, bass and bluegill, exposed to Cd through water (Cearley and Coleman 1974). Méndez-Armenta and Ríos (2007) reported that the behavior of rats was altered by a single intraperitoneal injection of Cd, although they did not find any behavioral disorder in the rats chronically exposed for 3–12 months. Nowadays, Cd is still an important industrial metal. It is extracted during the production of other metals, such as zinc, lead and copper and it is used in industrial and household products, mainly in batteries, pigments, metal coatings, plastics and some metal alloys (Méndez-Armenta and Ríos 2007). Human intoxication results mainly from cigarette smoking due to high concentrations of Cd in cigarettes but also from water, food and air contaminations (Bertin and Averbeck 2006).

Hardly any cadmium can get into the brain parenchyma because of the protection offered by the brain barrier system, i.e., the blood-brain and the blood-cerebrospinal fluid (CSF) barriers (Brunelli et al. 2011). However Shukla et al. (1996) reported that permeability through the blood-brain barrier is enhanced by chronic exposure to Cd. In animal studies, chronic exposure to cadmium has been found to cause a significant increase in Cd concentration in the brain, especially in the olfactory bulb (Suzuki and Arito 1967; Clark et al. 1985). In the case of intranasal instillation of cadmium, a significant amount of cadmium was taken up in the anterior parts of the olfactory bulb (Tjalve et al. 1996). On the other hand, there are only a few reports concerning the movement and action of Cd in brain extracellular space (Gottofrey and Tjalve 1991) and the precise mechanism by which cadmium alters behavior and causes CNS disorders remains to be elucidated.

Metals—neurotoxicity in fish

Numerous reviews have summarized the behavioral responses of fish to metal intoxication (Atchison et al. 1987). In addition to direct neurotoxic mechanisms, alterations in avoidance or attraction responses, activity patterns, critical swimming speed, respiratory behavior, intraspecific social interactions, reproduction, feeding, and predator avoidance can be attributed to direct damage to respiratory surfaces and interference with energy metabolism, osmoregulation, and endocrine function (Leblond and Hontela et al. 1999). Studies of chemoreception have quantified the extent to which metals attract or repel fish and the extent to which metals affect responses to endogenous chemical signals such as pheromones. Alterations in avoidance or attraction responses have been observed in response to a number of metals, including cadmium, copper, and mercury. In rainbow trout (*Oncorhynchus mykiss*), lake whitefish (*Coregonus clupeaformis*), Atlantic salmon (*Salmo salar*), and goldfish (*Carassius auratus*), copper induces avoidance behavior (Atchison et al. 1987). This avoidance behavior is attributed to the effects of copper on the olfactory bulb. Copper attenuates electrical responses of the olfactory bulb and receptor cells to excitatory compounds (Winberg et al. 1992). Furthermore, copper exposure causes degeneration of specific olfactory receptor cells, likely through oxidative-stress-mediated apoptosis (Chinni and Yallapragdo 2000).

Interestingly, oxidative stress may be partly responsible for the observed neurological effects in Wilson's disease, a genetic defect in copper metabolism leading to copper neurotoxicity in humans (Sindayigaya et al. 1994). In fish, Cd exposure has been correlated to changes in brain AChE activity, although these neuro-chemical changes have not been correlated with changes in swimming behavior in larval rainbow trout (Beauvais et al. 2001). Although many metals elicit an avoidance response, mercuric chloride and methylmercury attract fish (Atchison et al. 1987). Exposure of mercuric chloride and MeHg to the olfactory bulb and receptors of rainbow trout (*Oncorhynchus mykiss*) and Atlantic salmon (*Salmo salar*) depressed electrical responses (Baatrup et al. 1990). MeHg has also been found to preferentially accumulate in olfactory receptors and the olfactory nerve of Atlantic salmon following dietary exposure (Berntssen et al. 2003). Chronic dietary exposure reduced overall activity in the Atlantic salmon (*Salmo salar*) and caused preferential histopathological damage to the brain stem (Berntssen et al. 2003). Lead exposure caused a general increase in locomotor activity in mirror carp (*Cyprinus carpio*) (Stouthart et al. 1994). In zebrafish (*Danio rerio*) and fathead minnows (*Pimephales promelas*), Pb exposure reduced feeding ability, as evidenced by feeding miscues and increased prey-handling times. This reduction in feeding ability was attributed to psychomotor coordination, based on correlations between increased brain lead levels, increased serotonin and norepinephrine concentrations, and decreased feeding ability). Pb exposure induced similar increases in brain serotonin and decreases in GABA in walking catfish (*Clarias batrachus*) (Katti and Sathyanesan 1986). Although lead increased brain serotonin and norepinephrine levels, its exposure did not increase dopamine levels in fathead minnows.

Dou and Zhang (2011) examined neurological deficits caused by Pb during early embryonic stages in the zebrafish (*Danio rerio*) and further explored its potential molecular mechanism. Following Pb exposure (0.2 mM), embryos showed obvious neurotoxic symptoms with "sluggish" action, slow swimming movements and slow escape action. The TUNEL assay demonstrated that the reduction of nerve cells was due to increased apoptosis of neuron and glia cells. In conclusion, these findings identify that Pb-induced neurotoxicity can be caused by impaired neurogenesis, resulting in markedly increased apoptosis of special types of neural cells, neuron and glia cells.

The visual system is susceptible to the toxicological effects of metals (Feitosa-Santana et al. 2007; Barboni et al. 2009; Mela et al. 2012). A fish's eyes are adapted or modified for underwater vision, but they are not very different from human eyes (Groman 1982). The big difference between a human eye and the eye of a fish occurs in the lens. In humans it is fairly flat or dishlike; in fish, it is spherical or globular. Although the eye of a fish has rigid lens and its curvature is incapable of changing, it can be moved toward or away from the retina (like the focusing action of a camera). The retina, an important component of the CNS, is known to be a target for many toxicants. The retina is comprised of pigmented epithelium, which consists of photoreceptors (rod and cone cells), horizontal cells, bipolar cells, amacrine cells, and ganglion cells and nerve fibers that lead to the optic nerve. The optic nerve is comprised of four different types of fibers, most of which are afferent and project to the contralateral side of the brain (the optic tectum). Horizontal cells facilitate information flow between photoreceptors, and amacrine cells facilitate lateral flow of information among ganglion cells. Bipolar cells direct excitatory or inhibitory responses vertically to the ganglion cells, which subsequently send information to the central nervous system through the optic nerve (Bonci et al. 2006).

Varieties of rod and cone subtypes are observed in fish and suggest specialization in terms of discriminating wavelengths and intensity of light (Hawryshyn 1982). Fish may have a scotopic system, with input from the rods, that is responsible for achromatic, high sensitivity, low acuity vision, while a photopic system using the cones is responsible for color vision, low sensitivity and high acuity vision at higher light intensities (Kusmic and Gualtieri 2000). So, fish can distinguish colors. There are indications that some kinds of fish prefer one color to another and also that water conditions may make one color more easily distinguished than others. Freshwater fish are generally found to have porphyropsin as their main photosensitive pigment, while marine fish typically have rhodopsin in their retina (Costa et al. 2008). Glutamate is the putative neurotransmitter in the photoreceptors that mediates synaptic communication with bipolar cells, while GABA has been well-documented in horizontal and amacrine cells (Tanan et al. 2006). Glycine, dopamine, and acetylcholine have also been reported as neurotransmitters in the retina. A potential role of neuropeptides has yet to be elucidated.

In vivo studies in laboratory animals have shown that vision loss can result after exposure to mercury (Tessier-Lavigne et al. 1985). Methylmercury is able to cross the blood–retina barrier and induce changes in photoreceptors of *Hoplias malabaricus*, a neotropical fish, even under subchronic exposure (Fig. 4.1). According to Mela et al. (2012), this low level of MeHg is typically encountered by fish in their environments over long periods and the deterioration observed in photoreceptor layer is associated

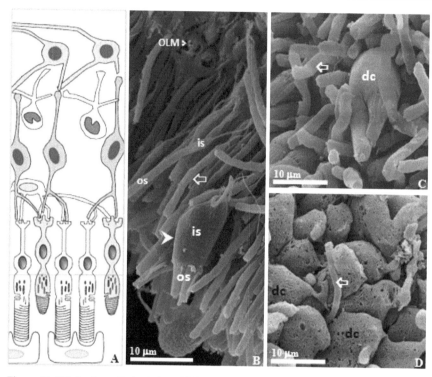

Figure 4.1. (A): A schematic depiction of fish retina. (B): Scanning electron micrograph showing the photoreceptor in retina of *Hoplias malabaricus*. Double cones (arrow-head), rods (open arrow) and outer limiting membrane (OLM) formed by prolongations of Müller cells. The photoreceptors have an inner segment (is) and an outer segment (os). (C): Scanning electron micrograph showing the photoreceptor in retina of *H. malabaricus*. Photoreceptors can be seen: (dc) double cone and a rod (open arrow). (D): Scanning electron micrograph showing the photoreceptor in retina of *H. malabaricus* after methylmercury exposure (0.075 mg MeHg g^{-1}) . Note the clear cell deterioration in the photoreceptors layer.

Schematic depiction of fish retina modified from: Seizure: European Journal of Epilepsy 2004; 13:113-128 (DOI:10.1016/S1059-1311(03)00082-7).

with cell death in the cones and rods. According to Evans and Garman (1980), one of the earliest signs of mercury poisoning is an impairment of scotopic (night) vision. This aspect of vision is mediated primarily by rod photoreceptor cells, which are unable to respond appropriately to light following exposure to mercuric compounds (Tessier-Lavigne et al. 1985). In addition, chronic exposure to mercury may result in a perturbation of peripheral vision, followed by a more severe loss of central vision (Saldana et al. 2006). Surprisingly, the retinal target sites due to metals exposure of vertebrates and the consequent neuronal damages have not been completely elucidated yet.

Lead exposure impairs visual functions in rodents, monkeys, as well as in children (Altmann et al. 1998). Fox (1992) demonstrated that selective rod apoptosis can be produced both *in vivo* and *in vitro* by lead exposure. Cadmium exposure gives rise to various visual defects in vertebrate species like ocular malformations such as microphthalmia in fresh water fish *Ambassis commersoni* Cuvier, small eyes in zebrafish (Cheng et al. 2001) and induces ocular anomalies in *Xenopus* embryos and damages the corneal endothelium of bullfrog. Cd exposure also causes apoptotic changes in photoreceptors and ganglionic cells mitochondrial function (Roozbehi et al. 2007). Using zebrafish as the model organism, Chow et al. (2009) showed that cadmium exposure induced microphthalmia, behavioral blindness, showing hyperpigmentation and loss of camouflage response to light. The number of retinal ganglion cells, was reduced, while photoreceptor cells, the last batch of retinal neurons to differentiate, were absent. Drastically reduced retinal ganglion cells axons and disrupted optic stalk showed that the optic nerves did not extend from the retina beyond the chiasm into the tectum. According to Erie et al. (2007), it raises the necessity of a solution/ further research concerning the issue of examining the potential mediating role of Cd-induced oxidative stress on the cells of the retina; it also raises the need for further research into molecular biomarkers. Together, these cited studies concentrate on the occurrence of ocular malformations that resulted from cadmium and lead exposure.

Neurotoxicity of Cyanobacteria

The occurrence of cyanobacterial waterblooms is a worldwide concern in environmental health, and the eutrophication phenomenon due to human activities, mainly waste production, has posed a serious danger to water resources. The toxins from cyanobacteria are known as cyanotoxin and comprise a large source of toxic natural products produced by these microorganisms.

Some of these toxins, which are characterized by the quick action, causing death by respiratory depression within minutes after exposure— have been identified as neurotoxic alkaloids or organophosphates.

In 1998, after several cases of major human intoxications in Australia, Europe and Brazil (Falconer and Humpage 2005), the World Health Organization (WHO) proposed a guideline value for the concentration in water of one of the most common cyanotoxin, microcystin-LR. This guideline for drinking water has now been adopted by sanitary authorities of many countries in the European community, and North (Canada) and South America (Brazil). However, the health risks for wild and domestic terrestrial vertebrates resulting from the exposure to cyanotoxins in water are still largely ignored, despite an increasing number of reports of poisoning in veterinary literature.

The risks for humans, especially through the consumption of drinking water, are discussed in the WHO book Toxic Cyanobacteria: A Guide to their Public Health Consequences, Monitoring and Management (Chorus and Bartram 1999).

Toxins from algae and cyanobacteria attain concentrations hazardous to human health through some sort of concentrating mechanism. As for the marine setting, the primary concentrating mechanism is accumulation in shellfish; for cyanotoxins in fresh water, high concentrations of toxins are primarily the result of surface scum formation, which to date has been the focus of most risk assessments. One reason for this lack of information is the difference between the most conspicuous accumulation mechanisms of marine phytotoxins, which is through shellfish, and of cyanotoxins, which is through scum formation. In consequence, exposure scenario assessments initially concentrated on cyanobacterial cell accumulation at bathing sites or drinking-water offtakes (Ibelings and Chorus 2007).

Toxins stabilize sodium channels by inhibiting inactivation of the sodium current during depolarization, thus keeping the sodium channel in the open state. This sodium channel stabilization causes a prolonged period of calcium influx through VSCCs, causing increased calcium-mediated secretions and contractions (Strichartz and Castle 1990). Toxins that stabilize sodium channels bind different sites compared to toxins that activate sodium channels; consequently, sodium-channel-stabilizer toxins can synergize with sodium-channel-activator toxins, causing larger membrane depolarizations at lower doses.

Other aquatic neurotoxins have mechanisms that do not center on ion regulation; for example, cyanobacteria of the genera Anabaena and Oscillatoria produce anatoxins that disrupt acetylcholine function at neuromuscular junctions. Anatoxin-a acts as an acetylcholine mimic, binding nicotinic acetylcholine receptors at vertebrate muscle endplates, and is reported to be eight times more potent than acetylcholine. The anatoxin-a(s) causes the same symptoms of neurotoxicity as anatoxin-a but by a different mechanism. Anatoxin-a(s) is a naturally occurring organophosphate that inhibits acetylcholinesterase activity in a manner similar to the organophosphate insecticides discussed previously (Bradbury et al. 2008).

Neurotoxins

Anatoxins are mainly produced by cyanobacteria in the *Anabaena* genus (Beltran and Neilan 2000), but also by other genera, such as *Planktothrix, Oscillatoria, MicrocystisAphanizomenon, Cylindrospermum* and *Phormidium.*

Reports of anatoxin poisoning are less frequent than microcystin toxicosis; however, poisoning has occurred worldwide (Beltran and Neilan 2000; Gugger et al. 2005). Anatoxins are neurotoxins and can generally be divided into anatoxin-a, homoanatoxin-a and anatoxina(s). Anatoxin-a is a secondary amine and has been detected in blooms worldwide. Homoanatoxin-a is a methyl derivative of anatoxin-a and has been identified in blooms in Japan (Namikoshi et al. 2004) and Ireland (Furey et al. 2003). Anatoxin-a(s) is a unique N-hydroxyguanidine methyl phosphate ester that has been detected in the Americas (Monserrat et al. 2001) and Europe (Henriksen et al. 1997). Anatoxin-a is a potent cholinergic agonist at nicotinic acetylcholine receptors in neurons and at the neuromuscular junctions (Thomas et al. 1993).

Anatoxin-a(s) is different from anatoxin-a and homoanatoxin-a. This neurotoxin has a unique chemical structure and is a natural occurring irreversible acetylcholinesterase inhibitor. The increased concentrations of acetylcholine in the synapse lead to persistent stimulation, followed by a neuromuscular block (Hyde and Carmichael 1991). The mechanism of toxic action is similar to that of carbamate or organophosphate insecticides. Animals poisoned with anatoxin-a(s) show a rapid onset of excessive salivation, lacrimation, diarrhea and urination. Clinical signs of nicotinic overstimulation include tremors, incoordination, and convulsions. Recumbency and respiratory arrest are most commonly observed in cases with a lethal outcome. The survival time of animals with anatoxin-a(s) poisoning is very short and animals may die within 30 minutes of exposure (Puschner and Humbert 2007).

Although microcystin and anatoxin poisonings make up the majority of cases reported in animals, other cyanotoxins are of concern. Saxitoxins and derived forms belong to the group of paralytic shellfish poisoning (PSP) and have been produced by a number of freshwater cyanobacteria, including *Aphanizomenon flos-aquae*, *Cylindrospermopsis raciborskii*, *Anabanea circinalis*, *Lyngbya wollei* and *Planktothrix* sp. (Carmichael et al. 1997; Sivonen and Jones 1999; Kaas and Henriksen 2000; Molica et al. 2005). In humans, saxitoxins are associated with the development of a neurological disease after the ingestion of shellfish contaminated with saxitoxins. All saxitoxin analogs have a high toxicity in mammals and act by binding to the voltage-gated sodium channel, followed by inhibition of the influx of sodium ions. This results in a blockade of neuronal transmission leading to respiratory arrest, neuromuscular weakness and cardiovascular shock. While saxitoxin poisonings have not been documented in animals, the risk for exposure certainly exists. Another cyanotoxin of concern to animals is cylindrospermopsin. This alkaloid has resulted in deaths in cattle (Saker et al. 1999) and has caused severe gastrointestinal disease in humans. Furthermore, this cyanotoxin is of particular concern because of its mutagenic and possible carcinogenic activities.

Neurotoxin—neurotoxicity in fish

Cyanobacterial toxins have been demonstrated to have a variety of harmful effects on fish species at different life stages (see Malbrouck and Kestemont (2005) for a review). Toxin contents depend on food consumption, which can be considerably less in carnivorous than herbivorous fish species (Gkelis et al. 2006). A further complicating factor is the time dependent nature of toxin accumulation vs. depuration in fish (and other biota). The concentrations found in fish are very much dependent on the short to medium- or long term history of exposure to cyanobacterial toxins (Kankaanpaa et al. 2002).

There are few reports of other aquatic neurotoxin effects on fish. Accumulations of saxitoxins and anatoxins in fish tissues have been documented primarily because they are monitored within the context of human health protection. Toxic endpoints of saxitoxins and anatoxins in fish, however, are rarely assessed. In laboratory studies.Anatoxin-a and anatoxin-a(s) elicit acetylcholine and acetylcholinesterase inhibition responses in electric rays (*Torpedo californica*) and in electric eels (*Electrophorus electricus*) (Hyde and Carmichael 1991; Swanson et al. 1991).

Neurotoxic Effects on Brazilian Fish Species

Neurological and behavioral activities of animals can be extremely sensitive to environmental contamination (Bretaud et al. 2000). The measurement of AChE activity is frequently used as a biomarker for exposure to certain groups of contaminants, such as organophosphorates and carbamates; however, other compounds also can influence the activity of this enzyme, which involve the use of many different toxins (Dutta and Arends 2003) and different compounds such as heavy metals (Payne et al. 1996).

Silva and collaborators (1993) studied the effects of Folidol 600 (methyl parathion) in *Callichthys callichthys* (Pisces: Callichththydae). The symptomatology developed by *C. callichthys* when treated with this agent follows the known sequence for poisoning in fish. The fish presented neurotoxic alterations including anxiety, agitation and loss of balance, followed by anorexia and an increase in the superficial respiratory rhythm with larger opercular amplitude. The muscular weakness and uncoordinated swimming movements as well as involuntary contractions seen as trismus on the fins were other symptoms observed in the same individuals after Folidol 600 exposure. In this way, about 90% of the plasma cholinesterase was inhibited in fish only four hours after the intoxication.

The brain tissue of *Corydoras paleatus* was exposed to three different pesticide classes—organophosphate (methyl parathion), carbamate (carbaryl) and pyrethoid (deltamethrin). *Corydoras paleatus* showed a inhibition of the AChE after 96 hours of exposure to concentrations of

7.2 mg.L^{-1} (84%) and 14.4 mg.L^{-1} (88%) of carbaryl and 0.5 mg.L^{-1} (74%), 1.0 mg.L^{-1} (99.7%) of methyl parathion, but no alteration was found in individuals exposed to deltamethrin. Only inhibition effects (92%) at the high concentration of methyl parathion were observed in the muscle AChE activity. In this study, AChE activity was inhibited in the brain of *C. paleatus* by carbaryl and methyl parathion, suggesting that this tissue is the more sensitive to pesticide exposure (Guiloski et al. 2012). In this study, deltamethrin did not disturb the brain and muscle AChE activity. The pollutants, which are not enzyme-specific, are required at high concentrations to cause enzyme inhibition (Sturm et al. 1999).

From the surrounding water, fish may absorb dissolved heavy metals that may accumulate in various tissues and organs and even be biomagnified in the food-chain/web (de la Torre et al. 2007). In the absorption process there are four possible routes for metals to enter a fish: the food ingested; simple diffusion of the metallic ions through gill pores; through drinking water; and by skin absorption (Sindayigaya et al. 1994).

In experiments with fish (*Cyprinus carpio*) exposed for four months to mercuric chloride, MeHg was accumulated significantly in brain and caused significantly increased levels of lipid peroxidation products and decreased the activities of superoxide dismutase (SOD) and glutathione peroxidase (Berntssen et al. 2003). According to these authors, compared with other organs, the brain is particularly susceptible to dietary Hg. It should be also noted that low dietary concentrations of mercury induced protective redox defenses in the brain evidenced by the induction of antioxidant enzyme SOD. Recently Mieiro et al. (2010) found that mercury exposure depleted antioxidant in fish. The total glutathione content and the activities of catalase, glutathione peroxidase, glutathione S-transferase and glutathione reductase significantly decreased. At the same time, fish exhibited unaltered lipid peroxidation levels, pointing out a higher propensity of mercury to inhibit enzymes than to oxidatively damage lipids. The authors concluded that mercury increased susceptibility of the fish brain to oxidative challenges. The exposure to HgCl$_2$ of freshwater fish matrinxã, *Brycon amazonicus* to sub-lethal concentration of mercury chloride for 96 hours in a static system increased the activities of superoxide dismutase, catalase, glutathione peroxidase, glutathione S-transferase and glutathione reductase (Monteiro et al. 2010). Collectively, these data suggest that oxidative stress in response to mercury exposure could be the main pathway of its toxicity induced by this metal in fish.

Neotropical fish (*Hoplias malabaricus*) exposures to Pb or MeHg for 70 days demonstrated the muscle cholinesterase (ChE) activity inhibition after 14 doses of methylmercury (Alves Costa 2007). In the same species, Rabitto et al. (2005) showed that *Hoplias malabaricus* exposed to Pb revealed a tendency for muscular ChE inhibition with a time-dependent decrease in activity (4 and 8 dietary doses), but a significant inhibition of ChE activity occurred after Pb^{2+} exposure to 14 dietary doses (21 mg g^{-1} w.w.). More recently, Rabitto et al. (2011) observed in *Cichla monoculus*, a predator species commercially important in Amazon human diet, that an unexpectedly significant increase in the cholinesterase activity occurred in muscle of individuals from the higher Hg concentration group.

The response of AChE was assessed in adult females of an indigenous teleost, *Cnesterodon decemmaculatus*, caught at a polluted site (heavy metals) and an important inhibition (between 35% and 38%) in specific brain AChE activity was observed (de la Torre et al. 2007). Linde-Arias et al. (2008) observed the effects of pollution in AChE activity in *Oreochromis niloticus* from a highly degraded Brazilian river. Low brain AChE activities were found in fish from the region with strong agriculture activity, showing the effects of metals from an industrialized and heavily degraded environmental area. Some studies on Cd toxicity have found an association with behavioral disturbances and cholinergic neurotransmission since an increase or a decrease in the AChE activity was verified in both animal models and humans that showed behavioral impairments after exposure to Cd (Pari and Murugavel 2007; Gonçalves et al. 2010).

The mechanism of AChE inhibition by metals is not clear. It is generally accepted that metals may deactivate enzymes by binding to their specific groups. A hypothesis for the inhibition is that the metal binds at the anionic site of ChEs (Guilhermino et al. 1998). Hence, with the metal at anionic sites, acetylcholine cannot bind properly to the enzyme and cannot be degraded. This hypothesis suggests that differences in the potential for some metals to inhibit AChE will be explained by properties such as ionic size, capacity of complex formation, electronegativity, and reduction potential (Alves Costa et al. 2007).

There are only a few studies describing the damage due to metals exposure on the visual system of fish. Pioneering these studies, Hawryshyn et al. (1982) showed behaviorally that the visual system of the rainbow trout is affected by exposure to mercury. In the last years, some studies investigated the effects of MeHg in the retina of a Brazilian fish species (*Hoplias malabaricus*) widely distributed in regions of Brazil's Amazon Rivers impacted by mercury (Bonci et al. 2006; Tanan et al. 2006; Mela et al. 2012). The experimentally exposed animals after acute and subchronic exposure to MeHg showed electrophysiological anomalous responses on horizontal cells (Tanan et al. 2006). In addition, Bonci (2006) observed in the retina of *H. malabaricus* a significant reduction in the number of amacrine cells located in inner nuclear layer after methylmercury exposure (0.075 mg Hg g^{-1}). According to the author, the reduction in the number of neural cells was due to induction of apoptotic process of cell death. Investigating the hypothesis of neural death by TUNEL technique in retinas of *H. malabaricus* intoxicated with methylmercury, Bonci et al. (2006) observed apoptotic cells in the inner nuclear layer and ganglion cell layer. A decrease in the number of amacrine and bipolar cells was also described by the same authors when individuals were exposed to 6 mg g^{-1} of methylmercury, or a reduction in the number of amacrine cells in the retina of *H. malabaricus* when exposed to 2 mg g^{-1} of methylmercury.

Mela et al. (2012) observed an abundant methylmercury deposits in the photoreceptor layer, in the inner plexiform layer and in the outer plexiform layer of *H. malabaricus*, demonstrating a dose-dependent bioaccumulation. According to Shrivastav et al. (1976) the primary effects of MeHg on synaptic transmission is the increase, followed by the decrease, of the spontaneous release of neurotransmitters, as well as a decrease in the release of neurotransmitters activated by the nerve impulses. Studies demonstrating the effect of mercury exposure on synaptic response of plexiform layers of fish retina have not yet been performed; however, Tanan et al. (2006) observed a dose dependent increase in the amplitude of the horizontal cell response, followed by a decrease, in the retina of *H. malabaricus* to the doses of 0.075 and 0.75 MeHg, respectively. According to Yuan and Atchison (1993), MeHg completely blocks the synaptic transmission and its effects are irreversible or only partially reversible. Due to the lack of data concerning the effect of mercury in the plexiform layers, Mela et al. (2012) suggest that the mercury deposits in these layers damage the retinal synaptic connections in *H. malabaricus* and impair the communication between the different retina cells. This implies that the action of MeHg on synaptic transmission may be responsible for, or partially contribute to, its neurotoxicity.

Besides, MeHg disturbs the integrity of the photoreceptors of *H. malabaricus* in very low concentrations. According to Mela et al. (2012), the ultrastructure analysis of retina revealed a cellular deterioration in the photoreceptor layer, morphological changes in the inner and outer segments of rods, structural changes in the plasma membrane of rods and double cones, changes in the process of removal of membranous discs and a structural discontinuity. These results lead to the conclusion that MeHg is able to cross the blood-retina barrier, accumulate in the cells and layers of retina and induce changes in photoreceptors of fish even under subchronic exposure.

As already noted, cadmium and lead are broad spectrum toxic agents capable of damaging a number of organ systems. The retina is a very important organ system that has received very little attention, as a possible target organ for cadmium or lead toxicity. However, there are not many studies of these metals in Brazilian species.

A study on *Hoplias malabaricus* exposure to saxitoxin (Silva et al. 2011) showed an increase in activity of SOD, suggesting a generation of superoxide radicals (O_2) in brain. Reactive oxygen species (ROS), such as

O_2, can be produced during the process of xenobiotic detoxification or due to the changing voltages in cellular membranes. Moreover, transition metals such as $Fe2\,\text{þ}$ can catalyze the formation of hydroxyl (OH-) radicals from hydrogen peroxide (H_2O_2) by reacting with the O_2 previously produced; the hydroxyl ion, in turn, is able to damage the structure of macromolecules such as DNA, proteins, carbohydrates, and lipids. The deficiency in the antioxidant system found in this study may have led to an increase in the hydroperoxide concentration, thus promoting LPO which can lead to the loss of cell-membrane integrity (Regoli et al. 2005) and the carbonylation of proteins, which can provoke cellular damage by forming adducts of proteins, concurrent with the generation of organic reactive species leading to oxidative stress.

In addition, the comet assay showed that STX can be a genotoxic substance that can lead to neurodegeneration (Thompson 2008). Moreover, the damages found in this study due to DNA fragmentation are an important result because they may induce cellular death by apoptotic processes (Fairbairn et al. 1995), and oxidative stress can lead to neurodegenerative disorders and provoke neuron pathology (Coyle and Puttfarcken 1993).

Conclusion

Despite a large number of studies, knowledge about the toxic mechanisms of metals, pesticides, toxins are still a challenge for toxicologists. Fortunately, science has made rapid strides in the last several decades, leading to an increased basic understanding of not only how cells live, grow and reproduce but also how and why cells become sick and die as a result of exposure to toxic pollutants. These basic insights, coupled with greatly improved analytical capabilities have lead to the development of the new biomarkers such as in the field of genomics, proteomics and metabolomics. Findings based on the use of these advances can lead to the identification of a coherent and consistent biological research pathway for biomarker development, validation and acceptance into public health practice. Such a pathway might include how disturbed cellular pathways are linked to more overt outcomes such as compromised organ function, death or cancer. Using these modern tools, it is now possible to look forward to the prospect of providing stronger scientific evidence and greater conviction in interpreting early health effects related to pollutant exposure. This would be especially important for biomarkers that are capable of finding effects from low levels of exposures, identifying sensitive subpopulations, and effects of some pollutants in the presence of other competing chemical exposure risks.

Acknowledgements

We acknowledge the financial support of CNPq- Brazil.

Keywords: neurotoxic effects, central nervous system, fish, oxidative stress, pesticides, metals, saxitoxin

References

Abdollahi, M., A. Rainba, S. Shadnia, S. Nikfar and A. Rezaie. 2004. Pesticide and oxidative stress: a review. Med. Sci. Monitor. 10: 141–147.

Abdullah, S., J. Muhammad and A. Javid. 2007. Studies on acute toxicity of metals to the fish (*Labeo roahit*). Int. J. Agri. Biol. 9: 333–337.

Almeida, L.C., L.H. Aguiar and G. Moraes. 2005. Methyl parathion effect in Matrinxã *Brycon cephalus* muscle and brain acetylcholinesterase activity. Ciência Rural. 35: 1412–1416.

Altmann, L., K. Sveinsson, U. Krämer, M. Weishoff-houben, M. Turfeld and H. Wiegand. 1998. Visual functions in 6-Year-Old children in relation to lead and mercury levels. Neurotoxicol. Teratol. 20: 9–17.

Alves Costa, J.R.M., M. Mela, H.C. Silva de Assis, E. Pelletier, M.A.F. Randi and C.A. Oliveira Ribeiro. 2007. Enzymatic inhibition and morphological changes in *Hoplias malabaricus* from dietary exposure to lead (II) or methylmercury. Ecotoxicol. Environ. Saf. 67: 82–88.

Anderson, H.R., J.B. Nielson and P. Grandjean. 2000. Toxicologic evidence of developmental neurotoxicity of environmental chemicals. Toxicology 144: 121–127.

Atchison, W.D. and T. Narahaski. 1984. Mechanisms of action of lead on neuromuscular junctions. Neurotoxicology 5: 267–282.

ATSDR (Agency for Toxic Substances, Disease Registry). 2002. Toxicological profile for DDT, DDE, and DDD. Atlanta, GA: U.S. Department of Health and Human Services, Public Health Services.

Atta, M.B., M.A. Noaman and H.E. Kassab. 1997. Effects of lead exposure on the physiology of neurons. Prog. Neurobiol. 24: 199–231.

Baatrup, E. and K.B. Døving. 1990. Histochemical demonstration of mercury in the olfactory system of salmon (*Salmo salar* L.) following treatments with dietary methylmercuric chloride and dissolved mercuric chloride. Ecotoxicol. Environ. Saf. 20: 277–289.

Bálint, T., Z. Szegletes, Z. Szegletes, K. Halsy and J. Nemcsók. 1995. Biochemical and subcellular changes in carp exposed to the organophosphorous metidation and the pyrethroid deltamethrin. Aquat. Toxicol. 33: 279–295.

Barboni, M.T.S., C. Feitosa-Santana, E.C. Zachi, M. Lago, R.Teixeira and A. Taub. 2009. Preliminary findings on the effects of occupational exposure to mercury vapor below safety levels on visual and neuropsychological functions. J. Occup. Environ. Med. 51: 1403–1412.

Beauvais, L., S.B. Jones, J.T. Parris, S.K. Brewer and E.E. Little. 2001. Cholinergic and behavioral neurotoxicity of carbaryl and cadmium to larval rainbow trout (*Oncorhynchus mykiss*). Ecotoxicol. Environ. Saf. 49: 84–90.

Beltran, E.C. and B.A. Neilan. 2000. Geographical segregation of the neurotoxin-producing cyanobacterium *Anabaena circinalis*. Appl. Environ. Microbiol. 66: 4468–74.

Berntssen, M.H.G., A. Aatland and R.D. Handy. 2003. Chronic dietary mercury exposure causes oxidative stress, brain lesions, and altered behaviour in Atlantic salmon (*Salmo salar*) parr. Aquati. Toxicol. 65: 55–72.

Bertin, G. and D. Averbeck. 2006. Cadmium: cellular effects, modifications of biomolecules, modulation of DNA repair and genotoxic consequences (a review). Biochimie. 88: 1549–1559.

Berzas Nevado, J.J., F.J. Bernardo, M. Jiménez, A.M. Herculano, J.L.M. do Nascimento and M.E. Crespo-López. 2010. Mercury in the Tapajós River basin, Brazilian Amazon: A review. Environ. Int. 36: 593–608.

Bhattacharyya, S. 1985. Toxicity of carbofuran and phenthoate in *Channa punctatus* and *Anabus testudineus*. J. Environ. Biol. 6: 129–137.

Bonci, D.M.O., S.M.A. de Lima, S.R. Grotzner, C.A. Oliveira Ribeiro, D.E. Hamassaki and D.F. Ventura. 2006. Losses of immunoreactive parvalbumin amacrine and immunoreactive a protein kinase C bipolar cells caused by methylmercury chloride intoxication in the retina of the tropical fish *Hoplias malabaricus*. Braz. J. Med. Biol. Res. 39: 405–410.

Botaro, D., J.P. Torres, O. Malm, M.F. Rebelo, B. Henkelmann and K.W. Schramm. 2011. Organochlorine pesticides residues in feed and muscle of farmed Nile tilapia from Brazilian fish farms. Food Chem. Toxicol. 49: 2125–2130.

Bradbury, S.P. and J.R. Coats. 1989. Comparative toxicology of pyrethroid insecticides. Rev. Environ. Contam. Toxicol. 108: 133–177.

Bradbury, S.P., R.W. Carlson, G.J. Niemi and T.R. Henry. 1991. Use of respiratory-cardiovascular responses of rainbow trout (*Oncorhynchus mykiss*) in identifying acute toxicity syndromes in fish. Environ. Toxicol. Chem. 10: 115–131.

Bradbury, S.P., R.W. Carlson, T.R. Henry, S. Padilla and J. Cowden. 2008. Toxic responses of the fish nervous system. *In:* R.T. Di Giulio and D.E. Hinton (eds.). The Toxicology of Fishes. CRC Press, Florida, USA, pp. 417–455.

Bretaud, S., J.P. Toutant and P. Saglio. 2000. Effects of carbofuran, diuron and nicosulfuron on acetylcholinesterase activity in Goldfish (*Carassius auratus*). Ecotoxicol. Environ. Saf. 47: 117–124.

Brocardo, P.S., P. Pandolfo, R.N. Takahashi, A.L.S. Rodrigues and A.L. Dafre. 2005. Antioxidant defenses and lipid peroxidation in the cerebral cortex and hippocampus following acute exposure to malathion and/or zinc chloride. Toxicology 207: 283–291.

Brunelli, E., A. Mauceri, M. Maisano, I. Bernabo, A. Giannetto, E. De Domenico, B. Corapi, S. Tripepi and S. Fasulo. 2011. Ultrastructural and immunohistochemical investigation on the gills of the teleost, *Thalassoma pavo* L., exposed to cadmium. Acta Histochem. 113: 201–213.

Carlson, R.W., S.P. Bradbury, R.A. Drummond and D.E. Hammermeister. 1998. Neurological effects on startle response and escape from predation by medaka exposed to organic chemicals. Aquat. Toxicol. 43: 51–68.

Carmichael, W.W. 1997. The cyanotoxins. Adv. Biochem. Res. 27: 211–256.

Casida, J.E. and G.B. Quistad. 1998. Golden age of insecticide research: past, present, or future? Annu. Rev. Entomol. 43: 1–16.

Cearley, J.E. and R.L. Coleman. 1974. Cadmium toxicity and bioconcentration largemouth bass and bluegill. Bull. Environ. Contam. Toxicol. 11: 146–152.

Cheng, S.H., P.K. Chan and R.S.S. Wu. 2001. The use of microangiography in detecting aberrant vasculature in zebrafish embryos exposed to cadmium. Aquat. Toxicol. 52: 61–71.

Chinni, S. and P.R. Yallapragda. 2000. Toxicity of copper, cadmium, zinc and lead to *Penaeus indicus postlarvae*: Effects of individual metals. J. Environ. Biol. 21: 255–258.

Chorus, I. and J. Bartram. 1999. Toxic cyanobacteria in water. A guide to public health consequence monitoring and management. London, UK.

Chow, E.S.H., C.W. Cheng and S.H. Cheng. 2009. Cadmium affects retinogenesis during zebrafish embryonic development. Toxicol. Appl. Pharmacol. 235: 68–76.

Clark, D.E., J.R. Nation, A.J. Bourgeois, M.F. Hare, D.M. Baker and E.J. Hinderberger. 1985. The regional distribution of cadmium in the brains of orally exposed adult rats Neurotoxicology 6: 109–114.

Clarkson, T.W. 1987. Metal toxicity in the central nervous system. Environ. Health Perspect. 75: 59–64.

Coats, J.R. 1990. Mechanisms of toxic action and structure-activity relationships for organochlorine and synthetic pyrethroid insecticides. Environ. Health. Perspect. 87: 255–262.

Costa, G.M., L.M. Anjos, G.S. Souza, B.D. Gomes and C.A. Saito. 2008. Mercury toxicity in Amazon gold miners: visual dysfunction assessed by retinal and cortical electrophysiology. Environ. Res. 107: 98–107.

Coyle, J.T. and P. Puttfarcken. 1993. Oxidative stress, glutamate, and neurodegenerative disorders. Science 262: 689–695.

De la Torre, F.R., A. Salibia and L. Ferrari. 2007. Assessment of the pollution impact on biomarkers of effect of a freshwater fish. Chemosphere 68: 1582–1590.

Dembele, K., E. Haubruge and C. Gaspar. 2000. Concentration effects of selected insecticides on brain acetylcholinestarase in the common carp (*Cyprinus carpio* L.). Ecotoxicol. Environ. Saf. 45: 49–54.

Dou, C. and J. Zhang. 2011. Effects of lead on neurogenesis during zebrafish embryonic brain development. J. Hazard. Mater. 194: 277–282.

Dringer, R. 2000. Metabolism and functions of glutathione in brain. Proc. Neurobiol. 62: 649–671.

Droge, W. 2002. Free radicals in the physiological control of cell function. Physiol. Rev. 82: 47–95.

Dutta, H.M. and D.A. Arends. 2003. Effects of endosulfan on brain acetylcholinesterase activity in juvenile bluegill sunfish. Environ. Res. 91: 157–162.

Ecobichon, D.J. 1995. Toxic effects of pesticides. *In:* C.D. Klaassen, M.O. Amdur and J. Doull (eds.). The basic science of poisons. McGraw-Hill, New York, USA, pp. 643–689.

Ecobichon, D.J. and R.M. Joy. 1974. Pesticides and neurological diseases. Boca Raton, USA.

Edwards, C.A. 1970. Persistent pesticides in the environment. CRC Crit. Rev. Environ. 12: 1–7.

Ek, J.C., M. Katarzyna, D. Mark and R. Norman. 2012. Barriers in the developing brain and neurotoxicology. NeuroToxicology. 33: 586–604.

Erie, J.C., J.A. Good, J.A. Butz, D.O. Hodge and J.S Pulido. 2007. Urinary cadmium and agerelated macular degeneration. Am. J. Ophthalmol. 144: 414–418.

Evans, H.L. and R.H. Garman. 1980. Scotopic vision as an indicator of neurotoxixity. *In:* W.H. Merigan and B. Weiss (eds.). Neurotoxicity of the visual system. Raven Press, New York, USA, pp. 135–47.

Falconer, I.R. and A.R. Humpage. 2005. Health risk assessment of cyanobacterial (Blue-green Algal) toxins in drinking water. Int. J. Environ. Res. Public Health 2: 43–50.

Fairbairn, D.W., L.P. Olive and K.L. O'Neil. 1995. The comet assay: a comprehensive review. Rev. Gen. Toxicol. 339: 37–59.

Feitosa-Santana, C., M.F. Costa, M. Lago and D.F. Ventura. 2007. Longterm loss of color vision after exposure to mercury vapor. Braz. J. Med. Biol. Res. 40: 409–414.

Fox, D.A. and M. Katz. 1992. Developmental lead exposure selectively alters the scotopic ERG component of dark and light adaptation and increases rod calcium content. Vis. Res. 32: 249–255.

Friberg, L. 1950. Health hazards in the manufacture of akaline accumulators with special reference to chronic cadmium poisoning. Acta Med. Scand. Suppl. 138: 240–249.

Furey, A., J. Crowley and A.N. Shuilleabhain. 2003. The first identification of the rare cyanobacterial toxin, homoanatoxin-a, in Ireland. Toxicon. 41: 297–303.

Ganjavi, M., H. Ezzatpanah, M.H. Givianrad and A. Shams. 2010. Effect of canned tuna fish processing steps on lead and cadmium contents of Iranian tuna fish. Food Chem. 118: 525–528.

Gibbins, I.L., C. Olsson and S. Holmgren. 1995. Distribution of neurons reactive for NADPH-diaphorase in the branchial nerves of a teleost fish, *Gadus morhua*. Neurosc. Lett. 193: 113–116.

Gkelis, S., T. lanaras and K. Sivonen. 2006. The presence of microcystins and other cyanobacterial bioactive peptides in aquatic fauna collected from Greek freshwaters. Aquat. Toxicol. 78: 32–41.

Gonçalves, J.F., A.M. Fiorenza, R.M. Spanevello, C.M. Mazzanti, G.V. Bochi, F.G. Antes, N. Stefanello, M.A. Rubin, V.L. Dressler, V.M. Morsch and M.R.C. Schetinger. 2010.

N-acetylcysteine prevents memory deficits, the decrease in acetylcholinesterase activity and oxidative stress in rats exposed to cadmium. Chem. Biol. Interact. 186: 53–60.

González-Doncel, M., E. de la Peña, C. Barrueco and D.E. Hinton. 2003. Stage sensitivity of medaka (*Oryzias latipes*) eggs and embryos to permethrin. Aquat. Toxicol. 62: 255–268.

Gottofrey, J. and H. Tjalve. 1991. Axonal transport of cadmium in the olfactory nerve of the pike. Pharmacol. Toxicol. 69: 242–252.

Groman, D.B. 1982. Histology of the Striped Bass. American fishiries Society. Bethsda, Maryland.

Grotto, D., J. Valentini, M. Fillion, C.J. Souza Passos, S.C. Garcia, D. Mergler and Jr. F. Barbosa. 2010. Mercury exposure and oxidative stress in communities of the Brazilian Amazon. Sci. Total Environ. 408: 806–811.

Gugger, M., S. Lenoir and C. Berger. 2005. First report in a river in France of the benthic cyanobacterium *Phormidium favosum* producing anatoxin-a associated with dog neurotoxicosis. Toxicon. 45: 919–928.

Guiloski, I.C., S.C. Rossi, C.A. Silva and H.C. Silva de Assis. 2013. Insecticides biomarker responses on a freshwater fish *Corydoras paleatus* (Pisces: Callichthyidae). J. Environ. Sci. Health, Part B. 48: 272–277.

Guilhermino, L., P. Barros, M.C. Silva and A.M.V.M. Soares. 1998. Should the use of inhibition of cholinesterases as a specific biomarker for organophosphates and carbamate pesticides be questioned? Biomarkers. 3: 157–163.

Hawryshyn, C.W., W.C. MacKay and T.H. Nilsson. 1982. Methyl mercury induced visual deficits in rainbow-trout. Can. J. Zool. 60: 3127–3133.

Henriksen, P., W.W. Carmichael and J.S. An. 1997. Detection of an anatoxin-a(s)-like anticholinesterase in natural blooms and cultures of cyanobacteria/blue-green algae from Danish lakes and in the stomach contents of poisoned birds. Toxicon. 35: 901–913.

Hyde, E.G. and W.W. Carmichael. 1991. Anatoxin-a(s), a naturally occurring organophosphate, is an irreversible active site-directed inhibitor of acetylcholinesterase (EC 3.1.1.7). J. Biochem. Toxicol. 6: 195–201.

Ibelings, B.W. and I. Chorus. 2007. Accumulation of cyanobacterial toxins in freshwater "seafood" and its consequences for public health: a review. Environ. Pollut. 150: 177–192.

Irie, K., M. Kawaguchi, K. Mizuno, J.-Y. Song, K. Nakayama, S.-I. Kitamura and Y. Murakami. 2011. Effect of heavy oil on the development of the nervous system of floating and sinking teleost eggs. Mar. Poll. Bull. 63: 297–302.

Ishikura, H. 1972. Impact of pesticide use on the Japanese environment. *In*: F. Matsumura, G.M. Boush and T. Misato (eds.). Environmental toxicology of pesticides. Academic Press, New York, USA, pp. 1–32.

Javed, M. and G. Mahmood. 2001. Metal toxicity of water in a strech of river Ravi from shahdera to baloki headworks. Pakistan J. Agric. Sci. 38: 37–42.

Kaas, H. and P. Henriksen. 2000. Saxitoxins (PSP toxins) in Danish lakes. Water Res. 34: 2089–2097.

Katti, S.R. and A.G. Sathyanesan. 1986. Lead nitrate induced changes in the thyroid physiology of the catfish *Clarias batrachus* (L.). Ecotoxicol. Environ. Saf. 13: 1–6.

Kavitha, P. and J.V. Rao. 2007. Oxidative stress and locomotor behaviour response as biomarkers for assessing recovery status of mosquito fish, *Gambusia affinis* after lethal effect of an organophosphate pesticide, monocrotophos. Pestic. Biochem. Phys. 87: 182–188.

Kostial, K. and V.B. Vouk. 1957. Lead ions and synaptic transmission in the superior cervical ganglion of the cat. Br. J. Pharmacol. 13: 219–222.

Kumar, A. and J.C. Chapman. 1998. Profenofos toxicity to the eastern rainbow fish (*Melanotaenia duboulayi*). Environ. Toxicol. Chem. 17: 1799–1806.

Kumaraguru, A.K., F.W.H. Beamish and H.W. Ferguson. 1982. Direct and circulatory paths of permethrin (NRDC 143) causing histopathological changes in the gills of rainbow trout, *Salmo gairdneri*, Richardson. J. Fish. Biol. 20: 87–91.

Kusmic, C. and P. Gualtieri. 2000. Morphology and spectral sensitivities of retinal and extraretinal photoreceptors in freshwater teleosts. Micron. 31: 183–200.

Lailson-Brito, J., P.R. Dorneles, C.E. Azevedo-Silva, T.L. Bisi, L.G. Vidal, L.N. Legat, A.F. Azevedo, J.P.M. Torres and O. Malm. 2012. Organochlorine compound accumulation in delphinids southeastern Brazilian coast. Sci. Total Environ. 433: 123–131.

Leblond, V.S. and A. Hontela. 1999. Effects of *in vitro* exposure to cadmium, mercury, zinc and 1-(2-chlorophenyl)-1-(4chlorophenyl)-2, 2-dichloroethane on steridogenesis by dispersed internal cells of rainbow trout (*Oncorhynchus mykiss*). Toxicol. Appl. Pharma. 157: 16–22.

Levin, E.D., E. Chrysanthis, K. Yacisin and E. Linney. 2003. Chlorpyrifos exposure of developing zebrafish: effects on survival and long-term effects on response latency and spatial discrimination. Neurotoxicol. Teratol. 25: 51–57.

Levin, E.D., H.A. Swain, S. Donerly and E. Linney. 2004. Developmental chlorpyrifos effects on hatchling zebrafish swimming behavior. Neurotoxicol. Teratol. 26: 719–723.

Linde-Arias, A.R., A.F. Inácio, C. Alburquerquehttp://www.sciencedirect.com/science/article/pii/S0048969708003422 - implicit0, M.M. Freirehttp://www.sciencedirect.com/science/article/pii/S0048969708003422 - implicit0 and J.C. Moreira. 2008. Biomarkers in an invasive fish species, *Oreochromis niloticus*, to assess the effects of pollution in a highly degraded Brazilian River. Sci. Total Environ. 399: 186–192.

Lukaszewicz-Hussain, J., J. Moniuszko-Jakoniuk and J. Rogalska. 2007. Assessment of lipid peroxidation in rat tissues in subacute chlorfenvinphos administration. Pol. J. Environ. Stud. 16: 233–236.

Luo, M. and R.P. Bodnaryk. 1988. The effect of insecticides on Ca–Mg-ATPase and the ATP dependent calcium pump in moth brain synaptosomes and synaptosome membrane vesicle from the bertha armyworm, Mamestra configurate W1K. Pestic. Biochem. Physiol. 30: 155–165.

Luo, W., D. Ruan, C. Yan, S. Yin and J. Chen. 2012. Effects of chronic lead exposure on functions of nervous system in chinese children and developmental rats. Neuro Toxicology. 33: 862–871.

Malbrouck, C. and P. Kestemont. 2006. Effects of microcystins on fish. Environ. Toxicol. Chem. 25: 72–86.

Malm, O., F.P. Branches, H. Akagi, M.B. Castro, W.C. Pfeizer and M. Harada. 1995. Mercury and methylmercury in fish and human hair from the Tapajós river basin. Braz. Sci. Total Environ. 175: 141–150.

Matsumura, F. 1972. Biological effects of toxic pesticidal contaminants and terminal residues. In: F. Matsumura, G.M. Boush and T. Misato (eds.). Environmental toxicology of pesticides. Academic Press, New York, USA, pp. 525–548.

Mela, M., M.A. Randi, D.F. Ventura, C.E. Carvalho, E. Pelletier and C.A. Oliveira Ribeiro. 2007. Effects of dietary methylmercury on liver and kidney histology in the neotropical fish *Hoplias malabaricus*. Ecotoxicol. Environ. Saf. 68: 426–435.

Mela, M., S. Cambier, N. Mesmer-Dudons, A. Legeay, S.R. Grotzner, C.A. Oliveira Ribeiro, D.F. Ventura and J-.C. Massabuau. 2010. Methylmercury localization in *Danio rerio* retina after trophic and subchronic exposure: A basis for neurotoxicology. NeuroToxicology. 31: 448–453.

Mela, M., S.R. Grotzner, A. Legeay, N. Mesmer-Dudons, J-.C. Massabuau, D.F. Ventura and C.A. Oliveira Ribeiro. 2012. Morphological evidence of neurotoxicity in retina after methylmercury exposure. NeuroToxicology 33: 407–415.

Méndez-Armenta, M. and C. Ríos. 2007. Cadmium neurotoxicity. Environ. Toxicol. Pharmacol. 23: 350–358.

Meyer, P.A., M.J. Brown and H. Falk. 2008. Global approach to reducing lead exposure and poisoning. Mutat. Res. 659: 166–175.

Mieiro, C.L., I. Ahmad, M.E. Pereira, A.C. Duarte and M. Pacheco. 2010. Antioxidant system breakdown in brain of feral golden grey mullet (*Liza aurata*) as an effect of mercury exposure. Ecotoxicology 19: 1034–1045.

Mittal, K., T. Adak and V.P. Sharma. 1994. Comparative toxicity of certain mosquitocidal compounds to larvivorous fish, *Poecilia reticulata*, Ind. J. Malariol. 31: 43–47.

Molica, R.J.R., E.J.A. Oliveira, P.V.V.C. Carvalho, A.N.S.F. Costa, M.C.C. Cunha, G.L. Melo and S.M.F.O. Azevedo. 2005. Occurrence of saxitoxins and anatoxin-a(s)-like anticholinesterase in a Brazilian drinking water supply. Harmful Algae. 4: 743–753.

Monteiro, D.A., F.T. Rantin and A.L. Kalinin. 2010. Inorganic mercury exposure: toxicological effects, oxidative stress biomarkers and bioaccumulation in the tropical freshwater fish matrinxã, *Brycon amazonicus* (Spix and Agassiz 1829). Ecotoxicology 19: 105–123.

Monteiro, V., D.G.S.M. Cavalcante, M.B.F.A. Vilela, S.H. Sofia and C.B.R. Martinez. 2011. *In vivo* and *in vitro* exposures for the evaluation of the genotoxic effects of lead on the Neotropical freshwater fish *Prochilodus lineatus*. Aquat. Toxicol. 104: 291– 298.

Moretto, A. and M. Lotti. 2006. Peripheral nervous system effects and delayed neuropathy. *In:* R.C. Gupta (ed.). Toxicology of organophosphate and carbamate compounds. Academic Press, Amsterdam, The Netherlands, pp. 361–70.

Monserrat, J.M., J.S. Yunes and A. Bianchini. 2001. Effects of Anabaena spiroides (cyanobacteria) aqueous extracts on the acetylcholinesterase activity of aquatic species. Environ. Toxicol. Chem. 20: 1228–35.

Namikoshi, M., T. Murakami, T. Fujiwara, H. Nagai, T. Niki, E. Harigaya, M.F. Watanabe, T. Oda, J. Yamada and S. Tsujimura. 1994. Biosynthesis and transformation of anatoxin-a in the ion channels. *In:* L. Chang (ed.). CRC Press, Florida, USA, pp. 609–655.

Narahashi, T. 2004. Role of ion channels in neurotoxicity. *In:* L. Chang (ed.). In Principles of Neurotoxicology, CRC Press, New York, USA, pp. 609–655.

Nielsen, J.B. and O. Andersen. 1991. Methyl mercuric chloride toxicokinetics in mice. II: sexual differences in whole-body retention and deposition in blood, hair, skin, muscles and fat. Pharmacol. Toxicol. 68: 208–211.

Nilsson, S. 2011. Brain and nervous system—autonomic nervous system of fishes. Academic press. University of Gothenburg, Gothenburg, Sweden.

Nitschke, I., F. Muller, J. Smith and W. Hopfenmuller. 2000. Amalgam fillings and cognitive abilities in a representative sample of the elderly population. Gerodontology 17: 39–44.

Oliveira Ribeiro, C.A. and H.C. Silva De Assis. 2005. AChE inhibition as a biomarker for pollutants contamination in tropical aquatic ecosystems. *In:* M. Parveen and S. Kumar (eds.). Recent trends in the acetylcholinesterase system. IOS Press, Amsterdam, The Netherlands, pp. 103–124.

Oudou, H.C., R.M. Alonso and H.C. Bruun Hansen. 2004. Voltammetric behavior of the synthetic pyrethroid lambda-cyhalothrin and its determination in soil and well water. Anal. Chim. Acta 523: 69–74.

Pari, L. and P. Murugavel. 2007. Diallyl tetrasulfide improves cadmium induced alterations of acetylcholinesterase, ATPases and oxidative stress in brain of rat. Toxicology 234: 44–50.

Payne, J.F., A. Mathieu, W. Melvin and L.L. Fancey. 1996. Acetylcholinesterase, an old biomarker with a new future? Field trials in association with two urban rivers and a paper mill in Newfoundland. Mar. Poll. Bul. 32: 225–23.

Penteado, J.C.P. and J.M. Vaz. 2001. The legacy of the polychlorinated biphenyls (PCBs). Quim. Nova. 24: 390–398.

Perger, G. and D. Szadkowski. 1994. Toxicology of pyrethroids and their relevance to human health. Ann. Agric. Environ. Med. 1: 11–17.

Puschner, B. and J-.F. Humbert. 2007. Cyanobacterial toxins. *In:* R.C. Gupta (ed). Veterinary Toxicology: Basic and Clinical Principles. Academic Press, New York, USA, 1224 pp.

Quintanar-Escorza, M.A., M.T. González-Martínez and J.V. Calderón-Salinas. 2010. Oxidative damage increases intracellular free calcium (Ca^{2+}) concentration in human erythrocytes incubated with lead. Toxicol. *in vitro* 24: 1338–1346.

Rabitto, I.S., J.R.M. Alves Costa, H.C. Silva de Assis, E. Pelletier, F.M. Akaishi, A. Anjos, M.A.F. Randi and C.A. Oliveira Ribeiro. 2005. Effects of dietary Pb(II) and tributyltin on neotropical fish, *Hoplias malabaricus*: histopathological and biochemical findings. Ecotoxicol. Environ. Saf. 60: 147–156.

Rabitto, I.S., W.R. Bastos, R. Almeida, A. Anjos, I.B.B. Holanda, C.F. Galvão Roberta, F. Filipak Neto, M.L. Menezes, C.A.M. Santos and C.A. Oliveira Ribeiro. 2011. Mercury and DDT exposure risk to fish-eating human populations in Amazon. Environ. Int. 37: 56–65.

Regoli, F., M. Nigro, M. Benedetti, D. Fattorini and S. Gorbi. 2005. Antioxidant efficiency in early life stages of the Antarctic silverfish *Pleuragramma antarcticum*: responsiveness to pro-oxidant conditions of platelet ice and chemical exposure. Aquat. Toxicol. 75: 43–52.

Roex, E.W.M., R. Keijzers and C.A.M. Van Gesterl. 2003. Acetylcholinesterase inhibition and increased food consuption in the zebrafish, *Danio rerio* after chronic exposure to parathion. Aquat. Toxicol. 64: 451–460.

Roozbehi, A. 2007. Effects of cadmium on photoreceptors and ganglionic cells of retinal layer in mice embryo—An ultrastructural study. Indian J. Exp. Biol. 45: 469–474.

Saker, M.L., A.D. Thomas and J.H. Norton. 1999. Cattle mortality attributed to the toxic cyanobacterium *Cylindrospermopsis raciborskii* in an outback region of North Queensland. Environ. Toxicol. 14: 179–182.

Saldana, M., C.E. Collins, R. Gale and O. Backhouse. 2006. Diet-related mercury poisoning resulting in visual loss. Br. J. Ophthalmol. 90: 1432–1434.

Sancho, E., C. Fernandez-Veiga, M. Sanchez, M.D. Ferrando and E. Andrev-Moliner. 2000. Alterations on AChE of the fish *Anguilla anguilla* as response to herbicide-contaminated water. Ecotoxicol. Environ. Saf. 46: 57–63.

Santovito, G., E. Piccinni, F. Boldrin and P. Irato. 2012. Comparative study on metal homeostasis and detoxification in two Antarctic teleosts. Comp. Biochem. Physiol. 155: 580–586.

Sayeed, I., S. Parvez, S. Pandey, B. Bin-Hafeez, R. Haque and S. Raisuddin. 2003. Oxidative stress biomarkers of exposure to deltamethrin in freshwater fish, *Channa punctatus* bloch. Ecotoxicol. Environ. Saf. 56: 295–301.

Scascitelli, M. and F. Pacchierotti. 2003. Effects of lindane on oocyte maturation and preimplantation embryonic development in the mouse. Reprod. Toxicol. 17: 299–303.

Shafer, T.J., D.A. Meyer and K.M. Crofton. 2005. Developmental neurotoxicity of pyrethroid insecticides: critical review and future research needs. Environ. Health Perspect. 113: 123–136.

Shrivastav, B.B., M.S. Brodwick and T. Narahashi. 1976. Methylmercury: effects on eletrical properties of squid axon membranes. Life Sci. 18: 1077–1082.

Shukla, A., G.S. Shukla and R.C. Srimal. 1996. Cadmium-induced alterations in blood-brain barrier permeability and its possible correlation with decreased microvessel antioxidant potential in rat. Hum. Exp. Toxicol. 15: 400–405.

Silva, H.C., H.S.G. Medina, E. Fanta and M. Bacila. 1993. Sub-lethal effects of the organophosphate folidol 600 (methyl parathion) on *Callichthys callichthys* (Pisces: Teleostei). Comp. Biochem. Physiol. 105: 197–201.

Silva de Assis, H.C., L. Nicareta, C. Klemz, J. H. Truppel and R. Calegari. 2009. Biochemical biomarkers of exposure to deltamethrin in freshwater fish. B. Arch. Biol. Technol. 6: 1401–1407.

Silva, C.A., E.T. Oba, W.A. Ramsdorf, V.F. Magalhães, M.M. Cestari, C.A. Oliveira Ribeiro and H.C. Silva de Assis. 2011. First report about saxitoxins in freshwater fish *Hoplias malabaricus* through trophic exposure. Toxicon. 32: 141–147.

Sindayigaya, E., R. Van Cauwenbergh, H. Robberech and H. Deelstra. 1994. Copper, zinc, manganese, iron, lead, cadmium, mercury and arsenic in fish from lake Tanganyika, Burundi. Sci. Total Environ. 144: 103–15.

Sivonen, K. and G. Jones. 1999. Cyanobacterial toxins. *In:* I. Chorus and J. Bartram (eds.). Toxic cyanobacteria in water: A guide to their public health consequences, monitoring and management. Spon Press, London, pp. 41–111.

Soderlund, D.M., J.M. Clark, L.P. Sheets, L.S. Mullin, V. J. Piccirillo, D. Sargent, J.T. Stevens and M.L. Weiner. 2002. Mechanisms of pyrethroid neurotoxicity: implications for cumulative risk assessment. Toxicology. 171: 3–59.

Solomon, K.R, G.R. Stephenson, C.L. Correa and F.A.D. Zambrone 2010. Praguicidas e o Meio Ambiente. 2010. ILSI Press Brazil.

Stouthart, A.J.H.X., F.A.T. Spanings, R.A.C. and S.E. Lock. 1994. Effects of low water pH on lead toxicity to early life stages of the common carp (*Cyprinus carpio*). Aquat. Toxicol. 30: 137–151.

Strichartz, G.R. and N. Castle. 1990. Pharmacology of marine toxins, effects on membrane channels. *In:* G.R. Hall and E. Strichartz (eds.). Marine Toxins. Origin, Structure and Molecular Pharmacology. American Chemical Society, Washington, USA, pp. 3–20.

Sturm, A., H.C. Silva de Assis and P.-D. Hansen. 1999. Cholinesterases of marine teleost fish: enzymological characterization and potential use in the monitoring of neurotoxic contamination. Mar. Environ. Res. 47: 389–98.

Suzuki, Y. and H. Arito. 1967. Effect of cadmium-feeding on tissue concentrations of elements in germ-free silkworm (*Bombyx mori*) larvae and distribution of cadmium in the alimentary canal. Ind. Health. 14: 875–879.

Swanson, K.L., R.S. Aronstam, S. Wonnacott, H. Rapoport and E.X. Albuquerque. 1991. Nicotinic pharmacology of anatoxin analogs. I. Side chain structure–activity relationships at peripheral agonist and noncompetitive antagonist sites. J. Pharmacol. Exp. Ther. 259: 377–386.

Syversena, T. and P. Kaur. 2012. The toxicology of mercury and its compounds. J. Trace Elem. Med. Biol. In press.

Tanan, C.L., D.F. Ventura, J.M. de Souza, S.R. Grotzner, M. Mela and A. Jr. Gouveia. 2006. Effects of mercury intoxication on the response of horizontal cells of the retina of trahira fish (*Hoplias malabaricus*). Braz. J. Med. Biol. Res. 39: 987–995.

Tessier-Lavigne, M., P. Mobbs and D. Attwell. 1985. Lead and mercury toxicity and the rod light response. Invest. Ophthalmol. Vis. Sci. 26: 1117–1123.

Thomas, P., M. Stephens and G. Wilkie. 1993. Anatoxin-a is a potent agonist at neuronal nicotinic acetylcholine receptors. J. Neurochem. 60: 2308–2311.

Thompson, L.M. 2008. Neurodegeneration: a question of balance. Nature. 452: 707–708.

Tjalve, H., J. Henriksson, J. Tallkvist, B. Larsson and N.G. SLindquist. 1996. Uptake of manganese and cadmium from the nasal mucosa into the central nervous systems via olfactory pathways in rats. Pharmacol. Toxicol. 79: 347–356.

Vahter, M., A. Akesson, C. Lidén, S. Ceccatelli and M. Berglund. 2007. Gender differences in the disposition and toxicity of metals. Environ. Res. 104: 85–95.

Ventura, D.F., A.L. Simões, S. Tomaz, M.F. Costa, M. Lago and M.T.V. Costa. 2005. Colour vision and contrast sensitivity losses of mercury intoxicated industry workers in Brazil. Environ. Toxicol. Pharmacol. 19: 523–529.

Vieira, C., S. Morais, S. Ramos, C. Delerue-Matos and M.B.P.P. Oliveira. 2011. Mercury, cadmium, lead and arsenic levels in three pelagic fish species from the Atlantic Ocean: Intra- and inter-specific variability and human health risks for consumption. Food Chem. Toxicol. 49: 923–932.

Vorobjeva, R.S.Z. 1957. Investigation of the nervous system function in workers exposed to cadmium oxide. In. Nevropat. Psikhiat. 57: 385–389.

Warfvinge, K. and A. Bruun. 2000. Mercury distribution in the squirrel monkey retina after in uterus exposure to Mercury vapor. Environ. Res. 83: 102–109.

Winberg, S., R. Bjerselius, E. Baatrup and K. Døving. 1992. The effect of Cu(II) on the electro-olfactogram (EOG) of the Atlantic salmon (*Salmo salar* L.) in artificial freshwater of varying inorganic carbon concentrations. Ecotoxicol. Environ. Saf. 24: 167–178.

Woolley, D.E. 1995. Organochlorine insecticides: neurotoxicity and mechanisms of action. *In:* L. Chang and R. Dyer (eds.). Handbook of Neurotoxicology. Marcel Dekker, New York, USA, pp. 475–510.

Yordy, J.E., R.S. Wells, B.C. Balmer, L.H. Schwacke and T.K. Rowles. 2010. Source of variation for persistent organic pollutant of common bottlenose dolphins (*Tursiops truncates*). Sci. Total Environ. 408: 2163–2172.

Yorifuji, T., T. Tsuda, S. Inoue, S. Takao and M. Harada. 2011. Long-term exposure to methylmercury and psychiatric symptoms in residents of Minamata, Japan. Environ. Int. 37: 907–913.

Yuan, Y. and W.D. Atchison. 1993. Disruption by methylmercury of membrane excitability and synaptic transmission of CA1 neurons in hippocampal slices of the rat. Toxicol. Appl. 23: 112–117.

Zheng, W., M. Aschner and J-F. Ghersi-Egea. 2003. Brain barrier systems: a new frontier in metal neurotoxicological research. Toxicol. Appl. Pharmacol. 192: 1–11.

Zohar, Y., J.A. Muñoz-Cueto, A. Elizur and O. Ka. 2010. Neuroendocrinology of reproduction in teleost fish. Gen. Comp. Endocrinol. 165: 438–455.

CHAPTER 5

Pollutants and Oxidative Stress

Francisco Filipak Neto

Introduction

Aerobic organisms continuously produce reactive oxygen species (ROS) during metabolism and eliminate these toxic substances through a very ancient antioxidant defense system. However, many chemicals can either potentiate ROS production through interactions with the systems that normally produce ROS at low levels, or by catalyzing and participating in some ROS-producing reactions. Other chemicals can impair antioxidant defenses, rendering cells more susceptible to oxidative damage. The presence of oxidative stress is commonly investigated in fish in biomonitoring and laboratory studies, since this process has a great effect on fish health.

Reactive Species and Oxidative Stress

Oxygen is essential for many metabolic processes that are vital to aerobic life such as ATP synthesis. However, reactive oxygen species are produced during aerobic metabolism, being prone to attack organic molecules from which cells are built. As a consequence of evolution and natural selection, several mechanisms that protect organisms from the toxic effects of

Universidade Federal do Paraná, Departamento de Biologia Celular, CX 19031, CEP 81.531-980, Curitiba–PR Brazil.

Email: filipak@ufpr.br

increased ROS production have been fine-tuned over time, allowing the benefits of aerobic metabolism as well as survival of organisms in oxygen-rich environments. In this context, the term "oxidative stress" can be defined as a state of imbalance toward the factors that generate ROS and away from the factors that protect cellular biomolecules from these reactants, leading to potential damage (Halliwell and Gutteridge 1999). Under conditions of oxidative stress, ROS and free radicals that are not reduced or removed from the cellular environment can cause damage to all cellular biomolecules including nucleic acids, lipids and proteins (Bokov et al. 2004). Fish are particularly sensitive to water contamination, because pollutants may impair many physiological and biochemical processes through xenobiotic-induced oxidative stress (Durmaz et al. 2006).

Whereas we define ROS as reactive species that contain oxygen, free radicals are atoms, molecules or ions with unpaired electrons on an otherwise open shell configuration (Halliwell and Gutteridge 1991). These unpaired electrons are usually highly reactive so that radicals are likely to take part in chemical reactions. Common examples of ROS include radicals such as molecular oxygen ($\cdot O{=}O\cdot$; the "dots" indicate the unpaired electrons), superoxide anion ($O_2\cdot^-$), hydroxyl radical ($\cdot OH$), and non-radical oxygen intermediates like hydrogen peroxide (H_2O_2) and singlet oxygen ($^1O_2$) (Lushchak 2011). Of the ROS, hydroxyl radical is the most reactive, being a very harmful byproduct of oxidative metabolism to cells (Cheng et al. 2002; Livingstone 2001).

Sources of Reactive Species

Endogenous ROS produced *in vivo* have extremely short half-lives and they are present at very low concentrations (usually not exceeding 10 nM) in part due to fine cellular control (Halliwell and Gutteridge 1991). But ROS concentration is a dynamic parameter; ROS are continuously generated and eliminated. This steady-state ROS concentration means that usually the amount of ROS produced is virtually equal to that eliminated. When the balance (redox status) is sufficiently disturbed, either oxidative or reductive stress is established (Lushchak 2011).

Mitochondrion is the major source of ROS in the eukaryotic cell since over 90% of oxygen consumed by organisms is utilized by this organelle (Papa and Skulachev 1997). In the mitochondrion, the reactions that generate most of the cell's ATP require energy-rich electrons from reduced substrates (NADH, $FADH_2$), to be passed along the electron transport chain up to molecular oxygen, generating water via a four-electron transfer. However, part of the electrons can leak (particularly from complexes I and III) and react with oxygen producing the free radical superoxide (Fig. 5.1). Superoxide anion is a significant mediator in numerous oxidative chain reactions, being

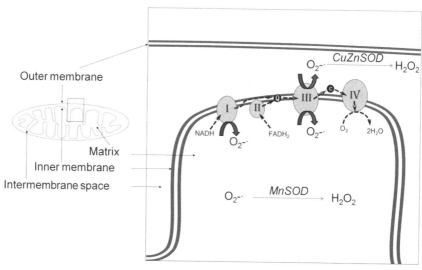

Figure 5.1. Mitochondrial pathway of ROS generation. Electrons from NADH or $FADH_2$ are transported through the transporting electron chain (Complex I/Complex II → Ubiquinone (Q) → Complex III → Cytochrome C (C) → Complex IV) until reacting with oxygen and protons to produce water. Dotted arrows indicate this pathway. However, a small percentage of electrons can be diverted from the normal transport pathway and react with oxygen before reaching complex IV, thus producing superoxide. Superoxide dismutase converts superoxide to hydrogen peroxide.

also a precursor to many other ROS (Halliwell and Gutteridge 1999). The degree of uncoupling depends on mitochondrial health status and many xenobiotics such as cadmium, chromium and mercury can uncouple or divert the "normal" electron flow, thereby increasing superoxide production (Stohs et al. 2001).

In cells such as hepatocytes, electron-transport chain of monooxygenases of the endoplasmic reticulum is the second most important source of ROS (Malhotra and Kaufman 2007). In this organelle, cytochrome P450 (CYP) activates molecular oxygen with the aid of additional enzymes and transfers it to a variety of endogenous and exogenous compounds. But, whenever the activated oxygen escapes from being added to the substrate, either superoxide or hydrogen peroxide is produced (Fig. 5.2). The level of ROS production depends on the CYP isoform, O_2 partial pressure and the type of substrate; some CYPs produce more ROS than others, and compounds such as polychlorinated biphenyls (PCBs), aromatic hydrocarbons (PAHs) and some organophosphates lead to particularly large production of ROS (Chambers et al. 2001; White 1991).

Certain amounts of ROS are produced in the plasma membrane, cytosol and peroxisomes by different enzymes, including peroxisomal oxidases, NADPH oxidases, nitric oxide synthetases (which produce nitric oxide,

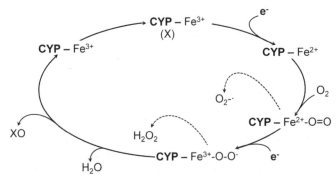

Figure 5.2. Cytochrome P450 (CYP) pathway of ROS generation. Electrons from NADPH are transferred to CYP thereby activating molecular oxygen, which is inserted in the substrate (X). However, activated oxygen can leak from the normal pathway and generate superoxide or hydrogen peroxide. Figure based on Boelsterli 2007.

a nitrogen reactive species) and lipoxygenases (Halliwell and Gutteridge 1999). Some cell types such as phagocytic cells of the innate immune system may promote localized environments with elevated oxidative stress. For example, macrophages can produce localized oxidative stress—"oxidative burst"—as part of the inflammatory response (Agnisola 2005; Federico et al. 2007) so that the production of some ROS are required to support the natural cellular function and regulate intracellular signaling (Perez-Matute et al. 2009). Note that activation of macrophages and recruiting of macrophage/monocytes may even affect redox homeostasis of organs with high local phagocytes population such as the liver, and potentiate oxidative stress damage.

Autooxidation of certain cellular components such as dopamine and δ-aminolevulinic acid, as well as redox cycling of many different compounds such as aromatic amines, metabolites of benzene and some pesticides may also be responsible for the production of substantial ROS amounts (Bonneh-Barkay et al. 2005; Lushchak 2011).

During redox cycling the xenobiotic takes up the electron from NADPH (an important antioxidant molecule) and if the redox potential allows, it transfers the electron to oxygen producing superoxide. This process can be cyclic as the xenobiotic can take up more electrons after transferring the previous one to oxygen. Redox cycling (Fig. 5.3) is therefore very problematic as a few xenobiotic molecules can lead to the production of large amount of ROS and consumption of NADPH (Bonneh-Barkay et al. 2005; Lushchak 2011).

Finally, some elements with changeable valences like iron, copper, chromium, cobalt and vanadium can generate hydroxyl radical through Fenton or Haber-Weiss reactions (Svensson 2008; Valko et al. 2005), as indicated.

$$Fe(III) + O_2^{\cdot -} \rightarrow Fe(II) + O_2 \tag{1}$$

$$Fe(II) + H_2O_2 \rightarrow Fe(III) + \cdot OH + OH^- \text{ (Fenton reaction)} \tag{2}$$

The overall reaction of the combined steps (1 and 2) is called Haber-Weiss reaction:

$$O_2^{\cdot -} + H_2O_2 \rightarrow O_2 + \cdot OH + OH^-$$

Considering the various sources of ROS (Fig. 5.4), the interactions of organisms and xenobiotics that cause or potentiate oxidative stress can be multiple. ROS and free radicals can cause both reversible and irreversible damage to biomolecules including proteins, nucleic acids and membrane lipids as well as initiate the expression of cytoprotective redox-sensitive transcription factors (Krivoruchko and Storey 2010). If ROS levels are not damaging, ROS signaling can have not only a protective role that increases the chance of surviving an insult related to oxidative stress (such as a hypoxic one), but can also precondition an animal to withstand further such insults (Krivoruchko and Storey 2010).

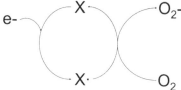

Figure 5.3. Redox cycling and ROS generation. During redox cycling, a substance (X) takes up one electron (X→X·) from NADPH and transfers it to molecular oxygen. During this process, the substance is initially reduced (receives the electron) and then oxidized (donates the electron), and superoxide is produced. The same substance can be utilized in additional rounds of ROS production, characterizing a devastating cyclic process.

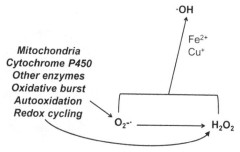

Figure 5.4. The cellular sources of ROS. ROS such as $O_2^{\cdot -}$ and H_2O_2 are produced during cell metabolism, particularly in mitochondria and endoplasmic reticulum (by cytochrome P450) through electron transport uncoupling, as well as by several enzymes, oxidative burst (in phagocytic cells), autooxidation of some cell components and redox cycling. Although $O_2^{\cdot -}$ and H_2O_2 are not very toxic *per se*, they can produce ·OH (by Fenton reaction) in the presence of some transition metals such as iron and copper at important sites for cell functioning (e.g., nucleus). Xenobiotics could, in principle, disturb any of these ROS sources.

Biological Effects

The disturbance of the dynamic equilibrium between production and elimination of ROS can lead to damage of cellular constituents such as lipids, proteins and nucleic acids.

Lipids

The reaction of ROS with lipids is considered one of the most prevalent mechanisms of cell damage (Halliwell and Gutteridge 1999). Some ROS can initiate lipid peroxidation, a self-propagating process in which a peroxyl radical (ROO·) is formed when an ROS has sufficient reactivity to abstract a hydrogen atom from an intact lipid (Fig. 5.5).

Lipid peroxidation (LPO) products are among the most common types of oxygenated molecules implicated in oxidative stress responses, and lipid peroxidation reactions are sub-categorized into two types: non-enzymatic and enzymatic. The oxidation rates of the non-enzymatic type are defined mainly by the number of double-bonds in lipids polyunsaturated fatty acid (PUFA) residues, and so this random process should affect all different (phospho)lipid classes with minimal dependence on the chemical nature of phospholipid polar head groups (Kagan 1988). Conversely, enzymatic oxygenations of free polyunsaturated fatty acid residues are controlled processes that generate many extra- and intracellular biological regulators that include a large diversity of eicosanoids, prostanoids, docosapentanoids and docosahexanoids (Pratico 2008; Serhan 2010).

Phospholipids bilayers form the structural basis of biological membranes that determine cell boundaries and enclose its organelles. The balance between membrane integrity and fluidity depends on the composition of saturated and unsaturated fatty acids. The alkenes in unsaturated fatty acids are susceptible to hydrogen abstraction by ROS to yield carbon-centered and peroxyl radicals (Radi et al. 1991), and unsaturated fatty acids in the mitochondrial membrane are especially vulnerable to peroxidation because of their intimate proximity to ROS generated by complexes I and III.

Oxidation of lipids may disturb biological membranes, leading to alteration of important properties such as permeability barrier and fluidity (Esterbauer 1993); both properties are essential for cell functioning and survival. No less important, lipid peroxidation may form lipid radical species that damage other cellular biomolecules. For example, lipid peroxidation end-products like malondialdehyde (MDA) and 4-hydroxynonenal (4-HNE) can react with both proteins and DNA (Hartley et al. 1997).

Oxidation of lipids is the most commonly used approach for oxidative stress determination in fish. Aquatic organisms contain high amounts of lipids with polyunsaturated fatty acid residues that are potential

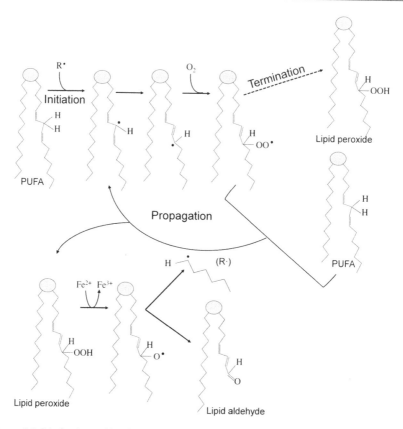

Figure 5.5. Mechanism of lipid peroxidation. In the *initiation* step, the reaction of a radical (R·) such as methyl (·CH$_3$) or hydroxyl (·OH) and an unsaturated fatty acid (PUFA) from a phospholipid produces a fatty acid radical. Since this radical is unstable, it reacts with molecular oxygen creating a peroxyl-fatty acid radical (ROO·/LOO·). LOO· reacts with itself creating cyclic peroxide or with another free fatty acid, producing one different fatty acid radical and one lipid peroxide (*propagation*). This cycle continues, as the new fatty acid radical reacts in the same way in a chain reaction mechanism. In living organisms, the *termination* of lipid peroxidation usually requires the play of an antioxidant molecule such as vitamin E, which can react with the lipid peroxide radical. In addition, lipid peroxides can produce lipid alkoxyl radical (through Fenton reaction) that can be fragmented into lipid aldehydes and other radicals. Cell defenses (e.g., CYP, glutathione S-transferase (GST), glutathione peroxidase (GPx)) can also neutralize lipid hydroperoxides avoiding the production of these harmful products. Adapted from Boelsterli (2007).

substrates for oxidation. Since lipids are oxidized usually through the formation of peroxides, the process of their formation has been called "lipid peroxidation". Again, in this case as well as in the ones described later (i.e., ROS-induced modification of proteins and nucleic acids), the monitored parameters most likely reflect the increase in steady-state ROS concentrations, rather than turnover of damaged molecules, and this may underestimate the real damage level.

Proteins

Proteins are susceptible to the attack of numerous forms of free radicals and ROS, leading to many different forms of protein oxidative modification (Berlett and Stadtman 1997; Stadtman 1992). Carbonylation of proteins due to direct action of ROS may cause loss of enzyme activity (Stadtman et al. 1991; Starke et al. 1987), and some lipid peroxidation products like 4-hydroxy-2-nonenal and other aldehydes may react with amino acid residues, impairing protein function (Stadtman and Levine 2000). Once oxidized, proteins must be either repaired or, if repair is not possible, degraded or cleared from the cell to minimize the potential negative effects of these damaged proteins. Almost all amino acids are susceptible to oxidative modification by one or more types of ROS. The sulfur-containing amino acids cysteine and methionine are unique in that there are specific enzymes to repair their oxidative damage, cysteine disulfides and methionine sulfoxides, respectively (Levine et al. 1999; Levine et al. 1996; Stadtman 1992). However, oxidation to other amino acids, or unresolved damage to cysteine and methionine, can result in oxidation moieties that cannot be repaired. In cases where repair is not possible, oxidized proteins are generally labeled with ubiquitin for degradation by the proteasome system or removed through autophagy. Despite the efficiency of clearance of oxidized proteins, certain damaged proteins can remain, accumulate and promote cellular dysfunction (Berlett and Stadtman 1997; Levine 1983; Pierce et al. 2008; Stadtman 1992).

Nucleic acids

Nuclear and mitochondrial DNA are both susceptible to oxidation that results in mutations and single-strand breaks along with the formation of 8-hydroxyguanosine (8-OHdG) (Marnett 2000). The oxidation product 8-OHdG is relatively stable and can be measured both in tissues and in excreted urine which accurately represents the amount of DNA oxidation/repair rate as a measure of DNA damage within the body as a whole (Williams et al. 1998; Wu et al. 2004). Oxidation of DNA has been strongly implicated in cellular senescence, apoptosis and the development of cancerous cell phenotypes. But, apart from ROS's direct damaging effects to coding sequences (Richards et al. 2011), ROS can also alter gene expression through epigenetic and genetic mechanisms. The epigenetic mechanism is concerned mainly with alterations in DNA-methylation patterns (Franco et al. 2008), whereas the genetic mechanisms rely on the activation of various redox-sensitive transcription factors (Allen and Tresini 2000).

Antioxidant Defense Systems and Aerobic Life

Considering that (1) oxygen in its molecular state is essential for aerobic organisms, (2) oxidative stress occurs when an organism's ability to detoxify ROS does not match the level of these potentially toxic substances, and (3) severe oxidative stress damages the cell structure (Fig. 5.6), the antioxidant defense system of aerobic organisms is expected to be very robust. This robust system is composed of low and high molecular antioxidants.

Low molecular mass antioxidants include compounds such as reduced glutathione (GSH), ascorbic acid (vitamin C), carotenoids, retinol (vitamin A) and α-tocopherol (vitamin E). They usually operate as free radical scavengers or serve as cofactors for antioxidant enzymes. For example, GSH is utilized by glutathione-dependent peroxidases and glutathione *S*-transferases, and α-tocopherol acts as a molecule that breaks the lipid peroxidation chain, with scavenging capacity for free radicals such lipid peroxyl, alkoxyl and hydroxyl radicals (Mukai et al. 1988).

High molecular mass antioxidants include the antioxidant enzymes superoxide dismutases, catalases, Se-dependent glutathione peroxidases and DT-diaphorase, as well as associated enzymes that provide needed cofactors—glutathione disulphide reductase, glutathione synthesizing enzymes and glucose-6-phosphate dehydrogenase. Non-specific high

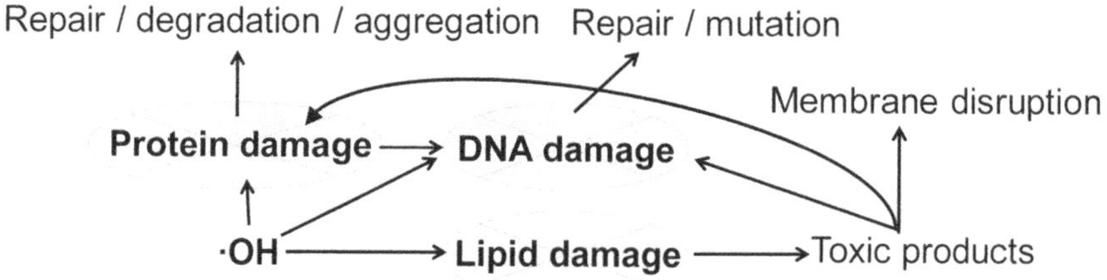

Figure 5.6. Oxidative stress and cell damage. ·OH is very reactive and can damage lipids (lipid peroxidation), proteins (protein oxidation/carbonylation) and DNA (DNA oxidation/breaks). Cells are endowed with a robust nuclear DNA repair system that is usually able to correct many types of alterations; in some minor cases either the damage remains as a mutation or epigenetic alteration, or the cell dies. Proteins repair system is modest as very few alterations are corrected; the unrepaired proteins are either degraded by proteosomes and lysosomes, or accumulate inside the cells as inclusion bodies. Lipids are not repaired at all. Lipid peroxidation toxic products are usually neutralized or removed by defense mechanisms; otherwise, membranes can lose their most fundamental function of selective permeability. Indirect damage is also possible. Lipid reactive products such as aldehydes can form adducts (covalent bounds) with DNA and proteins; DNA repair enzymes can be among the damaged proteins thereby impairing DNA repair (resulting in increased DNA damage levels), and protein-DNA crosslinks can be formed.

molecular mass antioxidants are represented by proteins that prevent ROS-induced damage by binding to transition metal ions (mainly iron and copper) such as metallothioneins and ferritin (Lushchak 2011). Some antioxidants like tocopherol and carotenoids are obtained by aquatic animals through food, but the majority is produced (Lushchak 2011).

In addition to the antioxidant defenses that prevent or control oxidative stress, there are many enzymes involved in the repair of oxidative damage, particularly for the DNA. ROS such as hydroxyl radical and breakdown products of lipid peroxides can damage both the DNA nucleobases and the sugar phosphate backbone, leading to a wide spectrum of lesions, including non-bulky (8-oxoguanine and formamidopyrimidine) and bulky (cyclopurine and etheno adducts) base modifications, abasic sites, non-conventional single-strand breaks, protein–DNA adducts, and intra/interstrand DNA crosslinks (Berquist and Wilson 2012). Cells devote extensive resources to prevent and correct DNA damage caused by ROS. Base excision repair (BER), nucleotide excision repair (NER), strand break (single- and double-stranded) repair, homologous recombination (HR), and interstrand crosslink (ICL) repair pathways all act to remedy ROS-induced DNA damage, maintain genetic information, and genomic stability (Berquist and Wilson 2012).

Fish oxidative stress, as well as its antioxidant potential, differs in relation to species, genetic variability, habitat, feeding behavior, nutritional state and age (Ahmad et al. 2004). Yet, tissue specificity is observed in vertebrates, which is a consequence of metabolic and antioxidant differences (Ognjanivic et al. 2008).

The exposure to chemical stressors can elicit pro-oxidant conditions. In response to it, an organism either aims to maintain the previous status by activation of corresponding protective mechanisms or goes to a new stable state. The adaptive response is usually realized via synthesis of new molecules of antioxidant enzymes and is essential for redox regulation.

Cell Signaling and Redox Regulation

The cellular redox status is finely regulated via biosynthesis and feedback mechanisms, in part because many ROS and NOS (nitrogen reactive species) are sensed by specific systems like Keap1/Nrf2, NF-κB, MAP-kinase and HIF-1α (Lushchak 2011). These regulatory proteins usually become activated by an oxidative signal in the cytoplasm, translocate into the nucleus and activate gene expression (Hansen et al. 2006; Sieprath et al. 2012). NF-erythroid 2-related factor 2 (Nrf2), for example, plays a key role in defense against oxidative stress. Under basal conditions, Nrf2 is anchored to the cytoplasm by Kelch-like ECH-associated protein 1 (Keap1; Itoh et al. 1999). However, when oxidative stress occurs, the cysteine residues of Keap1 become oxidized, releasing Nrf2 to enter the nucleus. There, it binds to the antioxidant response element (ARE) of many different genes, allowing transcription of antioxidants (Itoh et al. 1999; Tufekci et al. 2011). In this way, up-regulation of antioxidant enzymes is achieved through the alteration of gene expression, resulting in the synthesis of new molecules (Lushchak 2011). However, antioxidant enzymes are regulated by negative feedback from excess of substrate, and they are themselves sensitive to damage by ROS: superoxide dismutase by hydrogen peroxide, catalase by superoxide anion, glutathione disulphide reductase and glucose 6-phosphate dehydrogenase by both of them; glutathione S-transferase is easily inactivated by oxidants (Hermes-Lima and Storey 1993; Szweda and Stadtman 1992; Worthington and Rosemeyer 1976). Therefore, measured activities of the enzymes are a result of two processes: their production and inactivation (Bagnyukova et al. 2006).

Non-enzymatic Defenses

Although we will consider the individual aspects of some antioxidant molecules, it is essential to keep in mind that the non-enzymatic and enzymatic antioxidant defenses work together to regulate cell redox milieu. Among the non-enzymatic defenses, we will focus on general aspects of GSH and metallothioneins, since these are usually utilized as "biomarkers" in studies with fish, including the tropical ones.

Glutathione (GSH) is probably the most important antioxidant of organisms, with additional roles in different aspects of cell functioning such as protein synthesis, amino acid transport, DNA synthesis and cellular detoxification (Hasspieler et al. 1994; Sies 1999). GSH is a tripeptide (L-γ-glutamyl-L-cysteinyl-glycine) that represents the major non-protein thiol in the body; thiols are considered the first line of defense against oxidative stress in living organisms (Elia et al. 2006). It acts as a radical scavenger through two different mechanisms: first, it serves as a substrate for conjugation with electrophilic intermediates under the catalytic action of glutathione S-transferase (GST), and second, it directly reacts with free radicals through the oxidation of GSH to glutathione disulphide (GSSG). As indicated by the following reactions (1–3), GSH itself is converted into a radical (GS·, glutathyl), but two GS· react to form a GSSG molecule (4).

$$GSH + \cdot OH \rightarrow H_2O + GS\cdot \tag{1}$$

$$GSH + R\cdot \rightarrow RH + GS\cdot \tag{2}$$

$$GSH + ROO\cdot \rightarrow ROOH + GS\cdot \tag{3}$$

$$GS\cdot + GS\cdot \rightarrow GSSG \tag{4}$$

GSH is found in large quantities in organs exposed to toxins such as the kidney, liver, lung and intestine. A low level of oxidative stress may induce increase of GSH synthesis and detoxifying enzyme activity (adaptive mechanisms), while severe oxidative stress may cause the oxidation of GSH to GSSG and a reduction in antioxidant enzyme levels (Elia et al. 2006), impairing the adaptive mechanisms (Zhang et al. 2004). Thus, the calculation of the ratio of reduced glutathione to glutathione disulphide (GSH: GSSG) provides a useful and widely used means of assessing oxidative stress. If the level of oxidative stress is high, most of the glutathione will be in the disulphide form (Krivoruchko and Storey 2010). In addition, as reduced glutathione also acts as a reactant in conjunction with electrophilic substances, a change of GSH levels may also be an important indicator of the detoxification ability of an organism (Cheung et al. 2001).

Metallothionein (MT) is a cysteine-rich (20–30% of amino acid residues) low molecular protein (6–8 kda) involved in the homeostasis of essential metals such as copper, zinc and selenium as well as detoxification of xenobiotic metals like cadmium, mercury, silver and arsenic. The oxyradical scavenger property of MT can be predicted by the high sulfhydryl (from cysteine residues) content present in this protein. However, MT can protect the cells from oxidative stress not only by acting as an oxyradical scavenger, but through metal binding/release dynamics (Viarengo et al. 2000). MT expression can also be influenced by natural factors that may affect the bioavailability and accumulation of metals, such as salinity, and this constitutes an important factor to be considered when MT is measured and employed as a specific biomarker of metal pollution (Monserrat et al. 2007).

Enzymatic Defenses

Enzymatic antioxidant defenses are provided by several enzymes such as superoxide dismutase (SOD), catalase (CAT), glutathione peroxidase (GPx) and peroxiredoxins (Prx) that form a robust system in which the action of a particular enzyme can be partly replaced by the actions of other antioxidants (Bagnyukova et al. 2005).

The SOD–CAT system provides the first defense against oxygen toxicity due to the inhibitory effects on hydroxyl radical formation (Pandey et al. 2003). Superoxide dismutase dismutates superoxide anion ($O_2^{\cdot-} + O_2^{\cdot-} + 2\,H^+ \rightarrow H_2O_2 + O_2$), produced by mitochondria and other sources such as NADPH oxidases, to hydrogen peroxide, avoiding the formation of hydroxyl radical via the Haber–Weiss reaction in biological systems in the presence of H_2O_2 (Di Giulio et al. 1989). Manganese containing SOD (Mn-SOD) is found in the mitochondrial matrix while copper- and zinc-containing SOD (Cu, Zn-SOD) is found mostly in the cell cytosol (Miao and St. Clair 2009).

The hydrogen peroxide produced by SOD and other processes such as cytochrome P450 activity is further broken by catalase, glutathione peroxidase or peroxiredoxin. Otherwise, superoxide radicals can react with H_2O_2 in the presence of iron or some other metals and form highly reactive hydroxyl radicals as per classical Fenton reaction (Halliwell and Gutteridge 1999).

Catalases are heme-containing enzymes responsible for the conversion of H_2O_2, generated by several enzymes such as the acyl-CoA oxidase (involved in the fatty acid β-oxidation at peroxisomes), to water and oxygen ($2H_2O_2 \rightarrow 2H_2O + O_2$), thereby lowering the risk of hydroxyl radical formation. CAT is located primarily in peroxisomes, although it has been detected also in the cytosol of some organisms (Hiltunen et al. 2003).

The ROS-detoxifying role of superoxide dismutases and catalases depends on the redox properties of the metal group associated with the enzyme. Conversely, peroxidases use cysteine thiols at the active site to reduce inorganic and organic peroxides into the corresponding water/alcohols. According to the thiol electron donor, two classes of peroxidases are distinguished: glutathione peroxidases, which employ GSH, and thioredoxin peroxidases (also called peroxiredoxins), which usually employ thioredoxin as reductants (Brigelius-Flohe 2006).

There are two types of GPx (Brigelius-Flohe 2006). Classical GPx are multimeric, soluble and reduce H_2O_2 ($2GSH + H_2O_2 \rightarrow GSSG + 2H_2O$) and a wide range of organic hydroperoxides ($2GSH +$ lipid hydroperoxides$\rightarrow$ GSSG + lipid alcohols). Phospholipids hydroperoxides GPx are usually monomeric and membrane-associated, being involved in the reduction of soluble hydroperoxides and lipid hydroperoxides in membranes. This latter GPx type is therefore the main enzyme in repairing membrane lipid peroxidation (Halliwell and Gutteridge 2007; Helmy et al. 2000; Watanabe et al. 1997). The Se present in the active site of GPx contributes both to its catalytic activity and spatial conformation (Hamilton 2004; Rotruck et al. 1973); we shall see in later sections that Se supplementation lowers oxidative stress in tropical fish. GPx plays an especially important role in protecting membranes from damage due to lipid peroxidation (van der Oost et al. 2003), and GPx inhibition might reflect a possible antioxidant defense failure responsible for increase in lipid peroxidation levels.

Peroxiredoxins (Prx) reduce hydrogen peroxide, organic hydroperoxides (ROOH) and peroxinitrites with thioredoxins (trx) usually acting as electron donors ($ROOH + 2e^- \rightarrow ROH + H_2O$), although other reductants including GSH may be important in some cases for their activity (Rhee et al. 2005; Wood et al. 2003). Like glutathione peroxidases, peroxiredoxins are widely distributed in the cell compartments.

In addition to glutathione peroxidase which removes hydrogen peroxide and consumes lipid hydroperoxides using GSH, glutathione S-transferase (GST) is involved in the conjugation of compounds having reactive electrophilic groups such as many xenobiotics and lipid peroxidation end products with GSH (Grundy and Storey 1998). These enzymes are abundant in the cells, and usually generate less toxic and more hydrophilic molecules (Olsen et al. 2001). Although GST is primarily involved in biotransformation reactions, they also play an antioxidant role (Fig. 5.5) by neutralizing lipid hydroperoxides and electrophilic lipid peroxidation end products (Barata et al. 2005; Fernandes et al. 2008), thereby protecting the nucleophilic groups of macromolecules such as proteins and nucleic acids (Fournier et al. 1992). Other functions, not associated with detoxification, include repair of macromolecules oxidized by ROS, regeneration of S-thiolated proteins, and biosynthesis of physiologically important metabolites (Freitas et al. 2007).

Other antioxidant enzymes or enzymes with antioxidant functions include those involved in glutathione synthesis (γ-glutamylcysteine synthetase and glutathione synthetase) and GSSG reduction (GR, glutathione disulphide reductase: $\textbf{GSSG} + \text{NADPH} \rightarrow 2\,\textbf{GSH} + \text{NADP}^+$), NADP$^+$ reduction (G6PDH, glucose 6-phosphate dehydrogenase: glucose 6-phosphate + $\textbf{NADP}^+ \rightarrow$ 6-phosphogluconate + $\textbf{NADPH}$), and reduction of protein disulfide bonds (TrxR, thioredoxin reductase) and oxidized form of methionine in proteins (methionine sulfoxide reductases; Fig. 5.7).

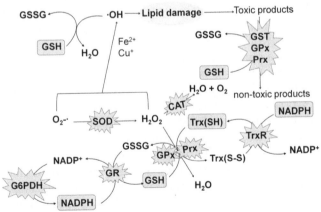

Figure 5.7. Cell antioxidant defense system. Due to the harsh outcomes of oxidative stress, a very robust antioxidant defense system is present in the cells. Primary antioxidant enzymes such as SOD and CAT are responsible for the metabolism of $O_2^{-\cdot}$ and H_2O_2, thus avoiding $\cdot$OH production. GPx and Prx are also involved in H_2O_2 metabolism, but they require cofactors such as GSH and Trx. Secondary enzymes such as GR, G6PDH, TrxR and those involved in GSH synthesis (not shown) are involved in providing reducing equivalents (cofactors) for the function of the primary antioxidant enzymes. In addition, antioxidants such as GSH and vitamin E can directly react with some inorganic and lipid radicals, and lipid toxic products such as lipid hydroperoxides can also be detoxified by Prx, GPx and GST activities.

Oxidative Stress Biomarkers

Most studies in ecotoxicology are designed to investigate the environmental condition concerning the presence of chemicals from different sources. In some cases, the main aim is to learn about the health of organisms inhabiting the investigated area, whereas in other cases it is to provide useful information about the risks for local biota and humans. Water quality, for example, is one of the main foci of many investigations, since water is a limited resource that is essential for life and human life-style.

Water quality can be determined using different physical, chemical and biological parameters. The physical and chemical parameters are very useful to evaluate pollution trends and sources (Alberto et al. 2001; Pesce and Wunderlin 2000), but they do not accurately predict changes in aquatic living systems, providing only partial information of the overall water quality (Hued and Bistoni 2005). Biological responses and effects (biomarkers) rather than quantitative estimations of mortality can provide the additional information that is needed for accurate assessment of environmental quality, since biomarkers response are the result of the physical/chemical-biological interactions that do not necessarily cause the death of the organism. In this way, biomarkers allow the determination of the adverse effects of sublethal concentrations of chemicals.

Cellular response to oxidative stress is usually related to two processes, the damage to cellular constituents and the up-regulation of antioxidant defenses. Actually, it is difficult to separate these two scenarios and the final results depend on many circumstances because these processes take place simultaneously (Kubrak et al. 2010).

Variation of a particular antioxidant is difficult to predict since it may be influenced by the class of chemicals, species sensitivity and several environmental and biological factors (Winston and Digiulio 1991). Due to the complexity of interactions between pro-oxidant factors and antioxidants, it appears that single responses cannot provide a general marker of oxidative stress. Thus, a particular antioxidant can be sensitive and specific but the response is difficult to predict. Antioxidants alterations are useful as "response biomarkers" indicating a varied pro-oxidant challenge and potentially important early warning signals (Regoli et al. 2002).

The evaluation of oxidative stress markers is a key question in the investigation of oxidative stress in organisms. In some cases, ROS level may be monitored by direct or nondirect methods. Although the direct registration of ROS is a very useful approach, it is virtually impossible to perform it *in vivo* due to technical reasons (Lushchak 2011). The induction of oxidative stress is usually monitored via registration of products of ROS-induced modification of cellular constituents. ROS-modified lipids, proteins and nucleic acids along with low and high molecular mass antioxidants and antioxidant potential constitute a battery of indices commonly used to describe oxidative stress (Lushchak 2011). However, due to specific nature of reactive species, there are no "ideal markers" of oxidative stress. At least several indices should be used to characterize oxidative stress development. In model experiments, the induction of oxidative stress has to be evaluated in two measures: dynamics of the process, and concentration effects. The dynamics is very important because different indices demonstrate varied time-courses (Lushchak 2011) or respond only at a specific period. For example, *Prochlodus lineatus* (curimbatá) exposed to atrazine (2, 10 and 25 μg l^{-1}) did not present alteration of either antioxidant enzymes activities (GST, SOD, CAT and GPx) or GSH concentration and lipid peroxidation (LPO) in the gills after acute exposure (48 h). However, at a subchronic exposure (14 days) to 10 μg l^{-1}, GST, SOD, CAT and LPO increased (Paulino et al. 2012). This is an extreme condition in which all evaluated biomarkers did not respond at short-term exposure, but increased at a longer exposure.

Authors usually combine different groups of oxidative stress biomarkers belonging to two main groups: the cell response biomarkers (ROS production and antioxidant defenses) and the stress effects biomarkers (damage to biomolecules). Biomarkers selection depends on the specific purpose of the investigation, and it requires minimally, a basic knowledge of the overall redox situation of cells (Fig. 5.8).

Evaluation of Antioxidant Defenses

Antioxidant responses are probably the most variable parameters to be evaluated in order to determine if the cell or organism is under oxidative stress or not. In some cases, the result is an increase or decrease of the biomarker; in other situations the parameter does not vary at all. To complicate it even more, the time-course responses can fluctuate greatly as we shall see in the following example.

P. lineatus, exposed to 1 or 5 mg L^{-1} of glyphosate-based herbicide (Roundup Transorb®), showed decreases of SOD, CAT and GST after six and 24 hours of exposure, but enzyme activities were reestablished after 96 hours. GPx increased only at the highest concentration. GSH decreased at 24 hours, but increased at the highest concentration at 96 hour-exposure (Table 5.1). LPO increased only at six hours, but it was similar to control level at 24 and 96 hours (Modesto and Martinez 2010).

According to the authors, the transitory decrease of liver GSH after 24 hours exposure, followed by a raise in GSH content after 96 hours exposure probably represents an "adaptation" following herbicide exposure. This general idea is corroborated by the transient increase of LPO at six hours followed by the return to control levels after 24 and 96 hour exposure. In addition, the authors interpret the finding inferring that the antioxidant

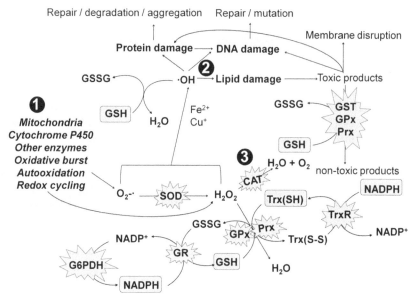

Figure 5.8. Oxidative stress and antioxidant defense systems. (1) The cellular sources of ROS. ROS such as $O_2^{-\cdot}$ and H_2O_2 are produced during cell metabolism, particularly in mitochondria and endoplasmic reticulum (by cytochrome P450) through electron transport uncoupling, as well as by other enzymes, oxidative burst (in phagocytic cells), autooxidation of some cell components and redox cycling. Although $O_2^{-\cdot}$ and H_2O_2 are not very toxic *per se*, they can produce $\cdot OH$ (by Fenton reaction) in the presence of some transition metals such as iron and copper. Xenobiotics can, in principle, disturb any of these ROS sources. (2) Oxidative stress and cell damage. $\cdot OH$ is very reactive and can damage lipids (lipid peroxidation), proteins (protein oxidation/carbonylation) and DNA (DNA oxidation/breaks). Cells are endowed with a robust nuclear DNA repair system that is usually able to correct many types of alterations; in some minor cases either the damage remains as a mutation or epigenetic alteration, or the cell dies. Proteins repair system is modest as very few alterations are corrected; the unrepaired proteins are either degraded by proteosomes and lysosomes, or accumulate inside the cells as inclusion bodies. Lipids are not repaired at all. Lipid peroxidation toxic products are usually neutralized or removed by defense mechanisms; otherwise, membranes can lose their most fundamental function of selective permeability. Indirect damage is also possible. Lipid reactive products such as aldehydes can adduct DNA and proteins; DNA repair enzymes can be among the damaged proteins, and protein-DNA crosslinks can be formed. (3) Cell antioxidant defense system. Due to the harsh outcomes of oxidative stress, a very robust antioxidant defense system is present in the cells. Primary antioxidant enzymes such as SOD and CAT are responsible for the metabolism of $O_2^{-\cdot}$ and H_2O_2, thus avoiding $\cdot OH$ production. GPx and Prx are also involved in H_2O_2 metabolism, but they require cofactors such as GSH and Trx. Secondary enzymes such as GR, G6PDH, TrxR and those involved in GSH synthesis (not shown) are involved in providing reducing equivalents (cofactors) for the function of the primary antioxidant enzymes. In addition, antioxidants such as GSH and vitamin E can directly react with some inorganic and lipid radicals, and lipid toxic products such as lipid hydroperoxides can also be detoxified by Prx, GPx and GST activities.

Table 5.1. Biomarkers response of liver of *P. lineatus* exposed to glyphosate-based herbicide.

Biomarker	SOD		CAT		GPx		GST		GSH		LPO	
mg/l	1	5	1	5	1	5	1	5	1	5	1	5
6 hr	=	↓	=	↓	=	=	↓	↓	=	=	↑	↑
24 hr	=	=	↓	↓	=	↑	=	↓	↓	↓	=	=
96 hr	=	=	=	=	=	↑	=	=	=	↑	=	=

Fish were exposed to 1 mg or 5 mg l⁻¹ of glyphosate-based herbicide during 6, 24 and 96 hr. Biomarkers responses are indicated as increase (↑), decrease (↓) or not altered (=). Source: Modesto and Martinez (2010).

defense at six hours of exposure was insufficient (decreased SOD, CAT and GST activities), leading to increased LPO as a function of the presence of the chemical. In longer experimental times, these defenses returned to basal levels and then they were apparently enough to combat the ROS, preventing the occurrence of this oxidative damage (Modesto and Martinez 2010).

Evaluation of Oxidative Stress Effects

Development of oxidative stress is evidenced by alterations of the levels of oxidative damage biomarkers; products of lipid and protein oxidation are most commonly utilized (Halliwell and Gutteridge 1999; Storey 2006).

Lipid peroxidation increases is frequently a quick response to ROS generation (Halliwell and Gutteridge 1999). However, products of lipid peroxidation are often unstable (Lushchak 2011), thereby removed at high rates (Dalle-Donne et al. 2003). Primary products include shortly lived species, frequently of radical nature. The secondary products consist of lipid peroxides, dienic conjugates and ketodienes, which can be measured by reliable, reproducible, non-expensive and easy to perform methods such as *Ferrous Oxidation/Xylenol Orange (FOX) Method* (Hermes-Lima et al. 1995). However, the most frequently used methods to monitor lipid peroxidation are based on measuring of the end products. Among others, malondialdehyde (MDA) is of particular interest because it is measured with thiobarbituric acid (TBA). But, according to Lushchak (2011), it should be noted that TBA reacts also with many types of compounds, such as different aldehydes, amino acids, and carbohydrates, and so we must refer to it as TBA-reactive substances (TBARS). Some relatively new approaches to measure the end products of lipid peroxidation were proposed recently. They are HPLC and immune techniques (Claeson et al. 2001), which are more specific than at TBARS measurement and can be recommended for research with aquatic organisms (Lushchak 2011).

Protein carbonylation can be measured by the spectrophotometric technique utilizing dinitrohenylhydrazine (Levine et al. 1990). According to Lushchak (2011), the parameter is reliable and can be recommended for a broad usage, but at least three issues should be taken into account in this case. Firstly, proteins always contain carbonyl groups and ROS-induced oxidation only adds new ones. Secondly, it is again a dynamic parameter because proteins, especially oxidized ones, can be catabolized. However, in some cases heavily oxidized proteins can be accumulated in the cell (Dunlop et al. 2009; Widmer et al. 2006). Thirdly, the set of cellular proteins with different carbonyl content may be changed at long-term studies, leading to apparent change in protein oxidation intensity (Lushchak 2011).

Reactive species also induce alterations in DNA. Because of the critical importance of DNA in storing information for the cell, it is very attractive to have a relevant method to evaluate its ROS-induced DNA modifications. Several techniques have been developed to address the issue. DNA strand breaks can be measured through alkaline precipitation assay (Olive 1988). The formation of oxidized bases, particularly 8-oxoguanine (8-OG), gives a powerful tool to evaluate this process. It can be measured by HPLC (Kelly et al. 2008) or immune (Bespalov et al. 1999) techniques. The comet assay, which in different modifications may reflect a variety of DNA damages, was successfully applied to fish (Caliani et al. 2009; Cavalcante et al. 2008; Toyoizumi et al. 2008), although the DNA breaks can have many different causes.

As we shall see in the next examples, biomolecule damages are commonly utilized to evaluate oxidative stress in tropical and subtropical fish.

Example 1. Lipid peroxidation (TBARS) and protein carbonylation increased in *Rhamdia quelen* (jundiá) exposed for 96 hours to sublethal concentrations of the pesticide methyl parathion (increase of LPO) and the fungicide tebuconazole (increase of LPO and PCO), but not for the glyphosate-based herbicide. The three agrichemicals led to decreased CAT and increased GST activity, reduced GSH and ascorbic acid concentrations (Ferreira et al. 2010).

Example 2. LPO (TBARs) and carbonyls increased in the liver and kidney of *Oreochromis niloticus* (Nile tilapia) exposed to a single oral dose of 120 µg microcystin-LR per fish and sacrificed after 24 hours. Hepatic GSH/GSSG ratio decreased. SOD, CAT and GST decreased in the liver, GPx and GR increased in the liver, but CAT and GR decreased and SOD increased in the kidney. GST and GPx were not altered in the kidney (Table 5.2; Atencio et al. 2009).

Table 5.2. Biomarkers response of liver and kidney of *O. niloticus* exposed to microcystin-LR.

Biomarker	SOD	CAT	GST	GPx	GR	GSH/GSSG	LPO	PCO
Liver	↓	↓	↓	↑	↑	↓	↑	↑
Kidney	↑	↓	=	=	↓	n.a.	↑	↑

Fish were exposed to 120 µg of microcystin-LR per fish and sacrificed after 24 hr. Biomarkers responses are indicated as increase (↑), decrease (↓) or not altered (=). n.a. = not analyzed. Source: Atencio et al. 2009.

These two studies exemplify what is generally expected and found on toxicological investigations: the increase of lipid peroxidation and protein carbonylation (as well as very variable responses of antioxidants). However, biomolecules damage biomarkers and do not respond as we expect them to in some cases.

Example 3. Teleost fish *Leporinus obtusidens* (piava) exposed during 90 days to 376 µg l^{-1} of clomazone or 1644 µg l^{-1} of propanil (both herbicides) presented decreased TBARS levels in the brain, muscle and liver.

Example 4. Rhamdia quelen exposed to glyphosate herbicide (Roundup®) during 96 hours presented increased TBARS production in the muscle, but decreased levels in the brain and no alteration in the liver (Glusczak et al. 2007).

Comparison of these additional examples (examples 3 and 4) with data on literature indicates that increases of LPO/PCO are more common at high but not excessive exposure levels, whereas at low levels (environmental concentrations) the response can be either an increase or a decrease. In addition, different organs may respond differently, some being the targets of toxicity whereas others have robust antioxidant defenses. Not least important, the exposure period also plays a role on the response as long-term exposure may allow fish accommodation through antioxidants regulation, as we discuss next.

Redox Deregulation and Physiological Accommodation

Chemical stress can deregulate redox balance, leading to biological damage of important cell structures. However, the cells and organisms are not passive to the aggression; they respond to it, increasing their defenses in order to avoid further injury or death. Thus, it is completely logical to find situations in which the measured biomarker had been altered at a specific period of time, but then has returned to "unaltered" (control/reference) levels. This kind of response usually occurs in the following scenario: a biological system is continuously exposed to a constant dose or concentration of a chemical during a period of time that allows the biomarker to reach a steady-state. We shall see this in the following two examples with tropical/subtropical fish.

Example 1. Boleophthalmus boddarti (mudskipper) exposed to a sublethal concentration of ammonia (8 mmol l^{-1} of NH$_4$Cl, pH 7.0) experienced a transient increase of lipid peroxidation (hydroperoxides), decreases of the contents of reduced glutathione (GSH) and of activities of glutathione peroxidase, glutathione reductase and catalase in the gills after 12 hours (Table 5.3). However, the majority of these parameters returned to levels similar to those of the control after 48 hours, which might contribute to the species' high ammonia tolerance. Likewise, there were decreases of the activities of GPx, GR, SOD and CAT in the brain after 12 hours; at 48 hours, the enzymatic activities returned to control levels (Ching et al. 2009).

Any investigation that has not considered GSH and CAT at 48 hour-exposure would not be able to detect a response in the gills due to physiological accommodation. Since the authors have measured several parameters (including GSH and CAT) at both exposure periods, they deduced that the gills experienced oxidative stress earlier than the brain, which corroborates with the fact that gills were directly exposed to ammonia (Ching et al. 2009).

Example 2. Clarias batrachus (Indian catfish) exposed to two nonlethal concentrations of arsenic (4.2 and 8.4 mg l^{-1}) for 1 to 10 days showed increase of lipid peroxidation during the beginning of the experiment, but then the LPO levels returned to control levels, with a biphasic pattern. The same pattern was observed for GSSG:GSH ratio. SOD and GPx increased but gradually returned to control levels (Bhattacharya and Bhattacharya 2007). Conversely, CAT increased on the first day; from day 2 to 10, CAT activity was lower than that of the control. This decrease associated with the increased activity of peroxisomal acyl co-A oxidase (ACOX, enzyme that participates in β-oxidation of very long chain fatty acids and produces H$_2$O$_2$) indicates that the peroxisomes may be potential targets of arsenic toxicity in fish (Bhattacharya and Bhattacharya 2007).

From these examples, it is evident that one must keep in mind that cells and organisms can modulate biochemical parameters with ease over time, so that studies that investigate only one period of time are limited in showing

Table 5.3. Biomarkers response of gills and brain of *B. boddarti* exposed to ammonia.

Biomarker	t	LOOH	TBARs	PCO	GSH	GSSG/GSH	GPx	GR	SOD	CAT
Gills	12 hr	↑	=	=	↓	=	↓	↓	n.d.	↓
	48 hr	=	=	=	↓	=	=	=	n.d.	↓
Brain	12 hr	=	=	=	=	=	↓	↓	↓	↓
	48 hr	=	↓	↑	↑	↑	=	=	=	↓

Fish were exposed to 8 mmol l^{-1} of NH$_4$Cl during 12 and 48 hr. Biomarkers responses are indicated as increase (↑), decrease (↓) or not altered (=); t = period of exposure; LOOH = hydroperoxides; n.d. not detected. Source: Ching et al. 2009.

these fluctuations. But biological responses are indeed complicated, since there are many other factors that interfere with the biomarkers.

Example 3. Carassius auratus (goldfish) exposed to chromium (Cr^{6+} and Cr^{3+} at 10 mg l^{-1}) over 24, 48 and 96 hours presented alteration of lipid peroxidation at the shortest exposure period and of protein carbonyls at intermediate and longest exposure periods (Kubrak et al. 2010). GSSG/Total GSH ratio increased at 24 hours and 48 hours of exposure, but returned to normal or even lower ratios at 96 hours. Antioxidant enzymes (SOD, CAT, GR, GST and G6PDH) activities were affected by the two forms of chromium at different times of exposure, but without apparent time-dependent pattern. Total GSH increased in the liver, but not in the other tissues that were analyzed (Kubrak et al. 2010).

Example 4. Cyprinus carpio (common carp) exposed to sublethal concentrations of the herbicide simazine (0.06 µg l^{-1}, 2 mg l^{-1} and 4 mg l^{-1}) for 14, 28 and 60 days showed increased production of ROS at the highest concentration after 60 days, but it did not reflect in increased lipid peroxidation (TBARs). SOD and CAT activities increased at 14 and 28 days, but decreased after 60 days exposure as compared with the control group (Stara et al. 2012). SOD was affected only in the muscle, whereas GPx decreased only in the liver. GR was not affected in any organ (brain, gill, muscle, intestine or liver). GSH levels increased in the brain for the three periods of exposure, but for the liver its increase was followed by a decrease at 60 days of exposure (Stara et al. 2012).

These last two examples illustrate some other aspects that are important when utilizing biochemical biomarkers to investigate the relationship between pollutants and redox unbalance/oxidative stress: they do not only respond differently, but some respond before others (dynamics of the process). In addition, the choice of the "possible target" must have a good biological support, since ecotoxicologists are interested in inferring the consequences of exposure or the health status of the model organism exposed to xenobiotics based on biomarkers analyses.

Pollutants and Redox Unbalance

Oxidative stress is a common mechanism of toxicity for many xenobiotics, although in many cases it is not clear whether it is the cause or the consequence of toxicity. Xenobiotics can affect the generation of ROS, antioxidant defenses or both. For example, mercury and cadmium uncouple mitochondrial respiration whereas iron, coper and chromium catalyze Fenton-like reactions; some organic compounds (polychlorinated biphenyls, polycyclic aromatic hydrocarbons, etc.) can induce and uncouple CYPs, whereas others (benzene, paraquat, etc.) can suffer redox cycling. In some cases it is the xenobiotic itself that affects the antioxidant defense, but in the majority of the cases, the ROS are involved.

Vertebrates have many cell types that are grouped in tissues and organs. Each cell type is endowed with specific levels of proteins that dictate cell structure (composition of membranes and other structures, number and size of organelles, shape and size of the cell, etc.) and command cell metabolism. These characteristics can render some cells more susceptible to xenobiotic accumulation and bioactivation than others. Thus, xenobiotic distribution and the level of ROS generation as well as the antioxidant capacity vary according to the organ and tissue, and so in many cases a xenobiotic has specific target organs.

Target Organs

The concept of biological targets is very important in ecotoxicology, since there are several situations in which one or more cell types, tissues or organs are more affected than others by xenobiotics. The reasons for the existence of a target are many. At molecular level, one tissue or cell type can express a particular molecule or class of molecules such as receptors, transcription factors, enzymes or transporters that are not present at the same levels in others (e.g., specific antioxidant defenses vary greatly according to the cell type). At tissue and organ level, one determined group of cells can be protected by additional biological barriers (e.g., hematoencephalic barrier that protects the nervous tissue). At organism level, a xenobiotic can bioaccumulate or be distributed unequally, being transported at first to a determined organ and then to the others (e.g., the liver and intestine are usually the first organs to interact with xenobiotics absorbed from food; other organs are exposed later).

For fish, laboratory and biomonitoring studies usually investigate one small group of possible target organs, such as the liver, brain, gills and muscle. In most cases, the variability of the responses and effects does not depend only on the xenobiotic, but also on the robustness of the antioxidant defenses in every organ or cell type. Antioxidant enzymes show tissue-specific differences in activities (Table 5.4) that reflect the functions of the tissues and the oxidative stress load that they experience (Lushchak et al. 2009a). However, compensation of low levels of some antioxidants by others (e.g., carp has low activities of SOD and GPx, but high levels of GSH; Table 5.4) may not effectively protect cells from some xenobiotics due to their different toxicity mechanisms.

The liver is the main organ involved in detoxifying xenobiotics and different toxic metabolites, hence possessing a very potent arsenal of antioxidants and biotransformation enzymes (Lushchak et al. 2005b) such as the GSTs (Halliwell and Gutteridge 1999; Hermes-Lima 2004). Whereas

Table 5.4. Comparative levels of antioxidants in different organs of fish.

Antioxidant defense	Organ	Fish species	Reference
SOD, GPx CAT GSH	L>I≈G>B L>I>B>G B>L>G>I	*C. carpio*	Hao and Chen 2012
SOD CAT GPx GR	I≥L≥M>B>G L>M>B≈I>G L>M≈I>G>B L>G>B≈I≈M	*C. carpio*	Stara et al. 2012
SOD, CAT, GPx	L>G	*C. carpio*	Xing et al. 2012
SOD CAT GPx	K>M≈G K>G>M K≈G>M	*C. carpio*	Oruç and Usta, 2007
SOD, CAT, L-SH GR, G6PDH	L>K>B K>L>B	*C. auratus*	Lushchak et al. 2009a
CAT, GPx, G6PDH, GST	L>K>B	*C. auratus*	Bagnyukova et al. 2007
SOD CAT GR G6PDH	L>K>B L>K>G>B L≈K>G≈B K>L>G>B	*C. auratus*	Kubrak et al. 2010
GPx, GR, G6PDH, GST, L-SH CAT	K>S K≈S	*C. auratus*	Kubrak et al. 2012
SOD CAT GSH GST	L≈G≈L L>G>H H>G>L L>H>G	*O. niloticus*	Thomas et al. 2009
SOD, CAT, GR, GST GPx	L>K K≥L	*O. niloticus*	Atencio et al. 2009
SOD CAT GPx, GSH GST	G>M>L L>M≈G M>G>L L>G>M	*B. cephalus*	Monteiro et al. 2009a
SOD, GPx CAT, GST	M>G>L L>G>M	*B. cephalus*	Monteiro et al. 2006
SOD, GR, L-SH CAT, GPx, GST G6PDH	L≈B≥M L>B>M B>L	*Perccottus glenii*	Lushchak and Bagnyukova 2007
SOD, GST CAT	L≈ G>I≈M L≈ G≈I≈M	*Fundulus heteroclitus*	Loro et al. 2012
SOD, CAT, GST GPx, GR, GSH	B>G G>B	*B. boddarti*	Ching et al. 2009
SOD CAT, GSH GPx	L>K≈S>H L>K>H>S L≥H≥K≥S	*O. mykiss*	Enis Yonar et al. 2011
SOD, GR, GSH	L>B	*O. mykiss*	Li et al. 2011

L-SH (low molecular mass thiols, represented mainly by GSH). B (brain), G (gills), H (heart), I (intestine), K (kidney), L (liver), M (muscle), S (spleen). "Basal/control levels" of antioxidants from the studies are presented in the table.

the liver receives xenobiotics that have been previously absorbed through the intestine, and at a lower extent through the gills, the gills are directly exposed to the water and have a large surface area and permeability, being the primary site for absorption of xenobiotics dissolved in the water (Sancho et al. 1997).

Differences on organ antioxidant capacities and responses can be exemplified for fish exposed to two very different classes of xenobiotics, diazinon (an organophosphate insecticide) and nickel (a non-essential metal).

Example 1. Analyses of antioxidants in *Cyprinus carpio* kidney, gills and muscle revealed that the kidney has the highest "basal" activities of SOD and CAT, and kidney is similar to the gills in terms of GPx activity (see Table 5.4). The muscle has the lowest GPx and CAT activities (Oruç and Usta 2007). Antioxidant responses are also different after exposure to sublethal concentrations of diazinon (0.0036, 0.018 and 0.036 ppb in water) for 5, 15 and 30 days. SOD activity increased at day 5, being more intense in the gills than in the kidney and muscle; SOD activity returned to control levels in the gills and kidney at long-term exposures. Gills response (increase) was also faster than that of muscle for GPx. CAT decreased dose-dependently in the muscle at day 5 and the activity remained lower than the control for the periods of 15 and 30 days. LPO generally increased at lowest and intermediate period of exposure, but then returned to control levels (Table 5.5) (Oruç and Usta 2007).

In general terms, antioxidants' results indicate that gills can, in principle, deal better with pollutants that deregulate redox milieu than the muscle, which is logical considering the anatomical localization and exposure to the outside environment of gills in comparison to muscle.

Table 5.5. Biomarkers response of gills, muscle and kidney of *C. carpio* exposed to diazinon.

Biomarker	[ ] ppb	SOD			GPx			CAT			LPO		
		5 d	15 d	30 d	5 d	15 d	30 d	5 d	15 d	30 d	5 d	15 d	30 d
Gills	0.0036	↑	=	=	=	=	↑	=	=	=	=	↑	↑
	0.018	↑	=	=	↑	=	↑	↑	=	=	↑	↑	=
	0.036	↑	=	=	↑	↑	↑		=	=	↑	↑	=
Muscle	0.0036	=			=	=	↑	↓	↓	↓	↑	↑	=
	0.018	↑	↑	↑	=	=		↓	↓	↓	↑	=	=
	0.036	↑	=	=	=	=	↑	↓	↓	↓	↑	↑	↑
Kidney	0.0036	↑	=	=	=	=	=	=	=	=	=	=	=
	0.018	↑	=	=	=	↓	=	=	=	=	↓	=	=
	0.036	↑	=	=				=	=	=	↓	↑	=

Fish were exposed to 0.0036, 0.018 and 0.036 ppb of diazinon during 5, 15 and 30 days. Biomarkers responses are indicated as increase (↑), decrease (↓) or not altered (=). Source: Oruç and Usta 2007.

Example 2. Nickel is a toxicologically relevant metal and the effects of waterborne Ni^{2+} (10, 25 and 50 mg l^{-1}) in the blood and blood-producing tissues (kidney and spleen) have been investigated in *C. autarus*. Ni accumulation increased renal iron content and resulted in elevated lipid peroxidation and protein carbonylation, accompanied by suppression of the activities of antioxidant enzymes (SOD, GPx, GR and G6PDH), indicating the development of oxidative stress in the kidney. In contrast to the kidney, the spleen presented increases of antioxidant enzymes (GPx, GR and G6PDH) and constant GSH levels. These responses indicate the presence of a powerful antioxidant defense system that allowed the spleen to deal with Ni-induced ROS production (Kubrak et al. 2012).

The nervous system is particularly susceptible to oxidative damage since the brain has a high oxygen consumption rate, its membranes are enriched with highly oxidizable polyunsaturated fatty acids, iron content is high and the activity of some first-line antioxidant enzymes (SOD, CAT and GPx; Table 5.4) is comparatively lower than in other tissues with high oxygen consumption (Verstraeten et al. 2008). Also, many neurotransmitters are autoxidizable molecules. For example, dopamine and noradrenalin can react with molecular oxygen to produce ROS and active quinones (Verstraeten et al. 2008); quinones have great potential to suffer redox cycling. A comparison of the brain with other organs reveals that the response can be quite different for some parameters. However, even when the brain responses point to the same direction (increase or decrease) of another organ, the importance may not be the same due to differences of tissue repair capacity; the brain's capacity is very low.

Note that in the following examples the brain can be considered an important target of two very different classes of chemicals (a herbicide and two metals). Although other organs can also be affected, the brain has a central function in organism survival and low capacity to repair itself.

Acute toxicity of herbicide glyphosate Roundup® (2.5, 5, 10 and 20 mg l^{-1}; 96 hours of exposure) to *C. auratus* was characterized by decreases of SOD, GR and G6PDH activities in the brain, liver and kidney, and increased CAT in the latter two organs. LPO was generally not affected and low molecular thiol levels (represented mainly by GSH) decreased in the liver and brain (Table 5.6). No concentration-dependent response was observed (Lushchak et al. 2009a). The same fish species exposed to sublethal concentrations of manganese (0.1 and 1 mM) for 96 hours presented increases of GPx and lipid peroxidation in the liver, kidney, gills and brain. CAT decreased in the liver and kidney, increased in the gills and was not altered in the brain. SOD increased in all the organs, except for the brain, where it decreased (Vieira et al. 2012). Yet, in another study with *C. auratus* exposed to Cr(III) (1.0–10.0 mg l^{-1}) for 96 hours, the authors sampled the brain and reported increase of lipid peroxidation (at 10.0 mg l^{-1}) and protein carbonyls (at 2.5, 5 and 10 mg

Table 5.6. Biomarkers response of brain, liver, kidney and gills of *C. autarus* exposed to different chemicals.

Chemical	Organ	[]	LPO	PCO	GSH	Total GSH	GSSG	GSSG/Total GSH	SOD	CAT	GPx	GR	G6PDH	GST
Glyphosate	Brain	2.5 mg/l	=	n.a.	↓	n.a.	n.a.	n.a.	↓	=	n.a.	↓	↓	=
		5.0 mg/l	=	n.a.	=	n.a.	n.a.	n.a.	↓	=	n.a.	↓	=	=
		10 mg/l	=	n.a.	=	n.a.	n.a.	n.a.	↓	=	n.a.	↓	↓	=
		20 mg/l	=	n.a.	=	n.a.	n.a.	n.a.	↓	=	n.a.	↓	↓	=
	Liver	2.5 mg/l	=	n.a.	=	n.a.	n.a.	n.a.	↓	↑	n.a.	↓	↓	↓
		5.0 mg/l	=	n.a.	=	n.a.	n.a.	n.a.	↓	↑	n.a.	↓	↓	↓
		10 mg/l	=	n.a.	=	n.a.	n.a.	n.a.	↓	=	n.a.	=	=	↓
		20 mg/l	=	n.a.	↓	n.a.	n.a.	n.a.	↓	=	n.a.	=	=	↓
	Kidney	2.5 mg/l	=	n.a.	=	n.a.	n.a.	n.a.	↓	↑	n.a.	↓	↓	=
		5.0 mg/l	=	n.a.	=	n.a.	n.a.	n.a.	↓	↑	n.a.	↓	↓	=
		10 mg/l	=	n.a.	=	n.a.	n.a.	n.a.	↓	↑	n.a.	↓	↓	=
		20 mg/l	↑	n.a.	=	n.a.	n.a.	n.a.	↓	↑	n.a.	↓	↓	=
Manganese	Liver	0.1 mM	↑	n.a.	n.a.	n.a.	n.a.	n.a.	=	↓	↑	n.a.	n.a.	n.a.
		1.0 mM	↑	n.a.	n.a.	n.a.	n.a.	n.a.	↑	↓	↑	n.a.	n.a.	n.a.
	Kidney	0.1 mM	=	n.a.	n.a.	n.a.	n.a.	n.a.	=	=	↑	n.a.	n.a.	n.a.
		1.0 mM	↑	n.a.	n.a.	n.a.	n.a.	n.a.	↑	↓	↑	n.a.	n.a.	n.a.
	Gill	0.1 mM	↑	n.a.	n.a.	n.a.	n.a.	n.a.	↑	=	=	n.a.	n.a.	n.a.
		1.0 mM	↑	n.a.	n.a.	n.a.	n.a.	n.a.	↑	↑	↑	n.a.	n.a.	n.a.
	Brain	0.1 mM	=	n.a.	n.a.	n.a.	n.a.	n.a.	=	=	↑	n.a.	n.a.	n.a.
		1.0 mM	↑	n.a.	n.a.	n.a.	n.a.	n.a.	↓	=	↑	n.a.	n.a.	n.a.
Chromium	Brain	1.0 mg/l	=	=	n.a.	↓	↓	=	=	↓	n.a.	=	=	↓
		2.5 mg/l	=	↑	n.a.	↓	↓	=	=	↓	n.a.	=	=	↓
		5.0 mg/l	=	↑	n.a.	↓	↓	=	=	↓	n.a.	↓	=	=
		10 mg/l	↑	↑	n.a.	↓	↓	↓	=	↓	n.a.	=	=	↓

Fish were exposed to glyphosate, manganese or chromium (III) during 96 hr. Biomarkers responses are indicated as increase (↑), decrease (↓) or not altered (=). n.a. = not analyzed. Sources: Lushchak et al. 2009a; Vieira et al. 2012; Lushchak et al. 2009b.

l^{-1}); decrease of total GSH and GSSG (at all concentrations), which reflected on decrease of GSSG/Total GSH only at the highest concentration, and CAT and GST decreases at almost all concentrations (Table 5.6) (Lushchak et al. 2009b). The authors did not observe a concentration-dependent response and explained the decreased glutathione concentration as a reflection of one of several processes—decreased synthesis or reduced transport into the brain (because most GSH is synthesized by the liver), accelerated degradation or enhanced export of oxidized form (Lushchak et al. 2009b).

Another organ with low SOD and CAT activities is the heart (Table 5.4). Indeed, during ischemic-reperfusion processes in humans, oxidative stress is involved in cardiomyocytes demise. For fish, there are some studies that analyzed the heart as a possible target of toxicity. Juvenile *O. niloticus* exposed to a sublethal concentration of the organophosphate trichlorfon for 96 hours presented increases of CAT, GST and GSH in the liver and gills, and SOD in the gills, preventing the increase of lipid peroxidation. In the heart, however, although SOD and GST activities increased, GSH decreased, resulting in LPO increase (Table 5.7). The heart was therefore considered the most sensitive organ when compared to the liver and gills (Thomaz et al. 2009).

The idea of differential sensitivities was also observed in another study. *Brycon cephalus* (matrinxã) exposed to 2 mg ml^{-1} of organophosphorus insecticide methyl parathion during 96 hours showed increased SOD, CAT and GST activities, and decreased GSH levels in the white muscle, gills and liver. GPx activity decreased and LPO increased in the former two organs, but both biomarkers were not altered in the liver. According to the authors, the gills and white muscle are more sensitive organs than the liver, since the antioxidant system of these tissues is not as efficient as that of the liver, which increases their vulnerability towards ROS. The increased GST activity was concomitant to the decreases in GSH content in all tissues analyzed (Monteiro et al. 2006).

Table 5.7. Biomarkers response of liver, gills and heart of *C. carpio* exposed to trichlorfon.

Biomarker	SOD	CAT	GST	GSH	LPO
Liver	↓	↑	↑	↑	=
Gills	↑	=	↑	↑	=
Heart	↑	=	↑	↓	↑

Fish were exposed to 0.5 mg l^{-1} of trichlorfon during 96 hr. Biomarkers responses are indicated as increase (↑), decrease (↓) or not altered (=). Source: Thomaz et al. 2009.

Dose/Concentration-response Studies

Two kinds of toxicological information are essential to understand the effects and risks of chemical exposure. The first is the variation of the response or effect over time. In principle, acute effects may be different from subchronic and chronic ones. Indeed, it is not hard to find different results even within these three general approaches, i.e., a 12 hour exposure may lead to different effects than a 24 hour exposure, although both are considered "acute exposures". The second information is the type and intensity of the response or effect. Laboratory studies deal with it through experiments utilizing three or more concentrations/doses of chemicals usually scaled at logarithmic steps. It may seem evident that the response increases with the dose. But it is usually not observed for many groups of biomarkers, in particular oxidative stress ones.

For example, *Cyprinus carpio* (common carp) exposed to the herbicide atrazine (4.28, 42.8 and 428 µg l^{-1}) for 40 days presented increased lipid peroxidation in function of the concentration (a clear concentration-response curve). However, SOD, CAT and GPx increased in the liver and gills at the lowest concentrations, but decreased at the highest one (Xing et al. 2012). The same response pattern was observed for *C. carpio* exposed to nano-ZnO (0.5, 5.0 and 50.0 mg l^{-1}) for 1, 3, 7, 10 and 14 days. In this case, SOD activity increased at the lowest concentration, but decreased at the intermediate and highest concentrations in the gills, liver, brain and intestine. CAT and GPx activities were not altered at the lowest concentration, increased at the intermediate and decreased at the highest. This type of curve (inverted "U") was not observed only for antioxidant enzymes; GSH concentration increased at the lowest and intermediate concentrations, but then decreased at the highest (Hao and Chen 2012). This type of response is not species-specific, as other fish species also present responses that do not obey an "expected" standard. *C. auratus* exposed to different concentrations (0.001, 0.005, 0.01, 0.05 and 0.1 mg l^{-1}) of pyrene for 10 days showed increased production of hydroxyl radical at concentrations of pyrene equal or greater than 0.01 mg l^{-1}, reduced GSH/GSSG ratios and SOD increase at all concentrations and increases of lipid peroxidation at 0.005–0.1 mg l^{-1} of pyrene. Although there is nothing different from what is expected for these biomarkers response in *C. auratus*, CAT and GST increased at lowest concentrations and then decreased at the highest ones (Sun et al. 2008).

These three studies illustrate what is usually found in ecotoxicology: whereas some biomarkers responses follow an expected pattern, others do not, at least in principle. The concept of *hormesis* can help defining these responses.

Hormesis is a dose-response relationship characterized by a biphasic (U- or inverted U-shaped) response, or in physiology and medicine, a non-monotonic behavior. At low doses, the endpoint response either increases or decreases from the baseline level; at high doses, the response changes its direction, forming a U- or inverted U-shaped curve. While non-monotonic biological response may have diverse mechanistic bases, hormesis is believed to occur as a result of overcompensation by the homeostatic control systems operating in biological organisms (Calabrese 2001; Calabrese 2008a,b, 2010). According to this idea, we could explain the results of those three studies as follows: the low exposure to the chemicals might have increased antioxidant defenses through increased expression of some genes coding for antioxidant enzymes (e.g., SOD, CAT, GPx) leading to increased activities. At higher exposures, either the gene expression might have been blocked by different mechanisms or the enzyme structure itself could be compromised, so that the compensatory mechanism would be overwhelmed. This logic may also be used to explain the results of other antioxidants, lipid peroxidation and so on, although the mechanisms may be different.

Oxidative Stress and *in vitro* Investigations

Most studies in ecotoxicology/toxicology of tropical fish aim to uncover the responses and effects of intact organisms to chemical exposure through the use of biomarkers. These approaches are very important in providing information on the organism's health as systemic interactions are present and chemicals can be biotransformed and transported through many organs and tissues that make up the living organisms. However, whenever the aim is to investigate the direct responses/effects of chemicals on single or very few cell populations, or the mechanisms underlying toxicity, isolated cells (fresh or cultured) are very useful. For this reason, this short section deals with *in vitro* approaches.

Hepatocytes, gills, macrophages and spermatozoa are usually utilized in toxicology studies with fish. Hepatocytes are particularly useful models since these cells are extraordinarily metabolic diverse, being responsible for most of the biotransformation reactions that occur in fish and other vertebrates. Oxidative stress biomarkers are commonly investigated on cultured hepatocytes. For example, *Prochilodus lineatus* (curimbatá) hepatocytes exposed to purified cylindrospermopsin toxin presented increases of G6PDH, reactive oxygen and nitrogen species and lipid peroxidation, although GSH, protein carbonylation and DNA damage have not been altered (Liebel et al. 2011). In most cases, cells are exposed to a single chemical as in this example, but in others, the authors use mixtures of chemicals. One interesting approach is to utilize tissue extracts of chemicals from fish captured in the natural environment. For example, hepatocytes from *Hoplias malabaricus* (traíra) exposed to a mixture of halogenated organic compounds extracted from the fat tissue of another fish (*Anguilla Anguilla*, European eel) from a contaminated site showed increases of CAT, GST and DNA damage, as well as decrease of GSH concentration and cell viability. According to the authors, the exposure of cells to a biological extracted mixture of chemicals may allow important interactions, even thought it makes impossible, the attribution of a specific effect to one causing agent in particular, and some unidentified chemicals (without analytical standards) can be present on the chemical mixture (Filipak Neto et al. 2007).

Interspecies comparisons can also be performed *in vitro*. For example, hepatocytes from *H. malabaricus* exposed to DDT and MeHg presented increases of ROS levels, CAT, G6PDH, lipid peroxidation and protein carbonylation, and decreases of SOD, GR and GSH (Table 5.8) (Filipak Neto et al. 2008). Conversely, hepatocytes from *Hypostomus commersoni* (cascudo) exposed to the same chemicals and in the same conditions showed only

Table 5.8. Biomarkers response of primary cultured hepatocytes of *H. malabaricus* and *H. commersoni* exposed to DDT, methyl mercury and their association.

	Fish	ROS	SOD	CAT	GST	GR	G6PDH	GSH	LPO	PCO
DDT	*H. malabaricus*	↑	↓	↑	↓	↓	=	↓	↑	=
	H. commersoni	↓	n.a.	=	↓	n.a.	n.a.	↓	=	n.a.
MeHg I	*H. malabaricus*	↑	↓	↑	=	↓	=	=	↑	=
	H. commersoni	↓	n.a.	=	=	n.a.	n.a.	=	=	n.a.
MeHg II	*H. malabaricus*	↑	↓	↓	=	↑	↑	↓	↑	↑
	H. commersoni	↓	n.a.	↑	=	n.a.	n.a.	↓	=	n.a.
Association	*H. malabaricus*	↑	=	↑	↑	↓	↑	↓	↑	↑
	H. commersoni	↓	n.a.	=	↓	n.a.	n.a.	↓	↑	n.a.

Hepatocytes were exposed *in vitro* to 50 nM of DDT, 0.25 (MeHg I) and 2.5 (MeHg II) µM of methyl mercury and to association of 50 nM of DDT and 0.25 µM of methyl mercury during 96 hr. ROS were quantified by the use of the fluorescent reagent dichlorofluorescein diacetate. Biomarkers responses are indicated as increase (↑), decrease (↓) or not altered (=). Sources: Bussolaro et al. 2010; Filipak Neto et al. 2008.

decrease of ROS and GSH levels (Table 5.8). The authors suggested that hepatocytes of herbivorous bottom fish *H. commersoni* may be somehow more resistant to DDT and MeHg exposure than those of the predator water column fish *H. malabaricus*, since antioxidant defenses and biomolecules integrity seemed to be less affected in the former species (Bussolaro et al. 2010).

Gill filaments and macrophage cultures from fish are also useful toxicology models. For example, short-term gill cultures from *Oncorhynchus mykiss* and *O. niloticus* have been utilized to investigate the effects of copper on apoptosis markers (Mazon et al. 2004; Monteiro et al. 2009b). Likewise, apoptosis have been studied in phagocyte-rich fraction (composed mainly by macrophages) isolated from head kidney of *Clarias batrachus* (Philippine catfish) and exposed to arsenic (Banerjee et al. 2011). Concerning oxidative stress biomarkers, macrophages isolated from fish *C. batrachus* showed increased production of ROS, translocation of the NADPH oxidase subunit p47[phox] to the plasma membrane (which is involved in NADPH oxidase activation) and increased LPO after exposure to arsenic at 0.50 µM (Datta et al. 2009a; Datta et al. 2009b).

One important question the researcher must deal with concerns the exposure situation *in vitro*, since many aspects determine the bioavailability of the chemical and consequently the cell response. For example, many chemicals are partitioned on lipids from fetal serum, whereas others react with cysteine and GSH present in the culture medium. In addition, the selection of the concentrations to be tested is a difficult task, since the *in vitro* situation is different from the *in vivo* one. Biotransformation, storage and mobilization, transport, removal of chemicals and metabolites play a role in determining the toxicity *in vivo*, whereas some of these processes do not even occur *in vitro*. But one model with individual cells does not seem to have most of these limitations, the spermatozoa. Most fish have external fecundation. In these fish, spermatozoa are usually released directly in the water. Water is thereby the ideal medium to be utilized for exposure purposes. Here is one example of a study that utilized this model.

Cyprinus carpio spermatozoa have been exposed *in vitro* to 0.2, 2.0 and 20 mg l^{-1} of carbamazepine (a human pharmaceutical commonly present in aquatic environment) for 2 hours. LPO increased in a concentration-dependent manner. PCO increased, but SOD, GR and GPx decreased at the highest concentrations (Li et al. 2010). The authors observed positive correlations between enzyme activity and motility, and negative correlations between damage to lipids/proteins and motility. They mentioned that LPO may be more sensitive than PCO to evaluate oxidative stress of fish spermatozoa (Li et al. 2010).

Common Studies and Data Interpretation

Laboratory/"field" exposures and biomonitoring are very common studies in ecotoxicology. In both cases, oxidative stress biomarkers have been utilized to characterize the health effects of chemical exposure or determine the "level" of chemical stress in one local (river, lake, etc.) in comparison to another. In this section, we will shortly discuss about the first group of studies.

Over this chapter, several examples of laboratory studies have been mentioned. In these studies, a biological model (fish species) has been kept in controlled conditions and exposed to a predetermined dose or concentration of chemical for a limited and known period of time. Variations of this strategy are to expose the fish to water from a reference/polluted site in the laboratory (under controlled conditions), or to place the fish on the sites of interest in fish-cages. For example, Almeida et al. sampled sediments collected from contaminated sites and utilized them to expose *Prochilodus lineatus* juvenile fish in the laboratory; oxidative stress data (CAT and GST activities, etc.) were utilized to classify some sampling sites in degrees of contamination (Almeida et al. 2005). In these cases, although the period of exposure can be controlled, the level of exposure and the composition of chemicals are usually unknown, since usually, very complex mixtures of chemicals are present in the natural environment. On both laboratory and "field" exposure approaches, fish species and age, nutrition (food supply), chemical species, dose/concentration, period and route of exposure are the main variables.

As we have seen, studies typically utilize high doses/concentrations of chemicals to characterize the biological responses/effects. These levels of exposure are difficult to be found in natural environments that are continuously impacted, unless in specific situations such as accidents (e.g., oil spills). When the results of these studies are compared with those utilizing environmental concentrations, some of the oxidative stress biomarkers seem to respond "oddly" in the latter. For example, the exposure to a chemical can lead to a lipid peroxidation decrease. Note that there is no ideal exposure design, as the doses/concentrations must be selected according to the purpose of the study. Likewise, the period of exposure can be quite different, some lasting a few hours whereas others for some months. Acute, subchronic and chronic exposures can also lead to very variable oxidative stress biomarkers responses.

Three are the common routes of exposure: trophic (feeding), waterborne or intraperitoneal injection. The choice of route depends on the purpose of the study and nature of the chemical, and can greatly influence the response. For example, *O. niloticus* exposed to the purified cyanotoxin cylindrospermopsin (200 µg kg^{-1}) through force-feeding (gavage) and intraperitoneal (i.p.) injection presented increases of NADPH oxidase activity, lipid peroxidation and protein carbonylation, as well as reduction of glutathione levels and γ-glutamylcysteine synthetase activity, the limiting enzyme in glutathione synthesis. No changes of DNA oxidation were observed. In fish sacrificed after five days, some of the biomarkers had recovered to their basal levels, which was not the case after 24 hours (Gutiérrez-Praena et al. 2011). Protein oxidation increased more in the kidney than in the liver. The two exposure routes led to different results, but there were no clear differences between them in terms of the severity of the toxic effects. Oral exposure led to higher alterations of NADPH oxidase activity and LPO in the liver, and i.p. exposure resulted in highest LPO and protein oxidation in the kidney and γ-glutamylcysteine synthetase activity in the liver (Gutiérrez-Praena et al. 2011).

Interpretation of oxidative stress data is usually very similar among the studies. Authors correlate the biomarkers among themselves and utilize mechanistic data available in literature to explain the importance or reason of the biomarker alteration. In some cases, the parameters are well-correlated, or the biomarkers respond as expected (antioxidant defenses decrease and biomolecules damage increase for example), making the interpretation easy. For example, Bagnyukova et al. have investigated the effects of ferrous sulphate on *C. auratus* L., and reported that G6PDH might be partially inactivated due to carbonylation, since negative correlations between protein carbonylation levels and the enzyme were observed. According to them, liver GST activity was inversely connected with TBARS levels, probably reflecting the involvement of liver GST in detoxification of aldehyde products of lipid peroxidation (Bagnyukova et al. 2006). In another study, juveniles *Rhamdia quelen* exposed to thorium for 30 days showed significant correlations between LPO levels and Th bioaccumulation (r^2 = 0.9046); GST and Th bioaccumulation (r^2 = 0.9789); and a negative correlation between the LPO levels and the GST activity in the gills. Exposure to Th led to LPO increase, CAT and SOD activities decrease, but no significant difference for GST (Kochhann et al. 2009).

Biomonitoring and Possible Interferences

Oxidative stress biomarkers are broadly utilized in studies that aim to establish the level of contamination of different sites, either compared to each other (if a pollution gradient is suspected) or to a reference site. However, any strong stress is usually accompanied by oxidative stress, as it may be responsible for physiological accommodation of organisms to a broad range of environmental stressors (Lushchak 2011). Particularly, freshwater fish must often cope with marked seasonal and daily changes in the temperature, oxygen, pH and pollutants, and these water physical–chemical parameters can induce distinct responses in fish antioxidant defenses (Wilhelm-Filho and Marcon 1996). Thus, the first part of this section is dedicated to some studies that utilized tropical/subtropical fish in biomonitoring; the final ones focus on some of the main interferences that must be clear when planning the study itself and interpreting oxidative stress data.

Oxidative stress and biomonitoring

Biomonitoring studies are essential for risk management, and oxidative stress biomarkers are usually integrated with other groups of biomarkers in multi-biomarkers evaluations. However, biological data are interpreted based on limited chemical data. Metals in tissues, water or sediments, and PAHs in bile are probably the most explored chemical data for interpretation due to low complexity and cost of analysis. See the following examples.

Example 1. O. niloticus captured during different months of the year in the Monjolinho River (SP, Brazil) had higher hepatic CAT and GST activities, metallothionein and LPO levels, and lower SOD and GPx activities than those from a reference site in the same river (Carvalho et al. 2012). According to the authors, the results are associated with the exposure of metals, and the reduced enzyme activities could indicate that the antioxidant capacity was surpassed by the amount of hydroperoxide products of lipid peroxidation. Yet, MT concentrations were higher at the months displaying the highest metal concentrations, supporting its usefulness in environmental monitoring even in complex environments where interference of other xenobiotics can be found (Carvalho et al. 2012).

Example 2. Oxidative stress biomarkers have been measured in the blood of three cichlid fish *O. niloticus* (Nile tilapia), *Tilapia rendalli* (Congo tilapia) and *Geophagus brasiliensis* (cará) captured during spring and autumn from two sites in Monjolinho River (SP, Brazil). SOD, GPx, GSH and LPO were, in general, highest in fish from the metal-contaminated site on both seasons, whereas CAT response were more variable (none, decrease or increase) among the three species when the sites were compared (Gaspar Ruas et al. 2008). SOD presented the most expressive variation among the enzymes considering the pollution situation. LPO was probably induced by the high levels of zinc, manganese and iron in the water in autumn, as well as of ammonia and copper in spring (Gaspar Ruas et al. 2008).

Example 3. Comparison among oxidative stress biomarkers in *Prochilodus lineatus* from different sampling sites along Salado River basin (a tributary of the middle Paraná River) indicated an increase of GST, GR and peroxidases activities and LPO in the fish from polluted sites in comparison to the reference site. Liver biomarkers response indicated chemical stress better than those in gills and kidney (Cazenave et al. 2009).

As we can see, data were interpreted according to metal, ammonia and a few other chemical parameters. Indeed, absence of chemical data from the major classes of xenobiotics is the main obstacle for interpretation. In some cases, authors do not have data about the chemicals and they utilize other information such as the main land uses and the localization of potential sources of pollutants (urban, industrial, and agricultural). However, it is very difficult to look at the oxidative stress biomarkers themselves and classify two or more sites as most or least impacted without additional information such as the "basal" antioxidant activities/levels or "basal" biomolecules damage levels. Thus, without a known "basal, normal or standard level", we can see differences between one parameter in two sites, but cannot be sure which (the highest or the lowest) is the "normal" and which is the "altered" one. To solve this limitation, a reference group is usually utilized to provide information on "what is the normal" for a species from a specific location. But, is the absence of a clear contamination source the only factor that must be considered to choose the reference site? As we shall see next, there are many factors that must be very similar between the site being studied and the reference one, so as to provide trustworthy data for making good comparisons.

Effects of temperature

Ectotherm organisms including fish are subject to varying temperatures during the seasons or even during the day in subtropical regions. In these animals, the metabolic rates of the processes that generate ROS vary according to the temperature. To compensate, adaptive responses by antioxidant defenses are necessary to guarantee homeostasis, particularly for long-term temperature variations (Bagnyukova et al. 2007).

The increase in temperature stimulates all metabolic processes in accordance with known thermodynamic principles. For example, it enhances oxygen consumption and, therefore, may increase ROS production as side products of intensified metabolism resulting in oxidative stress. In accordance with the abovementioned intensification of oxidative metabolism, one might expect a decrease in the risk of oxidative stress induction at decreased temperature (Lushchak 2011). But under some circumstances, the decrease in environmental temperature also may cause oxidative stress in fish (Malek et al. 2004). One possible consequence of temperature decrease is the increase of the percentage of polyunsaturated fatty acids on phospholipids to maintain the membrane fluidity. As we have seen, polyunsaturated fatty acids are the primary targets of lipid peroxidation.

An example of temperature effects on biomarkers were reported for *C. auratus* transferred from a low to a warm temperature. In these fish, lipid peroxidation increased and protein carbonyl content was reduced in the liver over the entire experimental course, but increased transiently in the kidney. The content of high-molecular mass thiols decreased by two-thirds in the brain and total low-molecular mass thiols (e.g., glutathione and others) increased transiently. SOD and CAT activities were generally unaffected, whereas glutathione-dependent enzymes (GPx and GST) were elevated in the brain and kidney. Notably, glutathione-dependent enzymes, but not the primary antioxidant enzymes (SOD and CAT), showed the most marked response to the transition from low to warm temperature, which may be necessary to establish long-term accommodation to temperature change. Hence, a short-term exposure to warm temperature disturbed several oxidative stress markers, but only slightly affected the activities of antioxidant enzymes (Bagnyukova et al. 2007).

Another example of the effects of temperature was reported for subtropical fish *Dicentrarchus labrax* (European seabass) juveniles maintained during 30 days to temperatures that reflect the average summer temperature (18°C and 24°C) and the temperature during heat waves (28°C). At t = 15 days, CAT and LPO was highest at 28°C. At t = 30 days, CAT and LPO was highest at 18°C. The highest levels of lipid peroxidation and catalase activity were observed in fish exposed to temperatures outside their thermal optimum, which means that thermal stress affects the cellular response of fish facing ROS (Vinagre et al. 2012). According to the authors, the oxidative stress response is not directly correlated to temperature as it is lowest at the optimal temperature (24°C) and increases outside the species' upper and lower thermal limits (Vinagre et al. 2012). Fish of the same species (*Dicentrarchus labrax*) but environmentally exposed to mercury presented ambivalent antioxidant responses, i.e., highest GR activity and lowest CAT activity in the warm period only, thus demonstrating that seasonal changes in environmental factors play a crucial role in regulating the antioxidant capacity of the brain of *Dicentrarchus labrax* (Mieiro et al. 2011).

Studies on fish mostly use antioxidant enzymes as biomarkers of environmental contamination. Although the response elicited by temperature is usually lower than that produced by contaminants, the variation induced by temperature may lead to confounding results when comparing sites with different thermal regimes like those located at different latitudes or at different depths (Vinagre et al. 2012). Thus, in order to assure that temperature is not a confounding factor, it is necessary to investigate the oxidative stress response in fish when exposed to temperatures that may realistically be present in areas commonly surveyed for pollution monitoring (Abele and Puntarulo 2004).

Effects of oxygen level

Oxygen level is an important physicochemical parameter that can greatly influence the response of oxidative stress biomarkers. The exposures to hypoxia as well as to hyperoxia induce alteration of the biomarkers in different fish species. For example, when fish are exposed to oxygen deficiency, the enhanced antioxidant potential may prevent the development of oxidative stress when oxygen supply is restored (Lushchak 2011). Indeed, large-scale mitochondrial ROS generation is usually associated with ischemia and reperfusion processes (Halestrap et al. 2007).

In *C. auratus*, both hyperoxia and heat shock increased lipid peroxidation within the first hours of treatment, but the return to initial conditions rapidly reduced levels of lipid peroxidation products, probably due to a compensation of GST (Lushchak et al. 2005a). Conversely, *C. carpio* maintained under hyperoxia conditions showed decreases of protein carbonylation in the kidney and muscle, and lipid peroxidation in the brain and liver, as well as increases of CAT and GPx in the brain (Lushchak et al. 2005b). Yet, hypoxia resulted in reduced SOD, CAT and GPx activities in juvenile *Piaractus mesopotamicus* (pacú) (Garcia Sampaio et al. 2008).

Effects of feeding

In their natural habitats, fish often experience periods of poor food supply in response to several factors (e.g., temperature, spawning migration, reproduction, etc.) and are well-adapted to long-term starvation (Furne et al. 2009; Perez-Jimenez et al. 2007; Rios et al. 2002; Van Dijk et al. 2005). However, the nutritional status has an important impact on antioxidant capacity since food nutrients are essential for the synthesis and function of antioxidants.

For example, *Brycon cephalus* (matrinxã) exposed to 2 mg L^{-1} of organophosphate methyl parathion for 96 hours presented increases of SOD, CAT and GST activities in the gills, white muscle and liver, but decreases of GPx activity in the white muscle and gills. Selenium supplementation for 8 weeks (0 or 1.5 mg Se kg^{-1}) prior to exposure reversed the pesticide effects of decreased GPx activity as it recovered the values presented by the control. GSH levels decreased in all tissues, and lipid peroxidation increased in the white muscle and gills exposed to the pesticide. Selenium supplementation completely avoided the pesticide-induced decreases of GSH in all tissues and increases in lipid peroxidation in the gills and white muscle (Monteiro et al. 2009a). According to the authors, Se supplementation had a protective effect promoting the maintenance of a high steady-state GSH level and a normal GPx activity after exposure. It also controlled lipid peroxidation in the gills and white muscle (Monteiro et al. 2009a).

Other authors observed similar effects of nutritional state on antioxidant capacity of fish. For example, the absence of changes in total glutathione after exposure to xenobiotics have been associated with supplementation with food, as feeding may be necessary to maintain the glutathione pools (Kubrak et al. 2010). According to these authors, the problem is that glutathione may be actively used by organisms especially under oxidative stress conditions. In this case, GSH reserves should be restored by *de novo* synthesis, which needs both energy input and certain amino acids. Therefore, fish starvation might affect the GSH pool (Kubrak et al. 2010) and this principle may be applied for other antioxidants.

Effects of age

Aging has been defined as the progressive accumulation of deleterious changes in the cells and tissues that increase the risk of death and disease with increasing age (Harman 2001). The three different aging patterns described in teleost fish are rapid (rapid senescence and sudden death after spawning—the organisms are semelparous), gradual (fish continue to grow, though at decreased rates, throughout their life span), and negligible (indeterminate growth with no increased mortality as a function of age) (Kishi 2004; Kishi et al. 2003; Patnaik et al. 1994).

The fish embryo or larva is generally considered the most sensitive stage in the life cycle of a teleost, being particularly sensitive to a range of low-level environmental stresses to which larvae may be exposed (Wiegand et al. 1999; Wiegand et al. 2000).

The age of fish used in the studies has been given little consideration, particularly in field studies. This can be problematic, especially in light of the fact that oxidative stress, damage products and antioxidant defenses can be age-affected in a large number of organisms (Beckman and Ames 1998). For instance, age-related changes in endogenous antioxidant systems were observed for SOD, CAT and GSH levels in *O. mykiss* (rainbow trout) and *Ameiurus melas* (black bullhead) (Otto and Moon 1996), and protein carbonylation is thought to play an important role in aging (Berlett and Stadtman 1997; Stadtman and Levine 2000). Here is one example for the subtropical fish *Salmo trutta*.

Farmed *S. trutta* in four different age groups (from five months to three years) had increased protein carbonyls and decreased 20S proteosome activity in the brain and liver as a function of increasing size and age. Total GSH levels in the liver declined as fish aged and the GSSG:GSH ratio increased (Carney Almroth et al. 2010). The authors reported a significant increase of protein carbonyls in the livers and decrease of 20S proteosome activity in the brain of 1 year old fish exposed to paraquat. No such increase was observed in the youngest fish, indicating that individuals of different ages respond differently to oxidative stress induced by paraquat (Carney Almroth et al. 2010). One important conclusion of the authors is that studies using oxidative stress biomarkers need to consider the ages of the fish (Carney Almroth et al. 2010).

Conclusion

Oxidative stress is an important mechanism of toxicity of many classes of chemical stressors so that different biomarkers can be utilized to evaluate fish health in tropical and non-tropical ecosystems. Antioxidant defenses and biomolecules damages are particularly useful for toxicity determination and risk assessment studies in the laboratory and the field. Whereas antioxidants biomarkers only make sense when analyzed together for interpretation purposes, individual parameters of cell damage (LPO, PCO and DNA oxidative damage) are more reliable to indicate the presence of oxidative stress if they are altered in a direction that makes sense with other toxicological information (biomarkers or chemical data). Oxidative stress parameters are very dynamic, being modulated by physicochemical and nutritional factors that are very relevant in biomonitoring studies, thus complicating comparisons between impacted and distant "reference" areas, or impeding comparisons between fish sampled in the field and fish acclimated in the laboratory (as a reference group). All these aspects must be analyzed by anyone interested in making use of oxidative stress as a useful tool for research in ecotoxicology.

Keywords: oxidative stress, ROS, antioxidant defenses, hormesis, GSH, catalase, superóxide dismutase, glutathione peroxidase, glutathione S-transferase, lipid peroxidation, protein carbonylation, liver, gill, brain

References

Abele, D. and S. Puntarulo. 2004. Formation of reactive species and induction of antioxidant defence systems in polar and temperate marine invertebrates and fish. Comp. Biochem. Physiol. Part A Mol. Integr. Physiol. 138: 405–415.

Agnisola, C. 2005. Role of nitric oxide in the control of coronary resistance in teleosts. Comp. Biochem. Physiol. Part A Mol. Integr. Physiol. 142: 178–187.

Ahmad, I., M. Pacheco and M.A. Santos. 2004. Enzymatic and nonenzymatic antioxidants as an adaptation to phagocyte-induced damage in *Anguilla anguilla* L. following *in situ* harbor water exposure. Ecotoxicol. Environ. Saf. 57: 290–302.

Alberto, W.D., D.M. Del Pilar, A.M. Valeria, P.S. Fabiana, H.A. Cecilia and B.M. De Los Angeles. 2001. Pattern recognition techniques for the evaluation of spatial and temporal variations in water quality. A case study: Suquia River basin (Cordoba-Argentina). Water Res. 35: 2881–2894.

Allen, R.G. and M. Tresini. 2000. Oxidative stress and gene regulation. Free Radic. Biol. Med. 28: 463–499.

Almeida, J.S., P.C. Meletti and C.B.R. Martinez. 2005. Acute effects of sediments taken from an urban stream on physiological and biochemical parameters of the neotropical fish *Prochilodus lineatus*. Comp. Biochem. Physiol. Part C Toxicol. Pharmcol. 140: 356–363.

Atencio, L., I. Moreno, Á. Jos, A.I. Prieto, R. Moyano, A. Blanco and A.M. Cameán. 2009. Effects of dietary selenium on the oxidative stress and pathological changes in tilapia (*Oreochromis niloticus*) exposed to a microcystin-producing cyanobacterial water bloom. Toxicon. 53: 269–282.

Bagnyukova, T.V., O.I. Chahrak and V.I. Lushchak. 2006. Coordinated response of goldfish antioxidant defenses to environmental stress. Aquat. Toxicol. 78: 325–331.

Bagnyukova, T.V., O.V. Lushchak, K.B. Storey and V.I. Lushchak. 2007. Oxidative stress and antioxidant defense responses by goldfish tissues to acute change of temperature from 3 to 23°C. J. Therm. Biol. 32: 227–234.

Bagnyukova, T.V., O.Y. Vasylkiv, K.B. Storey and V.I. Lushchak. 2005. Catalase inhibition by amino triazole induces oxidative stress in goldfish brain. Brain Res. 1052: 180–186.

Banerjee, C., R. Goswami, S. Datta, R. Rajagopal and S. Mazumder. 2011. Arsenic-induced alteration in intracellular calcium homeostasis induces head kidney macrophage apoptosis involving the activation of calpain-2 and ERK in Clarias batrachus. Toxicol. Appl. Pharmacol. 256: 44–51.

Barata, C., I. Varo, J.C. Navarro, S. Arun and C. Porte. 2005. Antioxidant enzyme activities and lipid peroxidation in the freshwater cladoceran Daphnia magna exposed to redox cycling compounds. Comp. Biochem. Physiol. Part C Toxicol. Pharmcol. 140: 175–186.

Beckman, K.B. and B.N. Ames. 1998. The free radical theory of aging matures. Physiol. Rev. 78: 547–581.

Berlett, B.S. and E.R. Stadtman. 1997. Protein oxidation in aging, disease, and oxidative stress. J. Biol. Chem. 272: 20313–20316.

Berquist, B.R. and D.M. Wilson. 2012. Pathways for repairing and tolerating the spectrum of oxidative DNA lesions. Cancer Lett. 327(1–2): 61–72.

Bespalov, I.A., J.P. Bond, A.A. Purmal, S.S. Wallace and R.J. Melamede. 1999. Fabs specific for 8-oxoguanine: Control of DNA binding. J. Mol. Biol. 293: 1085–1095.

Bhattacharya, A. and S. Bhattacharya. 2007. Induction of oxidative stress by arsenic in *Clarias batrachus*: Involvement of peroxisomes. Ecotoxicol. Environ. Saf. 66: 178–187.

Boelsterli, U.A. 2007. Mechanistic toxicology. The molecular basis of how chemicals disrupt biological targets. CRC Press Boca Raton.

Bokov, A., A. Chaudhuri and A. Richardson. 2004. The role of oxidative damage and stress in aging. Mech. Ageing Dev. 125: 811–826.

Bonneh-Barkay, D., S.H. Reaney, W.J. Langston and D.A. Di Monte. 2005. Redox cycling of the herbicide paraquat in microglial cultures. Mol. Brain Res. 134: 52–56.

Brigelius-Flohe, R. 2006. Glutathione peroxidases and redox-regulated transcription factors. Biol. Chem. 387: 1329–1335.

Bussolaro, D., F. Filipak Neto and C.A. Oliveira Ribeiro. 2010. Responses of hepatocytes to DDT and methyl mercury exposure. Toxicol. *In vitro*. 24: 1491–1497.

Calabrese, E.J. 2001. Overcompensation stimulation: A mechanism for hormetic effects. Crit. Rev. Toxicol. 31: 425–470.

Calabrese, E.J. 2008a. Hormesis: Why it is important to toxicology and toxicologists. Environ. Toxicol. Chem. 27: 1451–1474.

Calabrese, E.J. 2008b. U-shaped dose response in behavioral pharmacology: Historical foundations. Crit. Rev. Toxicol. 38: 591–598.

Calabrese, E.J. 2010. Hormesis is central to toxicology, pharmacology and risk assessment. Hum. Exp. Toxicol. 29: 249–261.

Caliani, I., S. Porcelloni, G. Mori, G. Frenzilli, M. Ferraro, L. Marsili, S. Casini and M.C. Fossi. 2009. Genotoxic effects of produced waters in mosquito fish (*Gambusia affinis*). Ecotoxicology 18: 75–80.

Carney Almroth, B., A. Johansson, L. Förlin and J. Sturve. 2010. Early-age changes in oxidative stress in brown trout, salmo trutta. Comp. Biochem. Physiol. Part B Biochem. Mol. Biol. 155: 442–448.

Carvalho, C.d.S., V.A. Bernusso, H.S.S.d. Araújo, E.L.G. Espíndola and M.N. Fernandes 2012. Biomarker responses as indication of contaminant effects in *Oreochromis niloticus*. Chemosphere 89: 60–69.

Cavalcante, D.G., C.B. Martinez and S.H. Sofia. 2008. Genotoxic effects of roundup on the fish *Prochilodus lineatus*. Mutat. Res. 655: 41–46.

Cazenave, J., C. Bacchetta, M.J. Parma, P.A. Scarabotti and D.A. Wunderlin. 2009. Multiple biomarkers responses in *Prochilodus lineatus* allowed assessing changes in the water quality of Salado River basin (Santa Fe, Argentina). Environ. Pollut. 157: 3025–3033.

Chambers, J.E., R.L. Carr, S. Boone and H.W. Chambers. 2001. The metabolism of organophosphorus insecticides. 2nd edition. *In:* R.I. Krieger (ed.). Handbook of pesticide toxicology. Academic Press, USA, pp. 919–928.

Cheng, F.C., J.F. Jen and T.H. Tsai. 2002. Hydroxyl radical in living systems and its separation methods. J. Chromatogr. B: Anal. Technol. Biomed. Life Sci. 781: 481–496.

Cheung, C.C.C., G.J. Zheng, A.M.Y. Li, B.J. Richardson and P.K.S. Lam. 2001. Relationships between tissue concentrations of polycyclic aromatic hydrocarbons and antioxidative responses of marine mussels, Perna viridis. Aquat. Toxicol. 52: 189–203.

Ching, B., S.F. Chew, W.P. Wong and Y.K. Ip. 2009. Environmental ammonia exposure induces oxidative stress in gills and brain of *Boleophthalmus boddarti* (mudskipper). Aquat. Toxicol. 95: 203–212.

Claeson, K., G. Thorsen and B. Karlberg. 2001. Methyl malondialdehyde as an internal standard for the determination of malondialdehyde. J. Chromatogr. B Biomed. Sci. Appl. 751: 315–323.

Dalle-Donne, I., R. Rossi, D. Giustarini, A. Milzani and R. Colombo. 2003. Protein carbonyl groups as biomarkers of oxidative stress. Clin. Chim. Acta. 329: 23–38.

Datta, S., D. Ghosh, D.R. Saha, S. Bhattacharaya and S. Mazumder. 2009a. Chronic exposure to low concentration of arsenic is immunotoxic to fish: Role of head kidney macrophages as biomarkers of arsenic toxicity to Clarias batrachus. Aquat. Toxicol. 92: 86–94.

Datta, S., S. Mazumder, D. Ghosh, S. Dey and S. Bhattacharya. 2009b. Low concentration of arsenic could induce caspase-3 mediated head kidney macrophage apoptosis with JNK-p38 activation in Clarias batrachus. Toxicol. Appl. Pharmacol. 241: 329–338.

Di Giulio, R.T., P.C. Washburn, R.J. Wenning, G.W. Winston and C.S. Jewell. 1989. Biochemical responses in aquatic animals—a review of determinants of oxidative stress. Environ. Toxicol. Chem. 8: 1103–1123.

Dunlop, R.A., U.T. Brunk and K.J. Rodgers. 2009. Oxidized proteins : mechanisms of removal and consequences of accumulation. IUBMB Life. 61: 522–527.

Durmaz, H., Y. Sevgiler and N. Üner. 2006. Tissue-specific antioxidative and neurotoxic responses to diazinon in Oreochromis niloticus. Pestic. Biochem. Physiol. 84: 215–226.

Elia, A.C., V. Anastasi and A.J. Dorr. 2006. Hepatic antioxidant enzymes and total glutathione of Cyprinus carpio exposed to three disinfectants, chlorine dioxide, sodium hypochlorite and peracetic acid, for superficial water potabilization. Chemosphere. 64: 1633–1641.

Enis Yonar, M., S. Mişe Yonar and S. Silici. 2011. Protective effect of propolis against oxidative stress and immunosuppression induced by oxytetracycline in rainbow trout (Oncorhynchus mykiss W.). Fish Shellfish Immunol. 31: 318–325.

Esterbauer, H. 1993. Cytotoxicity and genotoxicity of lipid-oxidation products. Am. J. Clin. Nutr. 57: S779–S786.

Federico, A., F. Morgillo, C. Tuccillo, F. Ciardiello and C. Loguercio. 2007. Chronic inflammation and oxidative stress in human carcinogenesis. Int. J. Cancer. 121: 2381–2386.

Fernandes, C., A. Fontainhas-Fernandes, M. Ferreira and M.A. Salgado. 2008. Oxidative stress response in gill and liver of Liza saliens, from the Esmoriz-Paramos coastal lagoon, Portugal. Arch. Environ. Contam. Toxicol. 55: 262–269.

Ferreira, D., A.C.d. Motta, L.C. Kreutz, C. Toni, V.L. Loro and L.J.G. Barcellos. 2010. Assessment of oxidative stress in Rhamdia quelen exposed to agrichemicals. Chemosphere. 79: 914–921.

Filipak Neto, F., S.M. Zanata, H.C. Silva de Assis, D. Bussolaro, M.V.M. Ferraro, M.A.F. Randi, J.R.M. Alves Costa, M.M. Cestari, H. Roche and C.A. Oliveira Ribeiro. 2007. Use of hepatocytes from Hoplias malabaricus to characterize the toxicity of a complex mixture of lipophilic halogenated compounds. Toxicol. In vitro. 21: 706–715.

Filipak Neto, F., S.M. Zanata, H.C. Silva de Assis, L.S. Nakao, M.A.F. Randi and C.A. Oliveira Ribeiro. 2008. Toxic effects of DDT and methyl mercury on the hepatocytes from Hoplias malabaricus. Toxicol. In vitro. 22: 1705–1713.

Fournier, D., J.M. Bride, M. Poirie, J.B. Berge and F.W. Plapp. 1992. Insect glutathione S-transferases—biochemical characteristics of the major forms from houseflies susceptible and resistant to insecticides. J. Biol. Chem. 267: 1840–1845.

Franco, R., O. Schoneveld, A.G. Georgakilas and M.I. Panayiotidis. 2008. Oxidative stress, DNA methylation and carcinogenesis. Cancer Lett. 266: 6–11.

Freitas, D.R.J., R.M. Rosa, J. Moraes, E. Campos, C. Logullo, I.D. Vaz and A. Masuda. 2007. Relationship between glutathione S-transferase, catalase, oxygen consumption, lipid peroxidation and oxidative stress in eggs and larvae of boophilus microplus (Acarina: Ixodidae). Comp. Biochem. Phys. A 146: 688–694.

Furne, M., M. Garcia-Gallego, M.C. Hidalgo, A.E. Morales, A. Domezain, J. Domezain and A. Sanz. 2009. Oxidative stress parameters during starvation and refeeding periods in Adriatic sturgeon (Acipenser naccarii) and rainbow trout (Oncorhynchus mykiss). Aquac. Nutr. 15: 587–595.

Garcia Sampaio, F., C. De Lima Boijink, E. Tie Oba, L. Romagueira Bichara dos Santos, A. Lúcia Kalinin and F. Tadeu Rantin. 2008. Antioxidant defenses and biochemical changes in pacu (Piaractus mesopotamicus) in response to single and combined copper and hypoxia exposure. Comp. Biochem. Physiol. Part C Toxicol. Pharmcol. 147: 43–51.

Gaspar Ruas, C.B., C.d.S. Carvalho, H.S. Selistre de Araujo, E.L. Gaeta Espindola and M.N. Fernandes. 2008. Oxidative stress biomarkers of exposure in the blood of cichlid species from a metal-contaminated river. Ecotoxicol. Environ. Saf. 71: 86–93.

Glusczak, L., D.d.S. Miron, B.S. Moraes, R.R. Simões, M.R.C. Schetinger, V.M. Morsch and V.L. Loro. 2007. Acute effects of glyphosate herbicide on metabolic and enzymatic parameters of silver catfish (*Rhamdia quelen*). Comp. Biochem. Physiol. Part C Toxicol. Pharmcol. 146: 519–524.

Grundy, J.E. and K.B. Storey. 1998. Antioxidant defenses and lipid peroxidation damage in estivating toads, Scaphiopus couchii. J. Comp. Physiol. B 168: 132–142.

Gutiérrez-Praena, D., A. Jos, S. Pichardo and A.M. Cameán. 2011. Oxidative stress responses in tilapia (*Oreochromis niloticus*) exposed to a single dose of pure cylindrospermopsin under laboratory conditions: Influence of exposure route and time of sacrifice. Aquat. Toxicol. 105: 100–106.

Halestrap, A.P., S.J. Clarke and I. Khaliulin. 2007. The role of mitochondria in protection of the heart by preconditioning. Biochimica Et Biophysica Acta-Bioenergetics 1767: 1007–1031.

Halliwell, B. and J.M.C. Gutteridge. 1991. Free radicals in biology and medicine, second edition: 540 pp. 1989.

Halliwell, B. and J.M.C. Gutteridge. 1999. Free Radicals in Biology and Medicine. 3th edn. Clarendon Press, Oxford.

Halliwell, B. and J.M.C. Gutteridge. 2007. Free radicals in biology and medicine. 4th edn. Oxford University Press, Oxford.

Hamilton, S.J. 2004. Review of selenium toxicity in the aquatic food chain. Sci. Total Environ. 326: 1–31.

Hansen, J.M., Y.M. Go and D.P. Jones. 2006. Nuclear and mitochondrial compartmentation of oxidative stress and redox signaling. Annu. Rev. Pharmacol. Toxicol. 46: 215–234.

Hao, L. and L. Chen. 2012. Oxidative stress responses in different organs of carp (*Cyprinus carpio*) with exposure to ZnO nanoparticles. Ecotoxicol. Environ. Saf. 80: 103–110.

Harman, D. 2001. Aging: Overview. Healthy aging for functional longevity 928: 1–21.

Hartley, D.P., D.J. Kroll and D.R. Petersen. 1997. Prooxidant-initiated lipid peroxidation in isolated rat hepatocytes: Detection of 4-hydroxynonenal- and malondialdehyde-protein adducts. Chem. Res. Toxicol. 10: 895–905.

Hasspieler, B.M., J.V. Behar and R.T. Digiulio. 1994. Glutathione-dependent defense in channel catfish (*Ictalurus punctatus*) and brown bullhead (*Ameiurus nebulosus*). Ecotoxicol. Environ. Saf. 28: 82–90.

Helmy, M.H., S.S. Ismail, H. Fayed and E.A. El-Bassiouni. 2000. Effect of selenium supplementation on the activities of glutathione metabolizing enzymes in human hepatoma Hep G2 cell line. Toxicology 144: 57–61.

Hermes-Lima, M. 2004. Oxygen in biology and biochemistry: role of free radicals. In: K.B. Storey (ed.). Functional Metabolism: Regulation and Adaptation. Wiley-Liss, Hoboken, USA, pp. 319–368.

Hermes-Lima, M. and K.B. Storey. 1993. In vitro oxidative inactivation of glutathione-s-transferase from a freeze-tolerant reptile. Mol. Cell. Biochem. 124: 149–158.

Hermes-Lima, M., W.G. Willmore and K.B. Storey. 1995. Quantification of lipid peroxidation in tissue extracts based on Fe(III)xylenol orange complex formation. Free Radic. Biol. Med. 19: 271–280.

Hiltunen, J.K., A.M. Mursula, H. Rottensteiner, R.K. Wierenga, A.J. Kastaniotis and A. Gurvitz. 2003. The biochemistry of peroxisomal beta-oxidation in the yeast Saccharomyces cerevisiae. FEMS Microbiol. Rev. 27: 35–64.

Hued, A.C. and M.D. Bistoni. 2005. Development and validation of a biotic index for evaluation of environmental quality in the central region of Argentina. Hydrobiologia. 543: 279–298.

Itoh, K., N. Wakabayashi, Y. Katoh, T. Ishii, K. Igarashi, J.D. Engel and M. Yamamoto. 1999. Keap1 represses nuclear activation of antioxidant responsive elements by Nrf2 through binding to the amino-terminal Neh2 domain. Genes Dev. 13: 76–86.

Kagan, V.E. 1988. Lipid peroxidation in biomembranes. CRC Press, Boca Raton, Fla.

Kelly, M.C., B. White and M.R. Smyth. 2008. Separation of oxidatively damaged DNA nucleobases and nucleosides on packed and monolith C18 columns by HPLC-UV-EC. J. Chromatogr. B: Anal. Technol. Biomed. Life Sci. 863: 181–186.

Kishi, S. 2004. Functional aging and gradual senescence in zebrafish. Strategies for engineered negligible senescence: Why Genuine Control of Aging May Be Foreseeable. 1019: 521–526.

Kishi, S., J. Uchiyama, A.M. Baughman, T. Goto, M.C. Lin and S.B. Tsai. 2003. The zebrafish as a vertebrate model of functional aging and very gradual senescence. Exp. Gerontol. 38: 777–786.

Kochhann, D., M.A. Pavanato, S.F. Llesuy, L.M. Correa, A.P. Konzen Riffel, V.L. Loro, M.F. Mesko, É.M.M. Flores, V.L. Dressler and B. Baldisserotto. 2009. Bioaccumulation and oxidative stress parameters in silver catfish (*Rhamdia quelen*) exposed to different thorium concentrations. Chemosphere. 77: 384–391.

Krivoruchko, A. and K.B. Storey. 2010. Forever young mechanisms of natural anoxia tolerance and potential links to longevity. Oxid. Med. Cell. Longev. 3: 186–198.

Kubrak, O.I., V.V. Husak, B.M. Rovenko, H. Poigner, M.A. Mazepa, M. Kriews, D. Abele and V.I. Lushchak. 2012. Tissue specificity in nickel uptake and induction of oxidative stress in kidney and spleen of goldfish *Carassius auratus*, exposed to waterborne nickel. Aquat. Toxicol. 118–119: 88–96.

Kubrak, O.I., O.V. Lushchak, J.V. Lushchak, I.M. Torous, J.M. Storey, K.B. Storey and V.I. Lushchak. 2010. Chromium effects on free radical processes in goldfish tissues: Comparison of Cr(III) and Cr(VI) exposures on oxidative stress markers, glutathione status and antioxidant enzymes. Comp. Biochem. Physiol. Part C Toxicol. Pharmcol. 152: 360–370.

Levine, R.L. 1983. Oxidative modification of glutamine-synthetase .1. Inactivation is due to loss of one histidine residue. J. Biol. Chem. 258: 1823–1827.

Levine, R.L., B.S. Berlett, J. Moskovitz, L. Mosoni and E.R. Stadtman 1999. Methionine residues may protect proteins from critical oxidative damage. Mech. Ageing Dev. 107: 323–332.

Levine, R.L., D. Garland, C.N. Oliver, A. Amici, I. Climent, A.G. Lenz, B.W. Ahn, S. Shaltiel and E.R. Stadtman. 1990. Determination of carbonyl content in oxidatively modified proteins. Methods Enzymol. 186: 464–478.

Levine, R.L., L. Mosoni, B.S. Berlett and E.R. Stadtman. 1996. Methionine residues as endogenous antioxidants in proteins. Proc. Natl. Acad. Sci. U.S.A. 93: 15036–15040.

Li, Z.-H., P. Li and T. Randak 2011. Evaluating the toxicity of environmental concentrations of waterborne chromium (VI) to a model teleost, *Oncorhynchus mykiss*: a comparative study of *in vivo* and *in vitro*. Comp. Biochem. Physiol. Part C Toxicol. Pharmcol. 153: 402–407.

Li, Z.-H., P. Li, M. Rodina and T. Randak. 2010. Effect of human pharmaceutical carbamazepine on the quality parameters and oxidative stress in common carp (*Cyprinus carpio* L.) spermatozoa. Chemosphere. 80: 530–534.

Liebel, S., C.A. Oliveira Ribeiro, R.C. Silva, W.A. Ramsdorf, M.M. Cestari, V.F. Magalhães, J.R.E. Garcia, B.M. Esquivel and F. Filipak Neto. 2011. Cellular responses of *Prochilodus lineatus* hepatocytes after cylindrospermopsin exposure. Toxicol. *In vitro*. 25: 1493–1500.

Livingstone, D.R. 2001. Contaminant-stimulated reactive oxygen species production and oxidative damage in aquatic organisms. Mar. Pollut. Bull. 42: 656–666.

Loro, V.L., M.B. Jorge, K.R. da Silva and C.M. Wood. 2012. Oxidative stress parameters and antioxidant response to sublethal waterborne zinc in a euryhaline teleost *Fundulus heteroclitus*: Protective effects of salinity. Aquat. Toxicol. 110: 187–193.

Lushchak, O.V., O.I. Kubrak, J.M. Storey, K.B. Storey and V.I. Lushchak. 2009a. Low toxic herbicide roundup induces mild oxidative stress in goldfish tissues. Chemosphere. 76: 932–937.

Lushchak, O.V., O.I. Kubrak, I.M. Torous, T.Y. Nazarchuk, K.B. Storey and V.I. Lushchak. 2009b. Trivalent chromium induces oxidative stress in goldfish brain. Chemosphere. 75: 56–62.

Lushchak, V.I. 2011. Environmentally induced oxidative stress in aquatic animals. Aquat. Toxicol. 101: 13–30.

Lushchak, V.I. and T.V. Bagnyukova. 2007. Hypoxia induces oxidative stress in tissues of a goby, the rotan *Perccottus glenii*. Comp. Biochem. Physiol. Part B: Biochem. Mol. Biol. 148: 390–397.

Lushchak, V.I., T.V. Bagnyukova, V.V. Husak, L.I. Luzhna, O.V. Lushchak and K.B. Storey. 2005a. Hyperoxia results in transient oxidative stress and an adaptive response by antioxidant enzymes in goldfish tissues. Int. J. Biochem. Cell Biol. 37: 1670–1680.

Lushchak, V.I., T.V. Bagnyukova, O.V. Lushchak, J.M. Storey and K.B. Storey. 2005b. Hypoxia and recovery perturb free radical processes and antioxidant potential in common carp (*Cyprinus carpio*) tissues. Int. J. Biochem. Cell Biol. 37: 1319–1330.

Malek, R.L., H. Sajadi, J. Abraham, M.A. Grundy and G.S. Gerhard. 2004. The effects of temperature reduction on gene expression and oxidative stress in skeletal muscle from adult zebrafish. Comp. Biochem. Physiol. Part C Toxicol. Pharmcol. 138: 363–373.

Malhotra, J.D. and R.J. Kaufman. 2007. Endoplasmic reticulum stress and oxidative stress: A vicious cycle or a double-edged sword? Antioxid. Redox Signal. 9: 2277–2293.

Marnett, L.J. 2000. Oxyradicals and DNA damage. Carcinogenesis 21: 361–370.

Mazon, A.F., D.T. Nolan, R.A.C. Lock, M.N. Fernandes and S.E.W. Bonga. 2004. A short-term *in vitro* gill culture system to study the effects of toxic (copper) and non-toxic (cortisol) stressors on the rainbow trout, *Oncorhynchus mykiss* (Walbaum). Toxicol. *In vitro*. 18: 691–701.

Miao, L. and D.K. St. Clair. 2009. Regulation of superoxide dismutase genes: Implications in disease. Free Radic. Biol. Med. 47: 344–356.

Mieiro, C.L., M.E. Pereira, A.C. Duarte and M. Pacheco. 2011. Brain as a critical target of mercury in environmentally exposed fish (*Dicentrarchus labrax*)—Bioaccumulation and oxidative stress profiles. Aquat. Toxicol. 103: 233–240.

Modesto, K.A. and C.B.R. Martinez. 2010. Effects of roundup ransorb on fish: Hematology, antioxidant defenses and acetylcholinesterase activity. Chemosphere. 81: 781–787.

Monserrat, J.M., P.E. Martínez, L.A. Geracitano, L. Lund Amado, C. Martinez Gaspar Martins, G. Lopes Leães Pinho, I. Soares Chaves, M. Ferreira-Cravo, J. Ventura-Lima and A. Bianchini. 2007. Pollution biomarkers in estuarine animals: Critical review and new perspectives. Comp. Biochem. Physiol. Part C Toxicol. Pharmcol. 146: 221–234.

Monteiro, D.A., J.A. de Almeida, F.T. Rantin and A.L. Kalinin. 2006. Oxidative stress biomarkers in the freshwater characid fish, *Brycon cephalus*, exposed to organophosphorus insecticide folisuper 600 (methyl parathion). Comp. Biochem. Physiol. Part C Toxicol. Pharmcol. 143: 141–149.

Monteiro, D.A., F.T. Rantin and A.L. Kalinin. 2009a. The effects of selenium on oxidative stress biomarkers in the freshwater characid fish matrinxã, *Brycon cephalus* exposed to organophosphate insecticide Folisuper 600 BR® (methyl parathion). Comp. Biochem. Physiol. Part C: Toxicol. Pharmacol. 149: 40–49.

Monteiro, S.M., N.M.S. Dos Santos, M. Calejo, A. Fontainhas-Fernandes and M. Sousa. 2009b. Copper toxicity in gills of the teleost fish, Oreochromis niloticus: Effects in apoptosis induction and cell proliferation. Aquat. Toxicol. 94: 219–228.

Moraes, B.S., V.L. Loro, A. Pretto, M.B. Da Fonseca, C. Menezes, E. Marchesan, G.B. Reimche and L.A. de Avila. 2009. Toxicological and metabolic parameters of the teleost fish (*Leporinus obtusidens*) in response to commercial herbicides containing clomazone and propanil. Pestic. Biochem. Physiol. 95: 57–62.

Mukai, K., Y. Kohno and K. Ishizu. 1988. Kinetic-study of the reaction between Vitamin-E radical and alkyl hydroperoxides in solution. Biochem. Biophys. Res. Commun. 155: 1046–1050.

Ognjanivic, B.I., J.G. Milovanovic, N.Z. Dordevic, S.D. Markovic, R.V. Zikic, A.S. Stajn and Z.S. Saicic. 2008. Parameters of oxidative stress in liver and white musile of hake (*Merluccius merluccius* L.) from the Adriatic Sea. Kragujevac J. Sci. 30: 137–144.

Olive, P.L. 1988. DNA precipitation assay: a rapid and simple method for detecting DNA damage in mammalian cells. Environ. Mol. Mutagen. 11: 487–495.

Olsen, T., L. Ellerbeck, T. Fisher, A. Callaghan and M. Crane. 2001. Variability in acetylcholinesterase and glutathione S-transferase activities in *Chironomus riparius* Meigen deployed *in situ* at uncontaminated field sites. Environ. Toxicol. Chem. 20: 1725–1732.

Oruç, E.Ö. and D. Usta. 2007. Evaluation of oxidative stress responses and neurotoxicity potential of diazinon in different tissues of *Cyprinus carpio*. Environ. Toxicol. Pharmacol. 23: 48–55.

Otto, D.M.E. and T.W. Moon. 1996. Endogenous antioxidant systems of two teleost fish, the rainbow trout and the black bullhead, and the effect of age. Fish Physiol. Biochem. 15: 349–358.

Pandey, S., S. Parvez, I. Sayeed, R. Haque, B. Bin-Hafeez and S. Raisuddin. 2003. Biomarkers of oxidative stress: a comparative study of river Yamuna fish *Wallago attu*. Sci. Total Environ. 309: 105–115.

Papa, S. and V.P. Skulachev. 1997. Reactive oxygen species, mitochondria, apoptosis and aging. Mol. Cell. Biochem. 174: 305–319.

Patnaik, B.K., N. Mahapatro and B.S. Jena. 1994. Aging in fishes. Gerontology 40: 113–132.

Paulino, M.G., N.E.S. Souza and M.N. Fernandes. 2012. Subchronic exposure to atrazine induces biochemical and histopathological changes in the gills of a Neotropical freshwater fish, *Prochilodus lineatus*. Ecotoxicol. Environ. Saf. 80: 6–13.

Perez-Jimenez, A., M.J. Guedes, A.E. Morales and A. Oliva-Teles. 2007. Metabolic responses to short starvation and refeeding in *Dicentrarchus labrax*. Effect of dietary composition. Aquaculture. 265: 325–335.

Perez-Matute, P., M.A. Zulet and J.A. Martinez. 2009. Reactive species and diabetes: counteracting oxidative stress to improve health. Curr. Opin. Pharmacol. 9: 771–779.

Pesce, S.F. and D.A. Wunderlin. 2000. Use of water quality indices to verify the impact of Cordoba City (Argentina) on Suquia River. Water Res. 34: 2915–2926.

Pierce, A., H. Mirzaei, F. Muller, E. De Waal, A.B. Taylor, S. Leonard, H. Van Remmen, F. Regnier, A. Richardson and A. Chaudhuri. 2008. GAPDH is conformationally and functionally altered in association with oxidative stress in mouse models of amyotrophic lateral sclerosis. J. Mol. Biol. 382: 1195–1210.

Pratico, D. 2008. Prostanoid and isoprostanoid pathways in atherogenesis. Atherosclerosis. 201: 8–16.

Radi, R., J.S. Beckman, K.M. Bush and B.A. Freeman. 1991. Peroxynitrite-induced membrane lipid-peroxidation—the cytotoxic potential of superoxide and nitric-oxide. Arch. Biochem. Biophys. 288: 481–487.

Regoli, F., S. Gorbi, G. Frenzilli, M. Nigro, I. Corsi, S. Focardi and G.W. Winston. 2002. Oxidative stress in ecotoxicology: from the analysis of individual antioxidants to a more integrated approach. Marine Environmental Research 54: 419–423.

Rhee, S.G., H.Z. Chae and K. Kim. 2005. Peroxiredoxins: A historical overview and speculative preview of novel mechanisms and emerging concepts in cell signaling. Free Radic. Biol. Med. 38: 1543–1552.

Richards, S.A., J. Muter, P. Ritchie, G. Lattanzi and C.J. Hutchison. 2011. The accumulation of un-repairable DNA damage in laminopathy progeria fibroblasts is caused by ROS generation and is prevented by treatment with N-acetyl cysteine. Hum. Mol. Genet. 20: 3997–4004.

Rios, F.S., A.L. Kalinin and F.T. Rantin. 2002. The effects of long-term food deprivation on respiration and haematology of the neotropical fish *Hoplias malabaricus*. J. Fish Biol. 61: 85–95.

Rotruck, J.T., A.L. Pope, H.E. Ganther, A.B. Swanson, D.G. Hafeman and W.G. Hoekstra. 1973. Selenium—Biochemical role as a component of glutathione peroxidase. Science 179: 588–590.

Sancho, E., M.D. Ferrando and E. Andreu. 1997. Inhibition of gill Na+, K+-ATPase activity in the eel, *Anguilla anguilla*, by fenitrothion. Ecotoxicol. Environ. Saf. 38: 132–136.

Serhan, C.N. 2010. Novel lipid mediators and resolution mechanisms in acute inflammation to resolve or not? Am. J. Pathol. 177: 1576–1591.

Sieprath, T., R. Darwiche and W.H. De Vos. 2012. Lamins as mediators of oxidative stress. Biochem. Biophys. Res. Commun. 421: 635–639.

Sies, H. 1999. Glutathione and its role in cellular functions. Free Radic. Biol. Med. 27: 916–921.

Stadtman, E.R. 1992. Protein oxidation and aging. Science 257: 1220–1224.

Stadtman, E.R. and R.L. Levine. 2000. Protein oxidation. Reactive Oxygen Species: From Radiation to Molecular Biology 899: 191–208.

Stadtman, E.R., C.N. Oliver and P.E. Starke-Reed. 1991. Implication of metal catalyzed oxidation of enzymes in aging, protein turnover, and oxygen toxicity. Korean J. Biochem. 23: 49–54.

Stara, A., J. Machova and J. Velisek. 2012. Effect of chronic exposure to simazine on oxidative stress and antioxidant response in common carp (*Cyprinus carpio* L.). Environ. Toxicol. Pharmacol. 33: 334–343.

Starke, P.E., C.N. Oliver and E.R. Stadtman. 1987. Modification of hepatic proteins in rats exposed to high oxygen concentration. FASEB J. 1: 36–39.

Stohs, S.J., D. Bagchi, E. Hassoun and M. Bagchi. 2001. Oxidative mechanisms in the toxicity of chromium and cadmium ions. J. Environ. Pathol. Toxicol. Oncol. 20: 77–88.

Storey, K.B. 2006. Reptile freeze tolerance: Metabolism and gene expression. Cryobiology 52: 1–16.

Sun, Y., Y. Yin, J. Zhang, H. Yu, X. Wang, J. Wu and Y. Xue. 2008. Hydroxyl radical generation and oxidative stress in *Carassius auratus* liver, exposed to pyrene. Ecotoxicol. Environ. Saf. 71: 446–453.

Svensson, E.P. 2008. Aquatic toxicology research focus. Nova Science Publishers, New York.

Szweda, L.I. and E.R. Stadtman. 1992. Iron-catalyzed oxidative modification of glucose-6-phosphate-dehydrogenase from Leuconostoc-Mesenteroides—Structural and functional-changes. J. Biol. Chem. 267: 3096–3100.

Thomaz, J.M., N.D. Martins, D.A. Monteiro, F.T. Rantin and A.L. Kalinin. 2009. Cardio-respiratory function and oxidative stress biomarkers in Nile tilapia exposed to the organophosphate insecticide trichlorfon (NEGUVON®). Ecotoxicol. Environ. Saf. 72: 1413–1424.

Toyoizumi, T., Y. Deguch, S. Masuda and N. Kinae. 2008. Genotoxicity and estrogenic activity of 3,3'-dinitrobisphenol a in goldfish. Biosci. Biotechnol. Biochem. 72: 2118–2123.

Tufekci, K.U., E. Civi Bayin, S. Genc and K. Genc. 2011. The Nrf2/ARE pathway: A promising target to counteract mitochondrial dysfunction in Parkinson's disease. Parkinsons Dis. 2011: 314082.

Valko, M., H. Morris and M.T.D. Cronin. 2005. Metals, toxicity and oxidative stress. Curr. Med. Chem. 12: 1161–1208.

van der Oost, R., J. Beyer and N.P.E. Vermeulen. 2003. Fish bioaccumulation and biomarkers in environmental risk assessment: a review. Environ. Toxicol. Pharmacol. 13: 57–149.

Van Dijk, P.L.M., I. Hardewig and F. Holker. 2005. Energy reserves during food deprivation and compensatory growth in juvenile roach: the importance of season and temperature. J. Fish Biol. 66: 167–181.

Verstraeten, S.V., L. Aimo and P.I. Oteiza. 2008. Aluminium and lead: molecular mechanisms of brain toxicity. Arch. Toxicol. 82: 789–802.

Viarengo, A., B. Burlando, N. Ceratto and I. Panfoli. 2000. Antioxidant role of metallothioneins: A comparative overview. Cell. Mol. Biol. 46: 407–417.

Vieira, M.C., R. Torronteras, F. Córdoba and A. Canalejo. 2012. Acute toxicity of manganese in goldfish *Carassius auratus* is associated with oxidative stress and organ specific antioxidant responses. Ecotoxicol. Environ. Saf. 78: 212–217.

Vinagre, C., D. Madeira, L. Narciso, H.N. Cabral and M. Diniz. 2012. Effect of temperature on oxidative stress in fish: Lipid peroxidation and catalase activity in the muscle of juvenile seabass, *Dicentrarchus labrax*. Ecological Indicators 23: 274–279.

Watanabe, T., V. Kiron and S. Satoh. 1997. Trace minerals in fish nutrition. Aquaculture. 151: 185–207.

White, R.E. 1991. The involvement of free-radicals in the mechanisms of monooxygenases. Pharmacol. Ther. 49: 21–42.

Widmer, R., I. Ziaja and T .Grune. 2006. Protein oxidation and degradation during aging: Role in skin aging and neurodegeneration. Free Radic. Res. 40: 1259–1268.

Wiegand, C., S. Pflugmacher, A. Oberemm and C. Steinberg. 2000. Activity development of selected detoxication enzymes during the ontogenesis of the zebrafish (*Danio rerio*). Int. Rev. Hydrobiol. 85: 413–422.

Wiegand, C., S. Pflugmacher, A. Oberemm, N. Meems, K.A. Beattie, C.E.W. Steinberg and G.A. Codd. 1999. Uptake and effects of microcystin-LR on detoxication enzymes of early life stages of the zebra fish (*Danio rerio*). Environ. Toxicol. 14: 89–95.

Wilhelm-Filho, D. and J.L. Marcon. 1996. Antioxidant defenses in fish of the Amazon. *In:* A.L. Val, V.M.F. Almeida-Val and D.J. Randall (eds.). Physiology and Biochemistry of the Fishes of the Amazon. INPA, Manaus, Brazil, pp. 299–312.

Williams, M.D., H. Van Remmen, C.C. Conrad, T.T. Huang, C.J. Epstein and A. Richardson. 1998. Increased oxidative damage is correlated to altered mitochondrial function in heterozygous manganese superoxide dismutase knockout mice. J. Biol. Chem. 273: 28510–28515.

Winston, G.W. and R.T. Digiulio. 1991. Prooxidant and antioxidant mechanisms in aquatic organisms. Aquat. Toxicol. 19: 137–161.

Wood, Z.A., E. Schroder, J.R. Harris and L.B. Poole. 2003. Structure, mechanism and regulation of peroxiredoxins. Trends Biochem. Sci. 28: 32–40.

Worthington, D.J. and M.A. Rosemeyer. 1976. Glutathione reductase from human erythrocytes. Catalytic properties and aggregation. Eur. J. Biochem. 67: 231–238.

Wu, L.L., C.C. Chiou, P.Y. Chang and J.T. Wu. 2004. Urinary 8-OHdG: a marker of oxidative stress to DNA and a risk factor for cancer, atherosclerosis and diabetics. Clin. Chim. Acta. 339: 1–9.

Xing, H., S. Li, Z. Wang, X. Gao, S. Xu and X. Wang. 2012. Oxidative stress response and histopathological changes due to atrazine and chlorpyrifos exposure in common carp. Pestic. Biochem. Physiol. 103: 74–80.

Zhang, J., H. Shen, X. Wang, J. Wu and Y. Xue. 2004. Effects of chronic exposure of 2,4-dichlorophenol on the antioxidant system in liver of freshwater fish *Carassius auratus*. Chemosphere. 55: 167–174.

CHAPTER 6

Genotoxicity and Mutagenicity

Marta Margarete Cestari

Introduction

Environmental damages or disturbances to genetic material are studied under the scope of several areas, such as toxicology, molecular biology, biochemistry, and genetics. These chemical or physical injuries on the DNA structure can promote changes and damages that are evaluated by genotoxic studies. Genotoxicity is the harmful action of a xenobiotic agent (chemical or physical) on DNA resulting in genetic mutation. This harm may result in tumors (somatic cells) or be transmitted from generation to generation (germline cells), affecting future generations developing diseases or malformations.

The results of damaging activities of these chemical agents range from single and double-strand breaks, cross-linking between DNA bases or between DNA bases and proteins, and formation of DNA adducts when a chemical compound is covalently linked to DNA (Preston and Hoffmann 2001). Contaminants may give rise to chain breaks in four different ways —direct binding of genotoxic compounds; through the effects of oxygen radicals; as a result of interaction with reactive metabolites; and as a consequence of enzymatic action of excision repair mechanism (Eastman and Barry 1992; Speit and Hartmann 1999). The damages described above

Universidade Federal do Paraná - Departamento de Genética, Centro Politécnico—Jardim das Américas - Cx Postal 19.071, CEP 81.531-990 Curitiba (Brazil).
Email: margaces@ufpr.br

are the precursors of DNA mutations, and may occur as point mutations in genes or chromosome mutations; both can affect somatic cells or germline cells. Gene mutations are the result of change on one or few base pairs, by replacement, addition, or deletion. Meanwhile, chromosomal mutations are classified by numerical and structural changes (Preston and Hoffmann 2001).

Mutations are also beneficial when they promote an increase in variability in individuals of a population, providing conditions to respond to environmental changes and allowing greater survival. These mutations are responsible for the evolution of species, but for our purposes this word is always linked to the idea that any change in genetic material is deleterious. Mutations and cancer are closely linked because both represent changes in permanent cells and can be inherited by daughter cells. For this reason, mutagenicity tests such as Ames test, genotoxicity test in fungi, etc. are recommended for pre-selection of chemicals to be evaluated for carcinogenic potential. Assays with compounds with mutagenic properties, therefore —potentially carcinogenic—can be performed with microorganisms and extrapolated to higher organisms. However, this extrapolation is frequently flawed because of the structural and metabolic simplicity of these microorganisms. Thus, the *in vivo* and *in vitro* tests performed with plants and vertebrates, capable of detecting gene mutations and chromosomal alterations, are more relevant (Rabello-Gay et al. 1991).

Assays with mutagenic compounds present in water or sediment could be performed under laboratory conditions with various types of animals, e.g., amphibians, mollusks, and fish, that can be used as biomarkers of water pollution (Minissi et al. 1996). To this end, the most suitable are fish as well as mammals. They undergo bioaccumulation and respond to low concentrations of mutagenic agents; they also activate, in response to pollutants, the cytochrome P450 system (diverse group of enzymes responsible for the oxidative metabolism of lipophilic compounds of endogenous or exogenous origin like steroids or environmental pollutants) (Goksoyr et al. 1991).

Among vertebrates or plants chosen for tests of mutagenicity or genotoxicity, the use of endemic organisms as environmental sentinels will always be preferable to the use of exotic species. In the Laboratory of Animal Cytogenetics and Environmental Mutagenesis (LCAMA in Portuguese) of UFPR, endemic Brazilian species are preferentially used (*Hoplias malabaricus, H. intermedius, Rhamdia quelen, Geophagus brasiliensis, Corydoras paleatus, Astyanax fasciatus, A. altiparanae, A. bimaculatus, A. serratus*), and the few occasions that exotic species are used (*Ciprinus carpio, Ctenopharyngodon idella, Tilapia rendali, Danio rerio*) are to compare the responses obtained to the same xenobiotics used in Brazilian fish.

Assays with living animals allow the compounds to be tested using several routes of administration, including oral, through food, water or by injecting directly into the animal. The effects on specific tissues or organs, in somatic or germline cells, as well as differences between acute and chronic contamination can be evaluated. These comparisons might provide a wide range of results that replicate in the best possible conditions, the human exposure to these compounds. It should be emphasized that due to the variety of possible effects of a xenobiotic, a single test is not enough to evaluate its action on a living being. One should consider that the effects may be restricted to the experimental species as well as to the synergistic or antagonistic effects to the compounds that may be present in the environment (Rabello-Gay et al. 1991).

For all these reasons, in experiments conducted at LCAMA, living specimens of endemic Brazilian species are used to analyze several tissues (blood, liver, kidney, brain, and gills) and different biomarkers (chromosome aberrations, piscine micronucleus, comet assay, DNA diffusion, frequency of normo- and polychromatic erythrocytes, and comet-FISH). Furthermore, on the bioassays already performed or those that are underway in LCAMA, different routes of administration of xenobiotics are tested as: a) trophic with *Hoplias malabaricus, Rhamdia quelen* e *Hoplias intermedius*; b) hydric with *Rhamdia quelen, Geophagus brasiliensis, Corydoras paleatus, Astyanax fasciatus, A. altiparanae, A. bimaculatus, A. serratus, Ciprinus carpio, Ctenopharyngodon idella, Tilapia rendali, Danio rerio*; c) by intraperitoneal injection with *Hoplias malabaricus, Rhamdia quelen* e *Astyanax serratus*.

Chromosomal Aberrations

Chromosomal aberrations are analyzed using fixed dividing cells at the metaphase stage. Ordinarily used are bone marrow cells of mice or rats, and the human lymphocyte cells. In fish, metaphase cells are obtained from kidney tissue, specifically from the anterior kidney of most species; less frequently, cells from the posterior portion are used. Although this test is commonly used in genotoxicity studies, Keshava et al. (1995) emphasize that this assay requires great experience and expertise in the material preparation to obtain high quality metaphase to a correct analysis.

In studies with human lymphocyte culture or of cell lines (human and mammals) it is less difficult to obtain metaphase cells because the culture conditions and protocols are well established in the laboratories. However, when applying the methodology to obtain *in vivo* metaphase cells, the strenuosness of obtaining cells for chromosomal analysis in suitable number and quality is far more cumbersome. This difficulty presents itself for various reasons such as: a) time of colchicine exposure varies from one fish species to another; b) stressed fish presents less kidney cells in cell division, and therefore, researchers in the field of cytogenetics should not process the material (fish) upon its arrival in the laboratory, but applying a mitogenic agent every 24 hours for two days in order to elicit an immunological response, and consequently obtain more cell divisions in the hematopoietic tissue.

When there is a need or wish to obtain metaphase cells from fish in bioassays or biomonitoring studies, colchicine or a mitogenic agent cannot be applied to them. The solution is to apply the short-term tissue culture methodology developed by Fenocchio et al. (1991) and modified by Cestari et al. (2004) and Ferraro et al. (2004). In this methodology, the anterior region of the kidney is removed and disaggregated into culture medium with colchicine. It is not injected directly into the fish, thus preventing further stress in the animal.

As a general rule, the chromosomal set of organisms, tissues or cells is fixed based on the shape and number of chromosomes. Changes in the shape and number thereof may occur spontaneously or artificially. When changes are observed and affect chromosomes, they are named chromosomal aberrations or may also be defined as chromosomal mutations (Lacadena 1996). Mutations are classified as genic, chromosomal numerical disorders (aneuploidy and euploidy), and chromosomal structural disorders (deletions, duplications, translocations, peri and paracentric inversions, breaks, gaps, rings, and breakdowns).

Gene mutations and chromosomal abnormalities are caused by many of the chemical compounds released in the environment. Some of these compounds (benzene, organochlorine pesticides, and metals as mercury) have aneugenic properties that result in altered chromosomal distribution during the cell division stage giving rise to aneuploidy. Aneuploidy in fish is hard to detect due to the large karyotipic variability within the same species. *Hoplias malabaricus* (trahira) and *Rhamdia quelen* (jundiá) are among the Brazilian endemic species that present different karyotypes and are used as bioindicators.

The trahira is considered from the cytogenetic point of view as a "complex of species" because it has seven distinct cytotypes (A, B, C, D, E, F, and G) (Bertollo 2007), without producing natural hybrids. *H. malabaricus* presents different diploid numbers because it has different types of sexual systems. The cytotypes A (2n=42), C (2n=40), E (2n=42), and F (2n=40) did not present heteromorphic sex chromosome but, differences in chromosomes showing metacentric, submetacentric, subtelocentric, and acrocentric shapes; cytotypes B ($\female$-2n=40,XX and $\male$-2n=40,XY), D ($\female$-2n=40,$X_1X_1X_2X_2$ and $\male$-2n=39,X_1X_2Y), and G ($\female$-2n=40,XX and $\male$-2n=41,XY_1Y_2), presented simple and multiple systems. Within the multiple system, there is a difference in diploid number between males and females (cytotypes D and G), what could be wrongly taken as a likely aneuploidy caused by a xenobiotic.

Since 1999, on LCAMA the species *H. malabaricus* has been studied cytogenetically and used to screen genotoxicity of xenobiotic agents in trophic bioassays (Ferraro 2003; Cestari et al. 2004; Ferraro et al. 2004; Lopez-Poleza 2004; Vicari et al. 2012) (Fig. 6.1), bioassays by intraperitonial injections (Ramsdorf et al. 2009), with cell cultures of hepatocytes (Filipak Neto et al. 2007), and brain cells (Silva et al. 2011). In order to establish the karyotype of the sample that will be used in bioassays, it is always necessary to assemble several karyograms of the control group specimens. Therefore, sample size recommended should not be less than 10 specimens, males and females in approximate number. This sample size number for both sexes is not always available—the batch of trahiras for bioassays must be offspring from the same pair of male and female; or from the batch that arrives in the laboratory, a reasonable number of specimens (♀ and ♂) should be karyotyped. For biomonitoring, all fish must have their karyotypes established and if more than one karyotype with numerical abnormalities occurs in the sample, manifested as aneuploidy, the test of chromosomal aberrations should not be used or the batch of fish should be excluded from the bioassay.

Figure 6.1. Specimen of *Hoplias malabaricus* preying on a freshly contaminated banded tetra (*Astyanax fasciatus*). Subchronic trophic contamination. Source: Tatiane Klingelfus (PhD student at PPGGEN-UFPR).

The species *Rhamdia quelen* (jundiá) has 49 synonyms (Silfvergrip 1996) and similar to trahira, is widely distributed throughout Brazil. This species has been intensively produced in fish farms because it has a good acceptance in the consumer market, high productivity in farm ponds, and shows high potential for commercialization. The jundiá also has a numerical karyotypic variability, but different from trahira, does not present heteromorphic sex chromosome. The karyotype variability with 2n ranging from 56 to 63 (Swarça et al. 2007) is most often caused by supernumerary chromosomes (Bs), and the small size of those chromosomes (less than 10 μm) (Hochberg and Herdtmann 1988; Fenocchio and Bertollo 1990, among others) are partially responsible for limitations in the studies of chromosomal aberrations in this species.

Currently, the species *Rhamdia quelen* is being much used in bioassays to test the genotoxicity of different xenobiotic agents. More frequently, the bioassays are by hydric exposure (Ferraro 2009; Piancini 2011; Ramsdorf 2011; Ghisi et al. 2011; Pamplona et al. 2011) seldom trophic (Costa 2011) (Fig. 6.2); more recently intraperitoneal injections have been used with nanoparticles of titanium dioxide (Master's degree of Tatiane Klingelfus, in progress). This demonstrates that the jundiá is being used in several routes of exposure in laboratory experiments, as well as being a target species in biomonitoring (Salvagni et al. 2011).

Figure 6.2. Specimen of *Rhamdia quelen* being fed through a diet gripper. Source: Paula Moiana da Costa—Doctoral work (2011)- PPGGEN-UFPR.

Some xenobiotics, referred to as clastogenic (metals, pesticides, petroleum products, among others), induce breaks and result in changes of chromosomal structure that can be detected by cytogenetic studies. These tests are indispensable to detect if mutagenic effects of a given compound are clastogenic or not. Overall, the structural chromosomal aberrations affect the sequence of genes in the chromosomes involving the gain or loss of different amounts of genetic material.

Al-Sabit (1991) described some structural chromosome aberrations which can be detected in chromosomes that were subjected to chemical and/or physical agents, stressing that ionizing radiations were those that resulted in more abnormalities. This book outlines some types of chromosomal aberrations such as: formation of isochromosomes; breaks of two chromatids of the same chromosome with the subsequent production of an acentric fragment; breaks of two chromatids of the same chromosome, one fragment joining the break point of the other chromatid, thus resulting in sister chromatids of different sizes; formation of ring chromosomes with or without centromere and the subsequent production of an acentric fragment; breaks near the telomeres of two nonhomologous chromosomes and subsequent fusion of them and formation of a dicentric chromosome and two acentric fragments; Robertsonian translocation; reciprocal translocation; pericentric and paracentric inversions; gap in one chromatid; break in one chromatid forming a simple fragment.

Among the chromosomal aberrations described by Al-Sabit (1991), Ale et al. (2004) found several types (isochromosome, break, ring, dicentric, gap, and some others) in specimens of *Oreochromis niloticus* following exposure to hydric conditions to $Pb(NO_3)_2$ at the concentrations of 100 and 300 µg Pb/L, for 96 hours. A significant increase of chromosomal aberrations number in the treated groups compared to the negative control.

Ferraro et al. (2004) exposed *Hoplias malabaricus* (2n=42) to subchronic trophic contamination observing chromosomal aberrations of only two types: breaks and gaps, similar to those described by Al-Sabit (1991). Apart from these chromosomal aberrations, the authors found loose strands in some chromatids that could be chromatin and named them as decondensations. Bioassays conducted by Ferraro et al. (2004) using subchronic trophic exposure (75 days—13 feeding cycles), at the concentrations of 0.3 µg/g body weight of Tributyltin (TBT) and of 21 µg/g body weight of inorganic lead (PbII), observed in both xenobiotics a significant increase of chromosomal aberrations in the treatment groups compared to control group. Additional methods, such as piscine micronucleus and comet assay were performed by the authors.

Another work with trophic contamination in *Hoplias malabaricus* (2n=42) was carried out by Cestari et al. (2004) that used two distinct periods referred as group A and group B. Group A was fed for 18 days (4 feeding cycles) and group B for 41 days (8 feeding cycles) with 21 µg/g body weight of $Pb(NO_3)_2$. Biomarkers, comet assay and chromosomal aberrations, were analyzed for group A and B. Statistical analysis showed that exposed groups (A and B) presented significant number of chromosomal aberrations when compared to control group, and group B presented significantly more chromosomal aberrations than group A. The aberrations identified in these groups were: gaps and breaks of single chromatid, decondensation of chromatin, acentric fragments, and pericentric inversions (Fig. 6.3). Also performed in was comet assay in erythrocytes.

Cestari et al. (2004) and Ferraro et al. (2004) identified one chromatin decondensation that was referred as a chromosomal aberration; at that time it had not been described in any previous work with fish. In addition to these works, this chromosomal aberration was found in *Rhamdia quelen* exposed to subchronic trophic contamination with aluminum sulfate (Costa 2011 Unpublisched data).

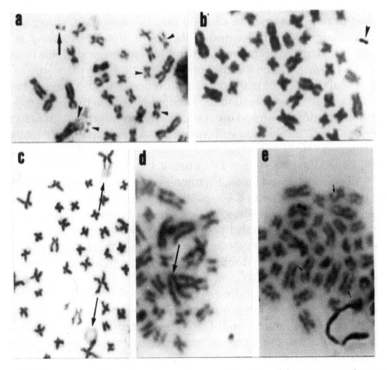

Figure 6.3. Chromosomal aberrations in specimens of *Hoplias malabaricus* exposed to trophic contamination at the concentration of 21 µg/g of body weight of $Pb(NO_3)_2$. a) arrow: acentric fragment and the chevron points to breads and chromatin decondensation; b) chevron: acentric fragments; c) arrows: chromatin decondensation; d) arrow: pericentric inversion; e) small arrows: breaks and gaps in one chromatid. Source: Priscila Maria Menel Lemos—(2009)—UFPR.

Ramsdorf et al. (2009) exposed *Hoplias malabaricus* (2n=40) to acute contamination (96 hours) by intraperitoneal injection of the $Pb(NO_3)_2$ at the concentrations of 7, 21, 63, and 100 µg/g body weight. There were no statistically significant results between treatments (7, 21, and 63 µg/g body weight) and control group, most likely due to the short exposure time, and low dosages. At the highest dosage (100 µg/g body weight) all fish died, demonstrating that at this concentration, intraperitoneal injection is lethal, although it has not been lethal by hydric exposure to fish *Oreochromis niloticus*. Intraperitoneal injection seems to be a more aggressive and direct form of exposure to a xenobiotic, or the endemic Brazilian species *Hoplias malabaricus* is more suceptible than the exotic species *Oreochromis niloticus*.

Eukaryotic chromosomes, exception of polytenes, under examination reveal a single and continuous structure of DNA associated with various proteins. Chromosome 1 of humans, the largest of all human chromosomes, has an approximate length of 7.5 cm. This molecule, in the interphase, is wrapped in fibrillar structure of 1,350 µm. In metaphase, chromosome 1 is approximately 10 µm. Because of its length, this DNA strand is more susceptible to damage of chemical, biological, and physical agents occurring in the ecosystems. A large number of damages that take place due to these agents reveal themselves in the onset of chromosome aberrations. The main types of damage that can lead to the emergence of structural chromosomal aberrations are double-strand breaks of DNA. Some agents such as ionizing radiation, specific antibiotics, and endonucleases are capable of causing this type of damage. However, while most genotoxic agents and UV radiation are not able to produce this type of DNA break, they result in some other DNA lesions, like single-strand break, thymine-thymine cyclobutane dimers, and some others, which, due to the repair and replication systems may result in double-strand breaks. Thus, double-strand breaks can arise from the accumulation of single-strand breaks in the DNA during the replication process, repair mechanisms, transpositions, mitotic recombination, and oxidative damage. Finally, double-strand breaks can lead to chromosomal aberrations (Obe et al. 2002).

Although the understanding of chromosomal structure is incomplete, several studies have pointed out that chromosomal abnormalities are a direct consequence of structural DNA damage. It is also recognized that chromosome losses, even partial ones, or aneuploidy arising from the problems of non-disjunction, are important factors in different types of cancer and in the aging process. These problems are related to defects in the spindle apparatus, centromere, or are connected to the level of chromosome condensation (Fenech 2000).

Aberrations cited here, such as breaks, deletions, ring and dicentric chromosomes, may result in acentric fragments that during telophase will not be in the daughter nuclei, but in a small nucleus named micronucleus. The same happens in the case of aneuploidy because the spindle fibers do not reach the kinetochores of one or more chromosomes and consequently, they will not be carried to the cell poles together with the other chromosomes to integrate in the daughter nuclei. These chromosomes that were outside of daughter nuclei also form micronuclei.

Piscine Micronucleus Test

Heddle (1973) and Schmid (1975) independently proposed a type of test that allowed the assessment of DNA damage by examining and quantifying the cytoplasmic structure known by hematologists as Howell-Joly corpuscle. They are found in populations of dividing cells. These structures received the name of micronucleus.

The principle of the test is based on the fact that during anaphase, the chromatids and acentric chromosomes fragments are not transported via spindle fibers to opposite poles, whereas fragments with centromeres are correctly transported. Upon telophase, chromosomes without damage are integrated into the nucleus of each daughter cell. Components that were not carried by the spindle can also be encompassed by the newly formed nuclei. However, some of these fragments, typically the very small ones, are not integrated into the newly formed nuclei, remaining in the cytoplasm making up structures identified as micronuclei (Schmid 1975).

Clastogenic or aneugenic events would give rise to micronuclei. The former refers to events that damage the chromosome directly or some of its components, especially DNA. The latter relates to events that damage the spindle apparatus and other components responsible for splitting the chromosomes (Albertini et al. 2000). Heddle et al. (1983) consider this test potentially sensitive to quantify the frequency of chromosomal abnormalities found in human chromosomes.

Micronuclei are considered those that are formed and are visibly separated from the main body of the cell having a size corresponding to 1/5 to 1/20 the size of the nucleus and yet not exceeding 1/3 the size of the main core. They must also possess distinguishable edges and with the same refringence of main nucleus. In the specific case of the fish, due to the usually small size of chromosomes, the size ratio ranges from 1/10 to 1/30 of nucleus size (Al-Sabti and Metcalfe 1995; Ayllon and Garcia-Vazquez 2000; Gustavino et al. 2001).

The process of micronuclei counting requires analysis between one and two thousand cells, all with cytoplasm and nuclear membranes undamaged, discarding the overlapping and damaged ones (Al-Sabti and Metcalfe 1995). Hooftman and De Raat (1982) using the micronucleus test originally developed by Schmid (1975) for bone marrow cells of mouse, adapted this test to blood cells of laboratory maintained fish. This modification of the original test became known as Piscine Micronucleus Test. Hooftman and De Raat (1982) have made observations on morphological changes in erythrocyte nuclei, usually elliptical, but presenting modified shapes in some cases.

The piscine micronucleus test has been extensively used to estimate the level of contamination exposure in many surveys since the 1980s. This test, measuring structural and numerical chromosomal damage, has been used to assess genotoxicity, and is thus a recommended indicator for environmental studies both in field and laboratory conditions (Belpaeme et al. 1998). The general consensus is that the counting of micronuclei during interphase is technically easier and faster when compared to the count of chromosomal aberrations during metaphase (Al-Sabti and Metcalfe 1995).

The piscine micronucleus test revealed great potential because it is quickly executed, inexpensive, and an excellent indicator of chemical contamination of fish (Hose et al. 1987). These authors, analyzing the frequency of micronucleus in circulating erythrocytes of two species of marine fish (*Genyonemus lineatus* and *Paralabrax clathratus*) collected from polluted and unpolluted regions of Southern California, found higher frequencies of micronuclei, four times higher in the former species and eleven times higher in the latter one. The authors describe several morphological changes in nuclei of circulating erythrocytes. The frequencies of nuclear morphological changes were significantly different between fish from the polluted area to fish from the unpolluted area; they argued that those morphological changes were indicators of genotoxic effects of chemicals present in the water. Furthermore, they considered that these changes should be included in the counts of micronuclei in the paper. They also concluded that the micronuclei test would be applicable to any fish species, regardless of its karyotypic features.

Carrasco et al. (1990), photographed and quantified the micronuclei, as well as the nuclear morphological changes found in their work. These changes were described and classified by the authors:

a) *Blebbed*: nuclei with a small evagination of the nuclear membrane, appearing to include euchromatin or heterochromatin (darker). The sizes of these evaginations are in the range of small protuberances to completely circumscribed structures, similar to micronuclei, but still connected to the main nucleus.

b) *Lobed*: nuclei with wider evaginations than those described for *Blebbed*. Its structure is not as clearly outlined as above. Some nuclei have several of these structures.

c) *Vacuoated*: nuclei that present a region resembling the vacuoles inside. These "vacuoles" are shown devoid of all visible material therein.

d) *Notched*: nuclei that have a well defined indentation in their surface. Generally with an appreciable depth in the nucleus. These grooves do not seem to possess any nuclear material and seem to be delimited by the nuclear envelope.

These abnormalities have been used by some authors as an indicator of cytogenetic damage in species of fish; however, Carrasco et al. (1990) found no significant association between changes in the morphology of the nucleus (including micronuclei) and chemical pollution levels in sediment or fish tissue revealing the shortcomings of the micronucleus test in fish. Its lack of sensitivity is probably due to the low and variable frequency of micronuclei in wild fish.

Ayllon and Garcia-Vazquez (2001) suggested that nuclear abnormalities should be included in the analysis of genotoxicity in fish based on the count of micronuclei, because the results would be more reliable and comprehensive. According to this study, the species *Oncorhynchus mykiss* presented different results to each of the tested compounds in the bioassay. Cyclophosphamide induced an increase in the frequency of micronuclei and morphological changes, mitomycin by itself induced an increase of frequency of nuclear changes, methyl methanesulfonate did not induce changes in the morphology or number of micronuclei, and the compounds N-ethyl-N-nitrosourea, acrylamide, and colchicine resulted in an increase in the frequency of micronuclei, but no morphological nuclear changes were observed.

Likewise, Grisolia and Cordeiro (2000) tested differences in responses between three species of fish (*Tilapia rendalli, Oreochromis niloticus,* and *Cyprinus carpio*) to four clastogenic compounds: bleomycin, cyclophosphamide, 5-fluorouacil, and mitomycin C and found that, in general, cyclophosphamide presented a higher clastogenic potential than the other compounds; the most susceptible species was *T. rendalli* and *C. carpio*, the most resistant.

Heddle et al. (1991) emphasize that micronuclei are short-term responses to a genotoxic compound, so that their expression depends on the intensity of exposure to pollution and probably is independent of exposure period. Minissi et al. (1996) also concluded that the micronucleus test in fish erythrocytes is representative of cytogenetic damage in the short-term exposure, since they were able to recover when the organisms were removed from the wild and kept in controlled laboratory conditions.

According to Al-Sabti and Metcalfe (1995) the highest induction of micronuclei usually occurs from one to five days of exposure, agreeing with the results of Grisolia and Cordeiro (2000), that observed the highest induction of micronucleus between two and seven days, post-treatment. According to Grisolia and Cordeiro (2000) the frequency of micronuclei decreased from the fourteenth day of exposure. A recent study by Pesenti (2012, unpublished data) showed that 30 days of depuration, *Astyanax fasciatus*, an endemic Brazilian species did not present decrease in nuclear morphological changes, demonstrating once again the susceptibility of this species.

The frequency of micronuclei within a cell population is highly dependent on the kinetics of cell proliferation. Such kinetics may vary according to the species of fish, tissue studied, and environmental changes, among other factors. Thus, it is not possible to establish an optimal time for micronucleus formation after exposure to genotoxic compounds without previous substantial work to standardize the testing procedures (Al-Sabti and Metcalfe 1995).

Like any laboratory technique, some factors should be considered in applying the micronucleus assay. The assay is not capable of detecting mitotic non-disjunction if it does not give rise to chromosome loss at anaphase; also, chromosome aberrations due to rearrangements, like translocations or inversion, would go undetected because they do not originate acentric fragments. In these cases, the test underestimates or lacks the sensitivity (Metcalfe 1989). In addition, the piscine micronucleus test (PMT) is performed on blood, more specifically with erythroid cells that have a mitotic index lower than hematopoietic tissue cell (kidney), resulting in very low counts of micronuclei. In several published studies with laboratory bioassays where fish contamination was controlled, PMT did not show enough sensitivity to establish the genotoxicity of xenobiotics (Ramsdorf et al. 2009; Erbe et al. 2011; Rocha et al. 2011; Ghisi et al. 2012; Vicari et al. 2012). Therefore, PMT alone is not enough to demonstrate whether a given xenobiotic is genotoxic or not. Thus, researchers have resorted to other experiments such as comet assay and chromosomal aberrations.

Besides all these caveats, there is a belief among researchers that PMT is a simple technique whose results could be analyzed by unpracticed students who have to make the micronuclei counts and differentiate them from technical artifacts, as stronger staining of cytoplasmic organelles and/ or dye precipitation (Giemsa) (Fig. 6.4). Experimental evidence accumulated since 1999, shows that in a universe of 2,000 erythrocytes of fish exposed to several xenobiotics, more than an average of two micronuclei is rarely found. For this reason, the nuclear abnormalities should be added to the few MN, as already done by the authors cited above. Figure 6.5 shows some nuclear anomalies found in experimental work of LCAMA of UFPR.

The use of micronucleus test *in vivo* is recommended to the members of European Economic Community for chemical analysis and quality control (Directive 92/69/EEC, Method B12). Unfortunately, Brazilian national legislation does not require such tests.

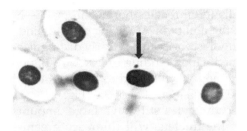

Figure 6.4. Erythrocytes of *Hoplias malabaricus* exposed to TBT. The arrow indicates a micronucleus. The other marks in the cytoplasm are not micronucleus. Source: Marcos Vinícius Ferraro 2003.

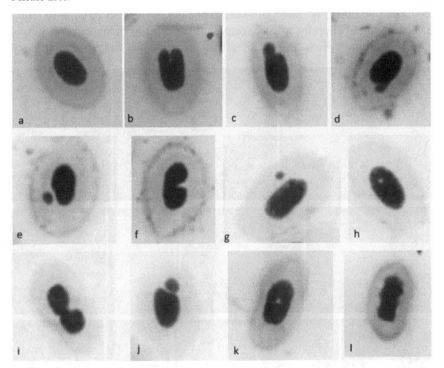

Figure 6.5. Several morphological alterations colored with Giemsa: a) normal; b and f) notched; c, d and j) blebbed; e and g) micronucleus; h and k) vacuolated; i) binucleous; l) lobed. Species *Astyanax fasciatus*. Source: Antonio E.L.M. Marques (Master's student at PPGGEN-UFPR).

Color image of this figure appears in the color plate section at the end of the book.

Erythrocyte Cells Stained with Acridine Orange

The micronucleus test with the use of acridine orange (AO) is a variation of normally executed piscine micronucleus test. The AO selectively stains young or immature erythrocytes (polychromatic erythrocytes—PCEs) unlike mature cells (erythrocytes normochromatic—NCEs). This option exists because of the ability of AO being able to bind to both DNA and RNA, but under fluorescence can be clearly distinguished with which one the AO is connected. Upon binding to DNA, AO fluoresces an intensely yellow-green color and when bound to RNA emits reddish color (Çavas and Gözükara 2005; McGahon et al. 1995) (Fig. 6.6).

The young erythrocytes still have large amounts of RNA in their cytoplasm—thus the nucleus will fluoresce greenish yellow and the cytoplasm reddish. However, mature erythrocytes will not present reddish color due to the absence of RNA in the cytoplasm (Çavas and Gözükara 2005).

The use of AO is especially useful because most fish have very small chromosomes. Thereby, any ocasional chromosomal break or even whole chromosomes, which are the origin of micronuclei may eventually go

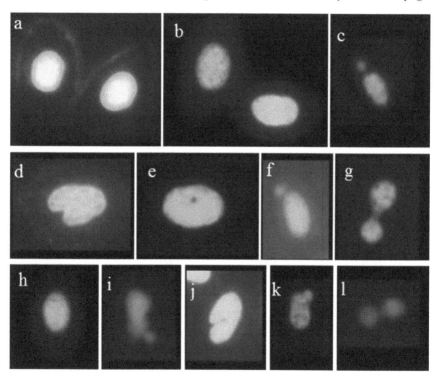

Figure 6.6. Mature erythrocytes (normochromatic), yellow greenish; imature erythrocytes (polychromatic) with reddish cytoplasm: a) erythrocytes normochromatics normals; b and h) erythrocytes polychromatics normals; c, f, i and k) blebbed; d) notched—erythrocyte polychromatic; e) vacuolated; g and l) binucleous; j) notched. Species *Astyanax fasciatus*. Source: Emanuele Cristina Pesente (Master's student at PPGGEN-UFPR).

Color image of this figure appears in the color plate section at the end of the book.

unnoticed when using Giemsa staining (Ueda et al. 1992). In addition, as highlighted above, the overestimated occurrence of micronuclei when they are only artifacts will be minimized, since AO stains only DNA as yellowish greenish. However, Giemsa staining will give artifacts (Fig. 6.7).

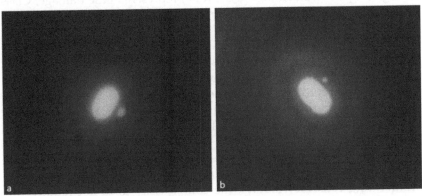

Figure 6.7. a) Normochromatic erythrocytes with one micronucleous; b) Normochromatic erythrocytes with one micronucleous. Species *Astyanax altiparanae*. Source: Galvan 2011 (Master in PPGGEN-UFPR).

Color image of this figure appears in the color plate section at the end of the book.

Comet Assay

The comet assay, also known as Single-Cell Gell Electrophoresis (SCGE) is a technique for detecting DNA damage in individual cells. Cells where DNA damage should be verified are embedded in low melting point (LMP) agarose and layered onto glass slides precoated with a thin layer of agarose. The slides are then subjected to a electrophoresis run (Speit and Hartmann 1999).

The DNA within the eukaryotic cells usually has several centimeters of length. Inside, the nucleus—between 5 and 10 µm of diameter—is arranged in several levels of compaction and condensation. Damage imposed on the DNA molecule causes a relaxation of this condensation and occasionally breaks the molecule (Rojas et al. 1999).

The principle of comet assay technique relies on the fact that the DNA in the cell that is not damaged will migrate together forming a head, and the damaged DNA will be constituted by fragments of various sizes. Smaller fragments tend to migrate more rapidly than larger ones. Under particular conditions that result in heavy damage in a cell, many chromosomal fragments of different sizes are formed and consequently, they will migrate at different speeds, forming a typical comet figure, head and tail (Olive et al. 1990).

There are several methodologies used to assess the extent of damage caused to DNA. One way to assess this damage is estimating the relationship between the radius of the head and the extension of the "tails" formed by DNA migration, where the damage 0 (zero) or Class 0 (zero) represents no damage to genetic material, Class 1 (one) represents little damage, Class 2 (two) average damage, Class 3 (three) great damage, and Class 4 (four) maximal damage (Fig. 6.8).

This assay is easy to perform, relatively inexpensive, with good reproducibility, fast execution, and sensitivity. In addition to the features mentioned, this test gives results from a few cells and data are obtained from individual cells and can be used to assess damage in cells from *in vivo* or *ex vivo* experiments (Monteith and Vanstone 1995; Sasaki et al. 1997; Rojas et al. 1999; Cestari et al. 2004; Ferraro et al. 2004; Filipak Neto et al. 2007, among others).

Unlike other types of testing as micronucleus (MN), chromosomal aberrations (CA), or sister chromatid exchanges (SCE) that require proliferating cells for viability, the comet assay does not require this condition and can be executed with virtually any cell type (Padrangi et al. 1995; Rojas et al. 1999).

Since genotoxic compounds very often are tissue specific, choosing comet assay has obvious advantages. As previously mentioned, this test does not require proliferating cells, for this reason the target tissue of the genotoxic compound can be analyzed directly and damaged cells quantified individually (Padrangi et al. 1995).

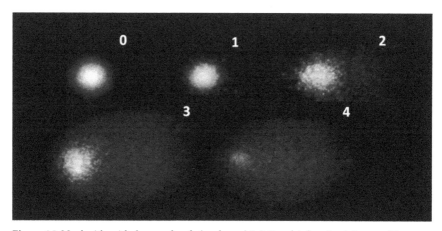

Figure 6.8. Nucleoids with damage levels (or classes) 1, 2, 3 and 4. Species *Astyanax altiparanae*. Source: Galvan 2011 (Master in PPGECO-UFPR).

Color image of this figure appears in the color plate section at the end of the book.

Rydberg and Johanson (1978) were the first to quantify the damage inflicted to DNA in the single cell. A cell suspension was embedded in agarose and it was spread onto a slide. This was placed in a lysis solution under mild alkaline conditions. They stained with acridine orange dye. Using a photometer they quantified the DNA damage. Double-strand breaks were shown in red color and single-strand breaks in green. The variation of tonalities between these two color indicated the amount of each break type.

Ostling and Johanson (1984) developed the gell electrophoresis on slide. It was named 'Comet Assay' due to the characteristic shape taken by the DNA during the migration process. In this technique, cells were embedded in low melting point agarose and layered onto a slide precoated with standard agarose. Then they were exposed to a lysis process using detergents and highly concentrated saline solutions. DNA would be released and then the slide would be submitted to electrophoresis in a neutral buffer. This technique had some limitations because it was able to detect only double-strand breaks in DNA and it could be mistaken with RNA present in the nucleus (Rojas et al. 1999).

Singh et al. (1988) aiming to verify DNA damage caused by X-ray and hydrogen peroxide (H_2O_2) have developed a variation of the above technique, using an electrophoresis buffer with a pH greater than 13. This change allowed detection of single-strand breaks of DNA, the labile alkali sites, and the delayed repair sites. This version is known as Single Cell Electrophoresis and for historical reasons, Comet Assay.

Olive et al. (1990) modified some parameters of Ostling and Johanson's (1984) technique, performing the cell lysis under alkaline conditions and electrophoresis in neutral pH or alkaline one (pH 12.3).

Although the techniques described by Olive et al. (1990) and Singh et al. (1988) are identical and similar in execution, the results obtained by the method developed by Singh et al. seems to be at least two times more sensitive to detect DNA damage (Rojas et al. 1999).

The DNA regions susceptible to alkylation process are more sensitive to degradation. These areas where depurination is increased become points of breaks in the DNA strand, therefore visible through the comet assay (Hank and Hock 1999).

The pH conditions of lysis solution and pH of the buffer during electrophoresis affect the relaxation of DNA molecule and influence the visualization of the damage, and consequently, the features of the resulting comet. The technique executed under pH close to neutral conditions (pH 7.5), comets have denser tails than those observed under alkaline conditions (pH greater than 10) which have dispersed tails. In alkaline conditions, those tails are shorter and more intensely stained, while in less alkaline conditions (pH between 10 and 11), they are longer and more faintly stained (Klaude et al. 1996).

Comet assay performed under alkaline conditions allows the detection of single-strand breaks of DNA, and non-denaturing conditions allow the detection of double-strand breaks of DNA (Olive et al. 1992; Olive and Banáth 1995). It is estimated that around two hundred breaks in the DNA strands in one cell is the sensitivity limit of the test. This sensitivity is higher than any other existing method for detecting DNA damage (Rojas et al. 1999). Olive et al. (1998) estimates that the sensitivity limit for single-strand breaks in diploid cells is around 50.

Single-strand breaks, detected by alkaline method (SCCGE), are the result of several reactions such as: nucleotide excision repair; direct excision of DNA structural proteins by chemical or physical agents; excision followed by intercalating chemicals in the DNA structure; action of endonuclease or topoisomerase among others (Horväthová et al. 1998; Mitchelmore and Chipman 1998).

The role of lysis in the comet assay is to remove cell contents, except nuclear material. The DNA remains well condensed due to the presence of a small amount of non-histone proteins. However, when placed in the electrophoresis solution (pH greater than 13), folded DNA begins to unwind from the break points, thus allowing them to be revealed by electrophoresis during the test (Yendle et al. 1997; Collins et al. 1997; Klaude et al. 1996).

Yendle et al. (1997) working with mouse keratinocytes and rodent hepatocytes observed that the time spent in the electrophoresis buffer affects the relaxation of DNA structure, with great influence in the analysis parameters of comet tails. Results of these researchers indicate that the agents affect only the proteins of DNA higher-order structure, and can increase the length of the tails, although not directly affecting the DNA. As a consequence, there would be two classes of agents that promote similar results in the comet assay, but they would have different implications as genotoxic compounds.

Collins et al. (1997) emphasize that DNA in the comet assay does not migrate as in the conventional electrophoresis, where the migration distance is inversely proportional to the fragment size. They also highlight that the distances between breaks detected by the test are in the order of 10^9 Da, beyond the detectable range of conventional electrophoresis. Detectable fragments in conventional electrophoresis are in the order or 1mm, whereas those in the comet tail are one hundredth of this value. The authors also emphasize that even after treatment with lysis and saline solution of electrophoresis, the nucleus and nucleoid retain their structure. This structure keeps the nucleosomes; therefore the formation of comet tail shape is the result of unwinding and loss of DNA higher-order structures due to damages inflicted by genotoxic compounds to this structure.

Comet assay requires individualized cells for analysis, therefore, it raises some practical limitations that must be considered. Separation and individualization of cells may require some grinding of the original tissues and use of trypsin, increasing the number of DNA breaks. Another consideration to be made is that increases in the number of breaks can be linked to stress (Mitchelmore and Chipman 1998).

In other words, when choosing the tissue or cells that will used for testing, these must be conveniently separated by means which do not damage them, but still allow their individualization. Blood cells may be diluted in fetal calf serum (Ramsdorf et al. 2009b) or physiological solution. Whichever means is being used, processing cells shall be performed under conditions to prevent any additional damage to DNA. Thus, cells should be handled in darkness to avoid damage to DNA by light.

Besides these facts, the mechanisms of DNA repair may be acting even before the analysis of the material has begun. Using cytosine-β-D-arabinofuranoside that inhibits the action of DNA polymerase as well as ligase, an increase of the breaks frequency in the test was detected (Mitchelmore and Chipman 1998).

It should be stressed that the DNA repair mechanisms in aquatic organisms and specifically in fish are slower than in mammalian cells (Espina and Weiss 1995).

Although the ideal conditions for increasing the sensitivity of the comet assay rely on the compound to be tested, the cells to be used, and the way of obtaining them, damage to DNA is detected even with short time exposure to genotoxic compounds.

Monteith and Vanstone (1995) detected DNA damage in rat hepatocytes exposed for a period of two hours to dimethylnitrosamine (DMNA) and then immediately analyzed.

Once single cells can be observed, it is possible, using this assay, to verify if all cells of a given population have the same amount of damage or, the mechanisms of DNA repair take place at the same rate. Thus, subpopulations of cells can be identified (Mitchelmore and Chipman 1998; Olive et al. 1990; Olive et al. 1992).

Compared with other genotoxicity assays, comet assay appears to be more sensitive. The sister chromatid exchange test (SCE) performed on ex-smokers showed discrepancies in the results obtained from smokers who had quit a few weeks before, but comet assay was more sensitive than the SCE in revealing the effects of smoking on these groups (Betti et al. 1995).

The comet assay originally developed to detect genotoxic effects in mammals, showed high efficiency in the identification of genotoxic compounds in aquatic organisms, besides the ample use in clinical studies of DNA repair in environmental and human monitoring.

Sasaki et al. (1997), using two marine snails with shells, *Patunopecten yessoensis* and *Tapes japonica*, demonstrated by the comet assay that these animals are excellent bioindicators of water pollution.

Using comet assay in fish to assess the effects of aflatoxina B1 (AFB1), Abd-Allah et al. (1999) demonstrated that this compound was responsible for extensive damage to DNA in different types of tissues analyzed in rainbow trout (*Oncorhyncus mykiss*). However, in the other species, catfish *Ictalurus punctatus*, no significant differences were found between treated and control fish.

Padrangi et al. (1995), using fish *Ameiurus nebulosus* and *Cyprinus carpio*, Silva et al. (2000) working with rodents and Navarrete et al. (1997) with a vegetable, *Allium cepa*, demonstrated the great efficiency of the comet assay in environmental monitoring.

In the specific case of comet assay performed with fish, blood proved to be a potentially useful tissue in biomonitoring. Firstly, it is obtained in a minimally invasive way, and secondly, its composition is ~97% of nucleated erythrocytes and only ~3% of leukocytes, what allows a considerable homogeneity in the sample (Mitchelmore and Chipman 1998; Theodorakis et al. 1994). Although leukocytes are more susceptible to damage by genotoxic compounds than erythrocytes, this susceptibility can be concealed by the large number of erythrocytes.

An additional advantage of using blood from fish is that minimal quantities are necessary for testing. The amount of 10 µl of whole blood diluted in 1ml of fetal calf serum, then using 10 µl of this dilution to mix with 120 µl of low melting point agarose is enough (Ramsdorf et al. 2009b). Using small amounts of blood allows the collection of multiple samples over time from the same fish, or to collect more than 10 µl of whole blood to perform swabs on a slide for piscine micronucleus analysis. Besides, from fish measuring over 15 cm more blood can be collected for additional hematological and biochemical analyses.

The tissues more frequently used in screening, in addition to blood tissue, are liver, because it is the main metabolism organ, gills due to their continuous contact with the aqueous phase, and kidney, because is a blood-producing organ in fish (Belpaeme et al. 1998). Since the comet assay analyzes individual cells, tissue breakdown has to be carefully performed. Cells may be split using gentle fragmentation processes or by enzymatic procedures. Cells may conveniently be separated by the least damaging method, while still allowing them to be individualized. Whichever methods are used, cell processing shall be performed without any additional damages being imposed on to DNA (Ferraro 2003). In the case of blood cells, as already explained above, these can be diluted in fetal calf serum or physiological solution. Ramsdorf et al. (2009b) solved the problem of blood storage, for further use in the comet assay, mainly in the context of field work, when there is no possibility of immediate execution, or when the bioassay has many specimens that need to be euthanized in the following days. They firmly established that fetal calf serum keeps the cells alive after 72 hours of collection, and DNA in the expected conditions for the comet assay when they are stored in darkness and refrigerated.

Regarding tissue cells of fish in comet assay, LCAMA is using them to compare the responses to xenobiotics in different compartments (tissues). Ramsdorf et al. (2009a) using *H. malabaricus*, have applied the comet assay technique in kidney cells collected after 96 hours following intraperitoneal injections at the concentrations of 7, 21, 64, and 100µg/g de $Pb(NO_3)_2$. The kidney tissue showed a significant difference between control and the doses of 21, 64, and 100µg/g de $Pb(NO_3)_2$; but the blood tissue was more sensitive than kidney, it presented significant difference between control and the lowest concentration of 7 µg/g de $Pb(NO_3)_2$. The authors concluded that these results occurred due to short exposure time and low concentrations of $Pb(NO_3)_2$.

Ghisi et al. (2011) applied the comet assay on gill cells of *Rhamdia quelen* following a long-term exposure of 60 days to Fipronil (0.05, 0.10, and 0.23 µg/L); they assumed the gills would be more sensitive because of their permanent contact with the aqueous contaminated environment, but using the comet assay technique or the hystopatology of gills there was no significant difference between the control and tested groups. In the same bioassay, Ramsdorf (2011) performed the comet assay in the blood observing differences between the control group and the concentrations of 0,10 and 0,23 µg/L, showed that fipronil under these concentrations and the same exposure period, blood suffered more damage than gill tissues. Fipronil at the concentrations tested was insufficient to result in damage to gills tissue, but the exposure time of 60 days was enough to present damage in blood. Erythrocytes have an average lifetime of 51 days in the bloodstream of fish (Tavares-Dias and Moraes 2004). Thus, virtually all red cells of jundiá had been replaced during the bioassay exposure time to the xenobiotic.

Benincá et al. (2012) peformed comet assay in blood and kidney tissues of fish in a biomonitoring study in the ponds of Santa Marta and Camacho (Santa Catarina State in Brazil). Damage observed in genetic material of erythrocytes was much higher than in kidney cells in all points and sampling times, even in the positive controls. These results once again show that kidney tissue is less susceptible to damage than the hemaotopoietic tissue under the same environmental conditions. In fish, kidney tissue has hematopoetic function, therefore, it may show some protection against xenobiotics that is not present in erythrocytes.

Costa (2011, unpublished data) in his recent doctoral work submitted *Rhamdia quelen* to subchronic trophic contamination to copper and aluminum sulfate applying the comet assay technique to liver, kidney, gill, brain, and blood tissues. The results pointed out that brain tissue was more susceptible to aluminum sulfate contamination than the others. It corroborates the scientific literature results that emphasize the neurotoxicity of aluminum sulfate for human beings. The liver and blood tissue also showed more DNA damage when compared to negative control, but less than the amount found in brain cells.

Frequently it is necessary to test several tissues and compare results because the information available in the scientific literature does not indicate the target tissues where the genotoxic compounds results in more DNA damage (Costa 2011; Galvan 2011, some additional works in progress at LCAMA). The way different tissues respond to the exposure to xenobiotics, and the technique reproducibility are important. At LCAMA a homogenizer Potter (1500 rpm, at lowest rotation) to break down the tissues is used. Assays should always run positive and negative controls to identify the problems related only to tissue breakdown.

DNA Diffusion Assay

The DNA diffusion assay is performed to check the amount of cells (nucleoids) that died, or in apoptosis or necrosis during exposure to xenobiotics, either in bioassays or when exposed to environmental impact. It is an important cytotoxic technique that should, whenever possible, be performed together with comet assay, because when the expected results of comet assay are lower than expected for a specific dose of xenobiotic, one that has already been tested in other species resulting in a much higher score. Often this low score is due to cell death and consequent scattering of DNA that may not have been seen in the analysis and therefore, not quantified. The lack of visualization occurs because the distinct "head" is missing and only a DNA "cloud" is visible, therefore this "cloud" does not count as damage because the "head" is absent. This cloud is also known as DNA ghost because in the absence of a concentrated DNA "head", they are not counted as damage when scoring the results of comet assay (Fig. 6.9). Some researchers read the comet assay slides for a second time, counting 100 nucleoids and identifying how many of them are DNA ghosts. This count must be done in all fish (control and contaminated ones). In the DNA diffusion test, apoptotic or necrotic cells that went through electrophoresis during the comet assay are visualized as ghosts because the DNA has been pulverized.

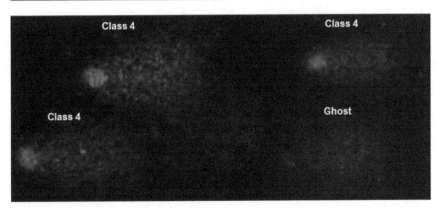

Figure 6.9. Nucleoides with damage levels (or class) 4 and one ghost. Species *Astyanax fasciatus*. (Ramsdorf 2011—Thesis PPGGEN-UFPR).

Color image of this figure appears in the color plate section at the end of the book.

The DNA diffusion assay is based on the comet assay but without the electrophoresis step. After the lysis step, the slides are placed in the electrophoresis buffer (pH>13.0) for 10 minutes. The duration of this step was set after several tests that lasted from five minutes up to one hour (Singh 2000; Díaz et al. 2009; Parolini et al. 2009). After this step, slides are neutralized and fixed.

This assay was first developed for human lymphocytes, seeking the quantification of apoptosis (Singh 2000). Apoptosis or programmed cell death in tissues of an organism is not associated with inflammation or scarring, unlike necrosis. Apoptosis is a normal event which occurs during and after development. It is an inevitable and important event in tissue remodeling during development and aging (Searle et al. 1982). This phenomenon occurs in cells damaged by given levels of toxic agents. It is also a crucial process for eliminating cancer cells (Guchelaar et al. 1997).

The nuclei of apoptotic cells are undefined, with no clear boundary due to diffusion of DNA in agarose, besides presenting a diameter three time greater than the average nuclear diameter. Nuclei of necrotic cells have a well-defined outer edge and a relatively homogeneous appearance (Singh 2000) (Fig. 6.10). The use of DNA diffusion assay to evaluate apoptosis was reviewed by Singh (2005) and its application in a low number of environmental studies have shown very promising results (Nigro et al. 2002; Frenzilli et al. 2004; Barga et al. 2006; Frenzilli et al. 2008; Rocco et al. 2010).

At LCAMA this technique has been employed to supplement the comet assay to assess nuclear abnormalities and the frequency of polychromatic erythrocytes. The works in progress are using the DNA diffusion assay in

Figure 6.10. Apoptotic and necrotic cells in the DNA diffusion test: a) normal nucleoid; b) apoptotic nucleoid; c) necrotic nucleoid. Species *Atherinella brasiliensis*. Source: Gustavo Souza Santos (PhD student at PPGECO-UFPR).

Color image of this figure appears in the color plate section at the end of the book.

fish erythrocytes of both freshwater and estuarine fish, demonstrating that is important to have more than one biomarker (cytogenetic or genotoxic) in environmental assessments as well as in bioassays.

Why is it Necessary to use More than One Genotoxic Biomarker?

As highligheted by Rabello-Gay et al. (1991) because of the wide range of possible effects of a xenobiotic, a single test is not enough to evaluate its effect on a living being; therefore several biomarkers should be used for biomonitoring and bioassays.

The first time that work was observed at LCAMA, the importance of using more than one genetic biomarker was realized during the work developed by Ferraro (2003) in her Master's thesis. In this work, the genotoxic effects of subchronic trophic contamination with inorganic lead (PbII) (21 µg Pb/g) and tributyltin (0.3 µg TBT/g de trahira) on bioindicator *Hoplias malabaricus* (trahira) was studied. Trahira presented the total score, mean, and median number of damages induced by TBT below the negative control, which made us question this xenobiotic genotoxicity (Table 6.1). The piscine micronucleus and chromosomal aberrations tests were performed, this time confirming the genotoxicity of TBT. Further analysis of comet assay slides made clear a high number of DNA ghosts that could not be added as damage because they lacked the "head" of a comet, resulting in mean and median numbers lower than negative control.

Generally in experiments and even in biomonitoring, the piscine micronucleus test proves to be less sensitive than comet assay to evidence genotoxicity. This has been demonstrated by several investigators (Metcalfe 1989; Ramsdorf et al. 2009; Erbe et al. 2011; Rocha et al. 2011; Ghisi et al. 2012; Vicari et al. 2012), who emphasized the difficulty of detecting in

Table 6.1. Total score of comet assay, means and medians found in trahira exposed to (PbII) and TBT.

Treatment	Total of Cells		Classes						
	Analysed	com comet	0	1	2	3	scores		
Control	100	16	84	16	0	0	16		
	100	6	94	5	0	1	8		
	100	23	77	22	1	0	24	mean	16,33
	100	2	98	2	0	0	2	median	16,00
	100	13	87	11	1	1	16		
	100	26	74	21	4	1	32		
Total	600	86	514	77	6	3			
Pb	100	17	81	16	0	3	25		
	100	66	34	50	16	0	82	mean	54,33
	100	58	42	48	8	2	70	median	62,00
	100	15	83	13	4	0	21		
	100	56	44	42	10	4	74		
	100	54	46	54	0	0	54		
Total	600	266	330	223	38	9			
TBT	100	5	95	5	0	0	5		
	100	0	100	0	0	0	0	mean	3,50
	100	1	99	1	0	0	1	median	4,00
	100	6	94	6	0	0	6		
	100	5	95	4	1	0	6		
	100	3	97	3	0	0	3		
Total	600	20	580	19	1	0			

Source: Ferraro 2003

this test, chromosomal aberrations such as translocations and para and pericentric inversions if they do not present the resulting fragments, thus underestimating the actual damage. Additional genetic biomarkers should be used as comet assay, to evidence several changes in DNA, as cited above, and also to demonstrate chromosomal aberrations which present chromosome changes that do not originate acentric fragments or chromosome loss during cell division.

Besides these techniques, FISH on chromosomes and comet-FISH, more sophisticated and specific cytogenetic techniques have been used recently at LCAMA to uncover changes in specific gene regions such as 18S, 5S, and telomeric.

Comparison of Genotoxicity between Endemic Brazilian and Exotic Species

In recent years, LCAMA has produced evidence of differences in responses between different species to the same xenobiotic. The studies are important when comparing exotic and endemic species under the same exposure conditions. Therefore, they were conducted comparing *Rhamdia quelen, Astyanas bimaculatus,* and *Cyprinus carpio* under exposure to hidric contamination to the herbicide Roundup and fungicide Folicur (Ferraro 2009, unpublished data). It was demonstrated that the endemic species (*Rhamdia quelen, Astyanas bimaculatus*) suffer more DNA damage than the exotic species under the same conditions of concentration and exposure time to xenobiotic (bioassays).

Recently Pesenti (2012, unpublished data) exposed *Astyanax fasciatus* (endemic) and *Ctenopharyngodon idella* (exotic) to hydric contamination of copper sulfate (0,039 μg/L) for 196 hours and two periods of decontamination. The endemic species presented more DNA damage than the exotic species using several genetic biomarkers. Therefore, these studies show that under the same experimental conditions and the species investigated, the endemic Brazilian species undergo more damage to their genetic material than the exotic species mentioned above. Thus fish present in the rivers of Brazil may suffer more DNA damage when exposed to xenobiotics than the exotic ones that have been and are being introduced in the aquatic ecosystems.

Acknowledgements

The author are grateful to the IBAMA (Instituto Brasileiro do Meio Ambiente) for authorizing the capture of specimens (IBAMA/MMA/ SISBIO license number: 0158/2006). This work was supported by CNPq (Conselho Nacional de Desenvolvimento Científico e Tecnológico), CAPES (Coordenação de Aperfeiçoamento de Pessoal de Nível Superior), Fundação Araucária (Fundacção Araucária de Apoio ao Desenvolvimento Científico e Tecnológico do Estado do Paraná) and also thank FURNAS—Empresa Eletrobrás fish donation to experiments. The author are specially grateful to the PhD Giovani Mocellin by revision of English.

Keywords: comet assay, chromosomal aberrations, piscine micronucleus, polychromatic erythrocytes, DNA diffusion assay, DNA cloud, Apoptosis, necrosis, endemic species

References

Abd-Allah, G.A., R. El-Fayoumi, J.M. Smith, R.A. Heckmann and K.L. O'Neil. 1999. A comparative evaluation of aflatoxin B_1 genotoxicity in fish models sing the omet assay. Mutat. Res. 446: 181–188.

Albertini, R.J., D. Anderson, G.R. Douglas, L. Hagmar, K. Hermminki, F. Merlo, A.T. Natarajan, H. Norppa, D.E.G. Shuker, R. Tice, M.D. Waters and A. Aitio. 2000. IPCS guidelines for the monitoring of genotoxic effects of carcinogen in human. Rev. Mutat. Res. 463: 111–172.

Ale, E., A.S. Fenocchio, M.C. Pastori, C. Oliveria Ribeiro and M.M. Cestari. 2004. Evaluation of the effects of $(NO_3)_2Pb$ on *Oreochromis niloticus* (Pisces, Cichlidae) by means of cytogenetic techniques. Cytologia 69(4): 453–458.

Al-Sabti, K. 1991. Handbook of Genotoxic Effects and Fish Chromosomes. Ljubljana, Yugoslavia.

Al-Sabti, K. and C.D. Metcalfe. 1995. Fish micronuclei for assessing genotoxicity in water. Mutat. Res. 343: 121–135.

Ayllon, F. and E. Garcia–Vazquez. 2000. Induction of micronuclei and other nuclear abnormalities in European minnow *Phoxinus phoxinus* and mollie *Poecilia latipinna*: an assessment of the fish micronucleus test. Mutation Research—Genetic Toxicology and Environmental Mutagenesis, Amsterdam 467: 177–186.

Barga, I. Del, G. Frenzilli and V. Scarcelli. 2006. Effects of algal extracts (*Polysiphonia fucoides*) on rainbow trout (*Oncorhynchus mykiss*): a biomarker approach. Mar. Environ. Res. 62 Suppl: 283–286.

Belpaeme, K., K. Cooreman and M. Kirsch-Volders. 1998. Development and validation of the *in vivo* alkaline comet assay for detecting genomic damage in marine flatfish. Mutat. Res. 415(3): 167–84.

Benincá, C., W. Ramsdorf, T. Vicari, C.A. Oliveria Ribeiro, M.I.M. Almeida, H.C. Silva de Assis and M.M. Cestari. 2012. Chronic genetic damages in *Geophagus brasiliensis* exposed to anthropic impact in Estuarine Lakes at Santa Catarina Coast–Southern of Brazil. Environ. Monit. Assess. 184: 2045–2056.

Bertollo, L.A.C. 2007. Chromosome evolution in the neotropical Erythrinidae fish family: An overview. In: E. Pisano, C. Ozouf-Costaz, F. Foresti and B.G. Kapoor (eds.). Fish Cytogenetics. Science Publishers, Enfield, USA, pp. 195–211.

Betti, C., T. Davini, N. Gianessi, N. Loprieno and R. Barale. 1995. Comparative studies by comet test and SCE analysis in human lymphocytes from 200 healthy subjects. Mutat. Res. 343: 201–207.

Carrasco, K.R., K.L. Tilbury and M.S. Myers. 1990. Assessment of the piscine micronucleus test as na *in situ* biological indicator of chemical contaminant effects. Can. J. Fish. Sci., Ottawa 47: 2123–2136.

Çavas, T. and S.E. Gözükara. 2005. Genotoxicity evaluation of metronidazole using the piscine micronucleus test by acridine orange fluorescent staining. Environ. Toxicol. Pharmacol. 19: 107–111.

Cestari, M.M., P.M.M. Lemos, C.A. Oliveira Ribeiro, J.R.M.A. Costa, M.V.M. Ferraro, M.S. Mantovani and A.S. Fenocchio. 2004. Genetic damage induced by trophic doses of lead in the neotropical fish *Hoplias malabaricus* (Characiformes, Erythrinidae) as revealed by the comet assay and chromosomal aberrations. Gen. Mol. Biol. 27(2): 270–274.

Collins, A.R., L.V. Dobson, M. Dusinská, G. Kennedy and R. Stétina. 1997. The comet assay: what can it really tell us? Mutat. Res. 375: 183–193.

Costa, P.M. 2011. Avaliação do efeito tóxico de sulfato de alumínio e sulfato de cobre em bioensaio de contaminação subcrônica via trófica no bioindicador *Rhamdia quelen* (Silurifome). M.S. Thesis, Universidade Federal do Paraná. Curitiba. Paraná. Brasil.

Diáz, A., S. Carro, L. Santiago, J. Estévez, C. Guevara, M. Blanco, L. Sánchez, N. López, D. Cruz, R. López, E.B. Cuetara and J.L. Fuentes. 2009. Estimates of DNA strand breakage in bottlenose dolphin (*Tursiops truncatus*) leukocytes measured with the comet and DNA diffusion assays. Gen. Mol. Biol. 32(2): 367–372.

Diretiva 92/69/CEE. Método B12. Disponível em < http:// europa.eu.int/eur-lex/ > Acesso em 01 de setembro de 2001.

Eastman, A. and M.A. Barry. 1992. The origins of DNA breaks: a consequence of DNA damage, DNA repair or apoptosis? Cancer Invest. 10: 229–240.

Erbe, M.C.L., W.A. Ramsdorf, T. Vicari and M.M. Cestari. 2011. Toxicity evaluation of water samples collected near a hospitalwaste landfill through bioassays of genotoxicity piscine micronucleus test and comet assay in fish *Astyanax* and ecotoxicity *Vibrio fischeri* and *Daphnia magna*. Ecotoxicology 20: 320–328.

Espina, N.G. and P. Weiss. 1995. DNA repair in fish from polluted estuaries. Mar. Environ. Res. 39: 309–312.

Fenech, M. 2000. The *in vitro* micronucleus tecnhique. Mutat. Res. 455: 81–95.

Fenocchio, A.S. and L.A.C. Bertollo. 1990. Supernumerary chromosomes in a *Rhamdia hilarii* population (Pisces, Pimelodidae). Genetica 81: 193–198.

Fenocchio, A.S., P.C. Venere, A.C.G. Cesar, A.L. Dias and L.A.C. Bertollo. 1991. Short term culture from solid tissues of fishes. Caryologia, Florence 44(2): 161–166.

Ferraro, M.V.M. 2003. Avaliação do efeito mutagênico do tributilestanho (TBT) e do chumbo inorgânico (PbII) em *Hoplias malabaricus* (Pisces) através dos ensaios: Cometa, Micronúcleo e de Aberrações Cromossômicas. Dissertation, Universidade Federal do Paraná, Curitiba, Paraná. Brasil.

Ferraro, M.V., A.S. Fenocchio, M.S. Mantovani, M.M. Cestari and C.A. Oliveira Ribeiro. 2004. Mutagenic effects of tributyltin (TBT) and inorganic lead (PbII) on the fish *H. malabaricus* as evaluated using the comet assay, piscine micronucleous and chromosome aberrations tests. Gen. Mol. Biol. 27(1): 103–107.

Ferraro, M.V.M. 2009. Avaliação de três espécies de peixes—*Rhamdia quelen*, *Cypr inus carpio* e *Astyanax bimaculatus*, como potenciais bioindicadores em sistemas hídricos através dos ensaios: cometa e dos micronúcleos. M.S. Thesis, Universidade Federal do Paraná. Curitiba. Paraná. Brasil.

Filipak Neto, F., S.M. Zanata, H.C. Silva de Assis, D. Bussolaro, M.V.M. Ferraro, M.A.F Randi, J.R.M. Alves Costa, M.M. Cestari, H. Roche and C.A. Oliveira Ribeiro. 2007. Use of hepatocytes from *Hoplias malabaricus* to characterize the toxicity of a complex mixture of lipophilic halogenated compounds. Toxicol. *In vitro* 21: 706–715.

Fisher, S.K., J.T. Lingenfelser, C.H. Jagoe and C.E. Dallas. 1995. Evaluation of the effects of cryopreservation of isolated erythrocytes and leukocytes of *Largemouth bass* by flow cytometry. J. Fish Biol. 46: 432–441.

Frenzilli, G., A. Falleni and V. Scarcelli. 2008. Cellular responses in the cyprinid leuciscus cephalus from a contaminated freshwater ecosystem. Aquatic Toxicol. 89(3): 188–196.

Frenzilli, G., V. Scarcelli and I. Del Barga. 2004. DNA damage in eelpout (*Zoarces viviparus*) from Göteborg harbour. Mutat. Res. 552(1–2): 187–95.

Galvan, G.L. 2011. Avaliação Genotóxica de efluentes químicos de laboratórios de instituição de ensino e pesquisa utilizando como bioindicador o peixe *Astyanax altiparanae* (Characidae). Dissertation, Universidade Federal do Paraná, Curitiba, Paraná. Brasil.

Ghisi, N.C., W.A. Ramsdorf, M.V.M. Ferraro, C.A. Oliveira Ribeiro and M.M.Cestari. 2011. Evaluation of genotoxicity in *Rhamdia quelen* (Pisces, Siluriformes) after sub-chromic contamination with Fipronil. Environ. Monit. Assess. 180: 589–599.

Ghisi, N.C. and M.M. Cestari. 2012. Genotoxic effects of the herbicide roundup in the fish *Corydoras paleatus* (Jenyns 1842) after short-term, environmentally low concentration exposure. Environ. Monit. Assess. (in Press).

Goksoyr, A., T. Anderson, D.R. Buhler, J.J. Stegeman, D.E. Willians and L. Forlin. 1991. Immunochemical cross-reactivity os β-naphthoflavone—inducible cytocrome P450 in liver microsomes from different fish species and rat. Fish Physiol. 9: 1–13.

Guchelaar, H.J., A. Vermes, I. Vermes and C. Haanen. 1997. Apoptosis: molecular mechanisms and implications for cancer chemotherapy. Pharm. World Sci. 19(3): 119–125.

Gustavino, B., K.A. Scornajenghi, S. Minissi and E. Ciccotti. 2001. Micronuclei induced in erythrocytes of *Cyprinus carpio* (teleostei, pisces) by X-Ray and colchicine. Mutat. Res. 494: 151–159.

Grisolia, C.K. and C.M.T. Cordeiro. 2000. Variability in micronucleus induction with different mutagens applied to several species of fish. Gen. Mol. Biol. 23: 233–239.

Hahn, A. and B. Hock. 1999. Assessment of DNA damage in filamentous fungi by single cell gel electrophoresis, comet assay. Environ. Toxicol. Chem. 18: 1421–1424.

Heddle, J.A. 1973. A rapid *in vivo* test for chromossomal damage. Mutat. Res. 18: 187–190.

Heddle, J.A., M. Hite, B. Kirkhart, K. Mavourin, J.T. Macgregor, G.W. Newell and M.F. Salamon. 1983. The induction of micronuclei as a measure of genotoxicity. Mutat. Res. 123: 61–118.

Heddle, J.A., M.C. Cimino, M. Hayashi, M.D. Romagna, J.D. Tucker, P.H. Vanprais and J.T. Macgregor. 1991. Micronuclei as a index of citogenetic damage: past, present and future. Environ. Mol. Mut. 18: 277–291.

Hochberg, V.B.M. and B. Erdtmann. 1988. Cytogenetical and morphological considerations in *Rhamdia quelen* (Pisces, Pimelodidae). The occurrence of β-chromosomes and polymorphic NOR regions. Brazil. J. Genet. 11: 563–576.

Hooftman, R.N. and W.K. de Raat. 1982. Induction of nuclear anomalies (micronuclei) in the peripheral blood erythrocytes of the eastern mudminnow *Umbra pygmea* by ethyl methanesulphonate. Mutat. Res. 104: 147–152.

Horváthová, E., D. Slamenová, L. Hlincíková, T.K. Mandal, A. Gábelová and A.R. Collins. 1998. The nature and origin of DNA single-strand breaks determined with the comet assay. Mutat. Res. 409: 163–171.

Hose, J.E., J.N. Cross, S.G. Smith and D. Diehl. 1987. Elevated circulating erythrocyte micronuclei in fishes from contaminated of Southern Califórnia. Mar. Environ. Res. 22: 167–176.

Katsumiti, A., F.X. Valdez Domingos, M. Azevedo, M.D. da Silva, R.C. Damian, M.I.M. Almeida, H.C. Silva de Assis, M.M. Cestari, M.A.F. Randi, C.A. Oliveira Ribeiro and C.A. Freire. 2009. An assessment of acute biomarker responses in the demersal catfish cathorops spixii after the Vicuña Oil Spillin a harbour estuarine area in Southern Brazil. Environ. Monit. Assess. 152: 209–222.

Keshava, C., T. Ong and J. Nath. 1995. Comparative studies on radiation-induced micronuclei and chromosomal aberrations in V79 cell. Mutat. Res. 328: 63–71.

Klaude, M., S. Eriksson, J. Nygren and G. Ahnströn. 1996. The comet assay: mechanisms and technical considerations. Mutat. Res. 363: 89–96.

Lacadena, J.R. 1996. Citogenética. 1ª ed. Madrid: Editorial Complutense.

Lopes-Poleza, S.C.G. 2004. Avaliação do efeito do metilmercúrio (CH$_3$Hg$^+$) em *Hoplias malabaricus* Através da freqüência de aberrações cromossômicas e dos ensaios cometa e micronúcleo. Dissertation. Universidade Federal do Parfaná. Curitiba. Paraná. Brasil.

McGahon, A.J., Seamus J. Martin, Reid P. Bissonnette, Artin Mahboubi, Yufang Shi, Rona J. Mogil, Walter K. Nishioka and Douglas R. Green. 1995. The end of the (Cell) line: Methods for the study of apoptosis *in vitro*. In: L.M. Schwrtz and B.A. Osborne (eds.). Cell Death. Academic Press, London, UK, pp. 172–173.

Metcalfe, C.D. 1989. Testes for predicting carcinogenecity in fish. Crit. Rev. Aquat. Sci. 1: 111–129.

Minissi, S., E. Ciccotti and M. Rizzoni. 1996. Micronucleus test in erythrocytes of *Barbus plebejus* (Teleostei, Pisces) from two natural enviroments: a bioassay for the *in situ* detection of mutagens in freshwater. Mutat. Res. 367: 245–251.

Mitchelmore, C.L. and J.K. Chipman. 1998. DNA strand breakage in aquatic organisms and the potential value of the comet assay in environmental monitoring. Mutat. Res. 399: 135–147.

Monteith, D.K. and J. Vanstone. 1995. Comparison of the microgel eletrophoresis assay and other assays for gemotoxicity in the detection of DNA damage. Mutat. Res. 345: 97–103.

Navarrete, M.H., P. Carrera, M. Miguel and C. de la Torre. 1997. A fast comet assay variant for solid tissue cells. The assessment of DNA damage in higher plants. Mutat. Res. 389: 271–277.

Nigro, M., G. Frenzilli, V. Scarcelli, S. Gorbi and F. Regoli. 2002. Induction of DNA strand breakage and apoptosis in the eel Anguilla anguilla. Mar. Environ. Res. 54(3–5): 517–20.

Obe, G., P. Pfeiffer, J.R.K. Savage, C. Johannes, W. Goedecke, P. Jeppesen, A.T. Natarajan, W. Martínez–López, G.A. Folle and M.E. Drets. 2002. Chromosomal aberrations: formation, identification and distribution. Mutat. Res. 504: 17–36.

Olive, P.L., J.P. Banáth and R.E. Durand. 1990. Heterogeneity in radiation-induced DNA damage and repair in tumor and normal cells measured using the "comet" assay. Rad. Res. 122: 86–94.

Olive, P.L., D. Wlodek, R.E. Durand and J.P. Banáth. 1992. Factors influencing DNA migration from individual cells subjected to gel electroforesis. Exp. Cell Res. 198: 259–267.

Olive, P.L. and J.P. Banáth. 1995. Sizing highly fragmented DNA in individual apoptotic cells using the comet assay and a DNA crosslink agent. Exp. Cell Res. 221: 19–26.

Olive, P.L., P.J. Johnston, J.P. Banáth and R.E. Durand. 1998. The comet assay: A new method to examine heterogeneity associated with solid tumors. Nature Med. 4: 103–105.

Ostling, O. and K.J. Johanson. 1984. Microelectrophoretic study of radiation-induced DNA damage in individual mammalian cells. Biochem. Biophys. Res. Com. 123(1): 291–298.

Padrangi, R., M. Petras, S. Ralph and M. Vrzoc. 1995. Alkaline single cell gel (comet) assay and genotoxicity monitoring using bullheads and carp. Environ. Mol. Mut. 26: 345–356.

Pamplona, J.H., E.T.O. Yoshioka, L.P. Ramos, T.A. Silva, W. Ramsdorf, M.M. Cestari, C.A. Oliveira Ribeiro, A.R. Zampronio and H.C. Silva de Assis. 2011. Subchronic effects of dipyrone on the fish species *Rhamdia quelen*. Ecotoxicol. Environ. Saf. 74: 342–249.

Parolini, M., A. Binelli, D. Cogni, C. Riva and A. Provini. 2009. Provini an *in vitro* biomarker approach for the evaluation of the ecotoxicity of non-steroidal anti-inflammatory drugs (NSAIDs). Toxicol. *In vitro* 23: 935–942.

Pesenti, E.C. 2012. Avaliação da genotoxicidade do sulfato de cobre e de reparo do DNA utilizando *Astyanax fasciatus* (Characidae) e *Ctenopharyngodon idella* (Cyprinidae) como bioindicadores. Dissertation, Universidade Federal do Paraná. Curitiba. Paraná. Brasil.

Piancini, L.D.S. 2011. Utilização de biomarcadores genéticos na avaliação aguda do efeito mutagênico dos contaminantes Atrazina e Cloreto de Cobre em*Rhamdia quelen* (Siluriformes, Heptapteridae). Dissertation, Universidade Federal do Paraná. Curitiba. Paraná. Brasil.

Preston, J.R. and G.R. Hoffmann. 2001. Genetic toxicology. *In*: Curtis D. Klaassen (ed.). *Casarett & Doull's*, Toxicology: The Basic Science of Poisons, 6. edn. McGraw-Hill Companies, New York, USA, pp. 321–350.

Rabello-Gay, M.N., M.A.R. Rodrigues and R. Monteleone-Neto. 1991. Mutagênese Teratogênese e Carcinogênese: métodos e critérios de avaliação. Editora da Sociedade Brasileira de Genética. Ribeirão Preto. São Paulo, Brasil.

Ramsdorf, W.A. 2011. Avaliação da toxicidade dos compostos fipronil, nitrato de chumbo e naftaleno em peixes. M.S.Thesis, Universidade Federal do Paraná. Curitiba. Paraná. Brasil.

Ramsdorf, W.A., M.V.M. Ferraro, C.A. Oliveira-Ribeiro, J.R.M. Costa and M.M. Cestari. 2009a. Genotoxic evaluation of different doses of inorganic lead (PbII) in *Hoplias malabaricus*. Environ. Monit. Assess. 158: 77–85.

Ramsdorf, W.A., F.S.F. Guimarães, M.V.M. Ferraro, J. Gabardo, E.S. Trindade and M.M. Cestari. 2009b. Establishment of experimental conditions for preserving samples of fish blood for analysis with both comet assay and flow cytometry. Mutat. Res. 673: 78–81.

Rocha, C., B. Cavalcanti, C.Ó. Pessoa, L. Cunha, R.H. Pinheiro, M. Bahia, H. Ribeiro, M.M. Cestari and R. Burbano. 2011. Comet assay and micronucleus test in circulating erythrocytes of *Aequidens Tetramerus* exposed to methylmercury. *In vivo* 25: 100–108.

Rydberg, B. and K.J. Johanson. 1978. Estimation of single strand break in single mammalian cells. *In* : P.C. Hanawalt, E.C. Friedbers and C.F. Fox (eds.). DNA Repair Mechanisms. Academic Press, New York, 465–468.

Rocco, L., G. Frenzilli, D. Fusco, C. Peluso and V. Stingo. 2006. Evaluation of zebrafish DNA integrity after exposure to pharmacological agents present in aquatic environments. Ecotoxicol. Environ. Saf. 73(7): 1530–6.

Rojas, E., M.C. Lopez and M. Valverde. 1999. Single cell gel electrophoresis assay: methodology and applications. J. Chromat. B 722: 225–254.

Salvagni, J., R.Z. Ternus and A.M. Fuentefria. 2011. Assessment of the genotoxic impact of pesticides on farming communities in the countryside of Santa Catarina State, Brazil. Gen. Mol. Biol. 31(1): 122–126.

Sasaki, Y.F., F. Izumiyama, E. Nishidate, S. Ishibahi, S. Tsuda, T. Shuji, N. Matsusaka, N. Asano, K. Saotome, T. Sofuni and M. Hayashi. 1997. Detection of genotoxicity of polluted sea water using shellfish and the alkaline single-cell gel electrophoresis (SCE) assay: a preliminary study. Mutat. Res. 393: 133–139.

Schmid, W. 1975. The micronucleus test. Mutat. Res. 31: 9–15.

Searle, J., J.F. Kerr and C.J. Bishop. 1982. Necrosis and apoptosis: distinct modes of cell death with fundamentally different significance. Pathol. Ann. 17(2): 229–259.

Silva, C.A., E.T. Oba, W.A. Ramsdorf, V.F. Magalhães, M.M Cestari, C.A. Oliveira Ribeiro and H.C. Silva de Assis. 2011. First report about saxitoxins in freshwater fish Hoplias malabaricus through trophic exposure. Toxicon. 57: 141–147.

Silva, J., T.R.O. Freitas, J.R. Marinho, G. Speit and B. Erdtmann. 2000. An alkaline single-cell gel electrophoresis (comet) assay for environmental biomonotoring with native rodents. Gen. and Mol. Biol. 23: 241–245.

Silfvergrip, A.M.C. 1996. A systematic revision of the neotropical catfish genus *Rhamdia* (Teleostei, Pimelodidae). Swedish Museum of Natural History, Stockholm.

Singh, N.P., M.T. McCoy, R.R. Tice and E.L. Schneider. 1988. A simple technique for quantitation of low levels of DNA damage in individual cells. Exp. Cell Res. 175: 184–191.

Singh, N. 2000. A simple method for accurate estimation of apoptotic cells. Exp. Cell Res. 256: 328–337.

Singh, N.P. 2005. Apoptosis assessment by the DNA diffusion assay. Met. Mol. Med. 111: 55–67.

Speit, G. and A. Hartmann. 1999. The comet assay (single cell gel test)—a sensitive genotoxicity test for the detection of DNA damage and repair. Met. Mol. Biol. 113: 203–212.

Swarça, A.C., A.S. Fenocchio and A.L. Dias. 2007. An update cytogenetic review for species of the families pseudopimelodidae, pimelodidae and heptapteridae (Pisces, Siluriformes). Suggestion of a cytotaxonomical classification. Caryologia. 60(4): 338–348.

Tavares-Dias, M. and F.R. de Moraes. 2004. Hematologia de Peixes Teleósteos. Ribeirão Preto. Brasil.

Theodorakis, C.W., S.J. D'Surney and L.R. Shugart. 1994. Detection of genotoxic insult as DNA strand breaks in fish blood cells by agarose gel electrophoresis. Environ. Toxicol. Chem. 13: 1023– 1031.

Ueda, T., M. Hayashi, N. Koide, T. Sofuni and J. Kobayashi. 1992. A preliminary study of the micronucleus teste by acridine orange fluorescent staining compared with cromossomal aberration teste using fish erythropoietic and embrionic cells. Wat. Sci. Technol. 25: 235–240.

Vicari, T., M.V.M. Ferraro, W.A. Ramsdorf, M. Mela, C.A. Oliveira Ribeiro and M.M. Cestari. 2012. Genotoxic evaluation of different doses of methylmercury (CH3Hgþ) in *Hoplias malabaricus*. Ecotoxicol. Environ. Saf. 82: 47–55.

Yendle, J.E., H. Tinwell, B. Elliot and J. Ashby. 1997. The genetic toxicity of time: Importance of DNA—unwinding time to the outcome of single-cell gel electrophoresis assays. Mutat. Res. 375: 125–136.

CHAPTER 7

The Use of Fish Biomarkers in the Evaluation of Water Pollution

Thiago E.M. Parente[1],* *and Rachel Ann Hauser-Davis*[2]

Introduction

The definition of biomarkers has been enunciated in many different forms as a reflection of their intense use in the environmental sciences over the past decades (NRC 1987; Schlenk 1999; Depledge et al. 1992; Adams 1987). In essence, a biomarker is any indicator of a stress agent that is somehow affecting an organism's ability to grow, reproduce, survive and adapt (or in other words, to live) in a given environment. In fact, the term biomarker is most accurately used to refer to an indicator at a sub-individual or, at most, an individual level of organization; this indicator may be alterations in molecular and biochemical processes, cellular structures and functions, tissue organization or mass and length ratios of individual organs or the whole body.

[1]Universidade Federal do Rio de Janeiro, Departamento de Bioquímica, Instituto de Química. Avenida Athos da Silveira Ramos 149. Centro de Tecnologia, Bloco A 5 andar sala 526. Cidade Universitária, Ilha, do Fundão, CEP: 21941-909, Rio de Janeiro, RJ, Brasil.
Email: parente@iq.ufrj.br (thiparente@gmail.com)
[2]Pontifícia Universidade Católica do Rio de Janeiro–PUC-Rio, Av. Marquês de São Vicente, 225, Departamento de Química, Laboratório de Bioanalítica, Gávea, CEP 22451-900, Rio de Janeiro, Brasil.
Email: rachel.hauser.davis@gmail.com
*Corresponding author

The correct identification of the stress agent, however, is one of the key points of biomarker use in order to monitor environmental health. Modifications in biomarker measurements can be due to contaminants in the environment, but often reflect other variables, such as organism health and seasonal variations (Fig. 7.1). In order to exclude, or at least minimize the possibility that biomarker responses are due to causes other than contaminants in the environment, the use of appropriate controls are fundamental in biomarker applications. In line with this, basic knowledge regarding the used biomarker is critical to understand which and how other variables might interfere with a specific biomarker response. With the use of these controls, a biomarker response will indicate the bioavailability of a certain chemical pollutant, which, in turn, is expected to trigger confirmatory analyses followed by environmental policies aiming to improve biomarker parameters and eventually restore environmental health (Fig. 7.2). The goal of this chapter is to analyze the foundations and weaknesses of the bridge connecting the use of biomarkers in fish species to water pollution, in particular in tropical species and ecosystems.

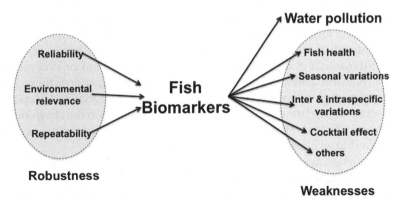

Figure 7.1. The robustness and weaknesses of the use of fish biomarkers in the evaluation of water pollution. The choice of which biomarkers to use is frequently based on their robustness (e.g., reliability, environmental relevance and repeatability). Biomarkers responses, however, might be altered by variables other than the presence of contaminants in the environment.

The Foundations of Fish Biomarker Research and Use

The foundations of biomarker research and use lie in a fundamental attribute of life, the capability of an organism to adapt to subtle changes in the environment. This adaptation is mechanistically based on alterations of molecular and biochemical processes within the cellular metabolism in response to environmental changes. The alterations in these processes, in

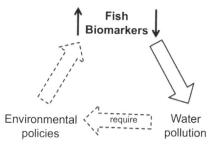

Figure 7.2. Role of fish biomarkers in environmental management. Biomarkers respond to the presence of contaminants in the environment indicating water pollution that, in turn, triggers environmental polices aiming to improve biomarker parameters and eventually restore environmental health.

turn, can lead to modifications in cellular structures, tissue organization and so on up the hierarchical level of biological organization. Since the early 1970s, researchers are looking for those processes and structures in order to establish a causative relationship between exposure to stress agents and fluctuations in these parameters.

However, it is important to discriminate between biomarkers and bioindicators. Both indicate the presence of contaminants in the environment, but while the former are measurements at the sub-individual, or, at most, at the individual level of organization, the latter are alterations at higher than individual levels of organization (e.g., populational or ecosystemic levels) (Hanson 2008). As with most arbitrary classifications, there is an overlapping region that often causes conflict, even in specialized literature. In this chapter, the presence or absence of a given species in the environment and stress indicators at higher than individual level of organization will not be referred to as biomarkers, but, instead, as bioindicators.

Biomarkers are defined as measurements in body fluids, cells or tissues indicating biochemical or cellular modifications due to the presence and magnitude of toxicants, or of host response (NRC 1987). This original definition was modified by Adams, specifically with aquatic organisms in mind. He included characteristics of organisms, populations, or communities that respond in measurable ways to changes in the environment (Adams 1987). Later, behavioral responses, latency and genetic diversity were added to this definition (Depledge et al. 1992).

In comparison to bioindicators, data that come from biomarkers are more reliable and repeatable, but less environmentally relevant (Fig. 7.3). Most of biomarkers' reliability lies in their specificity to a certain xenobiotic or, at least, to a given class of compounds. The more specific a biomarker is, the more confidently the stress agent can be identified. On the other hand, alterations in bioindicators can be caused by such a vast array of classes of substances that rarely can the stress agent can be identified. Repeatability is also higher in biomarkers than in bioindicators. As biomarker investigations

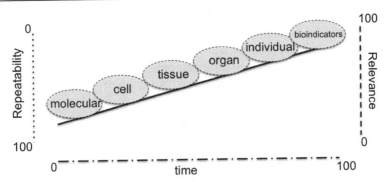

Figure 7.3. Hierarchical classification of biomarkers according to their repeatability, environmental relevance and time of response.

use less individuals than bioindicators, these studies can be replicated during short periods of time, while it is impractical to perform populational or ecosystemic surveys with such frequency. On the other hand, the gap from molecular, biochemical, tissual or organismic biomarkers to environmental pollution is much larger than the straightforward gap between bioindicators and pollution.

Three types of biomarkers were proposed in an attempt to classify responses as markers of exposure, effect and susceptibility (Schlenk 1999). As more biomarkers have been identified and characterized, it has become apparent that this tripartite definition has significant overlap, in that some biomarkers can be used in each of these capacities. As a review of the biomarker paradigm has described, an effect resulting from stressor exposure may be defined as an early adaptive non-pathogenic event or as a more serious altered functional event, depending on the toxicokinetics and mechanism of action of the stressor (Decaprio 1997). Likewise, biomarkers of exposure and effect may often be combined into a single classification with susceptibility occurring along any stage (Barret et al. 1997).

The foundational concept of the biomarker approach to assess adverse effects or stress is based on the hypothesis that the effects of stress are typically manifested at lower levels of biological organization before disturbances are realized at the population, community or ecosystem levels (Adams 1987). Such initial effects are observed at the molecular level primarily with the induction of cellular defense mechanisms, which may actually provide adaptive responses following exposure. However, if these defense processes fail or are overwhelmed, then higher level damage may occur, perhaps causing histological or physiological impairment, which again, may not be irreversible, depending on the capacity of the damaged system or organ. But if these processes are permanently affected or altered during vulnerable periods of organism development, reproduction and/or survival may be affected, eventually leading to changes at the population and possibly, community levels of biological organization (Schlenk 1999). In sum, effects at higher hierarchical levels are always preceded by earlier changes in biological processes, allowing the development of early-warning biomarker signals of effects at later response levels (Bayne et al. 1985).

Biomarkers of effect can be measured at any point along the continuum, from molecular to higher levels, and may vary tremendously in their specificity, repeatability and ecological relevance (Fig. 7.3). For example non-specific biomarkers of effect may include endpoints that can be altered by a number of stressors (i.e., HSP induction, adenylate levels, etc.) and primarily related to growth rate of an organism. However, the focus may be narrowed by using more specific biomarkers, which may be organ-specific. Examples of organ-specific markers may include enzymes released into the blood upon tissue damage such as various hepatic amino acid transaminases indicative of liver damage (Mayer et al. 1992). As will be discussed in the next section of this chapter, numerous agents may be responsible for the production of these effects; notwithstanding, these markers, in addition to other more specific markers, such as those for lipid peroxidation or oxidative stress, may provide more insight into the mechanism of the original damage observed from more general or non-specific measures. Indeed, acute stress protein production (HSPs, MT, SAP kinases) and/or cortisol release tend to be rapidly produced in response to acute stress, whereas histopathological changes such as membrane blebbing or neoplasia would be more appropriate for assessing chronic stress (Schlenk 1999). Because of the potential relationships between cellular and ecosystem level responses, and compensatory adaptive mechanisms, an effective suite of biomarkers of effect should provide some insight into the causality of observed higher level adverse effects (Schlenk 1999).

In an environmental context, biomarkers offer promise as sensitive indicators, demonstrating that toxicants have entered organisms, have been distributed between tissues, and are eliciting a toxic effect at critical targets (McCarthey and Shugart 1990). Good biomarkers are sensitive indices of both pollutant bioavailability and early biological responses (van der Oost et al. 2003). The use of biomarkers often increases the possibility of identifying the underlying causes behind toxic effects and provides information about bioavailability of pollutants and their potential ecological damage (Albertsson et al. 2007). For example, the use of specific biomarkers for estrogen exposure allied to chemical analyses played an important role in the identification of ethinylestradiol, the synthetic estrogen in contraceptive drugs, as a compound contributing to the feminization of fish downstream from sewage treatment works (Purdom et al. 1994; Sumpter and Jobling 1995; Routledge et al. 1998; Larsson et al. 1999). In general however, there is a lack of robust, sensitive and specific markers for the great majority of contaminants.

Biomarkers in Fish

Specifically regarding aquatic habitats, fish and molluscs appear as the major species where biomarkers have been applied to monitor water pollution (Viarengo et al. 2007). For several reasons, fish species have attracted considerable interest in studies assessing biological and biochemical responses to environmental contaminants (Powers 1989). Many fish species are amenable to both field and laboratory experiments and are easily raised and bred under laboratory conditions. Fish are particularly useful for the assessment of water-borne and sediment-deposited toxins where they may provide advance warning of the potential danger of new chemicals and the possibility of environmental pollution; also, they are particularly good models for studies involving biochemistry and comparative physiology, because they live in a variety of habitats and must adapt to environmental parameters and stress, which can be easily reproduced under laboratory conditions (Powers 1989).

Thus, fish are well recognized bioindicators of environmental changes, including chemical pollution (FAO/SIDA 1983; Espino 2000). Since fish span over a wide variety of feeding and living habits, they are exposed to chemical contamination from different food sources and water conditions. Fish can be found virtually everywhere in the aquatic environment and they play a major ecological role in the aquatic food-webs because of their function as energy carriers from lower to higher trophic levels. The understanding of toxicant uptake, behavior, and responses in fish may, therefore, have a high ecological relevance. They are also an important link between the environment and human populations through fisheries and consumption by local and other markets (FAO/SIDA 1983; Espino 2000).

Monitoring species should be selected from an exposed community on the basis of their relationship to the assessment endpoint as well as by following some practical considerations (Suter 1993). For the assessment of the quality of aquatic ecosystems, both criteria are met for numerous species of fish.

Most of the general biomarker criteria appear to be directly transferable to certain fish biomarkers (Stegeman et al. 1992). Between different fish species, however, considerable variation may become apparent in both the basic physiological features and the responsiveness of certain biomarkers towards environmental pollution (van der Oost et al. 2003).

In order to assess exposure to or effects of environmental pollutants on aquatic ecosystems, the following suite of fish biomarkers may be examined: biotransformation enzymes (phase I and II), oxidative stress parameters, biotransformation products, stress proteins, metallothioneins (MTs), MXR proteins, hematological parameters, immunological parameters, reproductive and endocrine parameters, genotoxic parameters, neuromuscular parameters, physiological, histological and morphological parameters (van der Oost et al. 2003).

Several of these fish biomarkers have been used in studies involving environmental risk assessments and aquatic ecosystem monitoring, including:

- *AChE activity*: Acetylcholinesterase (AChE), in nervous tissue, is responsible for acetylcholine degradation, one of the most important neurotransmitters in either the central or peripheral nervous system. AChE activity has, up to now, been used as an enzymatic biomarker of neurotoxicity caused by pesticides (see Chapter 4 for more details). However, an increasing number of studies provide evidence that the AChE activities in marine organisms may be affected by a wide range of contaminants other than pesticides, including heavy metals, petroleum-derived products such as polycyclic aromatic hydrocarbons (PAHs) and components of complex mixtures of contaminants (Kang and Fang 1997; Cunha et al. 2005; Oropesa et al. 2007; Chambers et al. 1978; Sheehan et al. 1991; Bocquene et al. 1995; Martinez Tabche et al. 1997; Mora et al. 1999; Akcha et al. 2000; Moreira et al. 2004). A more general use of this biomarker for the assessment of environmental quality is therefore under way (Matozzo et al. 2005; da Silva et al. 2005; Magni et al. 2006; Humphrey et al. 2007; Tsangaris et al. 2007).
- *Cytochrome P450 1A (CYP1A) activity:* One of the best studied parameters for measuring the effects of contaminants on aquatic organisms is cytochrome P450 1A dependent monooxygenase (CYP1A) activity, which is responsible for the biotransformation of a myriad of xenobiotic compounds such as polychlorinated biphenyls (PCBs), polycyclic aromatic hydrocarbons (PAHs), and polychlorinated dibenzo-pdioxins (PCDDs) and dibenzofurans (PCDFs), among others (van der Oost et al. 2003; Stegeman and Kloeppersams 1987; see more details in Chapter 2). Several field studies have identified correlations between CYP1A activity and contaminant concentrations (Galgani et al. 1991; Eggens et al. 1992). Two different approaches are normally used to quantify CYP1A induction: immunological measurements such as ELISA (Celander and Forlin 1991; Goksoyr 1991) and the measurement of catalytic activity with a highly specific model substrate such as 7-ethoxyresorufin in the 7-ethoxyresorufin-O-deethylation activity (EROD) assay (Burke and Mayer 1974).
- The use of the EROD activity as an environmental biomarker was suggested some 30 years ago (Burke and Mayer 1974), and from that time on the main objective of studies dealing with EROD activity has been to assess the effects of contamination by specific pollutants on target organisms in the marine and freshwater environments (Galgani

et al. 1991; Addison and Edwards 1988; Holdway et al. 1994; Parente et al. 2004). Because of the sensitivity of this biomarker in aquatic vertebrates, it has been widely used in biomonitoring studies for more than a decade (Galgani et al. 1992; Radenac et al. 2004; Parente et al. 2008).

- Therefore, CYP1A determinations may be used in several steps of environmental risk assessments, such as quantification of impact and exposure of various organic trace pollutants, environmental monitoring of organism and ecosystem 'health', identifying subtle early toxic effects, triggering of regulatory action, identification of exposure to specific compounds, toxicological screening and the research on toxic mechanisms of xenobiotics (van der Oost et al. 2003; Kang and Fang 1997), and it has been recommended that they should be measured simultaneously in studies of the effects of chemical contaminants on the aquatic environment (Vanderweiden et al. 1992).

- *Morphological parameters:* Several studies relate environmental contamination by metals and organic compounds to changes in morphological parameters that determine and describe environmental interferences in organisms, such as the Hepatossomatic Index (HSI) (more details in Chapter 12), that identifies possible liver disorders, and the Condition Factor (CF), a morphological parameter that evaluates the general health condition of the fish (Laflamme et al. 2000; Norris et al. 2000; Fernandes et al. 2008; Stephensen et al. 2000). These morphological parameters can, therefore, indicate changes in the fish's health state caused by environmental contaminants or stress.

- *Fish bile:* Biliary excretion can offer a way to analyze various contaminants in aquatic organisms, and fish bile has been extensively used as a biomarker for environmental contamination. The rapid metabolism and elimination of several contaminants by vertebrates, such as fish, result in low residual concentrations of these contaminants in muscle and liver tissues (Galgani et al. 1992). Chemical analysis of fish tissues therefore has limited usefulness as an indicator of environmental exposure to these contaminants, including PAH. Several species of fish living in habitats contaminated by PAH have been shown to contain high concentrations of biliary PAH metabolites (Krahn et al. 1987; Ariese et al. 1993; Kirby et al. 1999), which may therefore provide an alternative indicator of exposure for environmental monitoring and screening purposes (Norris et al. 2000). Fish bile has also been used as a screening method for metal exposure, for several studies indicate that many metals are excreted from the liver to the bile (Bunton and Frazier 1994; Dijkstra et al. 1996). Some examples are copper, which is excreted primarily in teleost fish by the bile, while other metals such as zinc can be excreted either by the bile or by the intestine. Recent studies have validated the use of certain metals excreted in bile as biomarkers for metal exposure (Hauser-Davis et al. 2012). Also, it has been discovered recently that certain metal binding proteins named metallothioneins are also excreted in bile, and that these proteins in bile follow the same trend as hepatic MT in situations of environmental exposure to metals (Hauser-Davis et al. 2012), making them interesting biliary biomarkers with regard to metal exposure.

The Weaknesses of Fish Biomarker Research and Use

The weaknesses of fish biomarker research and use consist of an array of physicochemical and biological variables capable of interfering with biomarkers responses, masking or emphasizing the presence of contaminants in the environment or their effects on the biota. These weaknesses include the fact that several biomarker responses may be modified in the environment due to effects of chemical speciation, kinetics and the adsorption of chemicals to sediment, accumulation through food chains and modes of toxic action which are not readily measured as short-term effects (McCarthey and Shugart 1990).

Fish are generally considered to be the most feasible organisms for pollution monitoring in aquatic ecosystems (McCarthey and Shugart 1990); however, some species are better suited than others, since migration patterns may often confuse biomarkers evaluation. Several fish species from temperate environments have been well characterized as suitable for their ecological habitats and metabolic adaptations. Hence, species such as trout (*Oncorhynchus* spp.), Atlantic cod (*Gadus morhua*), catfish (*Ictalurus punctatus*) and some others are being widely applied in aquatic toxicology. In tropical environments which display a much wider diversity of fish species, much less is known regarding the metabolism, adaptations and even the habits of tropical fish species. Also, certain biomarker responses established for one species may not necessarily be valid for another (ECETOC 1993). For example, certain species, such as the neotropical Pacu fish, have been known to exhibit the highest enzymatic GST values of any known vertebrate, due to its singular capacity for adapting to hypoxic or anoxic conditions as an adaption to the flooding and dry seasons that occur seasonally in the Amazon rainforest (Bastos et al. 2007). The ability of the Pacu fish to regulate the expression of GST genes in response to environmental pollutants and other variables are still unknown. Likewise, some species of tropical suckermouth catfish (*Hypostomus* spp. and *Pterygoplichthys* sp.) are unique vertebrates known to be unable to catalyze EROD activity despite accumulating CYP1A enzyme and inducing its gene expression upon exposition to aryl hydrocarbon receptor (AhR) agonists (Parente et al.

2009; Parente et al. 2011). Thus, biomarker responses in these species, and certainly in many other non- or poorly studied tropical fish, are very unique, making it quite dangerous to extend the standard knowledge of biomarkers responses obtained in temperate fish to species found in the tropics. Thus, the analyses of certain established biomarker responses, such as activities of xenobiotic biotransformation enzymes, like GST and CYP1A, would not be appropriate for tropical environmental contamination studies. Notably, Tilapia (*Oreochromis* spp.) is a species that has been emerging as the most used fish in biomonitoring surveys in tropical countries (Parente et al. 2004; Abdel-Moneim et al. 2012; Omar et al. 2012; Nogueira et al. 2011; Zagal and Mazmanci 2011; Pathiratne and Hemachandra 2010; Sun et al. 2009). Tilapia is originally from Africa but the species has been introduced in many countries for aquaculture purposes. Tilapias that escape from captivity or that was intentionally introduced in the field have colonized several natural rivers, lakes and even brackish marshes. The increasing use of Tilapia in environmental health studies and its massive use in aquaculture are building a substantial volume of basic information regarding genetics, metabolism and physiology that will be helpful in the validation of biomarkers responses (He et al. 2011; Soler et al. 2010; Guyon et al. 2012).

Fish condition at the time of analyses should also be taken into account: fish exhibit certain modifications when entering their reproduction stages, such as increments in body fat reserves, including hepatic changes, variations in feeding and in sexual hormones (van der Oost et al. 2003; Hauser-Davis et al. 2010) which may in turn influence biomarker responses due to extensive physiological changes that occur in fish during this period (Querol et al. 2002). Some organism modifications may also occur due to seasonality, since certain fish species change their feeding habits during different seasons, due to the availability of certain food items. Also, certain contaminants may be more bioavailable during the wet season (i.e., chlorinated contaminants such as PCBs and PBDEs, and metals), since increased water column and sediment dislocations are more frequent during this season, resuspending contaminants adsorbed by the sediment, making them more bioavailable to the biota.

Importantly, in the environment fish are many times exposed to complex contaminant mixtures, and chemical pollutants often interact with each other, leading to different biomarker responses than fish exposed to only one contaminant at a time, in what has been called "the cocktail effect" (van der Oost et al. 2003; Celander 2011). Easily transformed chemicals, such as organic compounds (PAH, PCBs) are likely not suitable as bioaccumulation markers for environmental exposure assessments, since the tissue levels of these contaminants will not accurately reflect levels in the surrounding environment (van der Oost et al. 1996).

The Gap between Fish Biomarker Responses to Water Pollution

The weaknesses in the use of fish biomarker in aquatic sciences represent a gap in the connection between biomarkers responses and the interpretation that the water is polluted. There is no absolute way to overcome this gap, as such weaknesses are an inherent issue of biomarkers and one can hardly guarantee that all weaknesses of the used biomarkers are known and were totally controlled during the experimentation process. However, this gap issue can be minimized to validate the obtained results and avoid misinterpretation. This minimization is achieved by using the appropriate controls, multiple biomarkers endpoints and different forms of exposure.

As in any field of experimental sciences, controls are of fundamental importance in the use of biomarkers to monitor water pollution. Multiple clean reference sites are desirable. The biomarkers responses at the clean sites are then used as reference values, or, in other words, the mean values expect to be found in fish inhabiting uncontaminated locations. Importantly, the use of multiple clean sites enables the estimation of the natural expected variance of a specific biomarker response. Values found to be out of this natural expected variance can be considered deviants and preliminarily seen as a reflection of water pollution. However, extra care must be taken when analysing the variation of control data. A variation is indeed expected. The length of this variation is unknown *a priori*. If all individuals of a specific control group or specific individuals of all groups are outliers in comparison to the other homogeneous group formed, these individuals should not be used as controls, as the control site might not be as clean as previously thought or the outliers might be reflecting other pathophysiological conditions. Unfortunately, very often, such clean sites are not available. In those situations, fish can be collected in several locations along the tested areas and the results compared among each other, identifying what could be considered priority areas or hot spots. This is particularly true when the purpose of the study is to monitor the effects of a known specific source of contaminants. In this case, samplings should be conducted at different distances from the source of discharge, up and downstream or better yet, before the source of contaminants begins operation. In this scenario, reference values would be the pre-operation or upstream biomarker responses.

The standardization of biological variables is also of key relevance to the correct interpretation of the results. Individuals used as controls must be about the same mean age, length and weight. It is also important that males and females are equally represented in control and tested groups. Control individuals should be also collected at the same season as the tested group. If tested groups are sampled in different seasons, different control groups should be collected accordingly. As discussed earlier in this chapter, all these variables can interfere with the responses of certain biomarkers.

A very interesting way to face this gap problem is to approach the question through different perspectives by using different species and forms of exposure. By using species with different ecological habits, the bioavailability of the contaminant is checked in different compartments of the environment and the risk of an unknown or uncontrolled variable interfering with the biomarker response is minimized. Most frequently, monitoring studies use feral fish. This is indeed an advantage as these fish have a life long history of exposure to the contaminants possibly present at the studied environment. On the other hand, feral fish are far more susceptible to the variables that constitute biomarker weaknesses, making it far more complicated to control the experiment condition. The use of caged fish, transplanted from control to test sites or vice-versa, is probably the best way to control most of the variables that can affect biomarker responses. If fish are transplanted from a control to test site, an impairment on the biomarker response would be expected due to the "new" exposure to contaminants, while in fish transplanted from a test to control site, the return of biomarker levels to the range of variation observed in unexposed fish due to the lack of environmental stimuli would be expected. The use of caged fish, however, has strong logistic and also ecological drawbacks. First, transporting fish around might not be the simplest task. Secondly, a cage full of fish is seen by any fisherman as a honey pot, and cases of fish (and cage) "disappearances" are not rare. Thirdly and most importantly, depending on the number/mass and of the species to be introduced, the experiment can unbalance the energy flow at the local environment. It is never recommended to introduce a species that is not already present in the environment; ideally native species should be used. Even using native species, all care must be taken to avoid fish from escaping to the field or to be preyed upon by other local species. Another good alternative is to "bring the environment to the laboratory" and expose model fish species to a series of dilutions of the water and sediment collected at test sites. While almost removing issues about the several kinds of fish variations, this approach introduces a key variation—that is, the conservation of the physicochemical parameters of the sample.

Conclusion

In this chapter we have presented key aspects that support the use of fish biomarkers, as well as their major weaknesses and the most used ways to overcome or, at least, to minimize those problems. The use of biomarkers in aquatic sciences has long been established as an early warning, predictive and relatively low cost tool to indicate water pollution. However, the knowledge acquired by academia has not been applied to its full potential by governmental regulatory agencies and private industries in monitoring the impacts of human activities on aquatic ecosystems. A great part of this resistance to the use of fish biomarkers is due to their weaknesses, which prevent a direct link between the biomarker responses to the definitive conclusion of water pollution.

The inability to directly connect the biomarkers responses to water pollution is not a constraint to researchers and scientists that follow the scientific method and, thus, do not prove their hypotheses but instead refuse alternative possibilities. Nonetheless, this is a huge concern among policy makers who urge for an ultimate proof of pollution. Though not a definite proof of pollution, biomarkers are indeed an excellent tool to indicate a possible contamination in aquatic ecosystems. Moreover, biomarkers indicate that this possible contamination would have been reaching the biota, causing changes in the metabolism or even harm to individuals.

Whether or not these metabolic changes and harmful effects are relevant for fish or environmental health is not the question addressed by the biomarkers applied in monitoring water pollution. These metabolic changes indicate that fish and other species of the community are exposed to chemical compounds that are naturally not present in their habitat and that, most likely, originate from anthropogenic activities. The presence of a contaminant in the ecosystem *per se* creates a risk for environmental health. Measuring or estimating this risk, however, is a different task that might include but is not limited to the use of biomarkers.

Acknowledgements

R.A.H.D. would like to thank Dr. Reinaldo Calixto (*in memoriam*) for his invaluable help throughout her academic career. T.E.M.P. thanks and is supported by a PEER Science grant (PGA-2000003446) from the U.S. Agency for International Development and the U.S. National Academies of Science.

Keywords: Xenobiotics, Biotransformation, Enzymes, CYP1A, EROD, GST, Bile metabolites

References

Abdel-Moneim, A.M., M. Al-Kahtani and O.M. Elmenshawy. 2012. Histopathological biomarkers in gills and liver of *Oreochromis niloticus* from polluted wetland environments, Saudi Arabia. Chemosphere 88(8): 1028–1035.

Adams, S.M. 1987. Status and use of biological indicators for evaluating the effects of stress on fish. *In*: S.M. Adams (ed.). Biological Indicators of Stress in Fish, Vol. 8. Am. Fish Soc. Symp. pp. 8–18.

Addison, R.F. and A.J. Edwards. 1988. Hepatic-microsomal mono-oxygenase activity in flounder *Platichthys flesus* from polluted sites in langesundfjord and from mesocosms experimentally dosed with diesel oil and copper. Mar. Ecol. Prog. Ser. 46: 51–54.

Akcha, F., C. Izuel, P. Venier, H. Budzinski, T. Burgeot and J.F. Narbonne. 2000. Enzymatic biomarker measurement and study of DNA adduct formation in benzo[a]pyrene-contaminated mussels, *Mytilus galloprovincialis*. Aquat. Toxicol. 49(4): 269–287.

Albertsson, E., P. Kling, L. Gunnarsson, D.G.J. Larsson and L. Forlin. 2007. Proteomic analyses indicate induction of hepatic carbonyl reductase/20 beta-hydroxysteroid dehydrogenase B in rainbow trout exposed to sewage effluent. Ecotox. Env. Saf. 68(1): 33–39.

Ariese, F., S.J. Kok, M. Verkaik, C. Gooijer, N.H. Velthorst and J.W. Hofstraat. 1993. Synchronous fluorescence spectrometry of fish bile—a rapid screening method for the biomonitoring of PAH exposure. Aquat. Toxicol. 26(3–4): 273–286.

Barret, J.C., H. Vainio, D. Peakall and B.D. Goldenstein. 1997. 12th Meeting of the Scientific Group on Methodologies for the Safety Evaluation of Chemicals: Susceptibility to Environmental Hazards. Environ. Health Perspect. 105(Supl. 4): 699–737.

Bastos, V.L.F.C., J.B. Salles, R.H. Valente, I.R. Leon, J. Perales, R.F. Dantas, R.M. Albano, F.F. Bastos and J. Cunha Bastos. 2007. Cytosolic glutathione peroxidase from liver of pacu (*Piaractus mesopotamicus*), a hypoxia-tolerant fish of the Pantanal. Biochimie 89(11): 1332–1342.

Bayne, B.L., D.A. Brown, K. Burns, D.R. Dixon, A. Ivanovici, D.R. Livingstone, D.M. Lowe and M. Moore. 1985. The effects of stress and pollution on marine animals. T Praeger Publishers, New York, USA.

Bocquene, G., C. Bellanger, Y. Cadiou and F. Galgani. 1995. Joint action of combinations of pollutants on the acetylcholinesterase activity of several marine species. Ecotoxicology 4(4): 266–279.

Bunton, T.E. and J.M. Frazier. 1994. Extrahepatic tissue copper concentrations in white perch with hepatic copper storage. J. Fish Biol. 45(4): 627–640.

Burke, M.D. and R.T. Mayer. 1974. Ethoxyresorufin—Direct fluorimetric assay of a microsomal O-dealkylation which is preferentially inducible by 3-methylcholanthrene. Drug Metab. Dispos. 2(6): 583–588.

Celander, M. and L. Forlin. 1991. Catalytic activity and immunochemical quantification of hepatic cytochrome-P-450 in beta-naphthoflavone and isosafrol treated rainbow-trout (*Oncorhynchus mykiss*). Fish Physiol. Biochem. 9(3): 189–197.

Celander, M.C. 2011. Cocktail-effects on biomarker responses in fish. Aquat. Toxicol. 105 Issues 3–4(3–4): 72–77.

Chambers, J.E., J.R. Heitz, F.M. Mccorkle and J.D. Yarbrough. 1978. Effects of crude-oil on enzymes in brown shrimp (*Penaeus* sp.). Comp. Biochem. Physiol. Part C Toxicol. Pharmacol. 61(1): 29–32.

Cunha, I., L.M. Garcia and L. Guilhermino. 2005. Sea-urchin (*Paracentrotus lividus*) glutathione S-transferases and cholinesterase activities as biomarkers of environmental contamination. J. Environ. Monitor. 7(4): 288–294.

da Silva, A.Z., J. Zanette, J.F. Ferreira, J. Guzenski, M.R.F. Marques and A.C.D. Bainy. 2005. Effects of salinity on biomarker responses in Crassostrea rhizophorae (Mollusca, Bivalvia) exposed to diesel oil. Ecotox. Environ. Safe. 62(3): 376–382.

Decaprio, A.P. 1997. Biomarkers: Coming of age for environmental health and risk assessment. Environ. Sci. Technol. 31(7): 1837–1848.

Depledge, M.H., J.J. Amaral-Mendes, B.R.S.H. Daniel, P. Kloepper-Sams, M.N. Moore and D.B. Peakall. 1992. The conceptual basis of the biomarker approach. *In*: D.B. Peakall and L.R. Shugart (eds.). Biomarkers: Research and Application in the Assessment of Environmental Health, Vol. 68. Berlin, Germany, pp. 15–29.

Dijkstra, M., R. Havinga, R.J. Vonk and F. Kuipers. 1996. Bile secretion of cadmium, silver, zinc and copper in the rat. Involvement of various transport systems. Life Sci. 59(15): 1237–1246.

ECETOC. 1993. Environmental hazard assessment of substances. European Centre for Ecotoxicology and Toxicology of Chemicals: Technical Report No. 51. Brussels, Belgium.

Eggens, M., F. Galgani, J. Klungsoyr and J. Everts. 1992. Hepatic erod activity in dab *Limanda limanda* in the German bight using an improved plate-reader method. Mar. Ecol. Prog. Ser. 91(1–3): 71–75.

Espino, G. 2000. Criterios generales para la elección de bioindicadores. *In*: G. Lanza Espino, S. Hernández Pulido and J.L. Carbajal Pérez (eds.). Organismos indicadores de la calidad del agua y de la contaminación (bioindicadores). Mexico, pp: 17–42.

FAO/SIDA. 1983. Manual de métodos de investigación del medio ambiente acuático. Parte 9. Análises de presencia de metales y organoclorados en los peces. pp: 35.

Fernandes, C., A. Fontainhas-Fernandes, D. Cabral and M.A. Salgado. 2008. Heavy metals in water, sediment and tissues of Liza saliens from Esmoriz-Paramos lagoon, Portugal. Environ. Monit. Assess. 136(1–3): 267–275.

Galgani, F., G. Bocquene, M. Lucon, D. Grzebyk, F. Letrouit and D. Claisse. 1991. Erod measurements in fish from the Northwest part of France. Mar. Pollut. Bull. 22(10): 494–500.

Galgani, F., G. Bocquene, P. Truquet, T. Burgeot, J.F. Chiffoleau and D. Claisse. 1992. Monitoring of pollutant biochemical effects on marine organisms of the French Coasts. Oceanol. Acta 15(4): 355–364.

Goksoyr, A. 1991. A semi-quantitative cytochrome P450IA1 ELISA: a simple method for studing the monooxygenase induction response in environmental monitoring and ecotoxicological testing of fish. Sci. Total Environ. 101: 255–262.

Guyon, R., M. Rakotomanga, N. Azzouzi, J.P. Coutanceau, C. Bonillo, H. D'Cotta, E. Pepey, L. Soler, M. Rodier-Goud, A. D'Hont, M.A. Conte, N.E.M. van Bers, D.J. Penman, C. Hitte, R.P.M.A. Crooijmans, T.D. Kocher, C. Ozouf-Costaz, J.F. Baroiller and F. Galibert. 2012. A high-resolution map of the Nile tilapia genome: a resource for studying cichlids and other percomorphs. BMC Genomics 13: 222.

Hanson, N. 2008. Does fish health matter? The utility of biomarkers in fish for environmental assessment. Ph.D. Thesis, University of Gothenburg, Göteborg, Sweedeen.

Hauser-Davis, R.A., F.F. Bastos, T.F. Oliveira, R.L. Ziolli and R.C. Campos. 2012. Fish bile as a biomarker for metal exposure. Mar. Pollut. Bull. 64(8): 1589–95.

Hauser-Davis, R.A., R.A. Gonçalves, R.L. Ziolli and R.C. Campos. 2012. A novel report of metallothioneins in fish bile: SDS-PAGE analysis, spectrophotometry quantification and metal speciation characterization by liquid chromatography coupled to ICP-MS. Aquat. Toxicol. 116–117: 54–60.

Hauser-Davis, R.A., T.F. Oliveira, A.M. Silveira, T.B. Silva and R.L. Ziolli. 2010. Case study: Comparing the use of nonlinear discriminating analysis and Artificial Neural Networks in the classification of three fish species: acaras (*Geophagus brasiliensis*), tilapias (*Tilapia rendalli*) and mullets (*Mugil liza*). Ecol. Inf. 5: 474–478.

He, A., Y. Luo, H. Yang, L. Liu, S. Li and C. Wang. 2011. Complete mitochondrial DNA sequences of the Nile tilapia (*Oreochromis niloticus*) and Blue tilapia (*Oreochromis aureus*): genome characterization and phylogeny applications. Mol. Biol. Rep. 38(3): 2015–2021.

Holdway, D.A., S.E. Brennan and J.T. Ahokas. 1994. Use of hepatic mfo and blood enzyme biomarkers in sand flathead (Platycephalus bassensis) as indicators of pollution in Port Phillip Bay, Australia. Mar. Pollut. Bull. 28(11): 683–695.

Humphrey, C.A., S.C. King and D.W. Klumpp. 2007. A multibiomarker approach in barramundi (*Lates calcarifer*) to measure exposure to contaminants in estuaries of tropical North Queensland. Mar. Pollut. Bull. 54(10): 1569–1581.

Kang, J.J. and H.W. Fang. 1997. Polycyclic aromatic hydrocarbons inhibit the activity of acetylcholinesterase purified from electric eel. Biochem. Biophys. Res. Commun. 238(2): 367–369.

Kirby, M.F., P. Matthiessen, P. Neall, T. Tylor, C.R. Allchin, C.A. Kelly, D.L. Maxwell and J.E. Thain. Hepatic EROD activity in flounder (*Platichthys flesus*) as an indicator of contaminant exposure in English estuaries. Mar. Pollut. Bull. 38(8): 676–686.

Krahn, M.M., D.G. Burrows, W.D. Macleod and D.C. Malins. 1987. Determination of individual metabolites of aromatic-compounds in hydrolyzed bile of english sole (*Parophrys vetulus*) from polluted sites in Puget-Sound, Washington. Arch. Environ. Contam. Toxicol. 16(5): 511–522.

Laflamme, J.S., Y. Couillard, P.G.C. Campbell and A. Hontela. 2000. Interrenal metallothionein and cortisol secretion in relation to Cd, Cu, and Zn exposure in yellow perch, *Perca flavescens*, from Abitibi lakes. Can. J. Fish. Aquat. Sci. 57(8): 1692–1700.

Larsson, D.G.J., M. Adolfsson-Erici, J. Parkkonen, M. Pettersson, A.H. Berg, P.E. Olsson and L. Förlin. 1999. Ethinyloestradiol—an undesired fish contraceptive? Aquat. Toxicol. 45(2–3): 91–97.

Magni, P., G. De Falco, C. Falugi, M. Franzoni, M. Monteverde, E. Perrone, M. Sgro and C. Bolognesi. 2006. Genotoxicity biomarkers and acetylcholinesterase activity in natural populations of *Mytilus galloprovincialis* along a pollution gradient in the Gulf of Oristano (Sardinia, Western Mediterranean). Environ. Poll. 142(1): 65–72.

MartinezTabche, L., B.R. Mora, C.G. Faz, I.G. Castelan, M.M. Ortiz, V.U. Gonzalez and M.O. Flores. 1997. Toxic effect of sodium dodecylbenzenesulfonate, lead, petroleum, and their mixtures on the activity of acetylcholinesterase of Moina macrocopa *in vitro*. Environ. Toxicol. Water Qual. 12(3): 211–215.

Matozzo, V., A. Tomei and M.G. Marin. 2005. Acetylcholinesterase as a biomarker of exposure to neurotoxic compounds in the clam tapes philippinarum from the Lagoon of Venice. Mar. Pollut. Bull. 50(12): 1686–1693.

Mayer, F.L., D.J. Versteeg, M.J. McKee, L.C. Folmar, R.L. Franey, D.C. McCue and B.A. Rattner. 1992. Biomarkers: Biochemical, physiological and histological marlers of anthropogenic stress. Lewis Publishing, Boca Raton, FL, USA.

McCarthey, J.F. and L.S. Shugart. 1990. Biomarkers of environmental contamination. Lewis Publishing, Boca Raton, FL, USA.

Mora, P., D. Fournier and J.F. Narbonne. 1999. Cholinesterases from the marine mussels *Mytilus galloprovincialis* Lmk. and *M. edulis* L. and from the freshwater bivalve Corbicula fluminea Muller. Comp. Biochem. Physiol. Part C Toxicol. Pharmacol. 122(3): 353–361.

Moreira, S.M., M. Moreira-Santos, R. Ribeiro and L. Guilhermino. 2004. The 'Coral bulker' fuel oil spill on the north coast of Portugal: Spatial and temporal biomarker responses in *Mytilus galloprovincialis*. Ecotoxicology 13(7): 619–630.

Nogueira, L., A.C. Rodrigues, C.P. Trídico, C.E. Fossa and E.A. de Almeida. 2011. Oxidative stress in Nile tilapia (*Oreochromis niloticus*) and armored catfish (*Pterygoplichthys anisitsi*) exposed to diesel oil. Environ. Monit. Assess. 180(1–4): 243–255.

Norris, D.O., J.M. Camp, T.A. Maldonado and J.D. Woodling. 2000. Some aspects of hepatic function in feral brown trout, Salmo trutta, living in metal contaminated water. Comp. Biochem. Physiol. Part C Toxicol. Pharmacol. 127(1): 71–78.

NRC. 1987. National Research Council Committee on Biological Markers—Biological markers in environmental health research.

Omar, W.A., K.H. Zaghloul, A.A. Abdel-Khalek and S. Abo-Hegab. 2012. Genotoxic effects of metal pollution in two fish species, *Oreochromis niloticus* and *Mugil cephalus*, from highly degraded aquatic habitats. Mutat. Res. 746(1): 7–14.

Oropesa, A.L., M. Perez-Lopez, D. Hernandez, J.P. Garcia, L.E. Fidalgo, A. Lopez-Beceiro and F. Soler. 2007. Acetylcholinesterase activity in seabirds affected by the Prestige oil spill on the Galician coast (NW Spain). Sci. Total Environ. 372(2–3): 532–538.

Parente, T.E., A.C.A.X. Oliveira and F.J.R. Paumgarttem. 2008. Induced cytochrome P450 1A activity in cichlid fishes from Guandu River and Jacarepaguá Lake, Rio de Janeiro, Brazil. Environ. Poll. 152: 233–238.

Parente, T.E., A.C.A.X. Oliveira, I.B. Silva, F.G. Araujo and F.J.R. Paumgarttem. 2004. Induced alkoxyresorufin-O-dealkylases in tilapias (*Oreochromis niloticus*) from Guandu River, Rio de Janeiro, Brazil. Chemosphere 54(11): 1613–1618.

Parente, T.E., M.F. Rebelo, M.L. da Silva, B.R. Woodin, J.V. Goldstone, P.M. Bisch, F.J. Paumgartten and J.J. Stegeman. 2011. Structural features of cytochrome P450 1A associated with the absence of EROD activity in liver of the loricariid catfish *Pterygoplichthys* sp. Gene 489: 111–118.

Parente, T.E.M., A.C.A.X. Oliveira, D.G. Beghini, D.A. Chapeaurouge, J. Perales and F.J.R. Paumgartten. 2009. Lack of constitutive and inducible ethoxyresorufin-O-deethylase activity in the liver of suckermouth armored catfish (*Hypostomus affinis* and *Hypostomus auroguttatus*, Loricariidae). Comp. Biochem. Physiol. Part C Toxicol. Pharmacol. 150: 252–260.

Pathiratne, A. and C.K. Hemachandra. 2010. Modulation of ethoxyresorufin O-deethylase and glutathione S-transferase activities in Nile tilapia (*Oreochromis niloticus*) by polycyclic aromatic hydrocarbons containing two to four rings: implications in biomonitoring aquatic pollution. Ecotoxicology 19(6): 1012–1018.

Powers, D.A. 1989. Fish as model systems. Science 246(4928): 352–358.

Purdom, C.E., P.A. Hardiman, V.J. Bye, N.C. Eno, C.R. Tyler and J.P. Sumpter. 1994. Estrogenic effects of effluents from sewage treatment works. J. Chem. Ecol. 8: 33–39.

Querol, M.V.M., E. Querol and N.N.A. Gomes. 2002. Fator de Condição gonadal, índice hepatossomático e recrutamento como indicadores do período de reprodução de *Loricariichthys platymetopon* (Osteichthyes, Loricariidae), bacia do rio Uruguai Médio, sul do Brasil, Iheringia. Serie Zoologia 92: 79–84.

Radenac, G., G. Coteur, B. Danis, P. Dubois and M. Warnau. 2004. Measurement of EROD activity: Caution on spectral properties of standards used. Mar. Biotechnol. 6(4): 307–311.

Routledge, E.J., D. Sheahan, C. Desbrow, G.C. Brighty, M. Waldock and J.P. Sumpter. 1998. Identification of estrogenic chemicals in STW effluent. 2. *In vivo* responses in trout and roach. Environ. Sci. Tech. 32(11): 1559–1565.

Schlenk, D. 1999. Necessity of defining biomarkers for use in ecological risk assessments. Mar. Poll. Bull. 39(1–12): 48–53.

Sheehan, D., K. Crimmins and G. Burnell. 1991. Bioindicators and environmental management. Academic Press, London.

Soler, L., M.A. Conte, T. Katagiri, A.E. Howe, B.Y. Lee, C. Amemiya, A. Stuart, C. Dossat, J. Poulain, J. Johnson, F. Di Palma, K. Lindblad-Toh, J.F. Baroiller, H. D'Cotta, C. Ozouf-Costaz and T.D. Kocher. 2010. Comparative physical maps derived from BAC end sequences of tilapia (*Oreochromis niloticus*). BMC Genomics 11: 636.

Stegeman, J.J., M. Brouwer, T.D.G. Richard, L. Forlin, B.A. Fowler and B.M. Sanders. 1992. *In*: R.J. Hugget, R.A. Kimerly, P.M. MJ and B.M. Bergman (eds.). Biomarkers: Biochemical, Physiological and Histological markers of Anthropogenic Stress. Lewis Publishers, Chelsea, MI, USA, pp. 235–335.

Stegeman, J.J. and P.J. Kloeppersams. 1987. Cytochrome-P-450 isozymes and monooxygenase activity in aquatic animals. Environ. Health Perspect. 71: 87–95.

Stephensen, E., J. Svavarsson, J. Sturve, G. Ericson, M. Adolfsson-Erici and L. Forlin. 2000. Biochemical indicators of pollution exposure in shorthorn sculpin (*Myoxocephalus scorpius*), caught in four harbours on the Southwest coast of Iceland. Aquat. Toxicol. 48(4): 431–442.

Sumpter, J.P. and S. Jobling. 1995. Vitellogenesis as a biomarker for estrogenic contamination of the aquatic environment. Environ. Health Perspect. 103: 173–178.

Sun, P.L., W.E. Hawkins, R.M. Overstreet and N.J. Brown-Peterson. 2009. Morphological deformities as biomarkers in fish from contaminated rivers in Taiwan. Int. J. Environ. Res. Public Health 6(8): 2307–2331.

Suter, G.W. 1993. Ecological risk assessment. Lewis Publishers, Boca Raton, FL, USA.

Tsangaris, C., E. Papathanasiou and E. Cotou. 2007. Assessment of the impact of heavy metal pollution from a ferro-nickel smelting plant using biomarkers. Ecotox. Environ. Safe. 66(2): 232–243.

van der Oost, R., A. Opperhuizen, K. Satumalay, H. Heida and N.P.E. Vermeulen. 1996. Biomonitoring aquatic pollution with feral eel (*Anguilla anguilla*). 1. Bioaccumulation: Biota-sediment ratios of PCBs, OCPs, PCDDs and PCDFs. Aquat. Toxicol. 35(1): 21–46.

van der Oost, R., J. Beyer and N.P.E. Vermeulen. 2003. Fish bioaccumulation and biomarkers in environmental risk assessment: a review. Environ. Toxicol. Pharmacol. 13(2): 57–149.

Vanderweiden, M.E.J., J. Vanderkolk, R. Bleumink, W. Seinen and M. Vandenberg. 1992. Concurrence of P450-1a1 induction and toxic effects after administration of a low-dose of 2,3,7,8-Tetrachlorodibenzo-P-Dioxin (Tcdd) in the rainbow-trout (*Oncorhynchus mykiss*). Aquat. Toxicol. 24(1–2): 123–142.

Viarengo, A., D.M. Lowe, C. Bolognesi, E. Fabbri and A. Koehler. 2007. The use of biomarkers in biomonitoring: a 2-tier approach assessing the level of pollutant-induced stress syndrome in sentinel organisms. Comp. Biochem. Physiol. Part C Toxicol. Pharmacol. 146(3): 281–300.

Zagal, A. and B. Mazmanci. 2011. Oxidative stress response in Nile tilapia (*Oreochromis niloticus*) exposed to textile mill effluent. Toxicol. Ind. Health. 27(1): 81–85.

CHAPTER 8

Blood Parameters of Estuarine and Marine Fish as Non-Destructive Pollution Biomarkers

R. Seriani,[1,]* D.M.S. Abessa,[1] C.D.S. Pereira,[2,3]
A.A. Kirschbaum,[4] L.D. Abujamara,[2] L.M. Buruaem,[5]
C. Félix,[6] G.C.R. Turatti,[6] L.R.G.B. Prado,[6]
E.C.P.M. Sousa[3] and M.J.T. Ranzani-Paiva[7]

Introduction

Blood is a tissue composed of circulatory cells in plasma that is responsible for carrying gases, proteins, and nutrients, as well as for maintaining an acid-base balance, removing metabolite waste from tissues, and actively participating in homeostasis. Because blood is continuously circulating among different tissues and organs, its characteristics may be altered due to stress or to any imbalance; thus, aspects of the blood may be considered a good indicator of overall health status in vertebrates (Aubin et al. 2001; Akinrotimi et al. 2010). Fish are certainly known to be in close relationship with the aqueous environment; hence, their blood will reveal conditions within their bodies long before there is any visible manifestation of

Authors' affiliations given at the end of the chapter.

disease (Musa and Omoregie 1999; Okechukwu et al. 2007). Due to this phenomenon, hematological indices are widely used by fish biologists and researchers the world over.

The use of hematological parameters in the assessment of fish physiology was first proposed by Hesser (1960). Since then, it has been employed in the detection of physiological changes, including those that result from the exposure to different stressful conditions such as handling, pollutants, medications, hypoxia, variations in pH and salinity, parasitism, feeding conditions, migratory activity, reproductive cycles, anesthetics, and acclimation (Blaxhall 1972; Duthie and Tort 1985; Ogbulie and Okpowasili 1999; Aubin et al. 2001; Ranzani-Paiva and Silva-Souza 2004; Tavares-Dias and Moraes 2004; Alwan et al. 2009; Seriani and Ranzani-Paiva 2012; Seriani et al. 2012). Thus, understanding hematological characteristics is an important tool that can be used as an effective and sensitive index for monitoring physiological and pathological changes in fish (Francesco et al. 2012), in field and laboratory conditions alike (Al-Sabti and Metcalfe 1995; Kirschbaum et al. 2009). Hematological parameters provide information about the health status of organisms, and they may also indicate abnormal environmental conditions (Seriani et al. 2011a). In this context, several authors worldwide have evaluated blood parameters in freshwater fish, particularly those with economic importance such as *Oreochromis niloticus* (Seriani et al. 2012), *O. mossambicus* (Nussey et al. 1995), *Clarias variepinus* (Van Vuren et al. 1994; Ololade and Oginni 2010), and *Ciprinus carpius* (Ajani and Akpoilih 2010; Vinodhini and Narayanan 2009), among others. On the other hand, knowledge of the hematology of marine fish is still scarce. According to Ranzani-Paiva and Silva-Souza (2004), hematological characteristics are known for only about fourteen marine teleosts; thus, these authors addressed the need for studies to describe the normal characteristics of the blood of healthy marine fish, as well as to understand how their blood responds to environmental contamination.

Blood Cells as Indicators of Fish Health

Ninety-seven percent of the cells in fish blood are erythrocytes and 3% are leukocytes. Thus, fish blood samples are relatively homogeneous (Mitchelmore and Chipman 1998). Leukocytes can also be represented by different cell types, including lymphocytes, neutrophils, monocytes, basophils and eosinophils (Ranzai-Paiva and Silva-Souza 2004) (Fig. 8.1).

The red blood cells (or erythrocytes) of fish are oval-shaped and have a central nucleus with acidophilic cytoplasm. They are nucleated and red-pigmented because they carry hemoglobin, the respiratory pigment that transports O_2 and part of CO_2 within the blood. Normally, these cells possess similar shapes and sizes, but in altered fish, abnormalities may occur, such

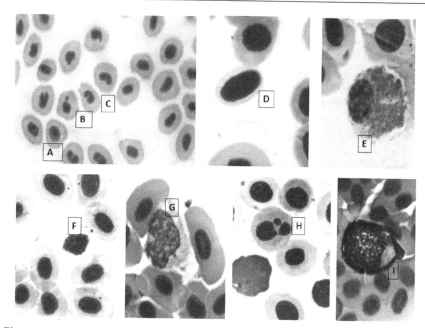

Figure 8.1. Peripheral cell types and nuclear abnormalities in erythrocytes from the tropical marine fishes studied. A: erythroblast, B: Micronucleus, C: nuclear abnormality, D: thrombocyte, E: eosinophil, F: lymphocyte, G: monocyte, H: neutrophil, I: basophil.

as the release of large quantities of young cells with gray colored cytoplasm. Erythrocytes are very susceptible to oxidative damage, because they metabolize reactive oxygen species (ROS) and are continuously exposed to high oxygen tensions. Consequently these cells are subject to changes to their permeability and antigenicity (Wagner et al. 1988; Tavares-Dias and Moraes 2004).

The hemoglobin concentration and the number of erythrocytes are both directly linked to physiological performance. Alterations to such factors may indicate diseases or health disorders. An increase in the number of erythrocytes is called erythrocytosis, and this phenomenon may be related to the proliferation of red cells by hematopoietic tissues or cancer. Conversely, anemia is characterized by a decrease in the amount of hemoglobin and is classified based on the number of erythrocytes and the concentration of hemoglobin. Anemia-inducing processes are attributed to the delay or inhibition of erythropoiesis, bleeding, hemolysis, or a significant parasite infestation (Ranzani-Paiva et al. 2000; Tavares-Dias and Moraes 2004; França et al. 2007; Seriani et al. 2009, 2010).

Fish thrombocytes (spindle cells) come in a variety of shapes, and they may be elliptic, rounded, oval-shaped, or fusiform (Ranzani-Paiva and Silva-Souza 2004) with acidophilic cytoplasm. Their nucleus is often large, and cytoplasm is scarce. Because of these features, they are sometimes confused with lymphocytes. These cells appear either as single cells or grouped together (Vázquez and Guerrero 2007). Even so, some controversy remains regarding the function and occurrence of thrombocytes. Some authors have obtained evidence that these cells have phagocytic functions (Stoskopf 1993; Matushima and Mariano 1996), but others have suggested their functions include blood clotting (Hrubec and Smith 1998) and organic defense (Matushima and Mariano 1996).

Leukocytes are non-pigmented, nucleated cells whose primary function is to combat infections and cellular debris. They have the ability to selectively migrate to and from vascular systems. Total and differential leukocyte counting may be used to evaluate the health status of fish after exposure to pollutants, because a reduction in the number of leukocytes (leukopenia) and lymphocytes is usually regarded as a stress response. Furthermore, an increase in the number of neutrophils has been described as an important indicator of acute stress that results from exposure to pollution (França et al. 2007; Seriani et al. 2009).

The differential counting of leukocytes, in which the different types of cells are identified and counted, is frequently performed, although distinguishing any specific type may be difficult. Lymphocytes are spherical, with a rounded nucleus and strongly basophilic cytoplasm. Monocytes are large cells, apparently with phagocytosis functions (Ranzani-Paiva and Silva-Souza 2004). Neutrophils are rounded cells, with a segmented nucleus. They may be frequent in the blood of some fish (and they exhibit phagocytic functions). Eosinophils and basophils are scarce and sometimes absent in the blood of fish, and their function is not well understood.

In general, there is enough data for authors to report that leukocytes are controlled by hormonal systems, and particularly by the release of cortisol during stressful episodes. According to Weys et al. (1998), cortisol changes the number of leukocytes, as well as their affinity to specific receptors, resulting in immune deficiency (Ellis 1981). Martins et al. (2000) describe this stress response in *Piaractus mesopotamicus*, in terms of the occurrence of lymphocytopenia (a reduction in the number of lymphocytes) and neutrophilia (an increased number of neutrophils).

The use of hematological analyses as stress biomarkers has an advantage, since it consists of a simple and non-destructive approach that allows for repeated measurements with the same experimental animals. When this approach is used, the test organisms are exposed to less stress as well. In addition, information about the existence, status and degree of possible sickness in fish can be rapidly obtained through use of hematological parameters (Francesco et al. 2012).

Hematological Changes in Tropical Fish *Centropomus parallelus*, *Micropogonias furnieri* and *Genidens genidens* from the Baixada Santista, São Paulo, Brazil

As previously mentioned, information on the hematology of marine fish is still scarce, and efforts have been dedicated not only to producing basic data on blood cell composition under normal conditions, but also to understanding how blood responds to natural and man-induced factors, including environmental contamination. This chapter presents the results of efforts made in Brazil to use hematology of estuarine and marine fishes as a tool to evaluate the effects of pollution, as well the implications of using this approach in environmental studies. More robust information has thus far been provided on *Centropomus parallelus* (fat snook), *Micropogonias furnieri* (whitemouth croaker), and *Genidens genidens* (catfish), after a joint effort between several research institutions.

The chosen species occupy different ecological niches in the estuarine ecosystem and also present economic and/or social importance. Fat snook meat is sold at good market prices (prime seafood), whitemouth croaker is a medium prize and is among the most consumed fish in many populations, and catfish is consumed by artisanal fishers.

Studies on the hematological responses of these fish to contamination have been conducted at three important estuaries situated along the central-south coast of São Paulo state (Fig. 8.2): the Santos Estuarine System (SES), the Itanhaem River Estuary (IRE) and the Cananeia-Iguape Estuarine Complex (CIEC).

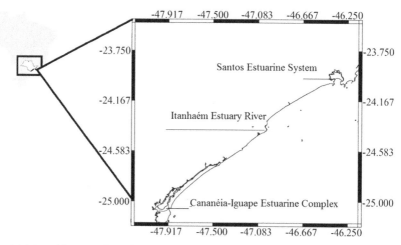

Figure 8.2. Map of the central-south coast of São Paulo showing the collection sites, the Santos Estuarine System (SES), the Itanhaém River Estuary (IRE) and the Cananéia-Iguape Estuarine Complex (CEIC).

The Santos Estuarine System (23°30'-24°S, 46°05'-46°30'W) is situated within the Baixada Santista Metropolitan Area (BSMA). The Baixada Santista is a unique region in both the civic sense and the ecological sense. It is comprised of nine cities located on the central coast of São Paulo, as well as the natural environments set on the lowland region between the littoral zone of the Atlantic Ocean and the escarpments and mountain ranges known as the Serra do Mar. The Santos Estuarine System is considered a highly polluted environment that multiple contamination sources contribute to, including a major industrial complex, the Port of Santos, discharge of untreated sewage from non-point sources and outfalls, the input of urban drainage waters, and leachates from both domestic and unregulated industrial landfills (Lamparelli et al. 2001). In this system, sediments have been described as toxic and contaminated by metals, poly-aromatic hydrocarbons (PAHs), poly-chlorinated biphenyls (PCBs), detergents, and other substances (Cesar et al. 2007; Hortellani et al. 2005; Abessa et al. 2008; Moreira et al. 2012), but effects on local biota are still unknown. Recent studies have shown biochemical effects on filter feeding bivalves (Pereira et al. 2012; Maranho et al. 2012) as well as the presence of bioaccumulation (Torres et al. 2012).

The other two estuaries have been used as reference areas. They are located further south and are subject to fewer anthropic impacts. The Itanhaém River Estuary is located in the south of the BSMA (23°50'-24°15'S, 46°35'00"W). Its surrounding area holds a local importance for commercial fishing activity. According to official data from a Sao Paulo state environmental agency (CETESB 2008), its estuarine waters present good quality throughout the year, and these results are supported by other investigations into the water and sediment (Abessa et al. 2006; Camargo and Biúdes 2006; Biúdes and Camargo 2006; Seriani et al. 2006, 2008). Moreover, the IRE basin is largely surrounded by well-preserved mangroves and Atlantic rainforest, and the mangrove area is considered part of a major Marine Protected Area.

The CIEC represents an important protected area comprised of broad mangrove and rainforest regions and representing a hotspot for biodiversity and economically exploitable living resources. It is also considered to be a Biosphere Reserve and Humanity Natural Heritage by UNESCO (Costa-Neto et al. 1997). It is for these reasons that the CIEC has been used as a reference site (Azevedo et al. 2009). Because the SES is considered to be a polluted region, comparisons were made between it and the IRE or CIEC, depending on the species. Young *M. furnieri* and *G. genidens* specimens were collected at the IRE and SES, whereas *C. parallelus* specimens were collected at the CEIC and SES. Sampling surveys were conducted during the summer and wintertime in order to consider seasonal variations, and qualitative-quantitative analyses were performed on erythrocytes and leukocytes.

After being captured, the fish were transferred to tanks containing water from the respective estuaries and anesthetized with benzocaine (3%). Next, blood samples were taken from the caudal vein using heparinized syringes, and biometric measures were then taken (weights, lengths). Blood samples were used for the red blood cell count (RBC) determinations that were performed using dilution in Natt and Herrick's solution, and counting took place in a Neubauer chamber. Hematocrit (Ht) was measured using the microhematocrit technique (Goldenfarb et al. 1971), while hemoglobin (Hb) concentration was evaluated using the cyanomethemoglobin method (Collier 1944). Mean corpuscular volume (MCV) and mean corpuscular hemoglobin concentration (MCHC) were calculated according to Wintrobe (1934). Approximately 100µl of blood were used for slides stained with May-Grunwald-Giensa (Rosenfeld 1947) for the total leukocyte (WBC) count and thrombocyte (TRB) count, according to the indirect method adopted by Hrubec and Smith (1998). Two thousand cells were analyzed per slide per specimen using optical microscopy (1000x).

Erythrocytes often are also analyzed for rates of DNA damage, since citogenotoxicity evaluations for micronuclei and nuclear abnormalities have been universally employed in the biomonitoring of coastal waters by using a variety of organisms, from invertebrates to fish (Rodriguez-Cea et al. 2003). The presence of micronuclei (MN) in peripheral blood erythrocytes has been used as a biomarker of genotoxicity since the 1970s (Schmidt 1975) (see more information in chapter 6). Micronuclei form from acentric chromosomal fragments or chromosomes that separate during their migration to the poles of the cell in anaphase (Heddle 1973; Schmidt 1975; Al-Sabti and Metcalfe 1995). According to Al-Sabti and Metcalfe (1995), the micronucleus can also be formed by apoptosis and by the action of physical agents and pollutants interacting with DNA and mitotic spindle. In addition to the micronuclei, other abnormalities in erythrocyte nuclei (Nuclear Abnormalities—NA) have been presented as citogenotoxicity biomarkers and are used as a complement to MN frequency. Ayllón and Garcia-Vasquez (2000), Kirschbaum et al. (2009) and Seriani et al. (2011) suggested that nuclear abnormalities as responses are primary and precede the formation of micronuclei, and can thus be considered more sensitive in assessing citogenotoxicity.

Moreover, both NA and MN biomarkers can be evaluated using the same slides prepared for leukocyte counting; thus, all of these analyses were employed together.

Hematological Responses of *Centropomus parallelus*

Hematological parameters from *C. parallelus* collected in the SES and CIEC were statistically different (Table 8.1). In the summertime, fish from the

Table 8.1. Hematological characteristics and nuclear alterations in erythrocytes of fat-snook *Centropomus parallelus* from the coast of Sao Paulo, Brazil.

Parameters	Winter								Summer							
	SES			CEIC					SES			CEIC				
	Mean		SD	Mean		SD			Mean		SD	Mean		SD		
Hematocrit (%)	36	±	5.1	29.55	±	3.4			45.67	±	4.8	32.1	±	7.85		
RBC (10^4/μL)	272.36	±	34.38	231	±	114.03			284.89	±	31.48	240	±	38.65		
MCV (fl)	134.22	±	26.28	127.71	±	39.26			161.31	±	18	132.63	±	20.12		
WBC (μL)	4.63	±	3.11	2.68	±	1.57			8.11	±	6.19	3.29	±	2.19		
Thrombocytes— Log10 (%)	4.7	±	0.27	4.13	±	0.4			4.84	±	0.51	4.95	±	0.58		

SES were found to have significantly higher values of Ht, MCV and TRB. During the wintertime, Ht, MCV and TRB values were significantly higher in fish from the SES.

It is recognized that temperature influences the metabolic activity of fish (Akinrotimi et al. 2010; Schmidt-Nielsen 2002). At lower temperatures, fish lower their metabolisms and, consequently, their oxygen consumption. Under these conditions, RBC production should be reduced, and reflected in the lower Ht and MCV values during the winter. In the summer, the significant increase in MCV in fish from the SES may represent the release of immature cells in the blood stream; because such young cells possess lower hemoglobin contents, the organism requires more cells to keep oxygen transport efficient in conditions of less dissolved oxygen contents in water.

Our results showed higher Ht values in *C. parallelus* from the Santos Estuarine System during both the winter and the summer. Many authors have reported increases of Ht levels in different fish species exposed to contaminants. This response was observed in *Pimelodus maculatus* and *Tilapia zilli* that had been exposed to sewage (Jeronimo et al. 2009; Saad et al. 1973), in *Pagothenia borchgrevinski* that had been exposed to hydrocarbons (Davison et al. 1992), in *O. mossambicus* that had been exposed to copper (Nussey et al. 1995), in *Rhamdia hilarii* that had been exposed to herbicide (Rigolin-Sá 1998), and in *Hoplias malabaricus* that had been exposed to methyl mercury (Oliveira-Ribeiro et al. 2006). Because the increases in Ht levels occur along with higher RBC values, these variations may have a variety of causes: i) increases in Ht levels could be a response to summer hypoxia (due to enhanced activity in higher temperatures) or to lower levels of dissolved oxygen in more saline waters; ii) these increases could be attributed to an attempt to optimize the process of detoxification of contaminants, which induces the production of erythrocytes and, consequently, the hematocrit; and iii) they could be the result of a stimulation by cell hemolysis caused by pollutants. Other potential causes need to be further investigated.

The thrombocyte numbers of *C. parallelus* collected in both study areas were higher during the summer. A similar phenomenon was observed by Rooted and Sing (1981) in a study on *Cirrhinus mrigala*, whereas Gardner and Yevich (1969) found that the morphology of thrombocytes of Ciprinodontiform fishes changed seasonally in temperate regions. Thus, thrombocytes of *C. parallelus* are subject to seasonal influences; however, mechanisms related to this response are still obscure.

The total number of leukocytes (WBC) was higher in *C. parallelus* from the SES, especially during the summer. This result indicates the presence of leukocytosis in the animals from this area. Leukocytosis in fish from the SES contrasts other studies that described a suppression of the immune system of fish exposed to pollutants (Vos et al. 1989; Weeks et al. 1992). Our study found that leukocytosis was directly related to environmental stress. A similar relationship was reported for the freshwater fish *Sarotherodon mossambicus* that had been exposed to sublethal concentrations of copper and endosulfan (Saravanan and Harikrishnan 1999), for *Heteropneustes fossilis* that had been exposed to nickel (Nanda 1997), for *Channa punctata* that had been exposed to lead (Hymavathi and Rao 2000), for *Clarias batrachus* that had been exposed to mercuric chloride (Joshi et al. 2002) and for *Labeo rohita* collected from polluted lakes in Bangalore(Karnataka, India) (Zutshi et al. 2010). In this context, the increase in leukocytes in the blood of *C. paralellus* would be the result of direct stimulation of its defense against diseases due to the presence of pollutants.

The causal relationships between the immunotoxicity of pollutants and fish diseases in the field remain unclear, as do the ecological significance of such effects. It is possible that immunological disruption caused by pollutants serves as a causal factor in the origin of fish diseases with multifactorial etiology (e.g., various skin diseases such as lymphocytosis, papilloma and skin ulcers (Vethaak 1993)). In certain situations, however, the opposite effect—protection of the fish against pathogens—may be observed after pollutant exposure (MacFarlane et al. 1986). But in cases of sub-acute or chronic exposure, hematological parameters are expected to decrease due to the depletion of hematopoietic function. Kirschbaum et al. (2009) reported the presence of tumors and citogenotoxicy effects in *C. parallelus* from the SES due to environmental contamination, especially by substances with carcinogenic, mutagenic and teratogenic potential such as PAHs and PCBs. Leukocytes, however, play an important role in nonspecific immunity, and their number in cell counting could be regarded as an indicator of the health condition of tropical marine fish (Misra et al. 2006). As for the phagocytes, they are directly involved in cell-mediated immune responses (Ballarin et al. 2004; Tavares-Dias and Moraes 2004; Tavares-Dias

2006). As in other vertebrates, white cell counts are affected by diseases, inflammation, stress, nutrition and physiological and environmental factors (Clauss et al. 2008).

Higher MN and NA rates were found in *C. parallelus* from the SES (Fig. 8.3), and these effects may be attributed to the complex contamination situation in this estuary—the contaminants originated from multiple sources (Cesar et al. 2006; Sousa et al. 2007; Abessa et al. 2008). Similar results were found in India (Mallick and Khuda-Bucksh 2003) in Centropomidae specimens from an estuary contaminated with PAHs and PCBs. An increase in NA frequency was found in CIEC organisms in the summer, but that response was considered a natural variation, as were similar results on Centropomidae from a non-polluted tropical estuary in India (Mallick and Khuda-Bukhsh 2003).

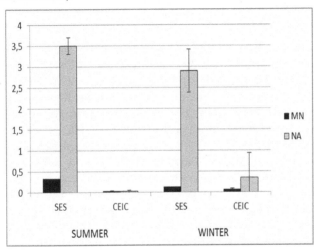

Figure 8.3. Rates of micronuclei (MN) and nuclear abnormalities (NA) in erythrocytes from *Centropomus parallelus* from the SES and CEIC, in the winter and the summer.

Hematological Responses of *Micropogonias furnieri*

In the case of *M. furnieri*, the results provided evidence of erythropoiesis inhibition among animals from the SES. This inhibition was characterized by high levels of Hb in erythrocytes, high MCHC values, and a lower number of erythrocytes (Table 8.2). Decreases in erythrocyte numbers, combined with increased concentrations of Hb and MCHC is recognized as a typical strategy for optimizing oxygen capture when erythropoiesis is inhibited and the environmental conditions are not favorable, whether due to natural factors (such as natural hypoxia or estuarine eutrophication) or to the presence of contaminants (Seriani et al. 2010).

Table 8.2. Hematological characteristics and DNA alterations in erythrocytes from whitemouth croaker *Micropogonias furnieri* from the coast of Sao Paulo, Brazil.

Parameters	Winter								Summer							
	SES			IRE			SES			IRE		
	Mean		SD	Mean		SD	Mean		SD	Mean		SD
Hematocrit (%)	29.0	±	4.2	31.3	±	2.6	28.6	±	5.8	34.4	±	3.7
RBC ($10^4/\mu L$)	2.7	±	0.5	1.4	±	0.5	2.4	±	0.4	1.9	±	0.4
MCV (fl)	108.6	±	27	225.6	±	45.9*	107.1	±	0.2	183.8	±	41.0
Hb (g/dL)	3.6	±	0.6	9	±	2.6	6.3	±	7.2	7.2	±	0.1
MCHC (g/dL)	12.7	±	2.9	28.8	±	7.8	12.3	±	33.9	10.9	±	0.7
WBC (μL)	85565	±	18154.2	53382.8	±	11710.4	68115.8	±	21172.1	43450	±	15436.1
Thrombocytes/ (μL)	3063	±	4258.3	2043.3	±	2123.2	11872.8	±	6512.5	9505	±	10451.8

An alternative hypothesis, which theorizes that the erythrocytes may have suffered hemolysis (Shah 2006; Adhikari et al. 2004; França et al. 2007), can be discarded. The SES fish exhibited high levels of MCHC and Hb. The reduction of RBC alone does not necessarily indicate hemolysis because in this case, there would be a reduction in Hb through leaching, which would reduce the concentration within erythrocytes (MCHC).

Changes to hematocrit responses have been reported as either acute or sub-acute effects (Ranzani-Paiva et al. 1997; Shah 2006). According to Fletcher and White (1986), hematocrit alterations can be splenomegaly, because the spleen acts as a phagocytic organ of erythrocytes. In our study, lower values were observed in hematocrit in the fish from the SES in both seasons. These changes to hematocrit are attributed to the increase in the volume of red cells, or to the decrease in plasma volume. MCV and MCHC also presented alterations resulting from ionic impairment, such as impaired ionic balance, which may trigger an increase in water retention that results in hemodilution. According to Allen (1994), MCV and MCHC may increase without changes to the number of erythrocytes, and when this occurs, the result will be the same amount of hemoglobin in a lower cell volume. When it came to seasonality, MCV and MCHC were found to decrease from the winter to the summer at both sites. This result suggests an effect of temperature in which the temperature increases cell production and releases these cells into the blood; however, these cells are smaller and carry lower amounts of hemoglobin.

The counting of leukocytes in *M. furnieri* exhibited a response that was similar to that of *C. parallelus*, with significantly higher numbers in specimens collected at the SES. This result indicates the occurrence of leukocytosis in these animals. The mechanisms involved in producing leukocytes are linked to cortisol metabolism (Weys et al. 1998). Under environmental stress conditions, cortisol release occurs and the expected consequence is inhibited production and the release of leukocytes (Wendelaar-Bonga 1997; Chen et al. 2002), as was previously mentioned for *C. parallelus*. However, because many pollutants found in the SES (including pharmaceuticals and organotins) may induce the activation of cytochromes in fish (Bainy et al. 1999; Pedrosa et al. 2001; Teles et al. 2003), the inhibitory action of the cortisol system could be suppressed by the stimulatory action of cytochromes, which itself may be induced by exposure to contaminants (Wilson et al. 1998). Another possibility is that the higher infestation of parasites in SES waters results in a greater leukocyte release in an attempt to combat infection. This area has received large amounts of untreated sewage (Abessa et al. 2005), and in a recent study (Pereira 2008), mussels from the SES were found to have high rates of parasitism that was attributed to sewage influence. The thrombocytes of *M. furnieri* were analyzed separately from the leukocytes, and we observed an increased number of thrombocytes in organisms from the SES compared to those from the IRE. Sub-lethal exposure to trace metals can increase the number of thrombocytes in fish. Changes to the total number of thrombocytes were observed by Lemly (2002). According to Pandey et al. (1996) and Shah and Altindaú (2005), this increase is often related to damage caused by metals in the gill, liver and kidney tissues. This response was not observed by Mazon et al. (2002) in *Prochilodus scrofa* that had been exposed to copper; however, the authors identified foci of hemorrhage resulting from the rupture of blood vessels in the gill lamellae of the fish. In both seasons, *M. furnieri* blood cells were found to have higher values of MN and NA when the specimens came from the SES. However, the differences were not significant, so results show a trend of more damaged DNA among organisms from the polluted region.

In addition, NA values in *M. furnieri* were significantly higher during the summer at clean and polluted sites alike (Fig. 8.4). This may be attributed to increased inputs of freshwater during the summertime (the rainy season), as well as to higher temperatures or other factors. This type of response was observed in other studies on estuarine fish (including Andrade et al. 2004). A similar result was observed by Kirschbaum (2010) in a study on *C. parallelus* from the Cananéia and Paranaguá estuaries. On the other hand, no significant differences in MN rates were detected between the winter and the summer. A similar situation was documented in a study on specimens from Lagoa dos Patos, Rio Grande do Sul, Brazil, the second largest lagoon in South America. In this study, croakers captured from this lagoon in the summer showed an increasing trend in weakening the DNA in the comet assay (Amado et al. 2006). The authors attributed this result to an increase

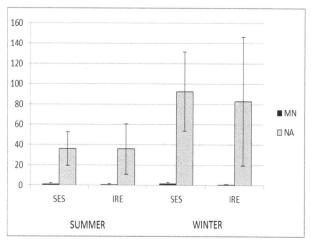

Figure 8.4. Rates of micronuclei (MN) and nuclear abnormalities (NA) in erythrocytes from *Micropogonias furnieri* from the SES and IRE, in the winter and the summer.

in the effectiveness of the DNA repair system, since the DNA weakening continued to occur; because the DNA was rapidly and more efficiently repaired, the formation of the MN erythrocytes of croakers decreased.

Hematological Responses of *Genidens genidens*

In the case of the catfish *G. genidens* from both the SES and the IRE, Ht, MCV, Hb and MCHC were not found to have any seasonal variations. Fish from the SES exhibited lower RBC numbers, higher values of MCV (reflecting macrocytic anemia), decreased hemoglobin levels, and increased MCH values during the winter (Table 8.3). In the summer, Hb was lower in the blood of animals from the SES. These results can be attributed to the delay or inhibition of erythropoiesis, bleeding, and hemolysis (Ranzani-Paiva et al. 2000; Tavares-Dias et al. 2002; França et al. 2007; Seriani et al. 2009, 2010). SES sediments possess high concentrations of metals (Lamparelli et al. 2001; Abessa et al. 2008), which may be related to the causes of anemia. Lead (Pb) is an important contaminant in the SES, and its relation to hemolysis and anemia processes is well understood. This element is capable of inhibiting the formation of hemoglobin by inactivating the enzyme ALA-D (aminolevulinic dehydratase acid delta) (Johansson-Sjobeck and Larsson 1979).

G. genidens specimens from the SES exhibited severe leukopenia and thrombocytopenia. Both leukocyte and thrombocyte numbers were clearly reduced in animals from the SES, indicating immunosuppression, which was possibly linked to the altered environment. Major effects occurred

Table 8.3. Hematological characteristics and DNA alterations in erythrocytes of the catfish *Genidens genidens* from the coast of Sao Paulo, Brazil.

Parameters	Winter								Summer							
	SES			IRE					SES			IRE			
	Mean		SD	Mean		SD			Mean		SD	Mean		SD	
Hematocrit (%)	30	±	2.17	30.69	±	6.26			25.375	±	4.083	28.6	±	1.858	
RBC/(10^4/µL)	114.8	±	30.74	243	±	44.9			189.5	±	36.596	245.9	±	52.7	
MCV (fl)	2.82	±	0.81	1.28	±	0.2			1.353	±	0.13	1.22	±	0.257	
Hb (g/dL)	7.2	±	0.063	7.7125	±	0.03			7.2	±	0.1	7.72	±	0.03	
MCHC (g/dL)	24.13	±	1.868	25.98	±	4.22			29.47	±	5.141	27.114	±	1.874	
WBC/(µL)	6051.5	±	2771.36	53373.28	±	15462.5			7976	±	2533	54149	±	40378.8	
Thrombocytes/(µL)	4483.75	±	3027.66	15022.5	±	8168.7			7028.75	±	6960.46	18866.25	±	11246.2	

in the winter. Thrombocytopenia and leukopenia may be responsible for damages to defense mechanisms, as well as for the increased susceptibility to opportunistic diseases. Pamplona (2011) identified necrosis in fish exposed to Dipyrone, associated with reductions in RBC and TRB, and leading to an impaired scar process and the removal of apoptosis cells under these conditions.

Thrombocytopenia can have devastating effects on fish. The thrombocytes are responsible not only for blood clotting, but also for controlling fluid loss from superficial wounds in fish. High levels of glucocorticoids decrease the number of thrombocytes and increase clotting time.

Immune responses in fish tend to be rapid because they are generally one of the initial responses that the organism employs to protect itself against stressor agents. As the first step, such a response occurs through the stimulation of the immune system and/or through the release of leukocytes from the reservoir tissue into the bloodstream. Thus, an increase in the number of white blood cells tends to occur rapidly, as was the case with *Satothedon melanotheron* that had been exposed to industrial effluents (Nte et al. 2011) and which was corroborated by other studies (Moharram et al. 2011; Datta et al. 2009). However, the long-term persistence of stressful conditions may lead to a suppression of leukopoietic centers, which, in turn, may cause a reduction in white blood cells, as was found in the *G. genidens* from the SES and studied herein.

The results show that *G. genidens* from the SES possess altered hematological parameters, which indicate a poor health status for this species when the specimens were compared to animals from the IRE. However, MN and NA rates in erythrocytes from *G. genidens* from the SES and the IRE were statistically similar (Fig. 8.5).

Comparisons of Hematological Responses of Estuarine Fish

When information on hematology of *G. genidens, C. parallelus* and *M. furnieri* are compared in qualitative terms, different responses become apparent. In general, fat snook and whitemouth croaker from polluted sites were found to have increased levels of MCV and MCHC, a reduced number of erythrocytes, and a higher number of leukocytes and thrombocytes. These fish thus presented with leukocytosis, thrombocytosis, and anemia. The catfish from the same polluted site, on the other hand, were found to have increased numbers of RBC, MCV and MCH, along with reduced levels of hemoglobin. These fish also exhibited lower numbers of leukocytes (leukopenia) and thrombocytes (thrombocytopenia). Such results show that hematological alterations in estuarine fish exposed to environmental contamination do occur, but they may be diverse depending on the species

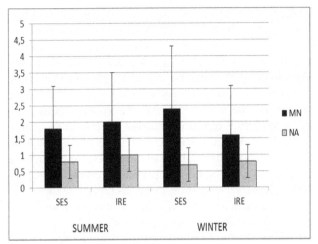

Figure 8.5. Rates of micronuclei (MN) and nuclear abnormalities (NA) in erythrocytes from *Genidens genidens* from the SES and the IRE, in the winter and the summer.

and their ecological characteristics (particularly dietary aspects and relationships to the bottom of the ocean), and physiological adaptations to naturally changing conditions.

In addition, some seasonality was observed, especially in the case of the snook and croaker. This seasonality suggests that natural fluctuations may interfere with the health status of local estuarine fish. Seasonal variations can significantly interfere with the immunological responses of fish because they are ectothermic organisms (Tavares-Dias and Moraes 2004). As for the genotoxic parameters, which were also determined using the blood samples, results of the MN and NA rates obtained for the three fish species showed that the three species exhibited different responses. The fat snook specimens from the polluted site experienced significant effects, whereas the catfish tended to have worse results when they lived in the SES. The whitemouth croaker did not exhibit significantly increased rates of MN and NA. The literature has reported varied results when these biomarkers were used: Carrasco et al. (1990) did not find any significant differences between the NA rates in fish from polluted sites and those in fish from clean sites, whereas Bombail et al. (2001) observed higher frequencies of MN and NA in *Pholis gunellus* when the specimens lived in a polluted area. Amado et al. (2006) observed high MN rates in *M. furnieri* from polluted sites of Lagoa dos Patos. Other studies demonstrated that fish from polluted areas exhibit significantly higher rates of MN and NA (Cavas and Ergene-Gözükara 2005; Rybakovas et al. 2009). NA rates have been found to be consistently higher than NA rates in studies on all three of these species. This result was widely discussed by Savage (1988), Kirschbaum et al. (2009), and Seriani et al. (2009). According to Savage (1988), this difference is due to the mechanism of formation of genotoxic damage, which becomes clear during cell division when broken DNA fragments are not transported by the spindle fibers to opposite poles of the cell during anaphase. They are then released, and because nuclear envelopes are formed during telophase, these elements may or may not be included (along the chromatin) in the nuclei of daughter cells, a fact that will depend on their location in the cell at random. If fragments are included within chromatin, nuclear abnormalities will develop, but if they remain excluded, micronuclei will form in the cytoplasm of the daughter cell. The fact is that, according to the author, there is a much higher probability of these elements being included rather than excluded from the nucleus, and this probability explains the higher rates of NA in relation to MN. According to Cavas (2008), nuclear anomalies are considered to be complementary to rates of micronuclei in routine DNA damage analyses involving the micronucleus.

Despite the initial data obtained for three Brazilian estuarine fish, much more information is necessary if we hope to develop a better understanding of how the blood of estuarine and marine fish responds to contamination. This promising approach is justified by its simplicity, speed, reliability, and ecological relevance, as well as by the wide diversity of tropical marine and estuarine fish species. Also, the lack of basic hematological data from most species should not be seen as a limitation; rather, it represents an opportunity for new research on the responses of fish to contaminants. Still, due to the simplicity and inexpensive nature of using blood parameters for environmental monitoring, further efforts should be dedicated to developing and implementing its use in environmental studies, and also to integrating them into other biomarker and ecotoxicological parameters. These methods may be spread and implemented in developing countries and on small fish farms (artisanal production), where, very often, there is no access to expensive equipment or technology. These methods are also very useful when cost is a limiting factor.

Acknowledgements

The authors would like to thank Mr. Bernardo Saraiva Ferracine for his help with images editing, the local fishers Mr Yu, Mr. Ricardo and Mr José Aparecido for their help during fish collections, the Lab Assistants, Biol. Sonia Aparecida Santiago Feliciano, Biol. Guilherme Franco Cavalheiro, Biol. Marcia Regina Gasparro and the NEPEA-UNESP staff for the assistance during field trips and laboratorial experiments.

Keywords: Blood, genotoxicity, fish, biomarker, environmental monitoring

References

Abessa, D.M.S., R.S. Carr, E.C.P.M. Sousa, B.R.F. Rachid, L.P. Zaroni, Y.A. Pinto, M.R. Gasparro, M.C. Bícego, M.A. Hortellani, J.E. Sarkis and P.M. Maciel. 2008. Integrative ecotoxicological assessment of a complex tropical estuarine system. *In:* T.N. Hoffer (Org.). Marine Pollution: New Research. Nova Science Publishers, New York, USA, pp. 279–312. Available at: https://www.novapublishers.com/catalog/product_info.php?products_id=15034

Abessa, D.M.S., F.V. Pinna, P. Romano, R. Seriani, F.L. Silveira and C. Magini. 2006. Water quality at the estuary of the Itanhaém River, SP, Brazil. *In:* COPEC [Org.] Proceeding of Environmental and Health World Congress, Santos, Brazil. Copec. pp. 42–45.

Adhikari, S., B. Sarkar, A. Chatterjee, C.T. Mahapatra and S. Ayyappan. 2004. Effects of cypermethrin and carbofuran on certain hematological parameters and prediction of their recovery in a freshwater teleost, *Labeo rohita* (Hamilton). Ecotoxicol. Environ. Saf. 58: 220–226. http://dx.doi.org/10.1016/j.ecoenv.2003.12.003

Ajani, E.K. and B.U. Akpoilih. 2010. Effect of chronic dietary copper exposure on haematology and histology of common carp (*Cyprinus carpio* L.). J. Appl. Sci. Environ. Manag. 14: 39–45. Available at: http://www.ajol.info/index.php/jasem/article/viewFile/63254/51138

Akinrotimi, O.A., B. Uedeme-Naa and E.O. Agokei. 2010. Effects of acclimation on haematological parameters of *Tilapia guineensis* (Bleeker 1862). Sci. World 5: 1–4. http://www.scienceworldjournal.org/article/view/8431

Allen, P. 1994. Changes in the haematological profile of the cichlid *Oreochromis aureus* (Steindachner) during acute inorganic mercury intoxication. Comp. Biochem. Physiol. C 108: 117–121. http://dx.doi.org/10.1016/1367-8280(94)90097-3

Al-Sabti, K. and C.D. Metcalfe. 1995. Fish micronuclei for assessing genotoxicity in water. Mutat. Res. 343: 121–135. http://dx.doi.org/10.1016/0165-1218(95)90078-0

Alwan, S.F., A.A. Hadi and A.E. Shokr. 2009. Alterations in haematological parameter of fresh water fish *Tilapia zilli* exposed to Aluminum. J. Sci. Its Appl. 3: 12–19.

Andrade, V.M., T.R.O. Freitas and J. Silva. 2004. Comet assay using mullet (*Mugil sp.*) and sea catfish (*Netuma* sp.) erytrocytes for the detection of genotoxicity pollutants in aquatic environment. Mutat. Res. 560: 57–67. http://dx.doi.org/10.1016/j.mrgentox.2004.02.006

Aubin, D.J.S., S. Deguise, P.R. Richard, T.G. Smith and J.R. Geraci. 2001. Hematology and plasma chemistry as indicators of health and ecological status in beluga whales, *Delphinapterus leucas*. ARCTIC 54: 317–331. http://arctic.synergiesprairies.ca/arctic/index.php/arctic/article/viewFile/791/817

Ayllon, F. and E. Garcia-Vazquez. 2000. Induction of micronuclei and other nuclear abnormalities in european minnow *Phoxinus phoxinus* and mollie *Poecilia latipinna*: an assessment of the fish micronucleus test. Mutat. Res. 467: 177–186. http://dx.doi.org/10.1016/S1383-5718(00)00033-4

Azevedo, J.S., A. Serafim, R. Company, E.S. Braga, D.I. Fávaro and M.J. Bebianno. 2009. Biomarkers of exposure to metal contamination and lipid peroxidation in the benthic fish *Cathorops spixii* from two estuaries in South America, Brazil. Ecotoxicology 18: 1001–1010. http://dx.doi.org/10.1007/s10646-009-0370-x

Bainy, A.C.D., B.R. Woodin and J.J. Stegeman. 1999. Elevated levels of multiple cytochrome P450 forms in tilapia from Billings reservoir—São Paulo, Brazil. Aquat. Toxicol. 44: 289–305. http://dx.doi.org/10.1016/S0166-445X(98)00084-8

Ballarin, L., M. Dall'oro, D. Bertotto, A. Libertini, A. Francescon and A. Barbaro. 2004. Haematological parameters in *Umbrina cirrosa* (Teleostei, Sciaenidae): a comparison between diploid and triploid specimens. Comp. Biochem. Physiol. A 138: 45–51. http://dx.doi.org/10.1016/j.cbpb.2004.02.019.

Bergmann-Filho, T.U. 2009. The use of fish biomarkers from an polluted estuary, SP, Brazil. M.S. Thesis. University of Aveiro. Aveiro, Portugal. 64p. Available at https://ria.ua.pt/bitstream/10773/858/1/2009001245.pdf

Biudes, J.F.V and A.F.M. Camargo. 2006. Changes in biomass, chemical composition and nutritive value of *Spartina alterniflora* due to organic pollution in the Itanhaém River Basin (SP, Brazil). Braz. J. Biol. 66: 781–789. http://dx.doi.org/10.1590/S1519-69842006000500003

Blaxhall, P.C. 1972. The haematological assessment of the health of fresh water fish. A review of selected literature. J. Fish Biol. 4: 593–604. http://dx.doi.org/10.1111/j.1095-8649.1972.tb05704.x

Bombail, V., D. Aw, E. Gordon and J. Batty. 2001. Application of the comet and micronucleus assays to butterfish (*Pholis gunnelus*) erythrocytes from the Firth of Forth, Scotland. Chemosphere. 44: 283–392. http://dx.doi.org/10.1016/S0045-6535(00)00300-3

Camargo, A.F.M. and J.F.V. Biudes. 2006. Influence of limnological characteristics of water in the occurrence of *Salvinia molesta* and *Pistia stratiotes* in rivers from the Itanhaém River basin (SP, Brazil). Acta Limnol. Bras. 18: 239–246. http://www.ablimno.org.br/acta/pdf/acta_limnologica_contents1803E_files/Artigo02_18%283%29.pdf

Campbell, T. and C. Ellis. 2007. Avian and exotic animal hematology and cytology. Wiley-Blackwell, New York. USA. http://www.lavoisier.fr/livre/notice.asp?ouvrage=1635180

Carrasco, K.R., K.L. Tilbury and M.S. Myer. 1990. Assessment of the piscine micronucleus test as an *in situ* biological indicator of chemical contaminant effects. Can. J. Fish. Aquat. Sci. 47: 2123–2136. http://www.nrcresearchpress.com/doi/abs/10.1139/f90-237#.UJ1YgYYzzNg

Çavas, T. and S. Ergene-Gözükara. 2005. Micronucleus test in fish cells: a bioassay for *in situ* monitoring of genotoxic pollution in the marine environment. Environ. Mol. Mutagen. 46: 64–70. http://dx.doi.org/10.1002/em.20130

Cesar, A., R.B. Choueri, I. Riba, M.C. Morales-Caselles, C.D.S. Pereira, A.R. Santos, D.M.S. Abessa and T.A. Delvalls. 2007. Comparative sediment quality assessment in different littoral ecosystems from Spain (Gulf of Cadiz) and Brazil (Santos and São Vicente Estuarine System). Environ. Int. 33: 429–435. http://dx.doi.org/10.1016/j.envint.2006.11.007

Clauss, T.M., A.D.M. Dove and J.E. Arnold. 2008. Hematologic disorders of fish. Vet. Clin. Exot. Anim. 11: 445–462. http://dx.doi.org/10.1016/j.cvex.2008.03.007

Collier, H.B. 1944. The standardization of blood haemoglobin determinations. Can. Med. Assoc. J. 50: 550–552. http://www.ncbi.nlm.nih.gov/pmc/articles/PMC1581573/pdf/canmedaj00573-0133.pdf

Costa-Neto, J.B., C.C. Maretti, C.F. Lino and J.L.R. Albuquerque. 1997. A Reserva da Biosfera da Mata Atlântica no Estado de São Paulo. SMA-SP/CETESB/UNESCO. São Paulo, Brazil. Available at: http://www.rbma.org.br/rbma/pdf/Caderno_05.pdf

Duthie, G.C. and L. Tort. 1985. Effects of dorsal artic carrulation on the respiration and haematology of the Mediterranean dog-fish *Segliorhinus canicula*. Comp. Biochem. Physiol. A 81: 879–883. http://dx.doi.org/10.1016/0300-9629(85)90923-5

Ellis, A.E. 1981. Inmunology of teleosts. *In:* R.J. Roberts (ed.). Fish Pathology, 2nd ed. Bailliere Tindall. London, UK, pp. 103–117. http://dx.doi.org/10.1002/9781118222942

Fletcher, T.C. and A. White. 1986. Nefhrotoxic and haematological effects of mercuric chloride in the plaice (*Pleuronectes platessa* L.). Aquat. Toxicol. 8: 77–84. http://dx.doi.org/10.1016/0166-445X(86)90054-8

França, J.G., M.J.T. Ranzani-Paiva, J.V. Lombardi, S. Carvalho and R. Seriani. 2007. Toxicidade crônica do cloreto de mercúrio (HgCl2) associado ao selênio, através do estudo hematológico em tilápia *Oreochromis niloticus*. Bioikos 21: 11–19.

Francesco, F., P. Satheeshkumar, D.S. Kumar, F. Caterina and P. Giuseppe. 2012. A comparative study of hematological and blood chemistry of Indian and Italian Grey Mullet (*Mugil cephalus* Linneaus 1758). HOAJ Biology vol. 1: 1–5. http://dx.doi.org/10.7243/2050-0874-1-5. Available at: http://www.hoajonline.com/journals/hoajbiology/content/pdf/5.pdf

Garner, R.G. and P.P. Yevich. 1969. Studies on the blood morphology of three estuarine cyprinodontiform fishes. J. Fish. Res. Bd. 26: 433–477. http://dx.doi.org/10.1139/f69-042

Goldenfarb, P.B., F.P. Bowyer, E.E. Hall and E. Brousius. 1971. Reproducibility in the hematology laboratory: the microhematocrit determination. American Journal of Clinic Pathogens 56: 59–9.

Harari, J., C.A.S. França and R. Camargo. 2008. Climatology and hidrography of Santos Estuary. *In:* R. Neves, J. Baretta and M. Mateus (eds.). Perspectives on Integrated Coastal Zone Management in South America. IST Press. Lisboa, Portugal, pp. 147–160. http://www.unisanta.br/nph/download/CoastalZoneManagementFinal.pdf

Heddle, J.A. 1973. A rapid *in vivo* test for chromosomal damage. Mutat. Res. 18: 187–90. http://dx.doi.org/10.1016/0027-5107(73)90035-3

Hesser, E.F. 1960. Methods for routine on fish haematology. Prog. Fish-Cult. 22: 164–171. http://dx.doi.org/10.1577/1548-8659(1960)22[164:MFRFH]2.0.CO,2

Hortellani, M.A., J.E.S. Sarkis, J. Bonetti and C. Bonetti. 2005. Evaluation of mercury contamination in sediments from Santos—São Vicente Estuarine System, São Paulo State, Brazil. J. Brazil. Chem. Soc. 16: 1140–1149. http://dx.doi.org/10.1590/S0103-50532005000700009

Hrubec, T.C. and S.A. Smith. 1998. Hematology of fish. *In:* B.F. Feldman, J.G. Zinkl and N.C. Jain (eds.). Schalm's Veterinary Hematology. 5th ed. W.W. Lippincott, Sydney, Australia, pp. 1120–1125.

Jeronimo, G.T., L.V. Laffitte, G.M. Speck and M.L. Martins. 2011. Seasonal influence on the hematological parameters in cultured Nile tilapia from Southern Brazil. Braz. J. Biol. 71: 719–725. http://dx.doi.org/10.1590/S1519-69842011000300005

Johansson-Sjobeck, M.J. and A. Larsson. 1979. Effects of inorganic lead on delta aminolevulinic acid dehydratase activity and hematological variables in the rainbow trout, *Salmo gairdneri*. Arch. Environ. Contam. Toxicol. 8: 419–431. http://dx.doi.org/10.1007/BF01056348

Kirschbaum, A.A. 2010. Citogenotoxicidade e determinação de metais, PAHs e organoclorados em tecidos de *Cathorops spixii* e *Centropomus parallelus* provenientes de três complexos estuarinos da costa brasileira. M.S. Thesis. University of São Paulo, São Paulo, Brazil.

Kirschbaum, A.A., R. Seriani, C.D.S. Pereira, A. Assunção, D.M.S. Abessa, M.M. Rotundo and M.J.T. Ranzani-Paiva. 2009. Cytogenotoxicity biomarkers in fat snook Centropomus parallelus from Cananéia and São Vicente estuaries, SP, Brazil. Genet. Mol. Biol. 32: 151–154. http://dx.doi.org/10.1590/S1415-47572009005000007

Lemly, A.D. 2002. Symptoms and implications of selenium toxicity in fish: the belews lake case example. Aquat. Toxicol. 57: 39–49. http://dx.doi.org/10.1016/S0166-445X(01)00264-8

Macfarlane, R.D., G.L. Bullock and J.J.A. Mclaughlin. 1986. Effects of five metals on susceptibility of striped bass to Flexibacter columnaris. T. Am. Fish. Soc. 115: 227–231. http://dx.doi.org/10.1577/1548-8659(1986)115<227:EOFMOS>2.0.CO,2

Mahiques, M.M., L. Burone, R.C.L. Figueira, A.A. Lavenere-Wanderley, B. Capellari, C.E. Rogacheski, C.P. Barroso, L.A.S. Santos and L.M. Cordero. 2009. Anthropogenic influences in a lagoonal environment: A multiproxy approach at the Valo Grande mouth, Cananéia-Iguape system (SE Brazil). Braz. J. Oceanogr. 57: 325–337. http://dx.doi.org/10.1590/S1679-87592009000400007

Mallick, P. and A.R. Khuda-Bukhsh. 2003. Nuclear anomalies and blood protein variations in fish of the Hooghly-Matlah River System, India, as an indicator of genotoxicity in water. Bull. Environ. Contam. Toxicol. 70: 1071–1082. http://dx.doi.org/10.1007/s00128-003-0092-3

Matushima, E.R. and M. Mariano. 1996. Kinetics of the inflammatory reaction induced by carrageenin in the swimbladder of *Oreochromis niloticus* (Nile tilapia). Braz. J. Vet. Res. Anim. Sci. 33: 5–10.

Mazon, A.F., E.A.S. Monteiro, G.H.D. Pinheiro and M.N. Fernandes. 2002. Hematological and physiological changes induced by short-term exposure to copper in the freshwater fish, *Prochilodus scrofa*. Braz. J. Biol. 62: 621–631. http://dx.doi.org/10.1590/S1519-69842002000400010

Misra, C.K., B.K. Das, S.C. Mukherjee and P.K. Meher. 2006. The immunomodulatory effects of tuftsin on the non-specific immune system of Indian major carp, Labeo rohita. Fish Shellfish Immun. 20: 728–738. http://dx.doi.org/10.1016/j.fsi.2005.09.004

Mitchelmore, C.L. and J.K. Chipman. 1998. DNA strand breakage in aquatic organisms and the potential value of the comet assay in environmental monitoring. Mutat. Res. 399: 135–147. http://dx.doi.org/10.1016/S0027-5107(97)00252-2

Musa, S.O. and E. Omoregre. 1999. Haematological changes in the mud fish *Clarias gariepinus* exposed to malachite green. J. Aquat. Sci. 14: 37–47. Available at: http://www.ajol.info/index.php/jas/cart/view/19971/17984

Nishigima, F.N., R.R. Weber and M.C. Bícego. 2001. Aliphatic and aromatic hydrocarbons in sediments of Santos and Cananéia, SP, Brazil. Mar. Pollut. Bull. 42: 1964–1072. http://dx.doi.org/10.1016/S0025-326X(01)00072-8

Nussey, G., J.H.J. Van Vuren and H.H. Du Preez. 1995. Effect of copper on haematology and osmoregulation of the Mozambique tilapia, *Oreochromis mossambicus* (Cichidae). Comp. Biochem. Physiol. C 111: 369–380. http://dx.doi.org/10.1016/0742-8413(95)00063-1

Ogbulie, J.N. and G.C. Okpokwasili. 1999. Haematological and histological responses of *Clarias gariepinus* and *Heterobranchus bidon* salis to some bacterial disease in Rivers State, Nigeria. J. Natl. Sci. Found. Sri. 27: 1–16. http://dx.doi.org/10.4038/jnsfsr.v27i1.2973

Okechukwu, E.O., J. Ansa and J.K. Balogun. 2007. Effects of acute nominal doses of chlorpyrifos-ethyl on some haematological indices of African catfish *Clarias gariepinus*. J. Fish. Int. 2: 190–194. Available at: http://docsdrive.com/pdfs/medwelljournals/jfish/2007/190-194.pdf

Oliveira-Ribeiro, C.A., F.F. Neto, M. Mela, P.H. Silva, M.A.F. Randi, I.S. Rabitto, J.R.M. Alves Costa and E. Pelletier. 2006. Hematological findings in neotropical fish *Hoplias malabaricus* exposed to subchronic and dietary doses of methylmercury, inorganic lead, and tributyltin chloride. Environ. Res. 101: 74–80. http://dx.doi.org/10.1016/j.envres.2005.11.005

Ololade, I.A. and O. Oginni. 2010. Toxic stress and hematological effects of nickel on African catfish, *Clarias gariepinus*, fingerlings. J. Environ. Chem. Ecotoxicol. 2: 14–19. Available at: http://www.academicjournals.org/jece/PDF/pdf2010/March/Ololade%20%20and%20Oginni.pdf

Omar, W.A., K.H. Zaghloul, A.A. Abdel-Khalek and S. Abo-Hegab. 2012. Genotoxic effects of metal pollution in two fish species, *Oreochromis niloticus* and *Mugil cephalus*, from highly degraded aquatic habitats. Mutat. Res. 746: 7–14. http://dx.doi.org/10.1016/j.mrgentox.2012.01.013

Pamplona, J.H., E.T. Oba, T.A. Silva, L.P. Ramos, W.P. Ramsdorf, M.M. Cestari, C.A.O. Ribeiro and A.R. Zampronio. 2011. Subchronic effects of dipyrone on the fish species *Rhamdia quelen*. Ecotoxicol. Environ. Saf. 74: 342–349. http://dx.doi.org/10.1016/j.ecoenv.2010.09.010

Pandey, A.K., K.C. George and M.P. Mohamed. 1996. Histopathological changes induced in gill of in estuarine mullet, *Liza parsia*, by sublethal exposure to mercuric chloride. Indian J. Fish. 43(3): 285–91. Available at: http://epubs.icar.org.in/ejournal/index.php/IJF/article/view/10140

Pedrosa, R.C., R. Gerenias, M.H. Silva, V. Figna, C. Locatelli and D.W. Filho. 2001. Biomonitoramento do Estuário do Ruio Itajaí-Açu. Utilizando a indução do citocromo p4501a e glutationa S-transferase de Bagres como biomarcadores. *In*: R. Moraes, M. Crapez, W. Pfeiffer, M. Farina, A.C.D. Bainy and V. Teixeira (eds.). Efeitos de Poluentes em Organismos Marinhos. Arte and Ciência Villipress. Brazil.

Pereira, C.D.S. 2008. Biomarcadores de exposição, efeito e bioacumulação de xenobióticos em mexilhões *Perna perna* (Linnaeus 1758) transplantados ao longo do litoral de São Paulo. Ph.D. Thesis. University of São Paulo, São Paulo, Brazil. Available at: <http://www.teses.usp.br/teses/disponiveis/21/21131/tde-24062008-142740/>. Accessed at 2012-09-19.

Pereira, C.D.S., M.L. Martín-Díaz, M.G.M. Catharino, A. Cesar, R.B. Choueri, S. Taniguchi, D.M.S. Abessa, M.C. Bícego, M.B.A. Vasconcellos, A.C.D. Bainy, E.C.P.M. Sousa and T.A. Del Valls. 2012. Chronic contamination assessment integrating biomarkers' responses in transplanted mussels—A seasonal monitoring. Environ. Toxicol. 27: 257–267. http://dx.doi.org/10.1002/tox.20638

Ranzani-Paiva, M.J.T. and A. Silva-Souza. 2004. Hematologia de peixes Brasileiros. *In:* M.J.T. Ranzani-Paiva, R.M. Takemoto and M.A.P. Lizama (eds.). Sanidade de organismos aquáticos. Varela, São Paulo, Brazil, pp. 89–120.

Ranzani-Paiva, M.J.T., A.T. Silva-Souza, G.C. Pavanelli and R.M. Takemoto. 2000. Hematological characteristics and relative condition factor (Kn) associated with parasitism in *Schizodon borelli* (Osteichthyes, Anostomidae) and *Prochilodus lineatus* (Osteichthyes, Prochilodontidae) from Paraná River, Paraná, Brazil. Acta Sci. 22: 515–521. Available at: http://periodicos.uem.br/ojs/index.php/ActaSciBiolSci/article/view/2940

Ranzani-Paiva, M.J.T., E.L. Rodrigues, M.L. Veiga, A.C. Eiras and B.E.S. Campos. 2003. Differential counts in "dourado" *Salminus maxilosus* Valenciennes, 1840, from Mogi-Guaçu River, Pirassununga, SP. Braz. J. Biol. 63: 517–525. http://dx.doi.org/10.1590/S1519-69842003000300018

Rigolin-Sá, O. 1999. Toxicidade do Herbicida Roundup (Glifosato) e do Acaricida Omite (Propargito) nas fases iniciais da ontogenia do bagre *Rhamdia hilarii* (Valenciennes, 1840) (Pimelodidae, Siluriformes). Ph.D. Thesis. Federal University of São Carlos, São Carlos, Brazil.

Rodriguez-Cea, A., F. Ayllon and E.Garcia-Vazquez. 2003. Micronucleus test in freshwater fish species: an evaluation of its sensitivity for application in field surveys. Ecotoxicol. Environ. Saf. 56: 442–448. http://dx.doi.org/10.1016/S0147-6513(03)00073-3

Rybakovas, A., J. Barsiene and T. Lang. 2009. Environmental genotoxicity and cytotoxicity in the offshore zones of the Baltic and the North Seas. Mar. Environ. Res. 68: 246–256. http://dx.doi.org/10.1016/j.marenvres.2009.06.014

Saad, M., A. Ezzat and A. Shabana. 1973. Effect of pollution on the blood characteristics of *Tilapia zillii* G. Water Air Soil Pollut. 2: 171–179. http://link.springer.com/article/10.1007%2FBF00655695?LI=true

Savage, J.R.K. 1988. A comment on the qualitative relationship between micronuclei and chromosomal aberrations. Mutat. Res. 207: 33–36. http://www.ncbi.nlm.nih.gov/pmc/articles/PMC1013369/pdf/jmedgene00309-0023.pdf

Schimidt-Nielsen, K. 2002. *Fisiologia animal: adaptação e meio ambiente.* Editora Santos, São Paulo, Brazil.

Schmidt, W. 1975. The micronucleus test. Mutat. Res. 31: 9–15. Available at: http://garfield.library.upenn.edu/classics1990/A1990EL74900001.pdf

Seriani, R. and M.J.T. Ranzani-Paiva. 2012. Alterações hematológicas em peixes: Aspectos fisiopatológicos e aplicações em ecotoxicologia aquática. *In:* A.T. Silva-Souza, M.A.P. Lizama and R.M. Takemoto (eds.). Patologia e Sanidade de organismos aquáticos. Massoni. Maringá, Brazil, pp. 221–242.

Seriani, R., F.V. Pinna, F.L. Silveira, P. Romano and D.M.S. Abessa. 2006. Toxicidade de água e sedimentos e estrutura da comunidade bentônica do estuário do Rio Itanhaém, SP, Brasil resultados preliminares. Mundo Saúde 30: 628–633. http://www.saocamilo-sp.br/pdf/mundo_saude/41/14_toxicidade_de_agua.pdf

Seriani, R., D.M.S. Abessa, C. Magini, F.V. Pinna, F.L. Silveira and P. Romano. 2008. Using bioassays and benthic community to evaluate the sediment quality at the estuary of Itanhaém River, SP, Brazil. Mundo Saúde 32: 294–301. http://www.saocamilo-sp.br/pdf/mundo_saude/63/294-301.pdf

Seriani, R., D.C. Dias, M.R.R. Silva, E.O. Villares, D.M.S. Abessa, L.B. Moreira, M.J.T. Ranzani-Paiva and D.H.R.F. Rivero. 2009. Hematological parameters of *Oreochromis niloticus* from a polluted site. The Proceedings of World Aquaculture Society. Veracruz, México.

Seriani, R., L.B.Moreira, D.M.S.Abessa, L.A.Maranho, L.D.Abujamara, N.S.B.Carvalho, A.A. Kirschbaum and M.J.T. Ranzani-Paiva. 2010. Haematological analysis in *Micropogonias furnieri* from two estuaries at Baixada Santista, São Paulo, Brazil. Braz. J. Oceanogr. 57: 1–8. http://dx.doi.org/10.1590/S1679-87592010000700011

Seriani, R., D.M.S. Abessa, A.A. Kirschbaun, C.D.S. Pereira, M.J.T. Ranzani-Paiva, A. Assunção, F.L. Silveira, P. Romano and J.L.N. Mucci. 2011. Relationship between water toxicity and hematological changes in *Oreochromis niloticus*. Braz. J. Aquat. Sci. Technol. 15: 47–53. http://www6.univali.br/seer/index.php/bjast/article/view/2129

Seriani, R., M.J.T. Ranzani-Paiva, A.T. Silva-Souza and S.R. Napoleão. 2011a. Hematological characteristics, micronuclei and nuclear abnormalities frequency in peripheral erythrocytes in fish from São Francisco Basin, Minas Gerais - Brazil. Acta Sci. Biol. Sci. 33: 107–112. http://dx.doi.org/10.4025/actascibiolsci.v33i1.7117

Seriani, R., D.M.S. Abessa, A.A. Kirschbaun, C.D.S. Pereira, M.J.T. Ranzani-Paiva, A. Assunção, F.L. Silveira, P. Romano and J.L.N. Mucci. 2012. Water toxicity and cyto-genotoxicity biomarkers in the fish *Oreochromis niloticus*. J. Braz. Soc. Ecotoxicol. 7(2): 79–84. http://dx.doi.org/10.5132/jbse.2012.02.010

Singh, N.N., V.J. Das and A.K. Srivastava. 2002. Insecticides and ionic regulation in teleosts: a review. Zoolog. Pol. 47: 21–36. http://www.biol.uni.wroc.pl/zoolog/pdf/vol_47_art_01.pdf

Sousa, E.C.P.M., D.M.S. Abessa, M.R. Gasparro, L.P. Zaroni and B.R.F. Rachid. 2007. Ecotoxicological assessment of sediments from the Port of Santos and the disposal sites of dredged material. Braz. J. Oceanogr. 55: 75–81. http://dx.doi.org/10.1590/S1679-87592007000200001

Stoskopf, M.K. 1993. Fish medicine. W.B. Saunders Company. Philadelphia, USA.

Tavares-Dias, M. and F.R. Moraes. 2004. Hematologia de peixes teleósteos. Ribeirão Preto, São Paulo, Brazil.

Teles, M., M. Pacheco and M.A. Santos. 2003. *Anguilla anguilla* L. liver ethoxyresorufin odeethylation, glutathione S-transferase, erythrocytic nuclear abnormalities, and endocrine responses to naphthalene and B-naphthoflavone. Ecotoxicol. Environ. Saf. 55: 98–107. http://dx.doi.org/10.1016/S0147-6513(02)00134-3

Van Vuren, J.H.J., M. Van Der Merwe and H.H. Du Preez. 1994. The effect of copper on the blood chemistry of *Clarias gariepinus* (Clariidae). Ecotoxicol. Environ. Saf. 29: 187–199. http://dx.doi.org/10.1016/0147-6513(94)90019-1

Vázquez, G.R. and G.A. Guerrero. 2007. Characterization of blood cells and hematological parameters in *Cichlasoma dimerus* (Teleostei, Perciformes). Tissue Cell 39: 151–160. http://dx.doi.org/10.1016/j.tice.2007.02.004

Vethaak, A.D. 1993. Fish disease and marine pollution, a case study of the flounder (*Platichthys flesus*) in Dutch coastal and estuarine waters. Ph.D. Thesis, University of Amsterdam, The Netherlands.

Vinodhini, R. and M. Narayanan. 2009. The impact of toxic heavy metals on the hematological parameters in common carp (*Cyprinus carpio* L.). Iran. J. Environ. Health. Sci. Eng. 6: 23–28. Available at: http://journals.tums.ac.ir/upload_files/pdf/_/12612.pdf

Vos, J., H. Van Loveren, P. Wester and D. Vethaak. 1989. Toxic effects of environmental chemicals on the immune system. Trends Pharmacol. Sci. 10: 289–292. http://dx.doi.org/10.1016/0165-6147(89)90031-X

Wagner, G.M., B.H. Lubin and D.T.Y. Chiu. 1988. Oxidative damage to red blood cells. *In:* C.K. Chow (ed.). Cellular Antioxidant Defense Mechanisms. CRC Press, Boca Raton, USA, pp. 185–195.

Weeks, B.A., D.P. Anderson, A.P. Dufour, A. Fairbrother, A.J. Goven, G.P. Lahvis and G. Peters. 1992. Immunologiocal biomarkers to assess environmental stress. *In:* R.J. Huggett, R.A. Kimerly, P.M. Mehrle and H.L. Bergman Jr. (eds.). Biomarkers: Biochemical, Physiological and Histological Markers of Anthropogenic Stress. Lewis Publishers, Chelsea, MI, USA, pp. 211–234.

Wilson, J.M., M.M. Vijayan, C.J. Kennedy, G.K. Iwama and T.W. Moon. 1998. Naphthoflavone abolishes interrenal sensitivity to ACTH stimulation in rainbow trout. J. Endocrinol. 157: 63–70. http://dx.doi.org/10.1677/joe.0.1570063

Wintrobe, M.M. 1934. Variations on the size and haemoglobin content of erythrocytes in the blood of various vertebrates. Folia Haematol. 51: 32–49.

[1]Núcleo de Estudos e Pesquisas em Ecotoxicologia Aquática–NEPEA–Universidade Estadual Paulista–UNESP–Campus do Litoral Paulista. Praça Infante Dom Henrique, s/n. São Vicente, SP, Brazil, 11330-900.

[2]Laboratório de Ecotoxicologia Caetano Bellibon–Universidade Santa Cecília–UNISANTA. Rua Doutor Oswaldo Cruz, 266, Santos, SP, Brazil, 11045-907.

[3]Centro de Ciências do Mar e Meio Ambiente. Universidade Federal de São Paulo. Avenida Almirante Saldanha da Gama, 89, Santos, SP, Brazil, 11030-400.

[4]Laboratório de Ecotoxicologia e Microfitobentos–Instituto Oceanográfico da Universidade de São Paulo. Praça do Oceanográfico, 191, São Paulo, SP, Brazil, 05508-900.

[5]Instituto de Ciências do Mar, LABOMAR, Universidade Federal do Ceará, Ceará, Av. da Abolição, 3207, Fortaleza, Ceará, Brazil, 60165-081.

[6]Pós Graduação *Lato Sensu*, Especialização em Gestão Ambiental. Universidade Paulista–UNIP–Campus Paraíso, Rua Vergueiro, 1211 – São Paulo, SP, Brazil, 01504-000.

[7]Centro de Pesquisas e Desenvolvimento de Peixes Ornamentais, Instituto de Pesca, São Paulo, SP. Avenida Francisco Matarazzo, 455, São Paulo, SP. 05001-970 Brazil.

*Corresponding author: robsonseriani@yahoo.com.br

CHAPTER 9

Histopathological Markers in Fish Health Assessment

Ciro Alberto de Oliveira Ribeiro[1],* and *Marisa Fernandes Narciso*[2]

Introduction

The concept of histopathology may be defined as the diagnosis and study of disease through the interpretation of cell and tissue samples under microscopy examination. When used as a biomarker to investigate the effects or the risk of a specimen's exposure to chemicals, histopathology is a powerful tool for evaluating morphological changes, including accessing the degree of pollution.

Considering the adverse effects of exposure to a complex mixture of chemicals (a more realistic situation), histopathological findings exhibit a high biological effectiveness. Two main factors can support this affirmation: (1) the high potential of bioaccumulation of large classes of chemicals to certain tissues; and (2) the complex relationship between synergic additive processes.

[1]Cellular Toxicology Laboratory, Department of Cellular Biology, C. Postal 19031, Federal University of Paraná, CEP: 81.531-980 Curitiba–PR Brazil.
Email: ciro@ufpr.br
[2]Department of Physiological Sciences, Federal University of São Carlos, Rodovia Washington Luis Km 235, CEP: 13565-905 São Carlos–SP Brazil.
Email: dmnf@ufscar.br
*Corresponding author

According to Adams et al. (1989), the advantage of histopathology as a biomarker lies in its intermediate location within the levels of biological organization. In fact, it functions as a medium-term response to sub-lethal stressors, or at the intermediate tissue level of biological organization, between molecular and individual levels. It is a particularly rapid method for detecting chronic effects. Experimentally, histopathology methods identify organs that are targets of toxicity, as well as the mechanism of action applied to fish risk assessment.

A target organ is a place in the organism where certain biological and physiological characteristics favor the greatest concentration of a specific class of chemicals. In general, a target tissue possesses appropriate conditions, or receptors, for a specific chemical or class of compounds, and in this target tissue, the effects may be more evident. Because this tissue may be more affected by a given chemical's concentration, the cellular responses are more easily measured, but the risk of exposure to the organism depends on the importance of the lesion, which may be classified as reversible or non-reversible. The consequences can be organ failure and the impairment of mechanisms that are vital to the organism's survival. Many morphological changes in response to injury are adaptive processes and are not related to cell survival. These processes include non-lethal and non-reversible damages. An important point is the possibility of distinguishing the progress of lesions among specimens if the exposure is not interrupted. This reversibility of determined change is biologically necessary to ensure a certain plasticity for the organism's survival. Otherwise, some cellular or tissue injury can become irreversible. Here, we include the irreversible changes that may lead to cell death with drastic consequences to tissue integrity and physiology. In these cases, the consequences may lead to population risk.

Water pollution induces pathological changes in fish that are frequently exposed to highly contaminated water. It produces adverse effects on exposed animals, especially in areas where the dilution rate of waste is low. When it comes to fish exposure to pollutants, their gills and livers are considered primary markers of aquatic pollution. The gills have a large surface area, and they risk direct contact with chemicals diluted in water; the liver is the first organ to have contact with chemicals after uptake, factors which make these tissues of additional interest for the investigation of environmental impacts.

Additionally, some morphological findings, such as neoplasms and pre-neoplastic lesions, play a major role in the monitoring programs because of their association with chemicals such as polycyclic aromatic hydrocarbons (PAHs), polychlorinated biphenyls (PCBs), DDT, dieldrin and chlordanes. Thus, some histopathological findings may be conclusive in establishing the main class of pollutants that the animals were exposed to.

Biomarkers as histopathological findings are intermediaries between the molecular level and the organism level and have been strongly recommended as a physiological approach to pollution investigation. According to Hinton et al. (1992), histopathological alterations occur earlier than reproductive changes and are more sensitive than growth or reproduction. As an integrative parameter, they may be excellent biomarkers for determining a correlation, environmental exposure, or adverse effects on an organism's health.

Measuring Histological Biomarkers

The concept of biomarkers has become increasingly utilized in the study of biological effects on organisms under chemical stress. The term "biomarker" is used to indicate and measure biochemical disorders or physiological and morphological injuries from either exposure to or the effects of chemicals at the sub-organismal or organismal level. In general, any measurable and reproductively biological evidence as a response to chemical exposure may be considered a biomarker. In this context, the histopathological findings in fish play an important role in investigating the risk of exposure to pollutants under experimental or natural conditions, particularly in the case of sublethal and chronic exposures. The liver is a site of high metabolism, biotransformation and excretion of xenobiotics. It is partly responsible for maintaining stable internal physiological conditions. Many studies show that the exposure of fish to contaminants induces lesions in the livers of different species, and these lesions have been found to be at distinct levels of severity, making this organ suitable for histological examination.

The standardization of sampling, preservation and laboratory techniques are well described for fish histology, but few studies involving the interpretation of histopathological findings have been developed. According to Bernet et al. (1999), in order to achieve a better understanding of histological findings after contamination exposure, the standard methods for describing and assessing these histological lesions between the organs from specimens or between specimens themselves need to be improved.

Many studies using distinct methods were employees in order to classify disorders in or changes to the liver after exposure to chemicals. Poleksić and Mitrović-Tutundžić (1994) proposed a method to interpret the histological details of gills in fish after chronic and sublethal exposure to pollutants. This method was later adapted for use on liver tissue by Maduenho and Martinez (2008), and it was then modified by Shiogiri et al. (2012), who also used it to evaluate the liver and gills. Another method was proposed by Bernet et al. (1999). It is based on the classification by Takashima and Hibidya (1995) and is in accordance with the recommendations of Sindermann (1979) and Susani et al. (1986). This method describes an assessment tool that can be applied to any given organ, standardizes the quantification of histopathological findings, and allows for a comparison between different studies and organs.

The method reported by Maduenho and Martinez (2008) and later by Shiogiri et al. (2012) consists of a semi-quantitative evaluation of tissue that considers the Degree of Tissue Change (DTC). The evaluation is based on the severity of the lesions and the possibility of tissue recovery in fish after exposure to chemicals. According to Maduenho and Martinez (2008), in the case of DTC calculation, tissue changes are classified into three progressive stages of impairment of tissue function: Stage I = changes that do not damage the tissue to such an extent that the organ cannot repair itself; Stage II = repairable changes that are more severe and affect the associated tissue function; and Stage III = changes that preclude the restoration of the structure of the tissue, even with an improvement in water quality. Although Shigiori et al. (2012) considered the same stages described above, the authors instead described four groups (G) to classify based on the type of damage and the location of the damage: (G1) hypertrophy and hyperplasia of cells and related changes; (G2) changes to specific tissue cells; (G3) blood vessel changes; and (G4) fibrosis and necrosis. According to the authors, the severity of the lesions can be evaluated using a Histopatological Index (HI). In order to calculate the severity of lesions (DTC or HI), the follow equation is used:

$$DTC \text{ or } HI = (1 \times \Sigma I) + (10 \times \Sigma II) + (100 \times \Sigma III)$$

Where:

ΣI, ΣII and ΣIII are the total number of alterations in each stage and the 1, 10 and 100 are integrated factors according to the severity of the damage.

After obtaining the DTC or HI, the median value is divided into five categories: 0–10, functionally normal tissue; 11–20, slightly to moderately damaged tissue; 21–50, moderately to heavily damaged tissue; 51–100, severely damaged tissue; and >100, irreparably damaged tissue. The method allows for a comparison to control groups and to tested groups of specimens.

The method proposed by Bernet and colleagues follows the classification described by Sindermann (1979) for the histopathological assessment of experimental studies. These classifications are also accepted by the National Oceanic and Atmospheric Administration (NOAA) quality assurance program on marine fish histopathology (Susani et al. 1986). The method describes the histopathological findings according to their importance factor, which means the relevance of the lesion depending on its pathological importance (i.e., how it affects organ function and the ability of the fish to survive). The importance factor varies on three levels:

(1) minimal pathological importance, the lesion is easy reversible; (2) moderate pathological importance, the lesion is reversible in most cases if the stressor is neutralized; and (3) marked pathological importance, the lesion is generally irreversible and leads to partial or total organ failure. Thus, the alterations are classified as regressive or progressive changes (for example, inflammation is regressive, while tumors are progressive). In the case of the liver, the regressive and progressive changes are evaluated at different sites or in different tissues from the organ as a functional unit of the tissue, including liver tissue itself, interstitial tissue, and the bile duct. In order to establish the Lesion Index for organs, the authors organized the respective pathological changes into five Reaction Patterns. Each reaction pattern includes several alterations that involve either a functional unit of the organ or an entire organ (Table 9.1). Every alteration is assessed using a score ranging from 0 to 6, depending on the degree and extent of the alteration: (0) unchanged; (2) mild occurrence; (4) moderate occurrence; and (6) severe occurrence. A detailed description of each alteration considered may be found in Bernet et al. (1999). According to the method, a lesion index of each organ may be obtained, and the sum of the lesion indices from different organs offers a lesion index of the specimen as a whole (total

Table 9.1. Histopathological findings and the respective importance factor in the liver to access the Lesion or Reaction Index. Adapted from Bernet et al. (1999).

Reaction pattern	Functional unit	Alteration	Importance factor
Circulatory disturbance		Hemorrhage and intercellular edema	1
Regressive changes	Epithelium	Architectural and structural alterations	1
		Deposits	1
		Nuclear alterations	2
		Atrophy	2
		Necrosis	3
	Supporting tissues	Architectural and structural alterations	1
		Deposits	1
		Nuclear alterations	2
		Atrophy	2
		Necrosis	3
Progressive changes	Epithelium	Hypertrophy	1
		Hyperplasia	2
	Supporting tissues	Hypertrophy	1
		Hyperplasia	2
Inflammation		Exudates	1
		Infiltration	2
Tumor		Benign	2
		Malignant	3

index). In both situations, an additional index (Reaction Index) may be calculated, where the quality of the lesion in each organ or in the specimen may be expressed.

To calculate the Organ Index (I_{org}), the following equation is used:

$$I_{org} = \sum rp \sum alt \left(a_{org\ rp\ alt} \times w_{org\ rp\ alt} \right)$$

Where:

org = organ (constant); rp = reaction pattern; alt = alteration; a = score value; w = importance factor

This index represents the degree of damage to an organ. A high index indicates a high degree of damage and allows for a comparison between specimens.

To calculate the Reaction Index ($I_{org\ rp}$), the following equation is used:

$$I_{org\ rp} = \sum alt \left(a_{org\ rp\ alt} \times w_{org\ rp\ alt} \right)$$

Where:

org, rp = constant

This index represents the details of the lesion in an organ by multiplying the importance factor and score values by the alterations of the corresponding reaction pattern. The sum of the five reaction indices in an organ is equivalent to the organ index (I_{org}) (Table 9.2).

To calculate the Total Index from one specimen if more than one organ is considered, the following equation can be used:

$$I_{total} = \sum org \sum rp \sum alt \left(a_{org\ rp\ alt} \times w_{org\ rp\ alt} \right)$$

Where I_{total} = Total Index

These indices represent the health status of the specimen based on the histopathological findings, and they make a comparison between specimens possible.

The same result can be obtained for the Total Reaction Index using the following equation, which represents the details of the lesion in all considered organs from one specimen and which also allows for a comparison between specimens:

Table 9.2. Index lesion of one individual where four organs were evaluated. Modified from Bernet et al. (1999).

Organ	Reaction Pattern					
	Circulatory disturbance (rp1)	Regressive changes (rp2)	Progressive changes (rp3)	Inflammation (rp4)	Tumor (rp5)	Σ
Org1	$I_{org1\,rp1}$	$I_{org1\,rp2}$	$I_{org1\,rp3}$	$I_{org1\,rp4}$	$I_{org1\,rp5}$	I_{org1}
Org2	$I_{org2\,rp1}$	$I_{org2\,rp2}$	$I_{org2\,rp3}$	$I_{org2\,rp4}$	$I_{org2\,rp5}$	I_{org2}
Org3	$I_{org3\,rp1}$	$I_{org3\,rp2}$	$I_{org3\,rp3}$	$I_{org3\,rp4}$	$I_{org3\,rp5}$	I_{org3}
Org4	$I_{org4\,rp1}$	$I_{org4\,rp2}$	$I_{org4\,rp3}$	$I_{org4\,rp4}$	$I_{org4\,rp5}$	I_{org4}
Σ	I_{rp1}	I_{rp2}	I_{rp3}	I_{rp4}	I_{rp5}	I_{total}

$$I_{.rp} = \sum_{org} \sum_{alt} \left(a_{org\,rp\,alt} \times w_{org\,rp\,alt} \right)$$

Where $I_{.rp}$ = Total Reaction Index

It is clear that a revision of the terms used to classify or describe histopathological findings in fish is needed when this biomarker is used to evaluate the risk of exposure to chemicals in both experimental studies in laboratories and biomonitoring programs. The diversity of alteration or damage descriptions will make it difficult to interpret or compare species or experiments. This problem is more evident or complex when the establishment of some Index of Lesion is not considered. Otherwise, before the application of any method, researchers must cluster or classify the largest possible number of alterations in reaction pattern (Bernet et al. 1999) or groups of damage classification (Shigiori et al. 2012), to consider other findings not described in either of the methods presented.

Fish Livers as Target Organs for Evaluating Exposure to Pollutants

The teleost liver is a relatively large organ. It is usually reddish brown in color in carnivorous species and lighter brown in color in herbivorous species. Different color aspects depend on the feeding habits, and on sexual or maturation stages, particularly in females. These characteristics change in individuals from aquaculture activities. The liver presents a high variability in size and position within the coleomic cavity depending on the species, but in general, it is located in the first portion of the cavity with processes extending the length of the abdomen.

The function of the liver as a digestive gland is to secret bile from the hepatic cells into the intracellular canaliculi, where it is carried into the extracellular bile canaliculi to form in the bile duct. Later, it joins with the hepatic duct and opens into the intestine. The hepatic duct possesses a branch in the majority of fish species. This branch leads into the gall bladder, where bile is stored. Bile plays an important role in digestion, especially of fatty foods in vertebrates, but it is also a vehicle of certain xenobiotic excretions. Thus, the chemical analysis of bile is also an important and complementary biomarker of exposure to organic pollutants, because it shows both acute and chronic exposure (depending on the biotransformation rate in the liver).

The tissue organization of the liver in fish differs from mammals in three main ways: (1) hepatocyte cords are not evident; (2) the apparent portal triads are not present; and (3) the pancreas can be observed with the hepatic parenchyma, particularly around vessels (Fig. 9.1). The sinusoids are presented but in irregular lining, with few Kupfer Cells. The endothelial cells forming the sinusoids are fenestrated, which limits the space of Disse with the hepatocyte border, which, in turn, presents microvilli at the surface. The polygonal hepatocytes are the most abundant cells in the liver. They possess a rounded and distinctive central nucleus and a prominent nucleolus. An evident ultra structure difference between fish and mammals is the peripheral arrangement of rough endoplasmic reticule within the cell, near the plasmatic membrane. In fish, this particular arrangement allows for the observation of cellular delimitation of hepatocytes, even under light microscopy. The loss of hepatocyte delimitation reveals disorders in organelle arrangement, which is an important finding in histopathological analyses (Fig. 9.2).

According to Boelsterli (2007), the liver is often a frequent target organ for toxicity. This is based on biological factors: (1) the organ receives a large amount of blood per unit of time; (2) the organ possesses a large variety of xenobiotic-metabolizing enzymes; (3) hepatocytes are polarized cells featuring a basolateral and apical cell domain and performing secretory and excretory functions, which increases the amount of membrane transporters also used by xenobiotics; (4) the liver is the tissue that is first exposed after oral exposure to xenobiotics; (5) hepatocyte cells do not have a basal membrane; (6) the contact of hepatocytes with the bloodstream is directly due to the fenested endothelial cells of sysusoids; (7) the liver presents immune cell populations as macrophages and T-Cells, which are targets for immune-cell-mediated toxicity. Other alterations not described or considered by Bernet at al. (1999) have been reported in different studies, as summarized in the Table 9.3. It can be seen that, depending on one's interpretation, many of these findings may be included in a Reaction Pattern, as described by Bernet et al. (1999), or in the groups of damage classifications presented by Shigiori et al. (2012).

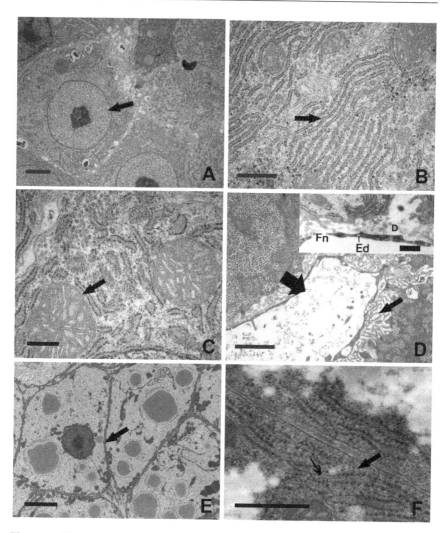

Figure 9.1. Ultrastructure of fish liver showing the normal aspect of hepatocytes. A. Observe the spherical nuclear shape (arrow) with evident nucleolus in *Astyanax* sp. (scale bar = 2 μm). B and C show the rough endoplasmic reticulum and the mitochondria (arrows), respectively, in *Astyanax* sp. (scale bar = 3 and 1 μm). D. Sinusoid (large arrow) in liver of *Astyanax* sp. The smaller arrow shows the space of Disse (scale bar = 2 μm). In detail (d), observe the endothelial cell (Ed), fenestration (Fn) and the space of Disse (D) (Scale bar = 0.5 μm). E. The arrow shows the limit between hepatocytes of *Hoplias malabaricus* after experimental exposure to MeHg due to the arrangement of rough endoplasmatic reticulum (smaller arrow) near the membrane (bigger arrow) as observed in F (scale bar –6 and 0.5 μm).

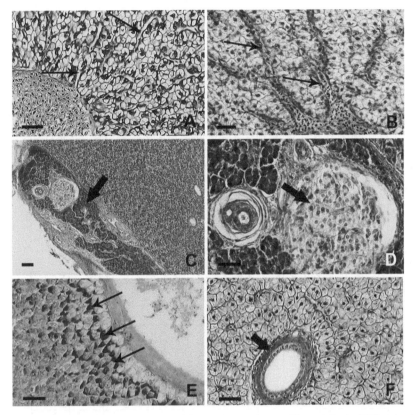

Figure 9.2. Cross section of a *Hoplias malabaricus* liver. Hematoxilin and Eosin (HE) stain. A. Observe the sinusoids and absence of cordonal organization (arrows) (scale bar = 50μ). B. The arrows show sinusoids formation (scale bar = 20 μm). C. The presence of pancreatic tissue (arrow) (scale bar = 100 μm), and in detail (D), observe the presence of endocrine tissue (islets of Langerhans) (arrow) (scale bar = 40 μm). E. The arrows show the glycogen accumulation after Shiffe's Reactive stain (scale bar = 50 μm). F. Biliar duct (arrow)(scale bar = 50 μm).

Liver Lesions as Biomarkers in Tropical and Neotropical Fish Species

Despite of a diversity of alterations and lesions found in the livers of fish used as biomarkers of effects of or exposure to pollutants, there are groups of damages that are more evident and frequent overall. This chapter will not present all lesions and alterations found in the literature (as shown above —in Table 9.3), but describes some histopathological findings and discusses them in order to clarify their physiological importance. Many findings still involve a certain degree of subjectivity, while in many cases, some of them are categorically ignored. The results show that researchers urgently need an

Table 9.3. Liver lesions described in the recent literature for fish species under experimental or natural exposure to different pollutants, not considered in the methods presented above.

Lesion (Bernet's Lesion Index classification)	Species	Contaminant	Exposure	Reference
Melanomacrophages centers (no classification)	*Prochilodus lineatus*	Diflubenzuron insecticide	Waterborne experimental	Pereira and Martinez (2008)
	Oreochromis mossambicus	^{60}Co gamma irradiation	Experimental	Buchari et al. (2012)
	Oreochromis niloticus	Wetland	Natural exposure	Abdel-Moneim et al. (2012)
	Clarias gariepinus	Freshwater	Natural exposure	Dyk et al. (2012)
	Barbus graellsii	Freshwater	Natural exposure	Raldúa et al. (2007)
	Alburnus alburnus	Freshwater	Natural exposure	Raldúa et al. (2007)
Piknotic nuclei (Nuclear alteration)	*Oncorhynchus mykiss*	Captan fungicide	Waterborne experimental	Boran et al. (2012)
	Oreochromis niloticus	Wetland	Natural exposure	Abdel-Moneim et al. (2012)
Vacuolation of hepatocytes or Steatosis (Deposits)	*Hydrocynus vittatus*	DDT-affected area	Natural exposure	McHugh et al. (2011)
	Channa Punctatus	Arsenic	Waterborne experimental	Roy and Bhattacharya (2006)
	Whitefish	Oil pipelines	Natural exposure	Lukin et al. (2011)
	Carassius auratus	Chromium (VI)	Waterborne experimental	Velma and Tchounwou (2010)
	Chanos chanos	Toxic metals	Natural exposure	Rajeshkumar and Munuswamy (2011)
	Oreochromis niloticus	Wetland	Natural exposure	Abdel-Moneim et al. (2012)
	Danio rerio	Phamaceutical solutions	Waterborne experimental	Madureira et al. (2012)
	Onchorynchus mykiss	Nonylphenol	Waterborne experimental	Uguz et al. (2003)
	Platichthys flesus	Estuary	Natural exposure	Koehler (2004)
Shrinkage of hepatocytes (Atrophy)	*Channa punctatus*	Cadmium VI	Waterborne experimental	Mishra and Mohanty (2008)
Hepatocyte hypertrophy or Hidropic change (Progressive change)	*Clarias gariepinus*	Freshwater	Natural exposure	Dyk et al. (2012)
Sinusoidal enlargement space (Circulatory disturbance)	*Cirrhinus mrigala*	Pyretroid pesticide	Waterborne experimental	Velmurugan et al. (2007)

Marker (classification)	Species	Matrix/compound	Exposure	References
Vitellogenin expression (No classification)	*Oryzias latipes and Danio rerio* / *Oryzias latipes*	HCH and Estradiol / Cianotoxins	Waterborne experimental / Waterborne and oral experimental	Wester et al. (2002) / Marie et al. (2012)
	Danio rerio / *Pimephales promelas*	Pharmaceutical solution / Freshwater (municipal wastewater)	Waterborne experimental / Natural exposure	Madureira et al. (2012) / Tetreault et al. (2012)
	Culaea inconstans	Freshwater (Municipal wastewater)	Natural exposure	Tetreault et al. (2012)
Karyolysis (Necrosis)	*Channa punctatus*	Captan fungicide	Waterborne experimental	Roy and Bhattacharya (2006)
Bile duct fibrosis (Architectural and structural alterations)	*Whitefish*	Oil pipelines	Natural exposure	Lukin et al. (2011)
Eosinophilic bodies or Hyaline degeneration (Necrosis)	*Solea senegalensis* / *Oreochromis niloticus*	Sediment / Wetland	Experimental / Natural exposure	Costa et al. (2012) / Abdel-Moneim et al. (2012)
Basophilic and eosinophilic foci (Tumour)	*Platichthys flesus* / *Platichthys flesus* / *Clarias gariepinus*	Estuary / Estuary / Freshwater	Natural exposure / Natural exposure / Natural exposure	Stentiford (2003) / Lang et al. (2006) / Dyk et al. (2012)
Difuse lipidosis or Fat degeneration (Necrosis)	*Salmo salar*	Soybean oil	Oral exposure experimental	Ruyter at al. (2006)
Fibrilar inclusions (Deposits)	*Platichthys flesus* / *Platichthys flesus* / *Clarias gariepinus*	Estuary / Estuary / Freshwater	Natural exposure / Natural exposure / Natural exposure	Stentiford (2003) / Lang et al. (2006) / Dyk et al. (2012)
Pleomorphic nuclei (Nuclear alterations)	*Platichthys flesus*	Estuary	Natural exposure	Stentiford (2003)
Bile stagnation (Architectural and structural alterations)	*Oreochromis niloticus*	Wetland	Natural exposure	Abdel-Moneim et al. (2012)
Perivascular and Peribiliary granulomatosis (Architectural and structural alterations)	*Clarias gariepinus*	Freshwater	Natural exposure	Dyk et al. (2012)
Celular fusion (Architectural and structural alterations)	*Colossoma macropomum*	Paraquat herbicide	Waterborne experimental	Salazar-Lugo et al. (2011)

intercalibration and standardization of the terms, descriptions, concepts and identification of a large amount of lesions or alterations. Another aspect that needs discussion or standardization is the physiological interpretation of the lesion or alteration, including its ecological significance to a population or community of fish.

Overall, despite the histopathological biomarkers mentioned previously, it can be concluded from the literature that liver histopathology in fish conducted according to standardized procedures is a useful tool to be incorporated in monitoring programs that assess the health of aquatic ecosystems. Because the liver is the main detoxification organ, and because it is responsible for metabolism and excretion of xenobiotic chemicals, multiple pathological liver alterations are to be expected in wild fish populations. Tropical and neotropical fish possess a more intense metabolism relative to species from more temperate climates. Because of this phenomenon, it is expected that the bioativation or biotransformation of organic molecules could be more accelerated than usual, which would promote increases in both effects and elimination. In short, tropical and neotropical animals are more sensitive to the same dose or concentration of chemicals.

Necrosis is a form of cell injury to living cells or tissues, such as infection, toxins or trauma, that result in unregulated membrane lesions and the digestion of cell components. In fish livers, necrosis is not necessarily due to specific pollutants, since little evidence links damage to specific organic or inorganic compounds (Rabitto et al. 2005).

The incidence of necrosis in the livers of fish after experimental and natural exposure is extensively described in the literature. This lesion is permanent in the liver and it can lead to organ failure depending on the extent of the damage. Although the occurrence of necrosis in the livers of tropical fish has been considered normal under some circumstances, these findings are still useful when establishing a diagnostic study because of their significance to fish health. Studies have reported on these lesions in *Hoplias malabaricus* after being experimentally exposed to cyanotoxin or naturally exposed to mercury in the Amazon region (Fig. 9.3), and they have described alterations to the shape of nuclei and necrotic cells (arrow) in *Hoplias malabaricus* after natural exposure to urban sludge and agricultural activities (Fig. 9.4). Also, the incidence of necrosis was high in *Rhamdia quelen* specimens after being experimentally exposed to BaP, and also in *Pimelodus maculatus* and *Oligosarcus hepsetus* specimens after natural exposure to an impacted reservoir (Fig. 9.3).

Steatosis, or fatty liver, can be a temporary or long-term condition, which is not harmful in itself, but it may indicate another type of problem. This kind of alteration is usually reversible once the cause of the problem is corrected. In fish, the liver is the organ responsible for changing fats from the diet to types of fat that can be stored and used for energy and new

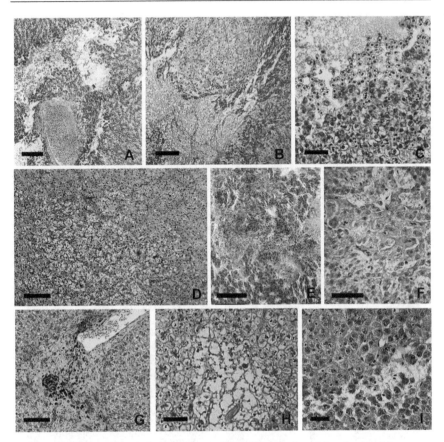

Figure 9.3. Incidence of necrosis in the livers of fish after experimental and natural exposures. HE stain. A, B, D and E. in *Hoplias malabaricus* after experimental exposure to cyanotoxin (scale bar = 100, 100, 50 and 20 μm). C. *In Rhamdia quelen* after experimental exposure to BaP (scale bar = 20 μm). F. In *Hoplias malabaricus* after natural exposure to mercury in the Amazon region (scale bar = 50, 20 and 20 μm). G and I. In *Pimelodus maculatus* after natural exposure to impacted reservoir (scale bar = 200 and 50 μm respectively). H. In *Oligosarcus hepsetus* after natural exposure to impacted reservoir (scale bar = 50 μm).

cell formation. The consequence is the deposit of large globules of fat in hepatocytes (more specifically, they are distributed within the cytoplasm). The exposure to toxic chemicals or low quantities of protein in the diet can contribute to the occurrence of fatty liver.

In fish, this finding is related to both acute and chronic responses to a variety of stressors. Lipodystrophy is believed to be a prenecrotic stage and has been observed in fish exposed to a myriad of pollutants. However, Biagianti-Risbourgt et al. (1997) described lipid storage as a mechanism of defense against the presence of liposoluble contaminants. Harbor activities contribute to an increase in many liposoluble chemicals, such as polycyclic

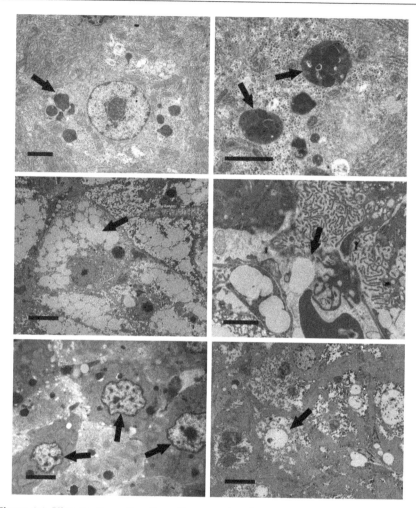

Figure 9.4. Ultrastructure alteration after natural and experimental exposure. A and B. Cytoplasmic inclusions such as cholestasis in *Micropogonia furnieri* (arrows) after natural exposure to urban sludge and agricultural activity (scale bar = 2 µm). C. Fatty liver disease (arrow) in *M. furnieri* (arrow) (scale bar = 5 µm) and in (D) alterations in sinusoid as the disorganization of membranes. Observe the damages to the endothelial cell (arrow) (scale bar = 2 µm). E and F show alterations to the shape of nuclei and necrotic cells (respectively) (arrow) in *Hoplias malabaricus* after natural exposure to urban sludge and agricultural activities (scale bar = 5µm).

aromatic hydrocarbons (PAHs) and organochlorines. This alteration was found in the liver of *Atherinela brasiliensis* after natural exposure to harbor activities, where a high incidence of steatosis was found (9.5). Although the mechanism is still unknown, the description of steatosis as a prenecrotic event has also been suggested by Abdel-Moneim et al. (2012) and was also

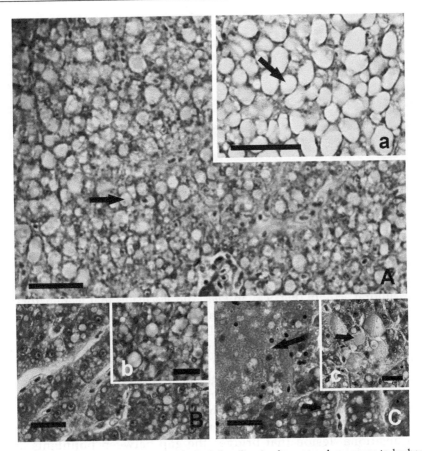

Figure 9.5. Steatosis in the liver of *Atherinela brasiliensis* after natural exposure to harbor activities. HE stain. A and a show high incidence of steatosis that is evident in the citoplasmatic vesicles (arrows) (scale bar= 50 μm). B and b. Lower incidence of steatosis (scale bar = 10 μm). C. Steatosis (smaller arrows) associated with necrosis (bigger arrow). (scale bar = 10 μm). In detail (c), observe some cells in necrosis (arrows) associated with the occurrence of steatosis (scale bar = 10 μm).

found in *A. brasiliensis* specimens that were naturally exposed to harbor activities and in *Micropogonia furnieri* after natural exposure to urban sludge and agricultural activity (Fig. 9.7). A nuclear alteration in hepatocytes of *A. brasiliensis* was associated with steatosis (Fig. 9.6). Typically, a diversity of studies with tropical species of fish has demonstrated the occurrence of this kind of lesion.

Intrahepatic cholestasis is characterized by episodes of liver dysfunction, and it is referred to as cholestasis when the hepatocytes present with a reduced ability to release bile leading to bile stagnation within the cell. The characteristic is the deposit of brownish-yellow granules in the

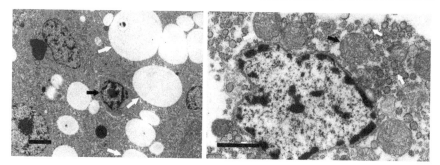

Figure 9.6. Nuclear alteration in hepatocytes of *Atherinela brasiliensis* (A and B), steatosis (A - white arrows) and rough endoplasmic reticulum disorganization (white arrows) (scale bar = 2 and 1 μm respectively).

cytoplasm (Fig. 9.4). With bile stagnation, many cytotoxic events, such as cytoskeleton disorders or vesicular fusion are related, and they have important consequences in terms of cellular metabolism. These alterations were found in *M. furnieri* after natural exposure to urban sludge and agricultural activity (Fig. 9.4), in *Pimelodus maculatus* that were naturally exposed to an impacted reservoir, and in *Hoplias malabaricus* that were naturally exposed to mercury and DDT (Fig. 9.7).

Physiologically, the cholestasis event reduces the absorption of fat in the body, which leads to an excess of fat in the feces (steatorrhea). This, in turn, leads to loss of appetite and weight. Long episodes of liver dysfunction occasionally develop into a more severe and permanent form of liver disease known as progressive intrahepatic cholestasis.

Neoplastic changes in the livers of fish are thought to represent a transitional lesion that bridges the gap between pollutants and hepatocellular malignant lesions, such as hepatocellular carcinoma (Hinton et al. 1992). In general, these lesions are associated with long-term exposure to carcinogenic compounds that results in the formation of adenomatous hyperplasia, which can develop into cholangioma (a benign tumor caused by the proliferation of bile ducts). Three different cell types are involved in carcinogens in the livers of fish: hepatocytes, bile epithelial cells and sinusoidal endothelial cells. Two or more common alterations found in the livers of fish are vacuolated and eosinophilic foci, which are precursor lesions of basophilic foci. The basophilic cell is a preneoplastic stage because this cell type persists during cancer progression, and it is the main evidence of malignant cancers (hepatocellular carcinomas) in fish. This kind of lesion is the most common preneoplastic event, and it first appears in the perisinusoidal and peribiliary zones.

Also, the chronic production of cellular oxidants from endogenous or exogenous sources is well described as a carcinogenic pathway due to damages to DNA. It may be attributed to chemicals that are non-DNA-reactive or indirectly DNA-reactive in cell growth.

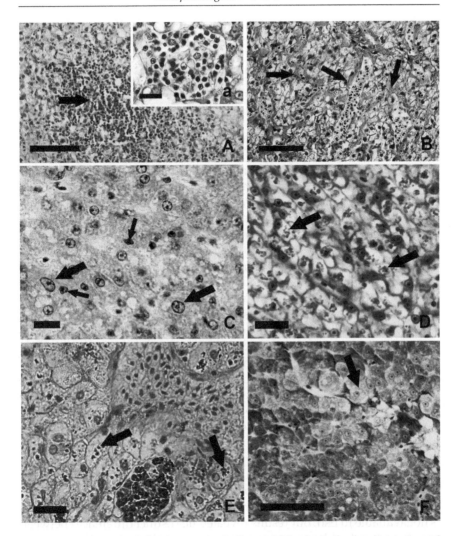

Figure 9.7. Tissue and cellular damages in the livers of fish after natural and experimental conditions. HE stains. A and a. Inflammatory response in *Rhamdia quelen* after experimental DDT exposure (arrow) (scale bar = 50 and 10 μm respectively). B. Tissue fibrosis in *R. quelen* after experimental exposure to TBT (arrows) (scale bar = 50 μm). C. Incidence of cell death (apoptosis) (smaller arrows) and pleomorphic nuclei in *Astyanax* sp. after natural exposure to urban sludge (bigger arrows) (scale bar = 10 μm). D. Intrahepatic cholestasis in *Pimelodus maculatus* that were naturally exposed to impacted reservoir (arrows); (E) in *Hoplias malabaricus* that were naturally exposed to mercury and DDT (arrows) (scale bar = 20 μm). F. Hepatocyte hypertrophy or hydropic change in *Astyanax* sp. that were naturally exposed to an impacted river (arrows) (scale bar = 50 μm).

Cholangioma in biliary epithelial layer is separated from the surrounding liver parenchyma by layers of connective tissue, but invasive cholangioma may be also observed. On the other hand, cholangio-carcinoma may be composed of progressively anaplastic biliary duct epithelial cells with high mitotic activity.

Benign neoplasms appear as a distinct separation of the adenomas from the surrounding tissues and also as compression of the surrounding tissues, but malignant neoplasms are characterized by several key features, such as atypical cellular morphology, loss of cellular polarity, lack of active macrophage centers and pancreatic tissue (as is the case with adenomas). The invasion into adjacent tissues with irregular borders, presenting satellites, depends on the tumor's aggressiveness. A lack of clear differentiation and nuclear and cellular pleomorphism are also symptoms of malignancy.

The occurrence of neoplasia in the livers of fish was found after natural exposure in the case of *Atherinela brasiliensis*, which was exposed to harbor activities, in *Oligosarcus hepsetus* from impacted reservoir, and in *Rhamdia quelen* after subchronic experimental exposure to DDT (Fig. 9.8). Among the lesions, eosinophilic focus, invasive acidophilic foci, invasive neoplastic tissue, non invasive neoplastic tissue and neoplastic tissue were observed (Fig. 9.8). Damages to the liver, including large areas of tissue differentiation in distinct species of fish after experimental and natural exposure were found in tropical species, and interpretations were often difficult due to the presence of similar alterations, such as neoplastic events. Tissue differentiation in *Rhamdia quelen* was found after intraperitoneal exposure to TBT and DBT, and vacuolated hepatocellular adenoma and hepatocellular carcinoma were found in *Atherinela brasiliensis* after natural exposure to harbor activities. A differentiated tissue was found in *Hoplias malabaricus* after intraperitoneal exposure to cyanotoxin (Fig. 9.9).

In the case of genotoxic and non-genotoxic carcinogens in fish, the age of the fish determines the development of idiopathic liver lesions and the appearance of neoplastic lesions. The gender specificity of tumor frequencies is not available to fish, but results revealed no distinct differences between female and male fish specimens that were naturally exposed to a myriad of contaminants. Additionally, studies with primary culture of hepatocytes from males and females exposed to carcinogens and oxyradicals apparently showed advantages in that there was a gender-specific response for males. This explanation may be combined with 17-β-stradiol, which led to a highly significant downregulation of NADPH production in females. NADPH is an important molecule needed for the majority of biotransformation pathways and for oxyradical scavenging (Koehler 2004). Controversially, according to the same author, 17-β-stradiol plays an important role as a tumor promoter in females due to the mitogen effect it has on the proliferation of initiated cells in toxically injured livers.

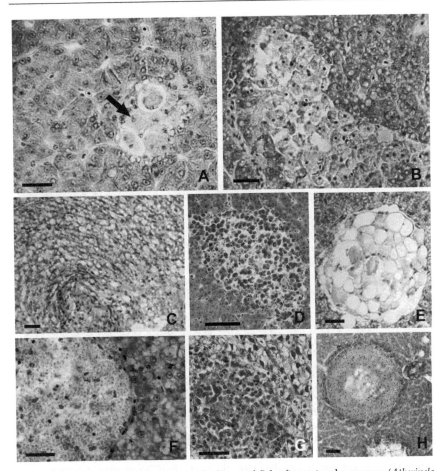

Figure 9.8. Occurrence of neoplasia in the livers of fish after natural exposure (*Atherinela brasiliensis* exposed to harbor activities and *Oligosarcus hepsetus* from impacted reservoir) and after experimental exposure to DDT (in the case of *Rhamdia quelen*). HE stain. A. Eosinophilic focus (arrow) in *A. brasiliensis* (scale bar = 50 μm). B and C. Invasive acidophilic foci in *A. brasiliensis* (scale bar = 10μm). D. Invasive neoplastic tissue in *O. hepsetus* (scale bar = 50 μm). E, F and H. Non-invasive neoplastic tissue in *A. brasiliensis* (scale bar = 10, 50 and 100 μm respectively). G. Invasive neoplastic tissue in *R. quelen* (scale bar = 50μm).

Melanomacrophages are the accumulation of pigmented macrophages involved in the storage of foreign material. Because of their prevalence and intensity, they have been proposed as a potentially useful biomarker of environmental degradations and pollution, but the occurrence under control conditions must be investigated further. Many studies have used the increase in melanomacrophage centers or aggregates as important non-specific hepatic lesions, including degenerative and necrotic processes, and the involvement of melanomacrophage centers with diseases has also

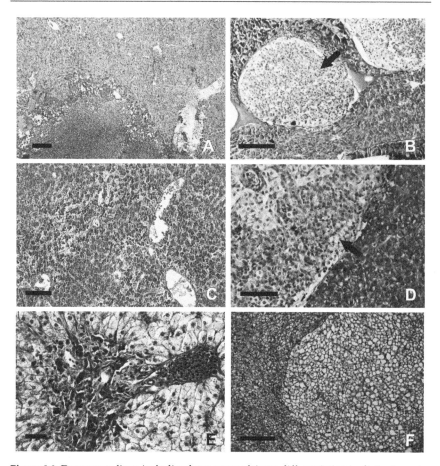

Figure 9.9. Damages to liver, including large areas of tissue differentiation in distinct species of fish after experimental and natural exposure. HE stain. A and C. Tissue differentiation in *Rhamdia quelen* after intraperitoneal exposure to TBT and DBT (scale bar = 200 and 50 µm respectively). B. Vacuolated hepatocellular adenoma (arrow) and (D) Hepatocellular carcinoma (arrow) in *Atherinela brasiliensis* after natural exposure to harbor activities (scale bar = 200 and 50 µm respectively). E and F in *Hoplias malabaricus* after intraperitoneal exposure to cyanotoxin (scale bar = 20 and 50 µm respectively).

been proposed. Starvation or chemical exposure may be the origin of this lesion, and it provides a sensitive indicator of stressful conditions in the aquatic environment. In general, these structures accumulate hemosiderin, lipofuscin and melanin pigments. In addition, normal storage, relocation and recycling of iron compounds from damaged red blood cells are all attributed to melanomacrophages.

In tropical fish, the occurrence of melanomacrophage centers and free macrophages in the liver after both experimental and natural exposure have been observed (Fig. 9.10). A high incidence of melanomacrophage centers,

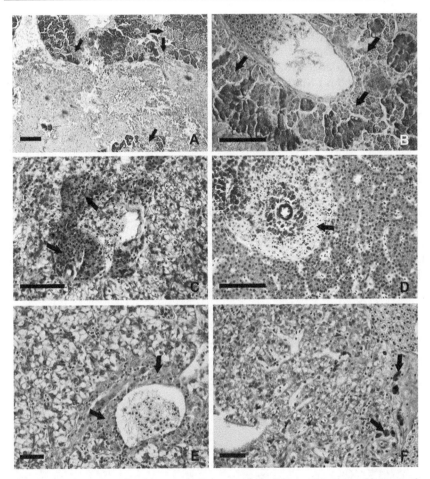

Figure 9.10. Occurrence of melanomacrophage centers and free melanomacrophages in the livers of fish after experimental and natural exposures. HE stain. A and B. High incidence of melanomacrophage centers, particularly around pancreatic tissue in *Pimelodus maculatus* that was naturally exposed to an impacted reservoir (arrows). C and D. Melanonomacrophage centers around damaged vessels and the biliary duct (arrows) in *Oligosarcus hepsetus* that was naturally exposed to an impacted reservoir. E and F. *Hoplias malabaricus* after experimental exposure to a mixture of BaP and DDT (arrows); and BaP, DDT and TBT respectively. In (F), observe the occurrence of free melanomacrophages (arrow) (scale bar = 50 μm).

particularly around pancreatic tissue, was found in *Pimelodus maculatus* that had been naturally exposed to an impacted reservoir. Melanonomacrophage centers around damaged vessels and the biliary duct were common in *Oligosarcus hepsetus* under the same exposure conditions. Experimentally, melanomacrophages were found in *Hoplias malabaricus* after subchronic and intraperitoneal exposure to a mixture (BaP + DDT and BaP + DDT + TBT). Additionally, the relationship between melanomacrophage centers and free macrophages has been used to establish the activation of immune response and also the failure of immune response.

Nuclear alterations or hypertrophy observed in the hepatic tissue may indicate intensive metabolic activity of the hepatocytes, as discussed by Maduenho and Martinez (2008), but they do not necessarily comprise the normal function of the organ. In general, this kind of alteration is very hard to confirm by light microscopy alone. Methods and parameters, such as nuclear and cellular deformation and its real role in cellular pathology, must be established. The incidence of cell death (apoptosis) and pleomorphic nuclei in *Astyanax* sp. after natural exposure to urban sludge were both described as nuclear alterations (Fig. 9.7).

Hemorrhagic congestion is a disorder involving the circulatory system. In some cases, it may be an important physiological biomarker. Hemorrhagic evidence in the livers of fish after experimental conditions was found in *Hoplias malabaricus* after cyanotoxin exposure, in *Rhamdia quelen* after both DDT and BaP exposure, and in *H. malabaricus* after methylmercury exposure (Fig. 9.11).

Inflammatory response in the livers of fish may be attributed to different factors, including nutritional stress, microbial infections or exposure to xenobiotics. In general, this kind of alteration represents a reversible finding that disappears after exposure to stimuli. It better represents an acute effect

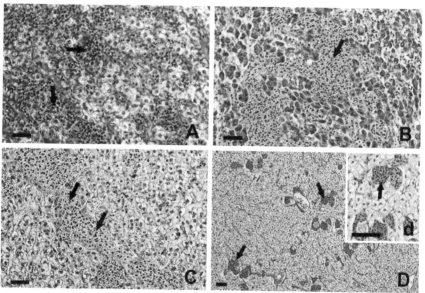

Figure 9.11. Hemorrhagic incidence in the livers of fish after experimental exposure. HE stain. A. In *Hoplias malabaricus* after cyanotoxin exposure (arrow). B and C. In *Rhamdia quelen* after both DDT and BaP exposure (arrows) (scale bar = 20 μm). D and d. In *H. malabaricus* after methylmercury exposure (arrow) (scale bar = 100 and 50 μm respectively).

than a chronic exposure. The presence of inflammatory response suggests the occurrence of cell death as a result of necrosis, and also that the immune system is not affected by the chemical. This data can be corroborated through the use of melanomacrophage counting. An inflammatory response in *Rhamdia quelen* was observed after experimental DDT exposure (Fig. 9.7). Overall, the occurrence of inflammatory response in the livers of fish is not necessarily a related effect, and it is found in both experimental and field studies. Recent studies using silver nanoparticles after oral exposure in *R. quelen* found an effect that involved inflammatory response. All exposed groups presented this kind of finding at higher or lower levels (author's note). The occurrence of parasites in the livers of fish is also indicative of immune response disorders, because the prevalence of this kind of finding may be related to the low regulation of immune response. In general, parasites are opportunists, and, at certain levels of infestation, may cause physiological damage to the immune system. Many toxic chemicals can affect immune response, such as organochlorine compounds. This kind of biomarker is more effective in biomonitoring programs, where the exposure to a complex mixture of compounds occurs under natural conditions. The parasites of the genus *calyptospora* are naturally present in the livers of many Brazilian species of fish, but the occurrence of a high infestation level reflects a low regulation of immune system, as found in *Astyanax* sp. that was naturally exposed to urban sludge. The incidence of multicellular parasites in the livers of fish represents more serious immune system disorders, as observed in *Pimelodus maculatus* when it was naturally exposed to an impacted reservoir, and in *Astyanax* sp. when it was naturally exposed to a polluted river (Fig. 9.12). The natural occurrence of parasites in some species of fish (particularly in the gills or on the skin) makes it difficult to treat this finding as a biomarker, but when high infestations are observed in the peritoneal cavity or within liver tissues, this data cannot be omitted.

Gills in Fish as Target Organ for Evaluating Exposure to Pollutants

Fish gills have numerous functions; they are the main respiratory organs of most teleost, and they play important role in osmotic and ionic regulation, acid-base equilibrium, and nitrogen excretion (Evans et al. 2005). Respiratory function depends on surface area and thickness of the lamellar epithelium, on gill blood perfusion, and on mucus density (Fernandes et al. 2007). Conversely, osmotic, ionic and acid-base homeostasis processes are affected by the functional surface of lamellae which favor the passive effluxes and/or influxes of ions. Acid-base equilibrium and N-excretion are both achieved by the dynamic net transepithelial fluxes of Na^+ and Cl^- and by electroneutral coupling of Na^+ influx and H^+/NH_4^+ efflux and Cl^- influx and HCO_3^- efflux (Dymowska et al. 2012). The predominant gill epithelial cell types (pavement

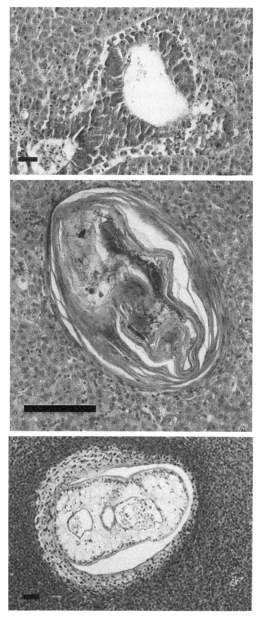

Figure 9.12. Occurrence of parasites in the livers of fish after natural exposure to pollutants. HE stain. A. Parasite of the genus *calyptospora* in *Astyanax* sp. that was naturally exposed to urban sludge (arrows) (scale bar = 10 μm). B and C. Multicellular parasite in *Pimelodus maculatus* that was naturally exposed to an impacted reservoir and *Astyanax* sp. that was naturally exposed to a polluted river. The arrow shows the inflammatory process around the parasite (scale bar = 50 μm).

cells (PVC), mitochondria-rich cells (MRC) and mucous cells (MC)) play an important role in these gill functions. Ionic uptake in freshwater fish depends on the number and the apical surface area of the MRCs (also called chloride cells); mucus produced by MCs and the glycocalix of PVCs on the epithelial surface helps to trap ions and create a ionic gradient close to the gill cell surface, which favors ionic uptake in freshwater fish (Perry and Laurent 1993; Moron et al. 2003; Sakuragui et al. 2003). Acid-base regulation during environmental hypercapnia or water acidification involves morphological changes in the apical surface of MRCs and PVCs (Goss et al. 1992; Perry and Laurent 1993; Dymowska et al. 2012).

The gill surface area is very and has thin distance between water and the blood. It is continuously in contact with water in the environment, which makes the gills the main route of contaminant uptake from water, particularly in freshwater fish. The accumulation of contaminants in these organs may directly or indirectly cause changes to the gills. Most direct changes result in some defense responses to keep the contaminant from getting into the gills and reaching the blood, such as increasing mucus release or cell damage, including necrosis. Indirect responses include a signal to activate the defense and/or compensatory responses to biochemical and physiological actions of toxic compounds. The activation of the hypothalamus-pituitary-interrenal axis lead to cortisol release that acts on the gill tissues in different ways: stimulating MRC proliferation and hypertrophy to maintain the water and ionic balance, as well as controlling the cardio-vascular system and energy mobilization (Wendelaar-Bonga and Van Der Meij 1989; Lock et al. 1994; Dang et al. 2000a,b). Both direct and indirect responses may impair one or more gill functions (Perry and Laurent 1993; Mazon et al. 2002; Sakuragui et al. 2003). On the other hand, the changes to gill morphology may also reflect systemic physiological changes of the organism in response to the temperature of the environment, or to hypoxia, hyperoxia, pH levels, or effects of toxic compounds (Mazon et al. 2002; Mitrovic and Perry 2009; Furukawa et al. 2011).

Filament and lamellar epithelial changes, PVC, MC, and MRC changes, blood vessel changes, tumors, necrosis, and fibrosis are the most common histopathological changes that can take place in the gills (Table 9.4 and Fig 9.13) (Poleksic and Mitrovic-Tutundzic 1994; Fernandes and Mazon 2003). The statistical review of histopathological changes to the gills of fish exposed to wide range of pollutants, which was performed by Mallatt (1985), revealed that most morphological changes take place in the lamellar structure and/or filament and lamellar epithelia and include epithelial hyperplasia, cellular hypertrophy, cell rupture and necrosis, edema and epithelial lifting, and leukocyte infiltration and aneurysm.

Hyperplasia of epithelial cells may result in the partial or complete fusion of two, several or total lamella. It may also reduce the protruding

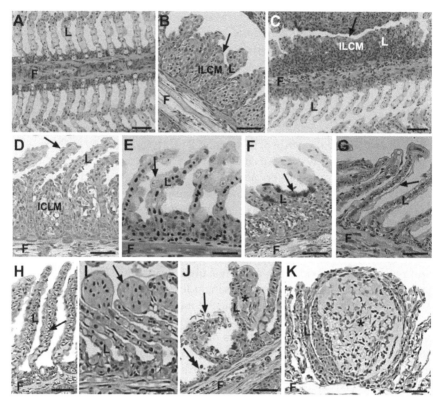

Figure 9.13. Most common histopatholological changes in the fish gills. A. Normal gills; B and C. Hyperplasia of epithelial cells. Note the expansion of interlamellar cell mass (ILCM), which causes partial lamellar fusion and reducing protrude lamella in B and A localized total fusion of several lamella in C; D. Hypertrophy of pavement cells (arrow); E. Hyperplasia and hypertrophy of mitochondria-rich cells (arrow); F. Hyperplasia of mucous cells (arrow); G. Epithelial lifting (arrow); H. Pillar cell system congestion (arrow); I. Lamellar telangiectasis (arrow); J. Lamellar aneurysm (*), epithelial rupture and hemorrhages (arrow), necrosis (double arrow). K. Tumor (*). Scale bar: 20 μm.

lamella and, consequently, the respiratory surface area (Fig. 9.13B, C). Such epithelial cell proliferation in response to environmental changes is reversible; the expansion and reduction of a cell mass between the lamella of the gills (interlamellar cell mass, ILCM) occurs over the course of hours to days and remodels the gills (Sollid et al. 2003; Nilsson 2007, 2012). Gill remodeling in response to oxygen needs during water normoxia/hypoxia, cold/warm temperature acclimation or resting/high swimming performance is well documented in crucian carp (*Carassius carassius*), goldfish (*Carassius auratus*), and eels (Tuurala et al. 1998; Sollid et al. 2003; 2005; Sollid and Nilsson 2006; Fu et al. 2011; Perry et al. 2012). The critical oxygen tension (PcO_2), which is the lowest water oxygen partial pressure

Table 9.4. Gill tissue changes induced by water contaminants and other environmental stressors and gill lesion types.

Gill Tissue Changes	Gill lesion types
General filament and lamellar epithelial changes	
Hyperplasia of epithelial cells	Thickening of interlamellar cell mass (ILCM) Focal partial or total fusion of two or more lamella Complete fusion of all lamella Decrease of interlamellar space Shortening of lamella
Hypertrophy of epithelial cell	Thickening of filament epithelium Thickening of lamellar epithelium Decrease of interlamellar space
Thinning of lamellar epithelium	
Edema	Lamellar epithelial lifting Rupture and peeling of lamellar epithelium Leukocyte infiltration of gill epithelium
Fibrosis	Scar tissue
Epithelial cell death	Focal epithelial necrosis Focal lamellar necrosis Complete necrosis of filament and lamella
Uncontrolled cell proliferation	Benign tumor Malign tumor
Mucous (MC) and mitochondria –rich (MRC) cell changes	
Hyperplasia of MC	MC present in the filament interlamellar region and/or lamellar epithelium Epithelium thickening
Hypertrophia of MC	Epithelium thickening
MC emptying or MC disappearance	
Hyperplasia of MRC	Increasing MRC density in the filament and/or lamella Epithelium thickening
Hypertrophia of MRC	Epithelium thickening
Blood vessels changes	
Pillar cell system constriction	Lamellar thinning
Pillar cell system congestion	Lamellar thickening
Lamellar telangiectasis	Apical lamellar thickening Epithelium rupture and hemorrages
Lamellar aneurysm	Entire lamellar thickening Epithelial rupture and hemorrages

required to maintain oxidative metabolism in resting fish, is low in fish with protruding lamella and, consequently, large surface area (Sollid et al. 2003; Fu et al. 2011). Epithelial cell proliferation occurs in numerous fish species exposed to low pH levels and toxic compounds such as ammonia, heavy metals and organic pollutants (Cerqueira and Fernandes 2002;

Monteiro et al. 2008; Shiogiri et al. 2012; Paulino et al. 2012b; Negreiros et al. 2011; Delunardo et al. 2013; Brito et al. 2012), and it is reversible after the environment is improved (Cerqueira and Fernandes 2002; Delunardo et al. 2013). Gill remodeling is not only related to oxygen needs but may also be considered to be a defense response employed to reduce the uptake of toxic compounds.

Cellular hypertrophy in general indicates an increase in cell activity, and it is a common response of cells in the outermost cell layer of the filament and lamellar epithelia. However, it can also be found in undifferentiated cells of the ILCM in the filament epithelium. The hypertrophy of the PVC (Fig 9.13D), which occupies more than 90%–95% of the outermost epithelial cell layer, increases the water-blood diffusion distance of lamellar epithelium and helps to reduce toxic uptake; conversely, it also reduces gases exchange (Moron et al. 2003; Sakuragui et al. 2003).

MRC density in the gill epithelia and the morphology of the MRC are biomarkers related to ionic regulation in fish. MRC hyperplasia and hypertrophy (Fig. 9.13E) indicate a compensatory response that increases the ion- transporting capacity of the gills and maintains the ionic balance, but such changes also increase the water-blood diffusion distance of the lamellar epithelium and reduce gas exchange (Bindon et al. 1994; Al-Ghanbousi et al. 2012). In histological sections, MRCs are identified in the filament epithelium by their large volume and round nuclei, with nucleoli in the outermost epithelial cell layer scattered among PVC in the filament epithelium close to the onset of lamella and in the filament interlamellar epithelium. MRCs are also found to be distributed throughout the lamellar epithelium in freshwater fish living in ion poor and soft waters and MRC density seems to be related to the fish's ability to retain body ions. The two erythrinid species, *Hoplias malabaricus* and *Hoplerythrinus unitaeniatus* living in the same soft water environment were found to have different MRC densities in the lamellar epithelium, which is evidence of different gill ionic permeability and varying capacities to maintain ionic balance (Moron et al. 2003).

Morphological studies of MRC using scanning electron microscopy showed that, in several species, there was an enlargement of the MRC fractional surface area (MRCFA) in the filament epithelium and also a positive relationship between the MRC fractional surface area in the filament epithelium and the whole-body Cl^- and Na^+ uptake rate in fish that are exposed to ion-poor and soft water or chronically injected with cortisol (Perry and Laurent 1993). The enlargement of the MRCFA in the filament epithelium was the result of increasing MRC density in contact with water in the environment and the MRC individual apical surface area. Na^+/K^+-ATPase immunocytochemistry under light and transmission electron microscopy used to identify the MRC in the gill epithelium revealed that cortisol increased MRC density as well as the volume of their tubular membrane system and also increased the Na^+/K^+-ATPase density in the tubular system (Dang et al. 2000b). Furthermore, two MRC populations were identified through the use of Na^+/K^+-ATPase immunocytochemistry: strongly stained cells (mature MRCs) and weakly stained cells which were found to be necrotic or apoptotic MRCs using transmission electron microscopy (Dang et al. 2000a).

Copper exposure induces cell proliferation while it enhances the MRC turnover, increasing the number of necrotic and apoptotic MRCs instead of mature cells (Dang et al. 2000a; Mazon et al. 2002). Aluminum exposure at low pH levels has a direct effect on the MRCs in that it reduces their number, Na^+/K^+-ATPase activity, and, consequently plasma ions (Monette and McCormick 2008; Camargo et al. 2009). Acute (48 h) and subchronic (14 d) exposure to low atrazine concentration (2 to 25 µg L^{-1}) causes slight to moderately gill damage (Paulino et al. 2012b). Paulino et al. (2012a) showed that, while subchronic exposure did not affect MRC density, acute exposure to 25 µg L^{-1} of atrazine decreased total MRC density in the filament epithelium, which was compensated by morphological adjustments on the epithelial surface that comes in contact with the water in the environment by either increasing emerged MRC density or increasing individual MRC apical surface area, which maintained ionic homeostasis (Fig. 9.14). Under realistic environmental conditions, a field study reported morphological changes in the PVC and MRC architecture and the MRCFA concomitant with the inhibition of Na^+/K^+-ATPase activity in fish from sites contaminated with organochlorine and metal. These results reveal the negative effect of multiple contaminants on these cells, which may cause ionic imbalance (Fernandes et al. 2013).

MCs are generally located on the gill filament border being rarely found in the lamellae, excepting in fish exposed to water contaminants (Banerjee and Chandra 2005). MC hyperplasia (Fig. 9.13F) and hypertrophy as well as increasing mucus production have been reported in several studies in which fish were exposed to mechanical, physical and chemical agents or potential pathogens infestation (Iger et al. 1994; Zaccone et al. 1989; Berntssen et al. 1997; Ferguson et al. 1992; Roberts and Powell 2003, 2005; Banergee and Chandra 2005; Banergee 2007; Diaz et al. 2005; Moron et al. 2009) however, MC decreasingwas reported in fish exposed to low temperature (Quiniou et al. 1998). Mucus layer cover the entire epithelial surface of gills being the first physical barrier against all stressors types; mucus contains large diversity of molecules such as immunoglobulin, lysozyme and proteolytic enzymes that are known to have antimicrobial and antiviral functions (Shephard 1994; Domeneghini et al. 1998; Roberts and Powell 2003, 2005). The polyionic nature of neutral and acidic glycoproteins released by MC as well as the glicocalix that cover the PVC attracts ions creating an ionic

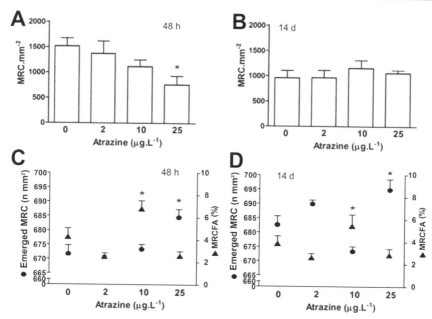

Figure 9.14. Mitochondria-rich cells (MRCs) (mean ± S.E.M.) in the filaments of the gills of *P. lineatus* exposed to atrazine. (A) Total MRC density after 48 hours of and (B) after 14 days of atrazine exposure identified using Na^+/K^+-ATPase immunocitochemistry; C and D. Emerged MRC density and MRC fractional area (MRCFA) in fish after 48 hours of (C) and 14 days of (D) atrazine exposure respectively, using scanning electron microscopy. (*) Indicates significant difference from the respective control group ($P < 0.05$). Modified from Paulino et al. (2012a).

gradient close to the gill surface that reduces ionic loss and favors ion uptake (Handy et al. 1989; Powell 2007). Hexoses and glycoproteins with sialic acids have low viscosities, helping to protect the epithelium from physical injuries, while sulfate acid mucosubstances have high viscosities, favoring the retention of suspended particles (Sibbing and Uribe 1985).

Glycoconjugate characterizations in MC using histochemistry techniques permits identify different MC types (Sabóia-Moraes et al. 1996; Paulino et al. 2012b). Contaminant exposure affects MC type density in the gills; per exemple, acute and chronic exposure to atrazine decreases the type 3 MCs in *P. lineatus* evidencing a mechanism to favor the washout of the toxic substance away from gill surface (Paulino et al. 2012b). The changes in the MC glycoconjugate may affect gill functions; blood O_2 tension was maintained in rainbow trout exhibiting MC hyperplasia (Powell et al. 1998) but CO_2 excretion through the gills was reduced by increasing mucus secretion after chemical irritation (Powell and Perry 1997). However, thick and extreme mucus secretion in response to any irritant affect the O2 uptake by increasing the water-blood diffusion distance (Ultsch and Gros 1979).

Blood vessel changes, in general, are pathologies related to the increasing of blood pressure and/or blood flow resistance through the lamella together with the fragility of the pillar cell junction system. Most blood vessel pathology, such as marginal channel dilatation (telangiectasis), blood congestion and aneurysms (Fig. 9.13H, I and J), occur in the lamella and may reduce the O_2 uptake by preventing the hemoglobin saturation of erythrocytes. Blood congestion and telangiectasis are progressive events if the level of contamination increases; they led to pillar cell system disintegration forming the aneurysms whichmay continuously increase resulting in epithelial rupture and hemorrhage (Fernandes and Mazon 2003). Lamellar epithelial lifting (Fig. 9.13G) can be a morphological expression of edema occurred between the inner and outermost cell layer of lamellar epithelium however, epithelial lifting may be a consequence of post-morten deterioration of lamellar structure (Munday and Jaisankar 1998) or represent an artifact occurred during sectioning of tissue embedding in paraffin medium. In general, lamellar epithelial lifting is absent or rarely present in the gills embedded in methacrylate resin.

Conclusion

Histopathological biomarkers have been largely used in both experimental and field studies to evaluate the effects of pollutants on fish. In general, there is no related effect, but some aspects may be associated with lesions or damages in target organs or tissues. Because the liver is a target organ for the majority of pollutants (and particularly for organic contaminants) due to its role in the biotransformation and detoxification processes, the histopathological studies on this tissue can provide a useful representation of the specimen in the evaluation of risk of exposure.

The high diversity of terms or concepts used to classify the pathological events in histopathology, including the difficultly in interpreting or describing a specific lesion or alteration, are important when showing the importance of this study, as well as the diversity of different cellular responses that are only revealed in histological images. Both acute and chronic exposure can be investigated using histological images and classified according to their severity. Additionally, histopathological findings show either the short-term or long-term effects, but in both cases, they represent damages at a higher level of biological organization, and they support noteworthy conclusions about risks to populations in natural ecosystems.

Despite these results, a standardization of the methods and techniques used for sample preservation, as well as for the identification and interpretation of lesions and alterations, is still strongly recommended. Thus, a broad revision is necessary in order to minimize the inappropriate use of terms, or to avoid an incorrect identification or interpretation of a lesion or damages to the liver and to aid in the comparative diagnoses among species.

Acknowledgements

Original work cited in this chapter was supported by Brazilian Agencies for Science and Technology supply (CNPq and CAPES), CNPq/INCT-TA and state Agencies for Science and Technology from São Paulo State (FAPESP) and Parana State (Araucaria Foundation). We also thank all students who gave their time to develop the projects.

Keywords: Histopathology, Biomarkers, Tropical fishes, Liver, Gills, Experimental Toxicology, Biomonitoring

References

Abdel-Moneim, A.M., M.A. Al-Kahtan and O.M. Elmenshawy. 2012. Histopathological biomarkers in gills and liver of Oreochromis niloticus from polluted wetland environments, Saudi Arabia. Chemosphere 88: 1028–1035.

Al-Ghanbousi, R., T. Ba-Omar and R. Victor. 2012. Effect of deltamethrin on the gills of Aphanius dispar: A microscopic study. Tissue and Cell 44: 7–14.

Banerjee, T.K. 2007. Histopathology of respiratory organs of certain air-breathing fishes of India. Fish Physiol. Biochem. 33: 441–454.

Banerjee, T.K. and S. Chandra. 2005. Estimation of zinc chloride contamination by histopathological analysis of the respiratory organs of the air breathing 'murrel' *Channa striata* (Bloch 1797) (Channiformes, Pisces). Vet. Arhiv. 75: 253–263.

Bernet, D., H. Schmidt, W. Meier, P. Burkhardt-Holm and T. Wahli. 1999. Histopathology in fish: proposal for a protocol to assess aquatic pollution. J. Fish Dis. 22: 25–34.

Berntssen, M.H.G., F. Kroglund, B.O. Rosseland and S.E. Wendelaar Bonga. 1997. Responses of skin mucous cells to aluminium exposure at low pH in Atlantic salmon (*Salmo salar*) smolts. Can. J. Fish. Aquat. Sci. 54: 1039–1045.

Bindon, S.D., K.M. Gilmour, J.C. Fenwick and S.E. Perry. 1994. The effects of branchial chloride cell proliferation on respiratory function in the rainbow trout, *Oncorhynchus mykiss*. J. exp. Biol. 197: 47–63.

Boran, H., E. Capkin, I. Altinok and E. Terzi. 2012. Assessment of acute toxicity and histopathology of the fungicide captan in rainbow trout. Exp. Toxicol. Pathol. 64: 175–179.

Brito, I.A., C.A. Freire, F.Y. Yamamoto, H.C.S. Assis, L.R. Souza-Bastos, M.M. Cestari, N.C. Ghisi, V. Prodocimo, F. Filipak Neto and C.A. Oliveira Ribeiro. 2012. Monitoring water quality in reservoirs for human supply through multi-biomarker evaluation in tropical fish. J. Environ. Monit. 14: 615–625.

Bukhari, A.S., H.E. Syed Mohamed, K.V. Broos, A. Stalin, R.K. Singhal and P. Venubabu. 2012. Histological variations in liver of freshwater fish Oreochromis mossambicus exposed to 60 Co gamma irradiation. J. Environ. Radioactiv. 113: 57–62.

Camargo, M.M.P., M.N. Fernandes and C.B.R. Martinez. 2009. How aluminium exposure promotes osmoregulatory disturbances in the neotropical freshwater fish *Prochilus lineatus*. Aquat. Toxicol. 94: 40–46.

Cerqueira, C.C.C. and M.N. Fernandes. 2002. Gill tissue recovery after copper exposure and blood parameter responses in the tropical fish *Prochilodus scrofa*. Ecotoxicol. Environ. Saf. 52: 83–91.

Costa, P.M., S. Caeiro, C. Vale, T.À. DelValls and M.H. Costa. 2012. Can the integration of multiple biomarkers and sediment geochemistry aid solving the complexity of sediment risk assessment? A case study. Environ. Poll. 161: 107–120.

Dang, Z., G. Flik, S.E. Wendelaar Bonga and R.A.C. Lock. 2000a. Cortisol increases Na+/ K+ ATPase density in plasma membrane of gill chloride cells in the freshwater tilapia, *Oreochromis mossambicus*. J. exp. Biol. 203: 2349–2355.

Dang, Z., R.C.A. Lock, G. Flik and S.E. Wendelaar Bonga. 2000b. Na^+/K^+-ATPase immunoreactivity in branchial chloride cells of *Oreochromis mossambicus* exposed to copper. J. Exp. Biol. 203: 379–387.

Das, S. and A. Gupta. 2012. Effect of cadmium chloride on oxygen consumption and gill morphology of Indian flying barb, Esomus danricus. J. Environ. Biol. 33: 1057–1061.

Delunardo, F.A.C., B.F. Silva, M.G. Paulino, M.N. Fernandes and A.R. Chippari-Gomes. 2013. Genotoxic and morphological damage in *Hippocampus reidi* exposed to crude oil. Ecotoxicol. Environ. Saf. 87: 1–9.

Diaz, A.O., A.M. Garcia and A.L. Goldemberg. 2005. Glycoconjugates in the branchial mucous cells of *Cynoscion guatucupa* (Cuvier 1830) (Pisces: Sciaenidae). Sci. Mar. 69: 545–553.

Domeneghini, C., R. Straini Pannelli and A. Veggatti. 1998. Gut lycoconjugates in *Sparus aurata* L. (Pisces, Teleostei). A comparative histochemical study in larval and adult ages. Histol. Histopathol. 13: 359–372.

Dymowska, A.K., P.-P. Hwang and G.G. Goss. 2012. Structure and function of ionocytes in the freshwater fish gill. Resp. Physiol. & Neurobiol. 184: 282–292.

Evans, D.H., P.M. Piermarini and K.P. Choe. 2005. The multifunctional fish gill: dominant site of gas exchange, osmoregulation, acid-base regulation, and excretion of nitrogenous wastes. Physiol. Rev. 85: 97–177.

Ferguson, H.W., D. Morrison, V.E. Ostland, J. Lumsden and P. Byrne. 1992. Responses of mucus-producing cells in gill disease of rainbow trout (*Oncorhynchus mykiss*). J. Comp. Pathol. 106: 255–265.

Fernandes, M.N. and A.F. Mazon. 2003. Environmental pollution and fish gill morphology. *In*: A.L. Val and B.G. Kapoor (eds.). Fish Adaptation. Science Publishers, Enfield, USA, pp. 203–231.

Fernandes, M.N., M.G. Paulino, M.M. Sakuragui, C.A. Ramos, C.D.S. Pereira and H. Sadauskas-Henrique. 2013. Organochlorines and metals induce changes in the mitochondria-rich cells of fish gills: An integrative field study involving chemical, biochemical and morphological analyses. Aquat. Toxicol. 126: 180–190.

Fernandes, M.N., S.E. Moron and M.M. Sakuragui. 2007. Gill morphological adjustments to environment and the gas exchange function. *In*: M.N. Fernandes, M.L. Glass, F.T. Rantin and B.G. Kapoor (eds.). Fish Respiration and Environment. Science Publishers, Enfield, USA, pp. 93–120.

Fu, S.J., C.J. Brauner, Z.D. Cao, J.G. Richards, J.L. Peng, R. Dhillon and Y.X. Wang. 2011. The effect of acclimation to hypoxia and sustained exercise on subsequent hypoxia tolerance and swimming performance in goldfish (*Carassius auratus*). J. exp. Biol. 214: 2080–2088.

Furukawa, F., S. Watanabe, M. Inokuchi and T. Kaneko. 2011. Responses of gill mitochondria-rich cells in Mozambique tilapia exposed to acidic environments (pH 4.0) in combination with different salinities. Comp. Biochem. Physiol. A 158: 468–476.

Goss, G.G., P. Laurent and S.F. Perry. 1992. Evidence for a morphological component in the regulation of acid-base balance in hypercapnic catfish (*Ictalurus nebulosus*). Cell Tissue Res. 268: 539–352.

Handy, R.D., F.B. Eddy and G. Romain. 1989. *In vitro* evidence for the ionoregulatory role of rainbow trout mucus in acid, acid/aluminum and zinc toxicity. J. Fish Biol. 35: 737–747.

Iger, Y., H. Abraham and S.E. Wendelaar Bonga. 1994. Responses of club cells in the skin of the carp *Cyprinus carpio* to exogenous stressors. Cell Tissue Res. 277: 485–491.

Koehler, A. 2004. The gender-specific risk to liver toxicity and cancer of flounder (Platichthys flesus (L.)) at the German Wadden Sea coast. Aquat. Toxicol. 70: 257–276.

Lang, T., W. Wosniok, J. Barsien, K. Broeg, J. Kopecka and J. Parkkonen. 2006. Liver histopathology in Baltic flounder (Platichthys flesus) as indicator of biological effects of contaminants. Mar. Poll. Bull. 53: 488–496.

Lock, R.A.C., P.H.M. Balm and S.E. Wendelaar Bonga. 1994. Adaptation of freshwater fish to toxicants: stress mechanisms induced by branchial malfunctioning. *In*: R. Müller and

R. Lloyd (eds.). Sublethal and Chronic Effects of Pollutants on Freshwater Fish. Fishing New Books, Oxford, UK, pp. 124–134.

Lukin, A., J. Sharova, L. Belicheva and L. Camus. 2011. Assessment of fish health status in the Pechora River: Effects of contamination. Ecotoxicol. Environ. Saf. 74: 355–365.

Maduenho, L.P. and C.B.R. Martinez. 2008. Acute effects of diflubenzuron on the freshwater fish Prochilodus lineatus. Comp. Biochem. Physiol. Part C 148: 265–272.

Madureira, T.V., M.J. Rocha, C. Cruzeiro, I. Rodrigues and R.A.F. Monteiro. 2012. The toxicity potential of pharmaceuticals found in the Douro River estuary (Portugal): Evaluation of impacts on fish liver, by histopathology, stereology, vitellogenin and CYP1A immunohistochemistry, after sub-acute exposures of the zebrafish model. Environ. Toxicol. Pharmacol. 34: 34–45.

Mallatt, J. 1985. Fish gill structural changes induced by toxicants and other irritants: A statistical review. Can. J. Fish. Aquat. Sci. 42: 630–648.

Mallatt, J., J.F. Bailey, S.J. Lampa, M.A. Evans and S. Brumbaugh. 1995. A fish gill system for quantifying the ultrastructural effects of environmental stressors: methylmerucy, Kepone®, and heat shock. Can. J. Fish Aquat. Sci. 52: 1165–1182.

Mazon, A.F., C.C.C. Cerqueira and M.N. Fernandes. 2002. Gill cellular changes induced by copper exposure in the South American tropical freshwater fish, *Prochilodus scrofa*. Environ. Res. 88A: 52–63.

McHugh, K.J., N.J. Smit, J.H.J. Van Vuren, J.C. Van Dyk, L. Bervoets, A. Covaci and V. Wepener. 2011. A histology-based fish health assessment of the tigerfish, Hydrocynus vittatus a DDT-affected area. Phys. Chem. Earth. 36: 895–904.

Mitrovic, D. and S.F. Perry. 2009. The effects of thermally induced gill remodeling on ionocyte distribution and branchial chloride fluxes in goldfish (*Carassius auratus*). J. exp. Biol. 212: 843–852.

Mittal, A.K., O. Fugimori, H. Ueda and K. Yamada. 1995. Carbohydrates in the epidermal mucous cells of a freswhater fish, *Mastacembelus pancalus* (Mastacembelidae, Pisces) as studied by electron-microscopic cytochemical methods. Cell Tissue Res. 280: 531–539.

Monette, M.Y. and S.D. McCormick. 2008. Impacts of short-term acid and aluminum exposure on Atlantic salmon (*Salmo salar*) physiology: a direct comparison of parr and smolts. Aquat. Toxicol. 86: 216–226.

Monteiro, S.M., E. Rocha, A. Fontaínhas-Fernandes and M. Sousa. 2008. Quantitative histopathology of *Oreochromis niloticus* gills after copper exposure. J. Fish Biol. 73: 1376–1392.

Monteiro, S.M., E. Rocha, J.M. Mancera, A. Fontaínhas-Fernandes and M. Sousa. 2009. A stereological study of copper toxicity in gills of *Oreochromis niloticus*. Ecotoxicol. Environ. Saf. 72: 213–223.

Moron, S.E., C.A. Andrade and M.N. Fernandes. 2009. Response of mucous cells of the gills of traíra (*Hoplias malabaricus*) and jeju (*Hoplerythrinus unitaeniatus*) (Teleostei: Erythrinidae) to hypo- and hyper-osmotic ion stress. Neotrop. Ichthiol. 7: 491–498.

Moron, S.E., E.T. Oba, C.A. Andrade and M.N. Fernandes. 2003. Chloride cells response to ion chalenge in two tropical freshwater fish, the erythrinids *Hoplias malabaricus* and *Hoplerythrinus unitaeniatus*. J. Exp. Zool. 298A: 93–104.

Munday, B.L. and C. Jaisankar. 1998. Postmortem changes in the gills of rainbow trout (*Oncorhynchus mykiss*) in freshwater and seawater. Bull. Eur. Ass. Fish Pathol. 18: 127–131.

Negreiros, L.A., B.F. Silva, M.G. Paulino, M.N. Fernandes and A.R. Chippari-Gomes. 2011. Effects of hypoxia and petroleum on the genotoxic and morphological parameters of *Hippocampus reidi*. Comp. Biochem. Physiol. C 153: 408–414.

Nilsson, G.E. 2007. Gill remodeling in fish—a new fashion or an ancient secret? J. exp. Biol. 210: 2403–2409.

Nilsson, G.E., A. Dymowska and J.A.W. Stecyk. 2012. New insights into the plasticity of gill structure. Resp. Physiol. & Neurobiol. 184: 214–222.

Olson, K.R., P. Fromm and W. Frantz. 1973. Ultrastructural changes of rainbow trout gills exposed to methyl mercury or mercury chloride. Fed. Proc. 32: 261.

Paulino, M.G., M.M. Sakuragui and M.N. Fernandes. 2012a. Effects of atrazine on the gill cells and ionic balance in a Neotropical fish, Prochilodus lineatus. Chemosphere 86: 1–7.

Paulino, M.G., N.E.S. Souza and M.N. Fernandes. 2012b. Subchronic exposure to atrazine induces biochemical and histopathological changes in the gills of a Neotropical freshwater fish, *Prochilodus lineatus*. Ecotoxicol. Environ. Saf. 80: 6–13.

Perry, S.F. and P. Laurent. 1993. Environmental effects on fish gill structure and function. *In*: J.C. Rankin and F.B. Jensen (eds.). Fish Ecophysiology. Chapman & Hall, London, UK, pp. 231–264.

Perry, S.F., C. Fletcher, S. Bailey, J. Ting, J. Bradshaw, V. Tzaneva and K.M. Gilmour. 2012. The interactive effects of exercise and gill remodeling in goldfish (*Carassius auratus*). J. Comp. Physiol. 182: 935–945.

Poleksic, V. and V. Mitrovic-Tutundzic. 1994. Fish gills as a monitor of sublethal and chronic effects of pollution. *In*: R. Muller and R. Lloyd (eds.). Sublethal and Chronic Effects of Pollutants on Freshwater Fish. Cambridge Univ. Press, Cambridge, UK, pp. 339–352.

Powell, M.D. and S.F. Perry. 1997. Respiratory and acid-base disturbances in rainbow trout blood during exposure to chloramines-T under hypoxia and hyperoxia. J. fish Biol. 50: 418–428.

Powell, M.D. 2007. Respiration in infectious and non-infectious gill diseases. *In*: M.N. Fernandes, M.L. Glass, F.T. Rantin and B.G. Kapoor (eds.). Fish Respiration and Environment. Science Publishers, Enfield, USA, pp. 317–339.

Powell, M.D., E. Haman, G.M. Wright and S.F. Perry. 1998. Respiratory responses to graded hypoxia of rainbow trout (Oncorhynchus mykiss) following repeated intermittent exposure to chloramines-T. Aquaculture 165: 27–39.

Quiniou, S.M.-A., S. Bigler, L.W. Clew and J.E. Bly. 1998. Effects of water temperature on mucous cell distribution in channel catfish epidermis: a factor in winter saprolegniasis. Fish & Sellfish Immunol. 8: 1–11.

Rajeshkumar, S. and N. Munuswamy. 2011. Impact of metals on histopathology and expression of HSP 70 in different tissues of Milk fish (Chanos chanos) of Kaattuppalli Island, South East Coast, India. Chemosphere. 83: 415–421.

Raldua, D., S. Diez, J.M. Bayona and D. Barcelo. 2007. Mercury levels and liver pathology in feral fish living in the vici of a mercury cell chlor-alkali factory. Chemosphere. 66: 1217–1225.

Roberts, S.D. and M.D. Powell. 2003. Comparative ionic flux and gill mucous cell histochemistry: effects of salinity and disease status in Atlantic salmon (*Salmo salar* L.). Comp. Biochem. Physiol. A 134: 525–537.

Roberts, S.D. and M.D. Powell. 2005. The viscosity and glycoprotein biochemistry of salmonid mucus varies with species, salinity and the presence of amoebic gill disease. J. Comp. Physiol. B 175: 1–11.

Roy, S. and S. Bhattacharya. 2006. Arsenic-induced histopathology and synthesis of stress proteins in liver and kidney of Channa punctatus. Ecotoxicol. Environ. Saf. 65: 218–229.

Sabóia-Moraes, S.M.T., F.J. Hernandez-Blazquez, D.L. Mota and A.M. Bittencourt. 1996. Mucous cell types in the branquial epithelium of the euryhaline fish Poecilia vivipara. J. Fish Biol. 49: 545–548.

Sakuragui, M.M., J.R. Sanches and M.N. Fernandes. 2003. Gill chloride cell proliferation and respiratory responses to hypoxia of the neotropical erythrinid fish Hoplias malabaricus. J. Comp. Physiol. B 173: 309–317.

Salazar-Lugo, R., C. Mata, A. Oliveros, L.M. Rojas, M. Lemusd and E. Rojas-Villarroel. 2011. Histopathological changes in gill, liver and kidney of neotropical fish Colossoma macropomum exposed to paraquat at different temperatures. Environ. Toxicol. Pharmacol. 31: 490–495.

Schwaiger, J., R. Wanke, S. Adam, M. Pawert, W. Honnen and R. Triebskorn. 1997. The use of histopathological indicators to evaluate contaminant-related stress in fish. J. Aquat. Ecosyst. Stress Recov. 6: 75–86.

Shephard, K.L. 1994. Functions for fish mucus. Rev. Fish Biol. Fish. 4: 401–429.

Shiogiri, N.S., M.G. Paulino, S.P. Carraschi, F.G. Baraldi, C. Cruz and M.N. Fernandes. 2012. Acute exposure of a glyphosate-based herbicide affects the gills and liver of the Neotropical fish, *Piaractus mesopotamicus*. Environ. Toxicol. Pharmacol. 34: 388–396.

Sibbing, F.A. and R. Uribe. 1985. Regional specialization in the oro-pharyngeal wall and food processing in the carp *Cyprinus carpio*. Netherlands J. Zool. 35: 377–422.

Sollid, J., E.R. Weber and G.E. Nilsson. 2005. Temperature alters the respiratory surface area of crucian carp *Carassius carassius* and goldfish *Carassius auratus*. J. exp. Biol. 208: 1109–1116.

Sollid, J. and G.E. Nilsson. 2006. Plasticity of respiratory structures-adaptive remodeling of fish gills induced by ambient oxygen and temperature. Resp. Physiol. Neurobiol. 154: 241–251.

Sollid, J., P. Angelis, K. Gundersen and G.E. Nilsson. 2003. Hypoxia induces adaptive and reversible gross-morphological changes in crucian carp gills. J. exp. Biol. 206: 3667–3673.

Stentiforda, G.D., M. Longshaw, B.P. Lyons and G. Jones. 2003. Histopathological biomarkers in estuarine fish species for the assessment of biological. Mar. Environ. Res. 55: 137–159.

Tetreault, G.R., C.J. Bennett, C. Cheng, M.R. Servos and M.E. Masterb. 2012. Reproductive and histopathological effects in wild fish inhabiting an effluent dominated stream, Wascana Creek, SK, Canada. Aquat. Toxicol. 110: 149–161.

Tuurala, H., S. Egginton and A. Soivio. 1998. Cold exposure increases branchial water–blood barrier thickness in the eel. J. Fish Biol. 53: 451–455.

Uguz, C., M. Iscan, A. Erguven, B. Isgor and I. Togan. 2003. The bioaccumulation of nonyphenol and its adverse effect on the liv of rainbow trout (Onchorynchus mykiss). Environ. Res. 92: 262–270.

Ultsch, G.R. and G. Gros. 1979. Mucus as a diffusion barrier to oxygen: possible role in O2 uptake at low ph in carp (*Cyprinus carpio*) gills. Comp. Biochem. Physiol. A 62: 685–689.

van Dyk, J.C., M.J. Cochrane and G.M. Wagenaar. 2012. Liver histopathology of the sharptooth catfish Clarias gariepinusas a biomarker of aquatic pollution. Chemosphere. 87: 301–311.

Velma, V. and P.B. Tchounwou. 2010. Chromium-induced biochemical, genotoxic and histopathologic effects in liver and kidney of goldfish, carassius auratus. Mut. Res. 698: 43–51.

Velmurugan, B., M. Selvanayagam, E.I. Cengiz and E. Unlu. 2007. Histopathology of lambda-cyhalothrin on tissues (gill, kidney, liver and intestine) of Cirrhinus mrigala. Environ. Toxicol. Pharmacol. 24: 286–291.

Wendelaar-Bonga, S.E. and C.J.M. Van Der Meij. 1989. Degeneration and death, by apoptosis and necrosis the pavement and cells the gills of the teleost *Oreochromis mossambicus*. Cell Tissue Res. 255: 235–243.

Wester, P.W., L.T.M. van der Ven, A.D. Vethaak, G.C.M. Grinwis and J.G. Vos. 2002. Aquatic toxicology: opportunities for enhancement through histopathology. Environ. Toxicol. Pharmacol. 11: 289–295.

Wobeser, G. 1975. Acute toxicity of methyl mercury chloride and mercuric chloride for rainbow trout (*Salmo gairdneri*) fry and fingerlings. J. Fish. Res. Board. Can. 32: 2005–2013.

Zaccone, G., S. Fasulo, P. Locasclo, L. Ainis, M.B. Ricca and A. Licata. 1989. Effects of chronic exposure to endosulfan on complex carbohydrates and enzyme activities in gill and epidermal tissues of the fresh water catfish *Heteropneustes fossilis* (Bloch). Arch. Biol. 100: 171–185.

Emerging Contaminants and Endocrine System Dysfunction

Daniele Dietrich Moura Costa

Introduction

There has been growing concern in recent years that a variety of natural and synthetic chemicals could be producing serious health effects in humans and other species by interfering in the actions of endogenous hormones. Public attention was originally drawn to the idea that environmental chemicals could disrupt the endocrine system of wildlife by the publication of Rachel Carson's "Silent Spring" in the 1960s. This book described deleterious reproductive effects of the then commonly used insecticide, dichlorodiphenyltrichloroethane (DDT), on birds and other wildlife. However, the first published evidence for environmental pollutants able to impact the endocrine system had actually appeared more than a decade earlier, when it was reported that consumption of a certain type of clover disrupted reproduction in sheep. Both of these reports involved chemicals that were able to impair the endocrine system. The organochloride pesticides, which were of greatest concern to Carson, were banned in Western Europe and North America in the 1970s, as were the polychlorinated biphenyls

Universidade Federal do Paraná, Setor de Ciências Biológicas, Departamento de BiologiaCelular, Laboratório de Toxicologia Celular, P.O. BOX 19031, ZIP CODE 81531-990, Curitiba, PR, Brazil.
Email: danidmc87@yahoo.com

(PCBs) widely used as electrical insulators. Both of these classes of chemical are extremely persistent, resistant to biodegradation, and in many areas their concentrations in fish tissues remain unchanged several decades after production has ceased. To these have been added a vast array of other chemicals which may reside in lake, river and ocean sediments for varying periods and that can be metabolized into a wide range of products of unknown toxicity (Kime 1999).

Nowadays there is a growing debate about the definition of endocrine disruptor compounds (EDCs). Originally, the concern over endocrine disruption was based almost entirely on perceived effects on the reproductive system and it was common to refer to the chemicals as estrogen mimics or estrogenic chemicals. Later, chemicals that could block estrogenic responses (anti-estrogens) or androgenic responses (anti-androgens) were found and it was soon recognized that chemicals could affect other elements of the endocrine system via interaction with hormone receptors other than those of the sex steroids (Philips and Harrison 1999). In order to establish consensus on the scope of the endocrine disrupter issue, to facilitate the identification of active chemicals and, ultimately, to underpin any future regulatory control, it is essential to agree to a precise definition of an endocrine disrupter (ED). The working definition considered herein is the same adopted in the final report of the US EPA's Endocrine Disruptor Screening and Testing Advisory Committee (EDSTAC): "An endocrine disruptor is an exogenous chemical substance or mixture that alters the structure or function(s) of the endocrine system and causes adverse effects at the level of the organism, its progeny, populations, or subpopulations of organisms, based on scientific principles, data, weight-of-evidence, and the precautionary principle". The US EPA also accept the definition adopted by Kavlock et al. (1996): An endocrine disrupting compound is "an exogenous agent that interferes with synthesis, secretion, transport, metabolism, binding action, or elimination of natural blood-borne hormones that are present in the body and are responsible for homeostasis, reproduction, and developmental process". Nowadays these two definitions are considered together and include not only the primary ways EDCs are thought to interfere with normal functioning of the endocrine system, but also some of the effects EDCs can have on an organism. The term endocrine disrupter is now preferred because it allows inclusion of health effects thought to result from interference with any part of the endocrine system, including thyroid, head, kidney and pituitary hormones. Thus, from a physiological perspective, an endocrine disrupting substance is a compound, either natural or synthetic, which, through environmental or inappropriate developmental exposures, alters the hormonal and homeostatic systems that enable the organism to communicate with and respond to its environment (Diamnati-Kandarakis et al. 2009; Pait and Nelson 2002).

The group of molecules identified as endocrine disruptors is highly heterogeneous and includes natural (17β-estradiol) and synthetic hormones (17α-athinylestradiol, 17β-trambolone, methyltestosterone), phytormones present especially in soy feeds (flavonas, genistein, coumestrol), chemicals used as industrial solvents/lubricants and their byproducts (polychlorinated biphenyls, polybrominated biphenyls, dioxins, plastics (bisphenol A)), plasticizers (phthalates), pesticides (methoxychlor, chlorpyrifos, DTT, endrin, lindane), fungicides (vinclozolin, ketoconazole), heavy metals (cadmium, mercury, lead, arsenic) and pharmaceutical agents (diethylstilbestrol, tamoxifen, fradazole, trilostane) (Diamnati-Kandarakis et al. 2009).

There are many pathways that may be impaired by endocrine disruptors. In vertebrates, different types of hormone systems could be main targets of environmental chemicals: reproductive hormones such as estrogens and androgens, thyroidal hormones, corticosteroids, growth hormone and their associated hypothalamus-pituitary releasing and stimulating hormones. However, research has to date focused nearly entirely on the interference with the steroidal reproductive hormones for obvious reasons. Firstly, reproductive hormones control one of the most important endpoints in the risk assessment of environmental chemicals, which is all aspects of reproductive development from sex differentiation to puberty, aspects of which are critical for population viability. Secondly, reproductive hormones are of vital importance during the early critical stages of embryonic development and sex differentiation. Because of the critical effects of estrogens on female and male reproduction and its ability to also induce abnormalities in male reproductive systems, studies in the endocrine disruptor area were initially centered on estrogenic effects of environmental chemicals. The subsequent identification of a very large number of chemicals that act, at least partially, by binding to the estrogen receptor, coupled with the widespread environmental distribution of certain of these chemicals, has resulted in a continuing focus on endocrine disruptors that act by altering signaling through the estrogen receptor. However, in recent years, there has been a rapid increase in information concerning the ability of various chemicals to mimic or inhibit the actions of other hormones (e.g., androgens, cortisol, thyroid hormone), or to inhibit enzymes related to synthesis, metabolism and clearance of hormones. It is now clear that, although endocrine disruptors that function as estrogens and/or anti-estrogens are the most common, endocrine disruption is a phenomenon that can potentially affect any hormonal system (Cooke et al. 2002; Scholz and Mayer 2008).

Except in cases of localized acute pollution, dead fish are now seen much more rarely, but there is increasing evidence that their health is being compromised by long-term low-level pollution. At one level this may lead to premature mortality, resulting in a decreased number of breeding seasons and thence fewer offspring. Accumulated evidence suggests, however, that even low levels of pollutants can disrupt the functioning of the endocrine system of fish, leading to decreased immune and stress responses, energy metabolism, osmoregulatory ability and reproductive function. Because of their exposure in the aquatic ecosystem, as the major repository of environmental pollutants, fish can provide an early warning of effects that may later become apparent in other wildlife and ultimately in humans themselves. Indeed, there is increasing evidence that some of the problems found in fish, including decreased fertility, genital abnormalities, altered behavior patterns and response to stress and disease, are now appearing in human population (Kime 1999; Toppari et al. 1996).

This chapter reviews the endocrine system of fish, the target organs and pathways that can be impaired by endocrine disruptors. In addition, the potential for these compounds to affect fish health, both in laboratory and field studies is discussed, as are potential future research directions in this area.

Fish Endocrine System

To fully understand how these pollutants can affect fish at very low levels, it is necessary to outline very briefly how the endocrine system works. It is a control system of the body which responds to internal and external signals to maintain the body's chemical equilibrium, to regulate sexual development and the seasonal reproductive cycles, and to evoke a stress response to external threats. At its core are the hypothalamus and pituitary, which respond to neural signals from the brain and convert them into hormone messengers who act on the individual glands such as the gonads, thyroid and the adrenocortical cells, present in the head kidney (Pait and Nelson 2002).

The primary function of an endocrine system is to transform various exogenous stimuli into chemical messengers—hormones—resulting at least in the expression of the appropriate gene and thus in the synthesis of proteins or in the activation of already existing tissue-specific enzyme systems. The endocrine system represents an important tool for the timely coordination of development (e.g., induction of spawning cycles or sexual maturity) and metabolism (e.g., glucose homeostasis). Exogenous stimuli like day length, temperature, light, or pheromones, as well as endogenous stimuli generally known as the "internal" clock, are processed in the central nervous system. After a complex chain of biochemical processes, the hypothalamus secretes releasing hormones or releases inhibition hormones that control the secretion of hormones from the pituitary gland. These glycoproteins, secreted by the pituitary gland, induce synthesis and release of tissue-specific hormones in the various glands (thyroid, adrenocortical cells, testes and ovary). Hormones secreted by these internal glands travel through the bloodstream to their target tissues and target cells, where they initiate a change in cellular activity by attaching to a receptor protein. This change is transmitted across the plasma membrane of a cell in different ways depending on the type of hormone. The cascade of different, interdependent physiological processes is regulated by complex mechanisms such as a negative feedback pathway that is turned on and off in response to fluctuating hormone levels: when hormone production of the glands peaks, the hormone acts as an inhibitor and causes the hypothalamus and/or pituitary to shut down the pathway producing the substance (Lintelmann et al. 2003).

The endocrine system in fish consists of various glands located throughout the body which synthesize and secrete hormones to regulate an array of biological processes (Table 10.1). The neural component of the endocrine system is the hypothalamus, which produces and secretes stimulant pituitary hormones. The pituitary gland in fish secretes a number

Table 10.1. Selected endocrine gland and hormonal action in fish.

Gland/Hormone	Target organ	Effects
Hypothalamus TRH CRH GnRH	Pituitary Pituitary Pituitary	Secretion of TSH Secretion of ACTH Secretion of LH and FSH
Pituitary TSH ACTH LH/FSH GH	Thyroid Adrenocortical cells Gonads Various	Stimulation of T_3/T_4 production Stimulation of cortisol production Stimulation of gonads Stimulation of growth
Thyroid Thyroxins (T_3/T_4)	Various	Adaptation to environmental, growth, development, metabolism, reproduction, etc.
Adrenocortical cells Cortisol	Gills, kidney, gut	Stress response, osmorreguation, energy metabolism, electrolytic balance, adipocit recruitment, larval development.
Gonads Estrogens (17βestradiol) Androgens (11KetoTestosterone)	Oocytes Liver Testis	Maturation Vitellogenin and Choriogenin production Spermiogenesis
Corpuscle of Stannius Hypocalcin	Gills	Calcium homeostasis
Pancreas Insulin Glucagon	All cells All cells	Increases glucose permeability Glycogen and lipid metabolism

of hormones which affect growth, osmoregulation, lipid metabolism, reproductive development and behavior, as well as controlling other endocrine glands (Bone et al. 1995; Pait and Nelson 2002). For example, the thyroid gland secretes the hormones thyroxine (T_4) and triiodothyronine (T_3), which are believed to aid fish in adapting to changes in temperature, osmotic stress, development and growth of larvae and embryos. The corpuscles of Stannius secrete hypocalcin which is thought to be involved in calcium homeostasis and may also be involved in controlling the ratio of calcium to sodium and potassium in the plasma (Bone et al. 1995; Pait and Nelson 2002). As in humans, the fish pancreas secretes insulin which aids in glucose permeability. Secretion of glucagon, also by the pancreas, enables increased glycogen and lipid metabolism. Although most of what is currently known about the effects of EDCs involves reproduction and reproductive behavior, other areas of the endocrine system, such as the thyroid, may also be targets for EDCs. Investigations of possible effects on other targets are just beginning.

Although many steps of this sensitive system can be influenced by different external stimuli, most effects of endocrine disruptors observed and explained until now are attributed to the function of the gonads, which control the development of sexual differentiation, secondary sex characteristics, and functioning of sex organs. Fish possess a reproductive endocrine system in that external cues are translated by the brain and hypothalamus into the release of gonadotrophin releasing hormone (GnRH). This in turn causes the pituitary gland, situated at the base of the brain, to release gonadotrophin which stimulates steroid synthesis in the gonads. At least some fish possess two gonadotrophins (FSH and LH), but some authors prefer to call FSH by GtH1 and LH by GtH2. FSH stimulates the ovary to produce estradiol which induces production of a yolk protein (vitellogenin) by the liver, while LH predominates just before spawning when it stimulates ovarian synthesis of a progestogen (17,20P-dihydroxy-4-pregnen-3-ona, usually abbreviated to 17,20PP or MIH—maturation inducing hormone), which induces maturation of the oocytes prior to ovulation. This progestogen may also play a role in sperm maturation. Male testis produces a major androgen—11-ketotestosterone—rather than testosterone. In fish, the gonads of both sexes synthesize testosterone, which may play an important role in feedback to the pituitary. The gonads of fish also have some of the properties associated with the liver of mammals in that they can convert steroid hormones into metabolites (Kime 1999; Pait and Nelson 2002).

In fish, the thyroid does not form discrete glands; it is scattered around the ventral aorta. This makes it very difficult to measure the changes induced in their structure by chemical pollutants. The hormones secreted by the thyroid are thyroxine (T_4) and triiodothyronine (T_3) and regulate general metabolic rate, growth and possibly embryonic development (Brown et al. 2003).

The adrenocortical cells are dispersed within the head kidney and secrete some hormones, specially cortisol a hormone involved in stress response, osmoregulation and energy metabolism. Growth in fish is continuous and not only dependent on secretion of growth hormone by the pituitary, but on age and the rate of metabolic activity and energy utilization as determined by both the thyroid and adrenocortical cells. Thyroid and adrenal activities are also involved in the osmoregulatory adaptations required by some species, such as salmon, which migrate between salt and fresh water (Hontela 2005).

In conjunction with the nervous and immune systems, the endocrine system forms the main regulatory mechanism that controls different pivotal functions in the animal body. The messengers of the endocrine system are hormones that are synthesized and excreted at very low quantities from specialized glands and transported to the target organ(s) via the bloodstream. Hormones are transported in the blood in free state or attached to carrier proteins and bind at the target organs to specialized hormone receptors on the cell surface or within the cell (nuclear receptors). This hormone-receptor complex then activates different cell or organ functions. The binding between hormone and receptor is based on steric complementarities comparable to the "key and lock" principle. The interaction between hormone, hormone receptor, and DNA shall be explained describing the estrogen receptor (ER). As already described, the lipophilic sex steroids released from the gonads into the bloodstream are transported to their target organs or tissues. They enter the cell by passive diffusion through the lipid membrane. The hormone then binds to the specific receptor protein, located within the cytosol or nuclei. The "free" receptor (i.e., without a ligand) is maintained in an inactive conformation through interactions with a number of associated proteins. Once activated, the receptor then forms homo or heterodimers, which seek out specific DNA motifs, termed "hormone response elements", located in the nucleus, upstream of hormone-responsive genes. Binding of the receptor complex to the hormone-response element of DNA results in chromatin rearrangement, usually allowing the cells' transcriptional machinery to access the promoter region of hormone-inducible genes, producing increased mRNA production followed by increased protein expression, resulting finally in observed effects such as increased growth in the reproductive tract organs and mammary glands (Lintelmann et al. 2003; Gillesby and Zacharewski 1998).

Hormones influence several essential regulatory, growth, developmental, and homeostatic mechanisms of the organism, such as reproduction, maintenance of normal levels of glucose or ions in the blood, blood pressure, general metabolism, and other muscle or nervous system functions. The balance of the hormones (homeostasis) in the organism is essential in order to prevent functional disorders. Therefore, the endocrine system includes a number of central nervous system-pituitary-target organ feedback mechanisms that enable the body to react very flexibly on internal or external changes of the hormone status. But this complex system is very sensitive upon disturbing influences that can severely impair the whole development of the organism (Lintelmann et al. 2003).

Mechanisms of Endocrine Dysfunction

The process by which hormones are produced in an endocrine organ, released into the circulation, transported to target tissues, and then bind to receptors in target cells to affect their subsequent activity is obviously extremely complex and tightly regulated. Each step of this process for any hormone signaling system is potentially vulnerable to disruption by an external agent. Indeed, there are known examples of endocrine disruption that result from effects at a wide variety of sites of action encompassing almost all aspects of hormone production, transport and action. For example, a single EDC may be both estrogenic and anti-androgenic. EDCs may be broken down or metabolized to generate subproducts with different properties. For instance, the estrogen agonist DDT is metabolized into the androgen antagonist DDE (Diamanti-Kandarakis et al. 2009; Rasier et al. 2007).

There are many pathways by which EDs can exert their effects. A chemical could have an effect on the hypothalamus, altering the production and secretion of hormones. A chemical can interact with pituitary cells and alter the levels of trophic pituitary hormones that regulate the activity of a steroidogenic organ (gonads and adrenocortical cells) or thyroid. Besides, toxic chemicals can cause the death of the hypothalamus neurons or the pituitary cells responsible for the production and secretion of a hormone or the pituitary cells, impairing hormone production by both organs. In both these events, a chemical has direct effects on the gland and alters hormone production and secretion. However, chemicals also could alter blood levels by altering and binding to transport proteins or affecting catabolism of the circulating hormone. EDCs can also mimic or antagonize the binding of the endogenous ligand to its receptor and could potentially affect other processes, such as the migration of the receptor from the cytoplasm to the nucleus. Chemicals could inhibit receptor dimerization or the binding of the ligand hormone and its receptor to the DNA or could alter the binding of cofactors and the initiation of transcription. Additional modes of action, such as altering the transportation or stability of the mRNA produced in response to hormonal stimulation or the translation of the mRNA into protein, are also possible (Cooke et al. 2002).

As described above, there are many possible mechanisms by which chemicals may interact with the endocrine system, of which the most important and studied are discussed below.

Direct Interaction with Hormone Receptors

In the initial phase of development of endocrine disruption research, attention was focused almost exclusively on so-called estrogen mimics: chemicals which induced biological responses normally associated with the action of natural estrogens. Since estrogens act by binding to specific receptors in target tissues, the ability of a chemical to bind to the same receptors was taken as sufficient evidence to define it as an estrogen mimic. With the expansion of the definition of endocrine disruption, EDCs now include chemicals which can interact with any hormone receptor, most notably androgen, thyroid and corticosteriod hormone receptors. Some compounds can bind to more than one receptor and a complex array of interactions is possible. Not only the hormone receptors, but also the enzymes involved in diverse pathways of synthesis, release, modification, metabolism and clearance of a hormone are liable to be impaired by EDCs.

Agonism

An exogenous agonist can be defined as a ligand that can bind to a receptor like the natural substrate and "turn it on". The activation of the hormone receptor then finally leads to the same effects that can be caused by endogenous hormone action. The potency of an exogenous agonist depends on its affinity to the receptor as well as on its ability to turn the receptor on. It should be mentioned that different species exhibit different structures of the hormone receptors. Therefore, ligand binding to a specific receptor does not automatically mean that this substance exhibits the same affinity for the respective hormone receptor of another species (Lintelmann et al. 2003).

The most frequently studied and best understood EDCs are those that mimic estrogens. Estradiol produced by the ovaries and transported via the circulatory system is passively taken up by the hepatocyte and then crosses the nuclear membrane. The unliganded estrogen receptor (ER) is maintained in an inactive conformation through interactions with a number of proteins, primarily heat shock proteins Hsp 70 and 90. Following the binding of estrogen to the receptor, the heat shock proteins dissociate allowing the ER to change its conformation to the active form. Once activated, the receptor forms a homodimer complex which seeks out specific DNA segments, in this case the estrogen response elements (EREs). Binding of the complex to the ERE results in a rearrangement of the chromatin and transcription of the gene, followed by production of the target proteins such as vitellogenin (vtg) and coriogenin (chg). A compound able to bind to the estrogen receptor in the cell might very well result in transcription and pleiotropic responses potentially affecting numerous functions within the organism (Gillesby and Zacharewski 1998; Pait and Nelson 2002).

Despite the most frequently studied and best understood type of EDCs are estrogens mimetics, there are an increasing number of studies with EDCs that mimic androgens, thyroid hormones and cortisol hormones among others.

Antagonism

Other EDCs can act on the hormone receptors via an antagonistic mechanism: an antagonist is a ligand that blocks or diminishes responses elicited by agonists because the receptor cannot be activated as usual. The inhibition of the receptor can be competitive (i.e., the endogenous agonist and exogenous antagonist compete for the same active binding site) or it can be noncompetitive (i.e., the inhibitor binds at the receptor or receptor-hormone complex, but not at the active binding site). Competitive inhibition can lead to total deactivation of the receptor; non-competitive inhibition can result in slower or reduced reactions performed by the receptor.

Typical antagonists for hormone receptors are the herbicides vinclozolin, and their metabolites that block the androgen receptor, or the pharmaceutical tamoxifen, which competes for binding sites at estrogen receptors (Kelce et al. 1995; Cook et al. 2003).

For both agonistic and antagonistic reactions between exogenous ligand and hormone receptor, the concentration of the ligand often plays an important role. Concentrations of endogenous hormones are normally very low. If concentrations of the xenobiotic in the organism are high, endocrine disruption effects can be evidenced even if the exogenous ligands exhibit only a low binding affinity to the receptor.

Indirect Interaction with the Hormone Receptors

Effects on hormone concentration, metabolism, synthesis, storage, release, transport and clearance

One of the most obvious mechanisms by which a compound may function as an endocrine disruptor is by having toxic effects on an endocrine organ(s) and subsequently altering the production of that hormone. A compound can also disrupt hormone synthesis by a specific effect on an endocrine organ, rather than by causing generalized toxicity. An example of this type of mechanism is the effects of some PCBs on steroid synthesis. Steroid hormones are synthesized from cholesterol by an elaborate enzymatic pathway, and compounds that either increase or decrease levels of steroidogenic enzymes would have the potential to produce alterations in the hormone production of those cells (Cooke et al. 2002).

There are many ways by which endocrine disrupters might affect the levels of hormones circulating in the bloodstream. For example, lipid soluble hormones, including sex steroids, thyroid hormones and glucocorticoids, are transported bound to carrier proteins and their effects are to some extent influenced by the levels of these proteins in the blood (Kime 1999). An influence on the transportation of the hormones via the bloodstream to the target tissues and organs can also lead to a disturbance of the endocrine system. Chemicals that compete with hormones for the binding sites of transport proteins may increase the level of free and, therefore, effective hormones (Lintelmann et al. 2003).

The levels of circulating natural hormones can be altered by chemical interference with the synthesis or breakdown of the hormone by mechanisms not necessarily mediated through hormone receptors. For example, the phytoestrogen sitosterol is able to reduce gonadal steroid biosynthesis by either affecting cholesterol availability or by altering the activity of P450-dependent enzyme. Tributyltin inhibits the conversion of androgens to estrogens by inhibiting P450 aromatase or by inhibiting testosterone metabolism and excretion. PCBs and dioxins are well known for their ability to induce certain CYP 1A1 and CYP 1B1 in liver. These enzymes have a key function in the synthesis and degradation of steroid hormones, and their production or activity can be influenced by various xenobiotics such as PCB congeners and dioxin (Safe and Krishnan 1995; Soontornchat et al. 1994). So it is possible that changes in rates of metabolism might disturb hormone levels.

Hormone receptor concentration

The responsiveness of a tissue to a hormone depends on the density of receptors within its component cells. The number of receptors is determined by their rate of synthesis and catabolism, which is itself controlled by complex feedback mechanisms involving hormone action. Some chemicals are known to interfere with this regulation. For example, TCDD can act to increase or decrease the expression of the estrogen receptor in fish (Kime 1999).

In receptor-mediated processes, both components, endogenous ligand and hormone receptor, own a key function. Every exogenous influence may, therefore, shift this sensitive balance. A so-called "down-regulation" of steroid hormones is discussed for some anti-estrogenic compounds, especially for 2,3,7,8-tetrachlorodibenzo-*p*-dioxin (TCDD). TCDD is an exogenous agonist for the arylhydrocarbon (Ah)-receptor. This receptor is not directly involved in hormone metabolism, but its activation can have different influences on the endocrine system by: (1) an increased degradation rate of estrogen receptors (down-regulation), (2) induction of estradiol metabolizing enzymes, and (3) inhibition of gene expression controlled by estradiol or growth promoters (Lintelmann et al. 2003).

The mechanisms of interactions between xenobiotics and the endocrine system presented in this chapter are only a small part of possible modes of action. The increased research in this field will surely elucidate more complex relations between exposed organisms, endocrine disruption, and xenobiotics.

Target Organs and Markers of Endocrine Dysfunctions

Biological effects in fish that have been attributed to the effects of endocrine disruptors include the inappropriate production of the blood protein vitellogenin (VTG; the female-specific and estrogen-dependent egg yolk protein precursor) in male and juvenile fish, inhibited ovarian or testicular development, abnormal blood steroid concentrations, intersexuality and/ or masculinization or feminization of the internal or external genitalia, impaired reproductive output, precocious male and/or female maturation, increased ovarian atresia (in female fish), reduced spawning success, reduced hatching success and/or larval survival, alteration in courtship behavior and alterations in early development (Jobling and Tyler 2003).

In addition to reproductive and developmental effects, fish exposed to endocrine disruptors may have a decreased response to stress or decreased growth and metabolism which can affect their ability to survive, or to defend themselves against predators. All of these factors can affect the ability of the species to survive and to reproduce itself in sufficient numbers to maintain the population viability (Kime 1999).

Endocrine disrupting effects are, sometimes, difficult to evaluate, because EDC are characterized by a delayed response, often measurable in years after exposure to low, physiologically relevant dosages during sensitive periods of organ development in embryo.

Hypothalamic-pituitary-gonadal Axis

Hypothalamic and pituitary dysfunction and abnormalities

The close neural relationship of the hypothalamus and the pituitary gland with the brain makes them particularly vulnerable to neurotoxins such as organophosphate pesticides and heavy metals such as lead and mercury. There are only few studies on these tissues, reflecting their very small size and inaccessibility.

The control of reproductive neuroendocrine function involves a group of neurons in the basal hypothalamus that synthesize and release the decapeptide GnRH. GnRH release drives reproduction throughout the life cycle, and this is the primary stimulus to the rest of the reproductive axis (the pituitary and gonads). GnRH release stimulates gonadotropin release from the anterior pituitary gland, which in turn activates steroidogenesis and gametogenesis in the ovary and testis. Steroid hormones produced by the gonad act on other target tissues that express estrogen, progestin, and/or androgens receptors, a concept that is fundamental to endocrine disruption because so many EDCs act to interfere with steroid hormone actions (Gore 2002). There is clear evidence that heavy metals and the organophosphate pesticides can damage the neurons of the hypothalamus which are responsible for GnRH release, leading to failure of ovaries and testes to produce yolky eggs and viable sperm. Organochlorine and organophosphate pesticides, cyanide, PAHs, PCBs, cadmium and mercury can all cause degeneration of the secretory cells of the pituitary gland and decrease its release of hormones. The feedback signal to the pituitary, which regulates plasma steroid hormone balance, can be disrupted by any xenobiotic that has hormone mimicking properties. Organochlorine pesticides such as lindane mimic natural estrogen and induce a negative feedback response in the pituitary (Kime 1999; Jobling and Tyler 2003).

Since the interaction between hypothalamus and pituitary is neural, the hypothalamic-releasing hormones, GnRH are not released in sufficiently high quantities to be detectable in peripheral circulation. Therefore, assays of hypothalamic function rely on hormone measurements of their corresponding pituitary hormones. If the pituitary sensitivity to hypothalamic output is compromised, then it is impossible to distinguish a primary hypothalamic or pituitary effect of an EDC. Another possible way to verify impairment in the hypothalamus/pituitary is the presence of necrotic and/or apoptotic cell in these tissues. Death of cells impairs the production and release of the hormones (Diamanti-Kandarakis et al. 2009).

Up until this moment, there are few studies demonstrating agonistic and/or antagonistic effects of xenobiotics in hypothalamus and pituitary receptors. Similarly, there are few reports of endocrine disrupting effects on synthesis, metabolism and release of hypothalamus-pituitary hormones. Nevertheless it is possible that xenobiotics exert their effects through these mechanisms. The effects of endocrine disruptors on hypothalamus and pituitary are almost always related to toxic and degenerative damage of the neurons and cells that produce the hormones, than with receptor interaction or metabolism of the hormones.

Male reproductive dysfunction and abnormalities

Abnormal function of the reproductive endocrine system in male and female fish can be caused by disruption of hypothalamic, pituitary or gonadal function, and by changes in the liver that affect the enzymes which deactivate steroid hormones. Both gonads and liver are fatty tissues and rapidly bioaccumulate both non-biodegradable organic pollutants and heavy metals that can be present at levels several orders of magnitude greater than either in surrounding water. In this section we will discuss the effects on male and subsequent female reproductive organs and functions. Impaired liver functions will be discussed in the following sections.

Endocrine dysfunction in the testis may be apparent by changes in its structure and secretion of hormones, in the activity of the enzymes that are necessary for steroid synthesis, in the quality and quantity of the sperm produced and in the hormone dependent behavior patterns of the male fish. Clear evidence for endocrine disruption of the testis is apparent when its steroid producing cells show increased or decreased activity, the proportions of sperm at different stages of development differ from those of unexposed fish or when there is a complete arrest in sperm production (Zhou et al. 2009). Testis structure and sperm quantity will be discussed in this section. Effects on the steroideogenic enzymes as well as in behavior and secondary sex characteristics are discussed in the following sections.

Although fish produce vastly more sperm than eggs, evidence suggests that even a small decrease in sperm quality or quantity can decrease the male's fertilizing ability. Viability of sperm in fish, as in other vertebrates, is dependent on the correct hormonal and nutritional environment during their development within the testis. This may in turn be affected by the internal hormonal environment during the early life stages in which the testis is differentiated. Endocrine disruption can therefore lead to abnormal development of the sperm and decrease its viability. This may take the form of either abnormal sperm structure or a decrease in its energy supply, both of which can alter its swimming ability and therefore its capacity to reach and fertilize the egg (Kime 1999).

The synthesis, release and binding of the maturation inducing hormone (MIH; 17α,20β-dihydroxy- 4-pregnen-3-one) is a possible target for EDCs in both male and female gonads. This hormone is considered a "maturation-inducing steroid" and is responsible for the final steps of maturation of sperm and oocytes. Interferences with releasing of trophic hormones are likely to be reflected in alterations in steroid hormones levels, and cause effects similar to direct interference (receptor binding or modulation of synthesis) with estrogens or androgens. Natural hormones, such as 17β-estradiol, as well as synthetic hormones, such as 17α-ethynilestradiol, alter male testis structure and function and decrease sperm quantity, quality and viability. Industrial chemicals, such as bisphenol A, nonylphenol, dioxins and PCBs, also impair the male reproductive tract (Scholz and Mayer 2008).

The effects of EDCs on male reproductive tract and sperm could be identified from testis histopathology (gonadal structure, GSI, gonadal sex ratio, feminization), hormone concentration (MIH and testosterone), enzymes activity (P450 aromatase, HSD) and reproductive impairment (spawning, fecundity, fertilization rate in offspring).

Female reproductive dysfunction and abnormalities

Xenobiotic induced disruption of female fertility follows essentially the same pattern as that of the male and can be caused by changes in pituitary-hypothalamic function, primary disruption of ovarian structure or hormone secretion, or changes in the rate of hormone deactivation. In addition, there may be changes in the synthesis of estrogen induced production of the yolk protein by the liver (vitellogenesis), which in turn can lead to failure in depositing sufficient yolk in the developing oocytes. Vitellogenesis provides a valuable biomarker for endocrine dysfunction in both sexes, but is more properly considered as part of the liver function. The main factor causing the increase in ovarian weight during the reproductive development of fish is the deposition of yolk into the developing eggs. Abnormal development of the ovary can be caused by lack of stimulation by pituitary hormones, failure of steroid synthesis or direct cellular damage. The most commonly observed effect in the ovary is a decrease in the numbers of large yolky eggs together with increased numbers of immature oocytes and/or atretic oocytes. The increased number of immature oocytes can be considered as an inhibition of ovarian development by a disruption in the hypothalamus-pituitary-gonadal axis (Kime 1999; Diamanti-Kandaraski et al. 2009).

Synthesis of the steroid hormones testosterone and MIH is identical in both ovary and testis, and dependent on the functional integrity of both the steroid producing enzymes and the receptors for pituitary gonadotrophins. Increase or decrease of female hormone levels such as 17β-estradiol in plasma also indicates dysfunction of hypothalamus-pituitary-ovaries and were reported in fish exposed to the effluent of pulp mills and other areas rich in PCBs and PAHs. Male and female gonads differ in that only the ovary has an active aromatase enzyme that converts testosterone into estradiol. Inhibition of this ovarian aromatase activity in female fish will decrease estrogen synthesis, and have consequential effects on the ability of its liver to synthesize yolk proteins, leading to retarded growth of the oocytes. Inhibition of aromatase activity leads to a reduction of 17β-estradiol plasma concentration resulting in reduced vitellogenin levels in female fish. Aromatase inhibitors could also be classified as androgenic compounds due to their associated masculinizing effects. These masculinizing effects are presumably provoked by accumulation of androgens since fish treated with the aromatase inhibitor fadrozole show elevated 11-ketotestosterone levels (Gray et al. 2002).

The effects of EDCs on female reproductive tract and oocytes could be identified from ovaries histopathology which indicates gonadal structure, gonadal sex ratio, masculinization, stage of developing oocytes, presence/absence of atretic oocytes, by GSI, hormone concentration (MIH, 17β-estradiol and testosterone), enzymes activity (P450 aromatase, HSD) and reproductive impairment (spawning, fecundity, fertilization rate in offspring).

Impairment of gonadal stereoideogenesis

In recent years, the interactions of EDCs with key enzymatic activities involved in both synthesis and metabolism of sex hormones has been investigated, since many compounds exert their effects though these enzymes.

Not only gonads, but also adrenocortical cells, possess enzymes related to stereoideogenesis process. However, in gonads these enzymes produce sexual steroids such as 17β-estradiol and 11keto-testosterone, while in adrenocortical cells these enzymes produce cortisol. In this section we will discuss the effects of EDCs on gonadal enzymes related to steroidogenesis; adrenocortical steroidogenesis will be discussed in the following sections.

Sex steroids have two classical functions in fish development: they act as morphogenic factors during sex differentiation and as activational factors during sexual maturation. Steroids also direct the development of germ cells and accessory glands and organs, as well as the modification of behavior, to ensure that sexual reproduction can take place. The interference of xenobiotics on the synthesis and clearance of key sex hormones may also alter bioavailable amounts of active hormones within the organism, and be a potential mechanism of endocrine disruption (Thibaut and Porte 2004).

Teleost fish produce several types of gonadal steroids, including estrogens, progestogens, androgens and numerous other steroids. A general picture of steroid production is shown in Fig. 10.1. Steroids are produced in specialized cells within the ovarian follicle (theca, granulosa) and the testis (Leydig cells). Steroideogenic cells possess enzymes that modify cholesterol and its derivatives into steroids hormones. These cells may also possess steroidogenic acute regulatory protein (StAR), which functions to transport cholesterol to the mitochondria, the true rate-limiting step in steroidogenesis. There are few reports of EDCs inhibiting the transportation

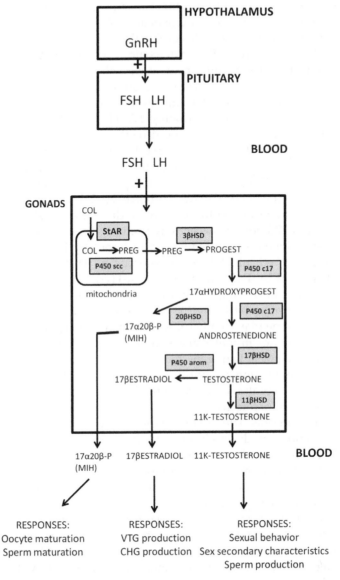

Figure 10.1. General mechanisms of gonadal production of steroid hormones. GnRH: Gonadotropin releasing hormone. FSH: follicle stimulating hormone. LH: luteinizing hormone. COL: cholesterol. PREG: pregnenolone. PROGEST: progesterone. 11αHYDROXYPROGEST: 11α-hydroxyprogesterone. 17α20β-P (MIH): 17α,20β-dihydroxy-4-pregene-3-one (maturation inducing hormone). StAR: steroidogenic acute regulatory protein. P450 scc: P450 cholesterol side chain cleaving enzyme. P450 c17: P 450 17α-hydroxylase. P450 c21: P450 21-hydroxylase. 3βHSD: 3β-hydroxylase. 11βHSD: 11β-hydroxylase. 17βHSD: 17β-hydroxylase. 20βHSD: 20β-hydroxylase. +: stimulus.

of cholesterol by StaR, although there are reports of chemicals that decrease the cholesterol availability (Young et al. 2005).

After the delivery of cholesterol to the inner mitochondrial membrane by StAR, the cytochrome P450 side-chain cleavage (P450scc), located at the inner mitochondrial membrane, converts cholesterol into pregnenolone. Pregnenolone, in turn, can serve as a substrate for cytochrome P450 17-hydroxylase (P450$_{C17}$) which catalyzes the hydroxylation of pregnenolone at C17 (17-hydroxylase) producing 17α-Hydroxypregnenolone. At the same time, pregnenolone can be directly converted into progesterone by 3β-HDS. 3β-HDS also converts17α-Hydroxypregnenolone into 17α-Hydroxyprogesterone. Progesterone is also converted into 17α-Hydroxiprogesterone by P450$_{C17}$. P450$_{C17}$ has dual enzymatic activity, it also catalyzes the removal of two-carbon (C20–21) acetic acid residues from 17-hydroxyprogesterone to yield the androgen androstenedione. Lyase activity and 17-hydroxylase activity occur in the same enzymatic pocket. Androsteridione is converted into testosterone by 17β-HSD. Testosterone can be converted into 11keto-testosterone by 11 β-HSD activity or into 17 β-Estradiol by P450 aromatase. 17α-Hydroxiprogesterone can also be converted into 17α,20β-dihydroxy-4-pregnen-3-one, the maturation inducing hormone (MIH) by 20β-HSD (Young et al. 2005).

Testosterone plays many roles in development of spermiogenesis and courting behavior in males and secondary sexual characteristics in both males and females. Estrogens have been mostly associated with female reproductive function, primarily as a central mediator of vitellogenesis in oviparous vertebrates. However, it has become evident that estrogens also play an important role in reproduction in males by stimulating proliferation of gonial stem cells. The MIH promotes the sperm and oocyte maturation and liberation into the environment (Thibaut and Porte 2004).

The fungicides ketoconazole and prochloraz have been reported to inhibit the P450 aromatase in fish. Ketoconazole also inhibits the P450scc and P450$_{C17}$ activities. 3β-HSD activity can be inhibited by the pharmaceutical trilostane (Ankley et al. 2009).

Xenobiotics that affect all these reported enzymes can impair the steroidogenesis and consequently the normal functions of the ovary and testis. Additionally, secondary sex characteristics and sexual behavior also depend on gonadal steroids and can be impaired by dysfunction in steroidogenesis process.

Steroidogenesis can be impaired in any level, from cholesterol uptake to its conversion into steroid hormones. The most reported targets of EDCs are the P450 enzymes (P450scc, P450 aromatase and P450$_{C17}$) and the HSD enzymes (3β, 17β, 11β and 20β-HSD). Although there are few reports about the issue, StAR transportation of cholesterol can also be impaired by EDCs.

Thyroid

The study of thyroid activity has undoubtedly been complicated by the difficulty of isolating the fish tissue which, unlike mammals, does not form a distinct gland. The thyroid tissue in most teleosts is scattered diffusely in the basibranchial region or around the ventral aorta, but, as in other vertebrates, consists of follicles formed of a single layer of epithelial cells surrounding an extracellular lumen containing proteinaceous colloid (Brown et al. 2004).

The hormones of the thyroid, in concert with those of the adrenal gland and growth hormone from the pituitary, regulate energy utilization, behavior and growth. Since thyroid hormones (THs) are incorporated into the developing eggs from maternal sources, they may also be involved in larval development, although their role in this is still unclear. In teleost fish, the thyroid hormones also play a pivotal role in the regulation of metamorphosis, in the transition from larval to juvenile stages (Brown et al. 2004; Power et al. 2001). THs also have a potential function in the control of both sexual differentiation and reproductive development (Cyr and Eales 1996). Furthermore, THs are also involved in the regulation of the immune system, as indicated by the suppression of immune function by hyperthyroidism (Yada and Nakanishi 2002). Besides, thyroid hormones are essential for normal brain development, for the control of metabolism, and for many aspects of normal adult physiology. Therefore, changes in the function of the thyroid gland or interference in the ability of the thyroid hormone to exert its action may produce effects on development, metabolism, or adult physiology and could severely compromise fitness and survival.

A general scheme on how thyroid hormones are produced in fish is shown in Fig. 10.2. The main thyroid hormones, thyroxine (T_4) and triiodothyronine (T_3), are tetra- and triiodinated derivatives of 4-hydroxydiphenyl ether. Thyroid function can be disrupted by action of pollutants at several sites, including production of thyrotrophin releasing hormone (TRH) from the hypothalamus, release of thyroid stimulating hormone (TSH) from the pituitary, synthesis (via thyroglobulin iodination, or deiodination of thyroxine to triiodothyronin) and metabolism (conjugation) of thyroid hormones, interference in the plasma binding proteins, or by binding to the thyroid receptor or by T_4 conversion in the thyroid and other responsive organs into T_3 (Brown et al. 2004; Scholz and Mayer 2008).

The thyroid is stimulated by thyroid stimulating hormone (TSH), a glycoprotein secreted by the pituitary thyrotrope cells. The thyrotrope cells are stimulated to secrete TSH by the TRH (thryrotrophin stimulating hormone) from the hypothalamus. No well-defined hypothalamic-pituitary portal system in fish exists and thyrotropes are innervated directly by

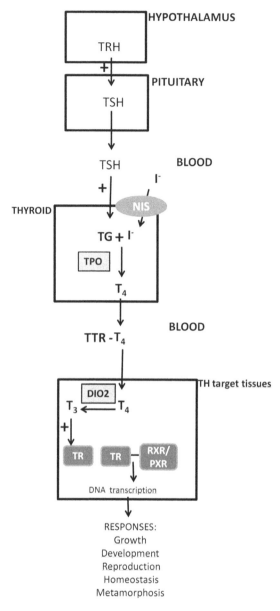

Figure 10.2. General scheme of thyroid hormone production and action on target tissues. TRH: thyrotropin releasing hormone. TSH: thyroid stimulating hormone. I-: iodide. NIS: sodium/iodide symporter. TG: thyroglobulin. TPO: thyroperoxidase. TTR: transthyretin. T3: triiodothyronine. T4: thyroxine. DIO: deiodinase. TR: thyroid hormone receptor. RXR: retinoid X receptor. PXR: pregnane X receptor. +: stimulus.

hypothalamic neurons. Circulating TH negatively influences the activity of the hypothalamus and pituitary and, when THs levels are reduced or "disrupted", the activity of the hypothalamus–pituitary increases so as to elevate THs production by the thyroid follicles and restore its circulatory set point levels, and vice versa. Thus, the activity of the HPT axis is regulated by negative feedback (Car and Patiño 2011).

Levels of TRH and TSH would be ideal endpoints to measure in studies of hypothalamus-pituitary-thyroid responses to potential thyroid-disrupting agents. However, there are several technical drawbacks that make these endpoints difficult to determine. First, TRH is released locally from nerve terminals innervating the pituitary gland and thus their levels in the peripheral circulation are very low. Also, because of species differences in the subunit primary structure of TSH, mammalian commercial antibodies typically do not recognize non-mammalian TSH and homologous antisera are not available for the majority of teleosts. Some studies have utilized quantitative PCR methods to examine changes in expression of the TSH-β subunit gene in pituitary. However, while data on gene expression can assist in hypothesis development regarding potential downstream effects on thyroid function, this information does not directly address the question of whether TSH secretion (and availability to the thyroid follicles) has been altered (Car and Patiño 2011).

Since thyroid is stimulated to produce THs by TSH, the first step in thyroid hormone synthesis is the uptake of iodide into the thyrocyte by the sodium/iodide symporter (NIS). Iodide is essential for thyroid hormone synthesis, and iodine deficiency is an important public health problem worldwide. Thus, chemicals that interfere with the NIS may interfere with thyroid hormone synthesis or may exacerbate problems of iodine deficiency (Zimmermann 2007).

Iodide, the form of iodine that enters the cell, must be oxidized to a higher oxidation state before it is transferred to the precursor of thyroid hormone, thyroglobulin. Of the known biological oxidizing agents, only H_2O_2 and O_2 are capable of oxidizing iodide. Organification of iodine is controlled by the enzyme thyroperoxidase (TPO), a heme-containing enzyme. A number of compounds, such as 6-propyl-2-thiouracil (PTU), a methylmercaptoimidazole, are known to block TPO. PTU is well known to reduce circulating levels of THs and to increase circulating levels of TSH. This reduction is caused by the ability of PTU to inhibit directly the function of the TPO enzyme (Taurog 2000; Miyazaki et al. 2004).

The main product of thyroid in fish is T_4. Once produced, T_4 is secreted into the blood and carried by specific proteins. In fish, most part of thyroid hormones is bound to transthyretin (TTR) and only a small fraction runs free. The role of serum binding proteins for thyroid hormone in thyroid homeostasis is not well understood. Some chemicals can bind to TTR and block the transportation of T_4 (Schussler 2000). Obviously, contaminants that affect the interaction between TTR can also affect the availability of TH to target tissues. TTR levels have been measured directly using ELISA in studies of contaminant effects in teleosts (Morgado et al. 2009). Quantitative PCR also can be used to gauge TTR gene expression (Shi et al. 2009) but these data may be difficult to interpret without information on changes in the actual levels of TTR protein.

Once in the serum, T_4 can be taken up into tissues by selective transporters to enter cells. There are a number of transporters that are likely to be important in the control of thyroid hormone uptake into various tissues and cells. However, little is known—or has been tested—about the ability of specific environmental or industrial chemicals to interfere with THs transporter function. Conceptually, disruption of TH transporter activity would interfere with TH availability to intracellular thyroid hormone receptors (TRs), but little is known about how potential EDCs may act upon these transporters (Köhrle 2007; Carr and Patiño 2011).

Inside the cell, T_4 is converted into T_3 by the type 1 or type 2 deiodinase. These outer-ring deiodinases are essential for thyroid hormone action. A number of environmental chemicals affect deiodinase activity, including PCBs (Kato et al. 2004). T_3 is the active form of THs and binds to the thyroid hormone receptor (TR). TRs and its ligands form homodimers and also heterodimers with other nuclear receptors, in particular with the retinoid X receptor (RXR) (Hahn et al. 2005). This complex (T_3-TR-TR/RXR) binds to thyroid hormone response elements (TRE) in DNA and promotes the gene's transcription. Some chemicals, such as PCBs, can act as antagonists of TR. These chemicals bind to TR and antagonize T_3, inhibiting TR-mediated gene activation (Hahn et al. 2005; Jugan et al. 2010).

Not only the TR activation can be disrupted by EDCS, but also many other pathways in the signaling cascade which culminates in gene expression. These other pathways are not yet well understood, but there are evidences that EDCs are able to activate receptor-dependent transcription of TH target genes by modulating upstream signaling without binding to the T_3-binding site of TRs. EDCs also could exert transcriptional effects by disrupting the recruitment/release of coactivators by TRs, by interfering with the expression of TR and their heterodimerization partner, or by interfering with the affinity between TR and TRE. All these potential targets need to be further investigated, as well as their physiological consequences (Jugan et al. 2010).

Many chemicals are known to decrease the serum half-life of T_4 by inducing liver sulfate and glucuronide conjugation pathways. Both of these pathways turn T_4 into T_4-sulfate or T_4-glucorinide. The enzyme involved in the T_4 glucuronidation is uridine diphosphate glucuronyl transferase (UGT) and the enzymes involved in T_4 sulfatation is sulfuryltransferase

(SULT). These enzymes can be induced by dioxin-like compounds acting on the AhR or through the pregnane X-receptor or constitutive androstane receptor nuclear receptors. T_4-sulfate or T_4-glucorinide can be converted into an inactive T_3 by deionidase type 1 which can be excreted via bile. In the liver, T_3 is also degraded by removal of one of its inner-ring iodines by deionidase type 3 to form the presumed inactive T_3 (Kretschmer and Baldwin 2005; Jugan et al. 2010).

THs are very well documented in diverse physiological process in adults. For many years, the notion that THs are important for teleost embryogenesis was based primarily on the knowledge that the hormone is maternally inherited in eggs and thus present way before its endogenous production, and that embryos contain the necessary mechanisms to regulate the availability and activity of THs (i.e., deiodinases and TRs) (Tindall et al. 2007, Walpita et al. 2007). For example, deiodinase activity seems to regulate the pace of development and pigmentation in *Danio rerio* embryos (Walpita et al. 2007), and offspring of hypothyroid zebrafish females (producing eggs with reduced TH levels) have malformed lower jaws (Mukhi and Patiño 2007). There are few reports suggesting embryo impaired development and exposure to EDCs.

Not only the embryo development, but also its transformation into larva and adult can be impaired by EDCs. Metamorphosis is characterized by major post-embryonic or post-hatch (in oviparous species) changes in body morphology (Bishop et al. 2006). In oviparous vertebrates, these changes generally result in the transformation of larval forms into juveniles. Among teleosts, (classical) TH-dependent metamorphic changes in external morphology range from the relatively mild, such as paired fin development in zebrafish (Brown 1997), to the radical, such as the extreme craniofacial rearrangement and loss of bilateral symmetry in flatfish (Schreiber 2006). The available information suggests that TH is involved in key aspects of teleost pre-hatch (embryogenesis) as well as post-hatch development (embryo-to-larva and larva-to-juvenile transformations) and embryos are possible targets for endocrine disruption.

Due to some difficulty in measuring levels of TRH, TSH, THs, deiodidase, UGT and SULT activities specially in small fish, thyroid histopathology has been a powerful tool to evaluate effects of EDCs on this organ. The use of thyroid histopathology is based on the premise that changes in the histological structure of the thyroid follicle cells, the appearance or content of the colloid, the degree of vascularization present within the thyroid gland or among thyroid follicles, and the overall size of the thyroid gland or tissue reflect changes in thyroid follicle activity in response to contaminant exposure. Thyroid follicle cell height is a widely used measure of thyroid gland function because it is directly proportional to the degree of stimulation by TSH. In some species, the degree of vascularization to the thyroid gland changes dramatically during goiter formation and can be measured independently as an endpoint of contaminant exposure (Mukhi et al. 2005). Finally, measurements of overall thyroid gland (or follicle tissue) volume can be used to determine goiter formation. Thyroid goiters may form due to hyperplasia and/or hypertrophy of thyroid follicle cells. Typically, goiter formation occurs in response to contaminants that inhibit TH synthesis thereby reducing the negative feedback regulation of TSH secretion. However, it is important to consider that increases in thyroid gland size (or thyroid follicle hypertrophy) also may reflect the effects of contaminants that activate the HPT axis at sites upstream of the thyroid gland or follicles (Carr and Patiño 2011).

Due to drawbacks in measuring TRH and TSH concentrations, measurement of TH plasma levels is the most commonly used assay as an endpoint for thyroid disruption in fish. However, without corroborating histopathological measurements, estimates of TH clearance (deiodinase, UGT and SULT activities), as well as activation or blockade of TR, TH plasma levels are difficult to interpret as they may reflect changes in TH secretion, metabolism, or deiodination.

Liver

The liver plays an important role in the endocrine system. The concentrations of hormones in plasma and the activity of the glands that secrete them are determined by the rate at which they are deactivated by the liver. Figure 10.3 shows how thyroid and steroid hormones can be metabolized in liver. The liver also has a major function in female reproduction, since it is the target tissue of ovarian estrogen, to which it responds by producing the yolk protein vitellogenin. Xenobiotics that affect any of these functions can therefore be considered to be potential endocrine disruptors.

Liver is an essential organ for steroid deactivation. It contains cytochrome P450-dependent mono-oxygenases, and reducing and conjugating enzymes that convert hormones into water-soluble products which can be more easily excreted. Cytochrome P450 enzymes of the liver are affected by copper, mercury and organotin, PCBs, PAHs, pulp mill effluent and municipal wastewater (Kime 1999; Jobling and Tyler 2002).

Constitutive androstane receptor (CAR) and pregnane X receptor (PXR), which are both members of the nuclear receptor family present in the liver, are activated by a large panel of xenobiotics, including endocrine disruptors, and induce the expression of drug transporters and xenobiotic metabolizing enzymes, including UDP-glucuronosyltransferases (UGTs) and sulfotransferases (SULTs). Hepatic SULTs and UGTs are involved in eliminating THs: the inactive glucuronide and sulfate derivatives of THs are eliminated in the urine and bile. Furthermore, T_4-sulfate is a

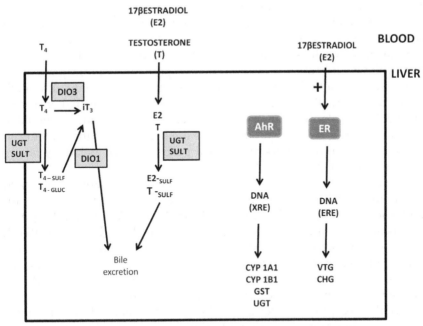

Figure 10.3. General mechanisms of steroid and thyroid hormone metabolization in liver, and activation of AhR and ER receptors. DIO: deiodinase enzymes. UGT: UDP-glucoronyltransferase. SULT: sulfotransferase enzymes. T_3: triiodothyronine. T_4: thyroxine. iT_3: inactive triiodothyronine. $T_{4\text{-SULF}}$: thyroxine sulfated. $T_{4\text{-GLUC}}$: thyroxine glucoronidate. E_2: 17β-estradiol. T: testosterone. $E_{2\text{-sulf}}$: 17β-estradiol sulfated. $T_{\text{-sulf}}$: testosterone sulfated. AhR: aryl hydrocarbon receptor. CYP: cytochrome P450 superfamily. GST: glutatione-S-transferase. ER: estrogen receptor. VTG: vitellogenin. CHG: coriogenin. ERE: estrogen responsive element. XRE: xenobiotic responsive element. DNA: deoxyribonucleic acid. +: stimulus.

considerably better substrate than T_4 for inner ring deiodination by liver type-I deiodinase, producing reverse-T_3, a biologically inactive metabolite of T_4. CAR was shown to mediate the induction of several isoforms of UGTs and SULTs involved in glucuronidation and sulfation of THs. Therefore, CAR/PXR activation by xenobiotics could indirectly result in disruption of TR activity (Gasiewicz and Park 2003; Hahn et al. 2005).

The fish liver also possesses receptors which specifically respond to estrogen (estrogen receptor) by synthesizing the yolk protein vitellogenin. Male, female and juvenile fish all possess these receptors in the liver, but under normal conditions only the female produces sufficient estrogen to activate the receptor and induce vitellogenin synthesis. This is usually a seasonal phenomenon and coincides with the need of the developing egg to incorporate yolk protein. Measurement of vitellogenin in plasma therefore provides both a useful indicator of the reproductive status of the female and a method of distinguishing the two sexes. Since vitellogenin is not normally detectable in significant amounts in male, juvenile or sexually regressed female fish, its presence in these animals is a clear indication of abnormal stimulation of the estrogen receptor by EDCs. The extremely high levels of vitellogenin in the plasma of estrogen stimulated fish, the ease with which it can be measured and the specificity of the response have made this one of the most widely used tests for endocrine disruption. In some cases, induction of vitellogenin occurs at levels at which there is also clear evidence of inhibition of testicular activity and steroid secretion due to EDCs exposure, while in other cases the high level of the protein can cause a range of non-endocrine related problems such as kidney failure (Hiramatsu et al. 2005).

Liver also possesses the Aryl hydrocarbon receptor (AhR), also called dioxin receptor (see more details in Chapter 3). This is a nuclear receptor and belongs to the basic helix-loop-helix (bHLH)/PAS (Period [Per]-Aryl hydrocarbon receptor nuclear translocator [Arnt]-Single minded [Sim]) family of heterodimeric transcriptional regulators. It is a ligand activated transcription factor known to mediate most of the toxic and carcinogenic effects of a wide variety of environmental contaminants such as dioxin (TCDD; 2,3,7,8-tetrachlorodibenzo-[p]-dioxin). The normal physiological role of this receptor is under discussion, but it is believed to be involved in cell proliferation and differentiation, in liver and immune system homeostasis and in tumor development. Molecular mechanisms for AhR activation in the absence of xenobiotics remain elusive in part because no definitive endogenous ligands have been identified. In the presence of some xenobiotics, especially TCDD, the AhR is activated and forms a dimer with ARNT (AhR nuclear translocator). The AhR-ARNT regulates gene expression by interacting with a specific DNA sequence, referred to frequently as a xenobiotic-response element, dioxin-response element or AhR-response element. These genes encode specially for CYP1A1, CYP1A2, CYP1B1, Glutatione-S-transferase, UDP-glucoronyltransferase and NADPH-quinone oxidoreductase. So the increased concentrations of these proteins or alteration in their activities are possible markers for endocrine dysfunction in liver (Barouki et al. 2007).

Diverse markers can be useful to evaluate endocrine disrupting effects in liver. The most widespread is the expression of vitellogenin by males and juvenile fish. The concentration of estrogen receptors, CAR receptors, PXR receptors and AhR can also be used to evaluate endocrine dysfunction in liver. All these receptors can activate DNA transcription and generate diverse proteins such as CYP1A1, CYP1A2, GSTs, UGTs, SULTs, among others. These proteins can be involved in xenobiotic and/or hormone metabolism and excretion. So these proteins are also possible targets for EDCs.

Adrenocortical Cells

Compared to the extensive knowledge on chemicals interfering with estrogen or androgen responses, the study of corticosteroids disruptors is an emerging field of research, and the identification of relevant xenobiotics and their underlying mechanisms of toxicity remains a major challenge.

The response of the organism, referred to as the stress response, activates the hypothalamo-pituitary-interrenal (HPI) axis and results in the release of corticoisteroid hormones (Fig. 10.4). The term interrenal rather than adrenal has been traditionally used in teleosts to describe the location of the steroid-producing tissue within the head kidney rather than to the kidney as in mammals. However, Norris (1997) proposed to use the term adrenocortical cells to identify, in teleosts and other non-mammalian vertebrates, the cells homologous to those found in the adrenal cortex of mammals which secretes cortisol.

Cortisol is the major corticosteroid hormone in fish and it has a role in osmoregulation and maintenance of electrolyte balance, regulation of metabolism, modulation of immune function and exerts antigonadal effects. There are still controversies concerning the specific role of cortisol in the physiological stress response. Despite this ongoing controversy, there is strong evidence that several important processes in life cycles of fish are dependent on cortisol that often facilitates actions of other hormones such as thyroid hormones or growth hormones. Thus, corticosteroid hormone-disrupting compounds could have severe impacts on adaptive responses with obvious implications for individual health and population development, even though these effects may be subtle and difficult to detect (Hontela 2005; Mommsen et al. 1999; Pankhurst and Van Der Kraak 2000).

Corticosteroid responses are regulated at various levels, i.e., biosynthesis, binding to serum proteins, cellular uptake, intracellular binding and metabolism, receptor binding, cellular export, degradation and excretion from the organism. Xenobiotics can potentially disrupt each step of corticosteroid action or they may act on multiple targets by mimicking the corticosteroid molecule. Thus, depending on the modified target and cell type or organ involved, xenobiotics may affect different physiological processes that are regulated by corticosteroids (Odermatt and Gumy 2008).

Secretion of corticosteroids in teleosts is under the control of the hypothalamus-pituitary axis, through the actions of corticotropin releasing hormone (CRH) synthesized by neurons that enter the pituitary and stimulate the secretion of ACTH from pituitary corticotropes. Although other trophic factors have been identified, ACTH is the most potent hormone capable of stimulating the adrenocortical cell to synthetizecortisol in teleosts (Hontela 2005).

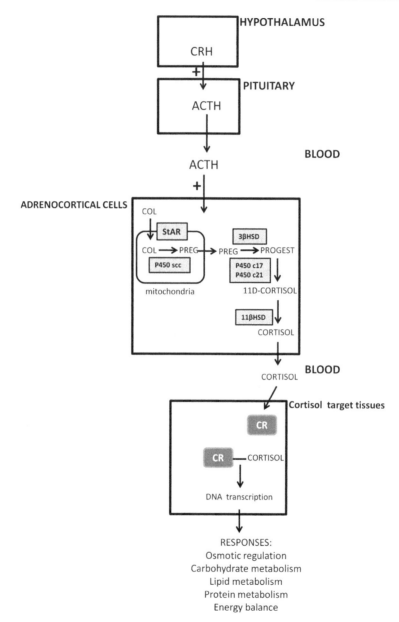

Figure 10.4. General scheme of cortisol production by adrenocortical cells and effects of cortisol on target cells. CRH: corticotropin releasing hormone. ACTH: adrenocorticotropic hormone. COL: cholesterol. PREG: pregnenolone. PROGEST: progesterone. 11D-CORTISOL: 11-deoxycortisol. StAR: steroidogenic acute regulatory protein. P450 scc: P450 cholesterol side chain cleaving enzyme. P450 c17: P 450 17α-hydroxylase. P450 c21: P450 21-hydroxylase. 11βHSD: 11β-hydroxylase. 3βHSD: 3β-hydroxylase. CR: cortisol receptor. +: stimulus.

Corticosteroids, similar to all steroid hormones, are synthesized from cholesterol in a series of reactions mediated by steroidogenic enzymes. Cholesterol first enters the mitochondria in a process facilitated by the steroidogenic acute regulatory (StAR) protein that transfers cholesterol from the outer mitochondrial membrane to the inner membrane. The first enzymatic step of steroidogenesis is the hydrolysis of the side chain of cholesterol by P-450scc (cholesterol side chain cleaving enzyme) yielding pregnenolone, a key precursor of corticosteroids. Pregnenolone is then transferred to the smooth endoplasmic reticulum where it is converted to progesterone by 3β-HSD and hydroxylated by 17α-hydroxylase (P-450$_{C17}$) and 21-hydroxylase (P-450$_{C21}$) yielding 11-deoxycortisol. Returning to the mitochondrion, 11-deoxycortisol is then modified by 11β-hydroxylase, also designated as P-450$_{C11}$, the enzyme that converts 11-deoxycortisol to cortisol (Stocco 2000; Conley and Bird 1997).

Plasma cortisol levels reflect the clearance rate from plasma and the rate of synthesis by adrenocortical cells, since the hormone is not stored in the cells but rather released into the circulation as it is synthesized. Corticosteroids, as well as other steroids synthesized in the body and xenobiotics absorbed from the environment, are metabolized mainly in the liver by Phase I and Phase II enzymes. These enzymes increase the solubility of the molecules, facilitating their excretion through feces or bile. It is important to note that the enzymes metabolizing corticosteroids are also the enzymes that metabolize other substrates, including numerous pollutants, and may detoxify or bioactivate the xenobiotics (Colby et al. 1994; Conley and Bird 1997).

Cortisol receptors (CR) have been characterized in liver, gut, adipose tissue, as well as gonads and osmoregulatory organs. The receptors for corticosteroids are present in the cytosol and following binding of the hormone to its receptor, the hormone-receptor complex is translocated to the nucleus. Metabolism of carbohydrates, protein and lipids is affected by cortisol in teleosts, and high plasma cortisol results in elevated plasma glucose levels. However, although the gluconeogenic actions of cortisol are well documented in mammals, the metabolic consequences of elevated plasma cortisol are still under some debate in teleost fish. Xenobiotics interfering with corticoesteroid availability or CR function are expected to disturb the cellular energy status by altering the expression of genes involved in carbohydrate, lipid and protein metabolism (Mommesen et al. 1999; Odermatt and Gumy 2008).

Various anthropogenic compounds, such as PCBs, can suppress corticosteroid levels and/or attenuate the corticosteroid response (Pottinger 2003). However, an elevation of cortisol levels has also been reported in fish following exposure to estrogenic compounds such as estradiol and 4-nonylphenol (Lerner et al. 2007). While suppression of the corticosteroid response could impair the organism's ability to adapt to changing environments or stress, a chronic or frequent hyperactivation may more seriously lead to growth suppression, reproductive dysfunction, and immunosuppression (Pottinger 2003). EDC compounds may also interfere with the early steps of steroid synthesis and thus affect reproductive steroid hormones as well (Aluru and Vijayan 2006).

Cortisol also suppresses the immune response, while the immune system itself comprises a range of intracellular messengers that can be affected by pollutants. Endocrine disruption of adrenal activity can therefore also affect both the ability of the fish to cope with natural stress and that induced by pollutants, and its susceptibility to disease (Kime 1999).

Osmoregulatory effects of cortisol are well known. Cortisol regulates sodium flux across membranes of gut, gills and kidney, and stimulates sodium transportation, both in and out in numerous teleost species subjected to various osmotic challenges. It stimulates the activity of $Na^+/K^+ATPase$ in osmoregulatory tissues and it seems to facilitate adaptation of teleosts to both sea and fresh water. It is generally accepted that the increase in cortisol facilitates adaptation to either sea water or fresh water during migration. EDCs that impair the cortisol production also impair osmorregulatory abilities (Hontela 2005).

There are many other pathways involved in cholesterol metabolism which are able to be disrupted by chemicals. Another classical example of a tissue-specific toxicity is the effect of o,p-DDD on the 11β-hydroxylase, a key steroidogenic enzyme present in the interrrenal tissue. Steroid hydroxylases contain heme-proteins, constituents that make them vulnerable to toxicants and oxidative stress. The adrenotoxic effects of DDT and its derivatives, and the sensitivity of the 11β-hydroxylase to DDT have been detected in numerous fish species (Horm 1989; Hontela 2005).

Recent studies in fish demonstrated an important connection between aryl hydrocarbon receptor (AhR) activation and the regulation of cortisol-mediated stress response. Fish exposed to the AhR agonist β-naphtoflavone showed impaired stressor-induced plasma and interrenal responses. β-naphtoflavone decreased ACTH-mediated cortisol production as a result of a reduced expression of StAR and P450scc, while 11β-hydroxylase expression was not affected. It was suggested that AhR activation might disrupt interrenal corticosteroidogenesis and target tissue responsiveness to corticosteroids stimulation in fish (Aluru and Vijayan 2006; Odermatt et al. 2006).

Potentially, disruption of glucocorticoid hormone action by environmental chemicals can occur at various steps: (1) regulation of the hypothalamus–pituitary–adrenocortical cells (HPA)-axis, (2) activity of enzymes with a role in steroidogenesis (StaR, P450scc), (3) binding capacity of serum proteins, (4) uptake into the target cell, (5) intracellular metabolism by 11β-HSD enzymes, (6) activation of the CR, (7) function of CR-associated proteins, and (8) degradation and excretion of the steroid hormone. Several recent reports provide evidence for the existence of chemicals that disturb different steps of glucocorticoid action, thus emphasizing the need for additional test systems to understand their mechanisms of action and to assess their health risks.

Other Kinds of Responses

Adipose tissue and energy metabolism

The adipose tissue works as an endocrine organ that secretes numerous hormones, growth factors, enzymes, cytokines, and complementary factors that participate in the body's feedback system regulating appetite and food intake.

PPARγ is a member of the nuclear receptor superfamily and constitutes a major regulator of adipogenesis. It is primarily expressed in adipose tissue, and its activation promotes adipocyte differentiation as well as the induction of lipogenic enzymes. Additionally, it contributes to maintenance of metabolic homeostasis through transcriptional activation of genes implicated in energy balance. During its activation, PPARγ forms a heterodimer with RXR (retinoic X receptor), and the complex binds to PPAR response elements in the regulatory regions (promoters) of target genes ultimately involved in the regulation of fatty acid storage and the repression of lipolysis (Migliarini et al. 2011).

Experimental evidence highlights that PPARγ is a molecular target for endocrine disruptors, in particular organotin compounds such as tributyltin (TBT) and triphenyltin. Kanayama et al. (2005) showed that TBT and triphenyltin functioned as agonists of PPARγ and RXR, acting as high-affinity ligands at levels comparable to known endogenous ligands.

It is possible that PPARγ signaling can interact with that of estrogen to influence adipogenesis. These findings have been recently reviewed (Grün and Blumberg 2006) and represent an important example of the mechanism by which environmental chemicals can interfere with body weight regulation. In addition, at high doses, TBT can inhibit aromatase enzyme activity in adipose tissue directly, leading to decreased estradiol levels and down-regulation of ER target genes. TBT at moderate to high doses inhibits the activity of 11β-hydroxysteroid dehydrogenase, resulting in decreased inactivation of cortisol. It has been hypothesized that the increased local corticosteroid levels could influence late stages in adipocyte differentiation and thus, metabolic regulation (Grün et al. 2006).

TCDD has been found to cause a dose dependent decrease in body weight, food intake, resting and total oxygen consumption, resulting in the so-called wasting syndrome. The reduction in body weight occurred via the inhibition of IGF-I circulating levels and of phosphoenolpyruvate carboxykinase (PEPCK) mRNA expression (Croutch et al. 2005).

In a very recent study conducted on zebrafish, natural mixtures of persistent organic pollutants POP (PBDE, PCB and DDT) extracted from freshwater ecosystems induced alterations in body weight gain and gene expression profiles associated with phenotypic changes in females. In particular, the gene expression changes concerned lipid metabolism, metabolic disorders and reproduction since PPARs and the ERs were the main targets of those EDs (Lyche et al. 2010).

Even though the researches in adipocyte tissue, adipogenesis and energy metabolism are still beginning, PPARγ activation and the expression of proteins related to lipid metabolism as well as body weight gain can be reasonable markers for EDCs exposure.

Behavior and secondary sex characteristics

Courtship rituals in fish can involve a complex communication between males and females via chemicals released into the surrounding water. These pheromones are, in many cases, the hormones secreted by the gonads and are probably a selective adaptation enabling males to detect females that are ready to spawn. Successful courtship is therefore dependent on the ability of the female to secrete the pheromone, the male to detect it, and for his neural and endocrine system to elicit a response that both triggers his production of sperm and his appropriate sexual behavior. Carbofuran and the organophosphate diazinon can both disrupt the ability of male fish to detect pheromones. Xenobiotics might also cause changes to other aspects of androgen dependent breeding behavior, particularly in species in which the male builds or guards a nest or spawning territory. This may be particularly important in species such as the stickleback which produces an androgen stimulated glue from the kidney for nest-building (Bell 2001; Dzieweczynski 2011).

Development of male secondary sexual characteristics, such as the red belly of sticklebacks or the hooked jaw in salmonids, is dependent on testicular secretion of 11-ketotestosterone, and any decrease in production of this steroid or effects on its receptors could inhibit his ability to attract a mate. Such changes can make a useful biomonitor for reduction in testicular function (Kime 1999).

Xenobiotics can also affect secondary sexual characteristics in species in which they are present. The most documented case is in mosquitofish and least, killifish, in which the anal fin of the male is modified to form a gonopodium that is used to transfer sperm to the female during internal fertilization. As in so many cases, the absence of a hormone can be just as disrupting as the presence of an inappropriate gonodopodium in females, which suggests a masculinization of females (Zhou et al. 2009).

Thomas et al. (2005) document that 17α-ethinylestradiol reduces the competitive reproductive fitness of the male guppy fish by significantly decreasing sexual coloration, sperm amount and frequency of courtship behavior. Male competition for mating and sperm competition are important factors affecting next generation quality in many species, and the consequences of feminization or demasculinization may be more serious (Evans and Magurran 2000).

Many EDCs are able to disrupt sexual behaviors and secondary sex characteristics when they are apparent. Breeding behavior, particularly in species in which the male builds or guards a nest or spawning territory, courtship behavior, and male competition for mating can be impaired by EDCs. Sperm production and liberation as also can be affected by EDCs. There has been an increase in researche in the field of behaviors that can be impaired by EDCs and some issues that are not understood now will soon be clear.

Development and growth

While fertilization and hatch rates have long formed the basis of many standard toxicity tests owing to the rapidity with which they can be carried out with small fish, they also provide valuable information on endocrine disruption. In many cases, abnormal fertilization and hatch rates as well as developmental deformities may be caused by the chemicals interfering with key developmental processes involving cellular signaling mechanisms, rather than by simple toxicity or chromosomal damage. Although the exact mechanism by which embryonic and larval development is controlled is not yet fully understood, it involves both the steroid and thyroid hormones and a range of intercellular messengers that are now covered by the broader definitions of endocrinology. As a result of the widespread use of early life stages for toxicity testing, there is a vast literature on the types of developmental abnormalities which can occur, but so far very little on how much of this is really due to endocrine disruption and how much is simply random toxicity or genetic damage which leads to deformity or mortality (Kime 1999; Zhou et al. 2009).

Many of the incidences of production of abnormal larvae may be a result of exposure during very short critical periods in early life, including exposure to contaminated yolk passed on from mothers who have accumulated high pollutant burdens during their whole life. Abnormal larvae development as a consequence of mothers' exposure can be considered a hereditary effect and will be discussed in the next section (Kinnbeg et al. 2003).

Heritability and transgenerational e ffects

Bioaccumulated EDCs inherited from the mother influence not only the reproductive development and physiology of the offspring, but also the offspring's reproductive behavior as adults, thereby magnifying the effects of EDCs though subsequent generations.

EDCs are able to decrease nutrient content of eggs and this decrease will result in smaller larvae with a decreased survival rate—the net result in both cases may be a decreased survival of the progeny. Furthermore, there is evidence that many xenobiotics can be passed on from the mother to the offspring by incorporation into the developing eggs. Although this off-loading of the pollutant burden will decrease its toxicity to the mother, it will inevitably increase the exposure of the developing offspring at critical stages of embryonic development. Since eggs rapidly become impermeable in water after spawning, it is probable that in most cases the embryo receives the majority of the toxicant from its mother rather than directly from the surrounding water (Crews et al. 2000).

By using multigenerational protocol to trace EDCs effects, one can better understand the cause-effect relationship between EDCs exposure and variety degeneration. The adverse effect of EDCs can be directly transferred from parents to their progeny or by mutations on gene and molecular targets. The effect of EDCs on these maternal/paternal behaviors could also affect the offspring's tability to cope with stress in future. Several reports have confirmed that EDCs have effects not only on F1 generation, but also on Fn (n≥2) generation in *Ozyrias latipes* (Patyna et al. 1999), *Perch fluviatilis* (Brown et al. 2005) and *Danio rerio* (Nash et al. 2004). Additionally, Raimondo et al. (2009) demonstrated that continued exposure to E_2 not only affects population growth rate of *Cyprinodont variegates*, but also alters their population structure and variability. The results were increased population risk of extinction and potentially decreased recovery time. These results indicated that EDCs exposure could result directly in genetoxicity and, probably, that it indirectly affects the genetic pool of the generation. However, linking EDCs and a variety of deterioration to multigeneration relevant impact on the organisms, with few exceptions, is an open challenge. Multigeneration toxicity testing for wildlife is still urgent if we want to elucidate the precise cause of this. In addition, most of the understanding about variety deterioration remains at the phenotypic level.

EDCs may affect not only the exposed individual but also its progeny and subsequent generations. Recent evidence suggests that the mechanism of transmission may, in some cases, involve the germline and be non-genomic (Anway and Skinner 2006). That is, effects may be transmitted not due to mutation of the DNA sequence, but rather through epigenetic modifications that regulate gene expression, such as DNA methylation and histone acetylation. There may be transgenerational effects of EDCs due to overt mutation or to more subtle modifications of gene expression independent of mutation (i.e., epigenetic effects). Epigenetic effects of EDCs include context-dependent transmission (e.g., the causal factor persists across generations) (Crews and McLachlan 2006) or germline dependent mechanisms (i.e., the germline itself is affected) (Anway and Skinner 2008).

Many of the known endocrine disruptors exert a complex pattern of toxicity, and they may interfere with more than one signaling pathway and with other proteins including steroid hormone metabolizing enzymes, transport proteins, binding proteins or proteins involved in signaling pathways. In contrast to chemicals disrupting estrogen and androgen action, relatively little is known about the interference of environmental chemicals with non-sex steroid hormone responses including corticosteroid and thyroid-mediated responses. Thus, there is a great need to investigate the impact of endocrine disruptors on these two neglected and many other endocrine organs such as Stanius corpuscles and ultimobranchial body, for which there are so far no studies available.

Studies on endocrine disruption in fish have focused almost exclusively on the reproductive steroids, and in less extent to thyroidal hormones and corticosteroids. However, these hormones regulate only a portion of the vital physiological processes under endocrine control. In principle, a number of different hormones and their corresponding physiological processes could be disrupted by environmental contaminants, provoking adverse effects. Potential targets could be those processes regulated by growth hormones, somatostatin, vasotocin, isotocin, melanin-concentrating hormones, endorphin, somatolacting, prolactin, atrial natriuretic peptide, and parathyroid hormones among others. In a few cases, potential target genes have been identified in the zebrafish (Hoshijima and Hirose 2007). However, it will be difficult to include this array of potential endocrine targets into present testing schemes. Hence, model species with extensive genetic characterization could be used for a comprehensive approach using modern techniques for screening whole genomes or proteomes. This type of screening may indicate relevant hormone regulated processes and pathways which could be integrated in targeted testing approaches.

Exposure to Endocrine Disrupting Compounds and the Risks for Fish Populations

Laboratory Studies

Natural and synthetic hormones

Steroid hormones are produced in male and female organisms and are lipophilic, fat-soluble molecules, which are mainly excreted as water-soluble glucoronates or sulphate conjugates. Natural hormones come principally from domestic and livestock sewage. 17β-estradiol (E_2), the most potent natural estrogen, induces vitellogenin syntheses in males of *Rhamdia quelen* as well as increases hepatic-somatic index (HIS) and impairs the liver homeostasis (Moura-Costa et al. 2010). E_2 induces vitellogenin expression and diminishes the aggressiveness of male fish to defend nest sites in *Pimephales promelas* (Shappell et al. 2010). In *Kryptolebias marmoratus*, the gonadsomatic index, the frequency of mature oocytes and the number of ovulated eggs decreased in individuals exposed to E_2 (Park et al. 2010). Males of *Cichlasoma dimerus* that expresses vtg in liver showed lower sperm production, presented immature germ cells in the lobular lumen of testes and some morphology changes in the hepatocytes due to the accumulation of vitellogenin after exposition to E_2 (Moncaut et al. 2003).

The fecundity of *Oryzias javanicus* exposed to E_2 was significantly lower than that of the control. The appearance of secondary sexual characteristics seemed to be inhibited and vitellogenin synthesis in males was induced by E_2 exposure. Male fish exposed also presented testis-ova (Imai et al. 2005). *Oryzias latipes* exposed to E_2 during embryo period showed significant differences in reproductive success and viability of offspring when they got to the sexual maturation. Larvae of the same species, exposed to E_2, presented significantly biased sex ratios once they got to sexual maturity. Further, intersex gonads were also observed in the species after estradiol exposure (Koger et al. 2000).

Poecilia reticulata exposed as adults to 17β-estradiol showed indications of blocked spermatogonial mitosis in males. The post-parturition ovaries of females exposed to E_2 showed effects that suggests an inhibited yolk deposition. Embryos of the species exposed to the same dose via the mother showed alterations in the histology of liver, revealing effects on the liver structure, and suggesting that some effect of maternal exposure to 17β-estradiol can be passed to the progeny (Kinnberg et al. 2003).

Not all actions of xenosteroids may be confined to the reproductive system. 17β-estradiol is emerging as a potent natural regulator of fish thyroid function at several levels in the thyroid cascade, and therefore xenoestrogens and possibly other steroids have the potential to influence fish thyroidal status (Cyr and Eales 1996). The influence of estrogens at different levels of the thyroid cascade appears to be species dependent. The E_2 enhances thyroid function in several species of tropical teleosts (Singh 1969; Bandyopadhyay et al. 1991), but depresses thyroid function in *Poecilia reticulate* (Sage and Bromage 1970). E_2 also induces *Channa gachua* thyroid to produce and release T_4 (Bandyopadhyay et al. 1991). *Dicentrarchus labrax* exposed to E_2 showed plasmatic cortisol decrease, suggesting that E_2 has effects on adrenocortical cells steroidogenesis (Teles et al. 2006).

Mature males of *Carassius auratus* exposed to E_2 showed severely affected reproductive behavior and physiology. The result also suggests that the hormone E_2 can act as an endocrine disrupting chemical, not only in physiological, but also in behavioral parameters. Mating pairs of *Oryzias latipes* showed impairment on reproductive success, while males presented altered sexual and nonsexual behaviors. The fecundities of the E_2-treated fish were reduced and the sexual behaviors (following, dancing, floating, and crossing) of male fish were suppressed (Oshima et al. 2003). The effects of E_2 on the sexual activity of *Gambusia holbrooki* showed adult males with a lower level of sexual activity, measured as the number of approaches and copulation attempts made toward non-exposed females. In addition, E_2-exposed males were less capable of impregnating females than the control males (Doyle and Lim 2005).

Phytoestrogens are naturally occurring substances with estrogenic activity found in plants. The most studied phytormones are the flavonoids which can be divided into several classes such as flavones, flavonols, anthocyanins, flavanones, and isoflavonoids. The main known phytoestrogens are the isoflavones, daidzein, genistein, equol, coumestrol, and β-sitosterol.

Betta splendens adults exposed to a range of concentrations of genistein, equol and β–sitosterol presented changes in spontaneous swimming activity, nest size and changes in the probability of constructing a nest. There was significant decrease in the intensity of aggressive behavior toward the mirror following exposure (Clotfelter and Rodriguez 2006). *Carassius auratus* exposed to β-sitosterol showed decreased plasma testosterone and 11-ketotestosterone levels in males and testosterone and 17 β-estradiol levels in females. Plasma FSH levels were elevated in male fish treated with β-sitosterol, suggesting that reduced plasma steroid levels were not due to effects on pituitary function. These data suggest that β-sitosterol reduces the gonadal steroid biosynthetic capacity through effects on the cholesterol availability or the activity of the side chain cleavage enzyme $P450_{SCC}$ (MacLatchy and Van Der Kraak 1995). *Carassius auratus* adult males exposed to β-sitosterol showed decrease of cholesterol in the mitochondria, suggesting that the phytoestrogen is impeding cholesterol transfer across the mitochondrial membrane (Leusch and MacLatchy 2003).

Danio rerio embryos at 24 hours post-fertilization exposed to genistein showed decreased heart rates, retarded hatching times, decreased body length, and increased mortality in a dose dependent manner. Embryos also presented effects related to endocrine dysfunction, such as pericardial edema, yolk sac edema, and spinal kyphosis (Kim et al. 2009). *Danio rerio* exposed continuously across three generations to phytosterol showed blood vitellogenin induction in all generations of exposed fish. The sterol also changed the sex ratio of the exposed fish: in F_1 generation, the predominant sex was male, while in F_2, female. This multigeneration test evidenced that phytosterols disrupt the reproduction system of zebrafish by changing the sex ratios and by inducing the vitellogenin production in the exposed fish (Nakari and Erkomaa 2003).

Waterborne exposures of *Oryzias latipes* to equol induced gonadal intersex (i.e., testis-ova) in males, while the ovaries of treated female fish showed delayed oocyte maturation, atretic oocytes, an enlarged ovarian lumen, proliferation of somatic stromal tissue, and primordial germ cells. Alterations of externally visible secondary sex characteristics also occurred: male (as identified by the gonadal phenotype) showed feminized secondary sex characteristics (Kiparissis et al. 2003).

Natural hormones like estradiol or progesterone are not suitable for oral applications—or only at higher dosage—because they are quickly metabolized (i.e., deactivated) and excreted. Therefore, mainly synthetic steroids are used for oral application. Ethylination or alkylation of the natural compound prevents metabolization and guarantees the desired effect. The group of synthetic hormones mainly consists of oral contraceptives (ovulation-inhibiting hormones) as well as steroids used for substitution therapy during menopause.

Males of *Pimephales promelas* exposed to 17α-ethynilestradiol (EE_2), the most common estrogen in contraceptive pills, have impaired ability to defend a spawning territory. Exposure to EE_2 impaired male's ability to compete and acquire territories compared to unexposed individuals. Higher levels of overall aggression (aggression rates of both fish) occurred in exposed fish relative to controls (Pawlowski et al. 2005). Breeding *Danio rerio* exposed to environmentally relevant concentrations of EE_2 showed that the reproductive hierarchies in breeding colonies were disrupted by exposure to EE_2. The effect was a reduction in the skew in male paternity and increased skew in female maternity. EE_2-exposed *Danio rerio* males also presented a decrease in 11-ketotestosterone plasma concentration (Coe et al. 2008).

Adult males of *Clarias gariepinus* exposed to EE_2 showed disappearance of spermatids/sperm from several testicular lumen/lobules. Immunocytochemical localization of GnRH and luteinizing hormone (LH) in preoptic area-hypothalamus and pituitary, respectively, revealed decreased immunoreactivity following EE_2 treatment in males. When females of the same species were treated with methyltestosterone, a precocious ovarian development was observed (Swapna and Senthilkumaran 2009).

A study with *Danio rerio* embryos exposed to EE_2 until three months post fertilization evaluated growth, development, gonad development and body vitellogenin (VTG) content. The results demonstrated a significant reduction in total body length and body weight, and an increase in morphological abnormalities for fish exposed to EE_2 as a function of exposure time. When the embryos got to maturity, a reduced number of spawning females and a reduced egg production were found for the female fish exposed to EE_2 (Van den Belt et al. 2003).

Eleven-month-old juvenile *Mugil cephalus* containing immature gonad phenotype were fed with EE_2 and showed abnormal values of serum vitellogenin and a premature oocyte development, suggesting that EE_2 accelerates ovarian differentiation (Aoki et al. 2011). When adult *Gobiocypris rarus* were exposed to EE_2, the growth and GST of fish were significantly reduced. Hepatosomatic indices (HSI) of male fish were significantly higher and plasma vitellogenin (VTG) induction could be observed in males after exposure to different concentrations of EE_2. Plasma VTG concentrations in females were also significantly higher than in controls. Also, feminization of male fish could be noticed and parts of males manifested the testis-ova phenomenon. Ovaries of female rare minnow exposed to EE_2 were degenerated (Zha et al. 2007).

Exposure of adult fathead minnows (*Pimephales promelas*) to the androgen 17α-methyltestosterone (MT) produces both androgenic and estrogenic effects, manifested as nuptial tubercle formation in females and vitellogenin production in males. In addition, aromatase activity was significantly decreased in ovarian microsomes and brain homogenates from exposed fish (Hornung et al. 2004). The influence of MT on growth responses, biological parameters and expression of genes involved in the GH–IGF pathway of the hypothalamic-pituitary-liver-gonadal axis were investigated in females and males of *Oreochromis niloticus* to evaluate the relationship between sex and MT-induced changes in these parameters. Female fish had a lower growth rate than male and MT increased growth performance and duodenal villi in females. Greater blood triglyceride levels indicated the vitellogenin process in female fish. MT exerted androgenic and, to some extent, estrogenic effects on several physiological parameters of this species (Phumyu et al. 2012).

17β-trenbolone (TB) is the active metabolite of an anabolic androgenic steroid used as a growth promoter in cattle and is a contaminant of concern in aquatic systems. In female gonads of *Pimephales promelas* exposed to TB, hydroxysteroid dehydrogenase, zona pellucida glycoprotein, and protein inhibitor of activated STAT were all down-regulated. In liver, a significant decrease in vitellogenin and brain aromatase mRNA expression was verified following TB exposure (Dorts et al. 2009).

Danio rerio exposed to environmentally relevant concentrations of TB from 0 to 60 days post-hatch (dph) showed a skewed sex ratio towards males. After the depuration period, no sign of reversibility was observed. Environmentally relevant concentrations of 17β-trenbolone cause a strong and irreversible masculinization of zebrafish, raising concern about the effects of androgenic discharges in the aquatic environment (Morthorst et al. 2010).

Natural estrogens, phytoestrogens and synthetic hormones can impair many endocrine functions in tropical fish. Reproductive parameters such as gonad morphology, stages of oocyte development and sperm viability are the most studied effects. However effects on steroidogenesis as well as on thyroid and adrenocortical cells have also been reported.

Pharmaceuticals

A pharmaceutical drug, also referred to as medicine, medication or medicament, can be loosely defined as any chemical substance intended for use in the medical diagnosis, cure, treatment, or prevention of disease. Some of these drugs have been reported for endocrine-disrupting effects on wildlife.

Fibrates are pharmaceuticals commonly used to control hypercholesterolemia in humans and frequently detected in the freshwater environment. Since cholesterol is the precursor of all steroid hormones, it is suspected that low cholesterol levels will impact steroidogenesis. *Danio rerio* exposed to bezafibrate (BZF) showed a time dependent decrease in the plasma cholesterol concentration. Plasma 11-KT also decreased significantly. Gonadal histology revealed the presence of germ cell syncytia in the tubular lumen of fish exposed to BZF and also an increased number of cysts containing spermatocytes, which indicate testicular degeneration (Velasco-Santamaría et al. 2011). These results show that bezafibrate exerts a hypocholesterolemic effect in adult male zebrafish and its potential as an endocrine disruptor due to its effect on the gonadal steroidogenesis and spermatogenesis. Another lipid regulator, gemfibrozil (GEM), reduces StaR and plasma testosterone in *Carassius auratus* by over 50% (Mimeault et al. 2005).

When mature *Danio rerio* were exposed to trilostane, an inhibitor of 3 β-hydroxysteroid dehydrogenase (3βHSD) used in the treatment of Cushing's syndrome, the drug had substantial impact on the transcriptional dynamics of zebrafish, as reflected by a number of differentially expressed genes, including transcription factors (TFs), and altered networks and signaling pathways. Expression of genes coding for 3βHSD and many of its transcriptional regulators remained unchanged, suggesting that transcriptional up-regulation is not a primary compensatory mechanism for this enzyme inhibition (Wang et al. 2011).

Tamoxifen is widely used in the treatment of breast cancer and can enter the aquatic environment via municipal wastewater. To evaluate potential effects on embryonic development of *Oryzias latipes*, fertilized eggs were exposed to tamoxifen. Adverse effects on hatchability and time of hatching were observed. Adults were also exposed and tamoxifen significantly increased plasma vitellogenin levels in males in a dose-dependent manner. Fecundity and fertility were detrimentally affected. Additionally, F_1 eggs were removed from tamoxifen-contaminated water to evaluate transgenerational effects. Hatchability was affected, but no morphological deformities were observed. A significant dose-dependent increase in the proportion of genotypic males occurred at all concentrations greater than 5 ug/l (Sun et al. 2007).

For *O. latipes*, ibuprofen affected several endpoints related to its reproduction, including induction of vitellogenin in male fish, fewer broods per pair, and more eggs per brood. Parental exposure to as low as 0.0001 mg/L of ibuprofen delayed hatching of eggs even when they were transferred to and cultured in clean water (Han et al. 2010). Delayed hatching is environmentally relevant because this may increase the risk of being predated.

Oryzias latipes exposed to diclofenac, one of the main concerns among pharmaceuticals and frequently found in sewage treatment plant (STP) effluents, presented alterations in the expression levels of cytochrome P450 1A, p53 and vitellogenin. These three biomarkers showed elevated expression levels, suggesting that diclofenac has the potential to cause cellular toxicity, p53-related genotoxicity and estrogenic effects (Hong et al. 2007). *Danio rerio* embryos were exposed to carbamazepine and diclofenac to evaluate the effects on embryo mortality, gastrulation, somite formation, tail movement and detachment, pigmentation, heartbeat, malformation of head, scoliosis, deformity of yolk, and hatching success. Specific effects on growth retardation were found for carbamazepine, on hatching, yolk sac and tail deformation. Diclofenac exerts effects on scoliosis and growth retardation (Brandhof and Montforts 2010).

Pimephales promelas exposed to the antiandrogen flutamide showed reduced fecundity and reduced embryo hatch. Qualitative histological assessment of ovaries from females indicated a decrease in mature oocytes and an increase in atretic follicles. Testes of males exposed to flutamide exhibited spermatocyte degeneration and necrosis. Increases in plasma testosterone and vitellogenin concentrations were observed in the females. Males exhibited elevated concentrations of β-estradiol and vitellogenin (Jensen et al. 2004).

Nowadays, there is increasing use of many medicines, from drugs for cancer to drugs for regulating physiological amended process. Among these drugs, many are able to impact the endocrine system of tropical fish. Most studies are focused on reproductive and developmental processes. Thus, there is a need to study the effects of these drugs on thyroid and adrenocortical physiology.

Phtalates

Phthalates are man-made chemicals used to improve the flexibility of plastics. They have been used in various products, such as toys, medical tubing, plastic bottles, packing cases, and cosmetics.

Reproductive and developmental effects of phthalates were evaluated in *Oryzias latipes*. Male fish showed a two-fold induction of testosterone and female fish expressed greater testosterone hydroxylase activity (Patyna et al. 2006). Females of *O. latipes* exposed to bis(2-ethylhexyl) phthalate (DEHP) showed decreased plasma vitellogenin, decreased gonad-somatic index (GSI) and oocytes in initial stages of maturation. Unlike female fish, no changes or adverse effects were observed in male fish (Kim et al. 2002). Effects of DEHP on the embryos of *O. latipes* include delayed hatching time, increased mortality and reduced body weight. Distortion of sex ratio was also observed after DEHP exposition (Chikae et al. 2004). *O. latipes*, exposed to DEHP from the time of hatching to three months of age showed distinct reproductive effects, such as decreased blood vitellogenin levels, decreased GSI and immature oocytes into ovaries (Kim et al. 2002).

Pimephales promelas, exposed to environmentally relevant concentrations of the plasticizer (DEHP) showed a reduction in plasma testosterone concentrations in males. There was also a significant reduction in 17β-estradiol among exposed animals. Contrary to what has been previously published for these two chemicals in mammals, the lower plasma testosterone concentrations in males exposed to DEHP was not a result of the inhibition of genes involved in steroidogenesis, nor due to an increase in the expression of genes associated with peroxisome proliferation. Rather, an increase in relative transcript abundance for CYP3A4 in the liver and androgen and estrogen-specific SULTs in the testes provides evidence that the decrease in plasma testosterone and E_2 may be linked to increased steroid catabolism. Feedback from the pituitary is not repressed, as the relative expression of follicle stimulating hormone β-subunit mRNA levels in the brain was significantly higher in DEHP exposed animals. In addition, luteinizing hormone β-subunit mRNA levels increased (Crago and Klaperb 2012).

DEHP also impairs the reproductive health of males of *Danio rerio*. Males treated with DEHP showed a significant increase in the hepatic-somatic index and levels of hepatic vitellogenin transcript. Females showed a reduction in fertilization success of spawned oocytes. Exposure to DEHP also causes alterations in the proportion of germ cells at specific stages of spermatogenesis in the testis, including a reduction in the proportion of spermatozoa and an increase in the proportion of spermatocytes, suggesting that DEHP may inhibit the progression of meiosis. These data demonstrated that exposure to high concentrations of DEHP disrupts spermatogenesis in adult zebrafish with a consequent decrease in their ability to fertilize oocytes spawned by untreated females (Uren-Webster et al. 2010).

DEHP was found to inhibit the growth of *Poecilia reticulate*in terms of body weight and body length. Fish DEHP exposure induced significantly reduced body length and weight as a consequence of lipid metabolism alterations (Zanotelli et al. 2009).

Phthalates showed, in almost all cases, an anti-estrogenic effect in fish. This chemical has no intrinsic anti-thyroid activity in humans, but a by-product generated by gram-negative bacteria is an inhibitor of thyroperoxidase, an enzyme strongly involved in TH synthesis. A limited number of studies suggest that exposure to some phthalates may be associated with altered thyroid function, but fish data remain insufficient.

PCBs

Polychlorinated biphenyls (PCBs) are synthesized via chlorination of biphenyl and two hundred and nine congeners are possible. PCBs were used as dielectric fluids in transformers and large capacitors, as pesticide extenders, plasticizers in sealants, heat exchange fluids, hydraulic lubricants, cutting oils, flame retardants, and in plastics, paints, adhesives, and carbonless copy paper. Use of these compounds has been proscribed since the 1980s, but PCBs are still frequently detected in fish tissues, such as blood, muscle and adipose tissue.

Danio rerio exposed to a mixture of PCBs demonstrated a bioaccumulation of PCBs in males and females as well as a maternal transfer to the eggs. Several reproductive traits were altered after PCB exposure, including a reduction in the number of fertilized eggs per spawn as well as an increase in the number of poorly fertilized spawns. Ovary histology revealed a decrease in maturing follicles and an increase in atretic follicles in the ovaries of females exposed to PCBs (Daouk et al. 2011). *D. rerio* exposed to a relevant environmental concentration of PCB mixture showed a decrease in late vitellogenic follicle stages. In addition, proliferation of granulosa cells was decreased in exposed fish. This was accompanied by increased apoptosis of granulosa cells. Vitellogenin was not detected in both male and female fish, suggesting an anti-estrogenic effect of PCB mixture (Kraugerud et al. 2012).

Exposure of *Micropogonias undulates* to Aroclor 1254, a commercial mixture of PCBs, has been shown to impair reproductive neuroendocrine function in this species. In addition, hypothalamic tryptophan hydroxylase (TPH), the rate-limiting enzyme in serotonin synthesis, has been shown to be a target of PCB neuroendocrine toxicity. Aroclor 1254 significantly inhibited hypothalamic TPH activity and gonadal growth (Khan and Thomas 2006). *M. undulatus* exposed to Aroclor 1254 or one of three individual congeners (planar PCB 77, ortho-substituted PCB 47 and PCB 153) showed impairment of thyroid function. Aroclor 1254 decreased plasma T_3 levels. Exposure to PCB 153 lowered both T_4 and T_3, while PCB 47 had no effect on thyroid hormone levels. PCB 77 increased T_4 levels (LeRoy et al. 2006). *Sebastes schlegeli* exposed to PCB 153 showed alteration in 17α-hydroxyprogesterone in females and males. Plasma concentrations of E_2 and vitellogenin also increased in females and males exposed to PCB 153 (Jung et al. 2005). PCB153, a non-coplanar PCB lacking dioxine-like activity, exerts effects in the vitellogenin induction in young individuals of *Spaurus aurata*, while PCB-126, a coplanar PCB prototypical AhR agonist, showed an anti-estrogenic activity in the same species (Calò et al. 2010).

Aroclor 1254 also affects the peripheral thyroid hormone metabolism and thyroid hormone plasma levels in *Oreochromis niloticus*. Decreased plasma T_4 and reverseT_3 levels were observed in tilapia exposed to Aroclor 1254. Hepatic type I deiodinase (D1) activity was depressed and hepatic type III (D3) activity was increased by Aroclor 1254 exposure (Coimbra et al. 2005).

The hypothalamus–pituitary–interrenal (HPI) axis of *Oreochromis mossambicus* is compromised by short-term exposure to PCB 126, showing decreased plasma cortisol levels. This suggests an impaired ability to acutely activate interrenal steroidogenesis in PCB treated tilapia. Adrenocorticotropic hormone (ACTH) and *in vitro* cAMP-stimulated cortisol release from head kidneys were lower in tissues from tilapia exposed to PCB 126 than in tissues from control animals indicating direct toxic effects on the interrenal cells (Quabius et al. 1997).

All these data demonstrate that exposure to PCBs at environmentally realistic concentrations can have profound effects on the thyroid, steroideogenic and reproductive status of the species. These data and the persistence of this chemical class in the environment increase the probability of other endocrine disrupting targets.

Dioxins

Dioxins are not intentionally manufactured, but are typically formed and released through industrial activities such as chlorine bleaching at pulp and paper mills, chlorination at waste and drinking water treatment plants, and from municipal solid waste and industrial incinerator emissions. They are chemically stable, lipophilic, hydrophobic, and resistant to biological degradation. They tend to associate to organic matter and readily bioaccumulate in aquatic food webs. Management plans, legislation and technological changes have reduced their release, substantially reducing environmental levels over the past 25 years. However, considerable amounts still circulate in contaminated aquatic environments.

Dioxins are a reproductive toxicant and endocrine disruptor, yet the mechanisms by which they cause these reproductive alterations are not fully understood. Adult *Danio rerio* exposed to 2,3,7,8- tetrabromodibenzo-p-dioxin (TBDD) or to a mixture of brominated dioxins showed alterations in spawning success, gonad morphology, decreased hepatic vitellogenin gene expression, and offspring early life-stage development. All brominated dioxins spiked to the feed were detected in female fish and transferred to eggs. Exposure to the dioxin mixture and TBDD clearly induced AhR-regulated genes and EROD activity. Exposure to TBDD reduced spawning success, altered ovarian morphology and reduced hepatic vitellogenin gene expression (Haldén et al. 2011). *D. rerio* exposed to 2,3,7,8-tetrachlorodibenzo-p-dioxin (TCDD) showed a dose-related reduction of egg numbers and lethal anomalies of their offspring (edema and malformations of the notochord). Histology of the ovaries revealed severely impaired development of previtellogenic to vitellogenic oocytes (Wannemacher et al. 1992). *Sparus aurata* exposed to TCDD showed significant increases in the EROD activity, as well as in AhR gene expression in liver (Abalos et al. 2008).

Oryzias latipes adult males exposed to TCDD showed differential gene expression in liver, brain and testes. Of the 42 altered genes, cytochrome P450 1A (CYP1A) mRNA was the only transcript significantly higher in TCDD-exposed brain, whereas 12 transcripts (including CYP1A and AhR) were significantly higher in TCDD-exposed liver and testes. In addition, TCDD-treated males showed adverse histopathological changes in the brain, and glycogen depletion was observed in the liver. Significant histological changes also occurred in the testis, and included disorganization of spermatogenesis at the testis periphery, disruption of the interstitium and Leydig cell swelling Volz et al. 2005).

While TCDD induces several overt toxicities in adult fish, it also exerts detrimental effects on reproductive success. Reproductive toxicity of TCDD varies among different fish species, and certain species are more vulnerable than others. Acute TCDD exposure impairs female reproduction via reduced steroid hormone secretion, depressed capacity for ovarian steroid biosynthesis, altered ovarian growth and development, fecundity with age, age to maturation, secondary sexual characteristics and egg and larval size, reduced egg production and early life stage toxicity of offspring (Giesy et al. 2002; Palstra et al. 2006; Wu et al. 2001). Less is known regarding TCDD-induced reproductive toxicity in male fish. However, it has been demonstrated in medaka fish that TCDD causes lesions in the testis (Volz et al. 2005).

P. flesus adults exposed to TCDD showed depressed plasma total T_4 levels, but no changes in free T_4 or plasma total T_3 levels. TCDD could interfere with the development because flounder metamorphosis is regulated by THs and chemically induced hypothyroidism inhibits metamorphosis beyond early stages (Schreiber and Specker 1998).

Adult fish are less susceptible to TCDD-induced toxicity compared to earlier life stages, requiring considerably higher body burdens to elicit adverse effects. TCDD exerts toxic effects in embryos of *Pimephales promelas*, *Ictalurus punctatus*, *Coregonus artedii*, *Oryzias latipes*, *Catastomus commersoni* and *Danio rerio*. Signs of TCDD toxicity, including edema, hemorrhaging, and craniofacial malformations were observed in these species (Eolen et al. 1998). Using *Danio rerio* and *Oryzias latipes* as fish models, 13 dioxin isomers, including TCDD, induced the expression of AhRs and CYP1A1 in the early life stages of embryos, indicating that dioxin isomers modulate AhR activity during the early stages of embryos (Hano et al. 2010).

Effects of dioxin exposure on reproductive success of fish are a growing concern. While this has not been extensively studied, some results suggest that sublethal exposure to TCDD impairs gonad development, ovulation and survival of offspring. Overall, it was shown that exposure to sublethal concentrations of brominated dioxins may impair reproductive physiology in fish and induce AhR-regulated genes.

Heavy metals

Heavy metals are trace elements and occur naturally in the earth's crust, but are introduced and concentrated into the environment through mining and manufacturing processes.

Arsenic is used in metallurgic applications and in wood preservatives, herbicides, pharmaceuticals, and glass. Smelting and refining industries also release it into the environment.

A recent study by Shaw et al. (2007) suggested a role for arsenic in the disruption of acclimation of fish to seawater, a GR-dependent process. Arsenic reduces the expression of certain stress-related genes, without affecting plasma cortisol levels or hepatic cortisol receptors mRNA expression. Disruption of GR function by arsenic seems to be indirect, involving altered posttranslational modifications or impaired interaction with receptor-associated factors (Odermatt and Gumy 2008). *Pangasianodon hypophthalmus* males exposed to a mixture of Pb, Mo, Rb and As revealed all developmental stages of germ cells with necrotic spermatogonia in testis. Vacuolization and hypertrophy of Sertoli cells were also observed after mixture exposition (Yamaguchi et al. 2007). The testicular architecture of *Colisa fasciatus* showed degenerative changes in lobules after arsenic oxide exposure. In addition to the degenerative changes in the lobules, the interstitial Leydig cells (steroid-secreting cells) presented significant reduction in diameter. Varying degrees of necrosis and pyknosis suggested reduced secretory levels. Females of *C. fasciatus* showed decreased development of oocytes, reduced number and diameter of nucleolus and increased number of atretic follicles (Shukla and Pandey 1984).

Cadmium enters the aquatic environment mainly from atmospheric fallout and in effluents from smelting and refining industries. Exposure of fish to cadmium chloride ($CdCl_2$) reduced thyroid epithelial cell height and lowered plasma THs concentrations in *C. batrachus* (Gupta et al. 1997). In *C. punctatus*, $CdCl_2$ decreased thyroidal T_4 content but not iodine peroxidase activity (Bhattacharya et al. 1989). Adult males and females of *Oryzias latipes* exposed to Cd showed alterations in reproductive endpoints, including plasma vitellogenin, hepatic estrogen receptor, plasma steroids, gonadal-somatic indice (GSI), and gonadal steroid release. Gonadal steroid release was significantly decreased in males and females and female plasma estradiol levels were significantly altered in Cd-exposed individuals (Tilton et al. 2003). Exposure of adult females of *Carassius auratus* to Cd inhibited growth and altered sexual behavior. The gonad-somatic index (GSI) decreased, plasma LH levels increased and ovulation did not occur in females exposed to Cd. Cd also caused an ovarian recrudescence (rebuilding of ovaries) and a decrease in the number of ovulating females (Szczerbik et al. 2006).

Oryzias latipes exposed as embryos to Cd showed impairment in the reproductive fitness, steroids concentration and vitellogenin expression when they got sexually mature. The number of total eggs ovulated, percentage of fertilized eggs and size of the eggs were smaller than in animals not exposed to Cd. In addition, the exposure elevated estradiol concentration and hepatic vitellogenin in males (Foran et al. 2002). Sexually immature *Clarias gairepinus* exposed to cadmium showed a very high 20αHSD enzyme activity in sperm. This enzyme converts 17-hydroxy progesterone (17P) substrate to 17,20α-dihydroxy progesterone (17, 20αP) product and the rate of enzyme activity is related to substrate (17P) concentrations. The results showed that 20αHSD enzyme activities in fish sperm may be used as indicators of water contamination with heavy metals (Ebrahimi 2007).

Clarias batrachus exposure to cadmium chloride causes adverse effects on the pituitary ACTH cells, gonadotropin- and thyrotropin-secreting cells, thyroid gland and gonads. $CdCl_2$ caused a significant increase in the ACTH cell nuclear indices, whereas the thyrotropin- and gonadotropin secreting cells showed inactivation and accumulation of secretary products. The epithelial height of the thyroid follicles was also significantly reduced as compared to that of the untreated control fish. In female and male fish the gonad-somatic index had a significant reduction. Ovarian maturation seems to have become arrested at perinucleolar stage and spermatogenesis at spermatocyte stage (Jadhao et al. 1994).

Lead enters the environment from its mining, but mainly from the refining and smelting of Pb and other metals. In aquatic ecosystems, Pb is generally tightly bound to particulates and sediments. Pb caused thyroid epithelial cell hypertrophy, reduced thyroid colloid content, decreased plasma THs levels and deiodinase activity and inhibited thyroid I⁻ uptake in *C. batrachus* (Katti and Sathyanesan 1987). $PbNO_3$ reduced plasma T_3 levels and liver deionidase activity in *Heteropneustes fossilis* (Chaurasia and Kar 1999). $Pb(NO_3)_2$ elicited biphasic effects on estradiol, testosterone and cortisol: stimulatory at lower concentrations and inhibitory at higher concentrations in *H. fossilis*. In contrast, progesterone, 17-hydroxyprogesterone, 17,20β-dihydroxyprogesterone, corticosterone, 21-deoxycortisol and deoxycorticosterone were inhibited in a dose-dependent manner in the same species (Chaube et al. 2010).

Mercury cycles in the environment as a result of both natural and anthropogenic activities. Its release into the environment has increased with industrialization mainly due to emissions from combustion of waste and fossil fuels. Once in the aquatic environment, Hg can be methylated and then accumulates in predators from aquatic food webs. In *C. punctatus* exposed to $HgCl_2$, plasma T_4 levels and thyroidal iodoperoxidase activity decreased (Bhattacharya et al. 1989). Exposure of *C. punctatus* (Ram and Sathyanesean 1984) or *L. parsia* (Pandey et al. 1993) to $HgCl_2$ caused thyroid follicular epithelial cells to become columnar and the colloid to exhibit varying degrees of vacuolization. Hg-based compounds caused thyroid epithelial cell hypertrophy, reduction in colloid content, and inhibition of thyroid iodide uptake in *C. batrachus* (Kirubagaran and Joy 1989). *Stizostedion vitreum* exposed to methylmercury showed significantly impaired both growth and gonadal development in males, which was apparent as reduced fish length, weight, and gonadosomatic index. Testicular atrophy was observed in fish; mercury also suppressed plasma cortisol (Friedmann et al. 1996).

The study of the effects of $HgCl_2$, emisan (a methoxy-ethyl mercury fungicide), and methyl mercuric chloride (CH_3HgCl) on the adrenocortical-pituitary activity of the *Clarias batrachus* showed that adrenocortical cells were highly stimulated by all Hg forms and became hyperplastic. Hg also induces necrotic changes in some areas of the head kidney. These areas had an extensive infiltration of lymphocytes, with localized sites of necrosis and hyperplasia. The ACTH cells in the pituitary were hypertrophied and degranulated in the Hg-treated groups, suggesting increased secretion of ACTH. The plasma cortisol level decreased significantly in Hg-exposed groups. These results suggest that Hg impairs the adrenocortical-pituitary activity of the catfish (Kirubagaran and Joy 1991).

Heavy metals concentration in aquatic ecosystems increased in the last decades due to human activity. There are sufficient data showing that heavy metals impair endocrine functions such as thyroid, gonadal and adrenocortical functions in tropical fish, suggesting that populations can be impacted by these chemicals.

Pesticides

Pesticides are used to control a wide variety of insect and plant pests. Pesticides are usually applied as a formulation containing the active ingredient, along with other materials, such as solvents, wetting agents or carriers. Some active ingredients have the potential to impact the endocrine system, as do some of the surfactants used in the formulations.

Crain et al. (1997) showed that atrazine has the ability to stimulate production of the enzyme aromatase which converts androgens in estrogens, and presumably could interfere with sexual differentiation and development. *P. promelas* exposed to atrazine showed a decrease in total egg production and reduced numbers of spawning events. Gonad abnormalities, such as delayed maturation, were observed in both males and females of atrazine-exposed fish (Tillitt et al. 2010). *Gobiocypris rarus* exposed to atrazine showed impairment of adrenocortical cells. Histological analyses showed lesions such as extensive expansion in the lumen, degenerative and necrotic changes of the tubular epithelia, shrinkage of the glomerulus as well as increase of the Bowman's space in kidney. The expressions of Na^+,K^+-ATPase, cortisol receptors, hsp70 and hsp90 in the kidney were significantly decreased in exposed animals (Yang et al. 2010a). Sexually mature males of *Carassius auratus* exposed to atrazine showed effects on the concentrations of gonad and plasma sex steroids, plasma vitellogenin (VTG) and gonad histopathology. Atrazine induced suppression of both testosterone and 11-ketotestosterone, and increased 17β-estradiol levels. Atrazine exposure also induced structural disruption in the testis and elevated the levels of atresia in females (Spanò et al. 2004). *Carassius auratus* exposed to atrazine or diuron revealed alterations on some behavioral endpoints related to swimming and social activities. Both herbicides decreased grouping behavior and atrazine also increased surfacing activity (Saglio and Trijasse 1998).

P. promelas exposed to environmentally relevant concentrations of the herbicide linuron, showed a reduction in plasma testosterone (T) and 17β-estradiol concentrations in males. Rather, an increase in relative transcript abundance for CYP3A4 in the liver and androgen and estrogen-specific sulfotransferases in the testes provides evidence that the decrease in plasma testosterone and E_2 may be linked to increased steroid catabolism (Crago and Klaperb 2012).

DDT [1,1,1-trichloro-2,2-di(4-chlorophenyl)ethane], a derivative of diphenylethane, was produced for the first time almost 60 years ago. Since then, it has been intensively used as an insecticide especially against insects communicating diseases like malaria or sleeping sickness. Technical DDT is a mixture of mainly *p,p'*-DDT and *o,p'*-DDT. All these compounds are highly lipophilic, nearly insoluble in water and very persistent, with half-life periods for microbial degradation of 3 to 20 years. The dehydrochlorination of DDT—the main step during biotic as well as abiotic transformation—leads to the main DDT metabolite, DDE. *Sarotherodon mossambicus* exposed to DDT showed greater thyroid epithelial cell height and nuclear diameter and these changes were reversed upon transfer of fish to clean water (Shukla and Pandey 1986). *Liza parsia* exposed to DTT showed a decrease in thyroid epithelial cell height, degeneration of epithelial cells, and depletion of colloid (Pandey et al. 1995).

The effects of o.p'-DTT in *Oreochromis niloticus* include vitellogenin induction by direct interaction between the pollutant and the estrogen receptor and decrease in plasma estradiol and testosterone concentrations (Leaños-Castañeda et al. 2007). *Danio rerio* exposed to DDT showed increased alteration in sperm release and activity as well as the life span of their trails (Njiwa et al. 2004). *Heteropneustes fossilis* exposed to DDTs showed decrease in GSI and plasma levels of E_2 (Singh and Singh 2007).

Oryzias latipes exposed to *o,p'*-DDT showed that the pollutant affects not only the adults, but also embryos via maternal transfer of the compound. In adults, an intersex condition of the gonad was observed in males exposed at early life stages, while the females showed more advanced development of oocytes, indicating that this estrogen agonist can alter gonadal development when exposure occurs continuously over the period of gonadal differentiation. Vitellogenin was also induced in males by exposure to DDT during early life stages. The maternal transfer of the pollutant showed delayed hatching time of the offspring (Metcalfe et al. 2000).

The expression of androgen, estrogen and cortisol receptors and hsp70 and hsp90 were significantly suppressed following p,p-DDE exposure in *Grobiocypris rarus* (Yang et al. 2010b). o,p'-DDE and heptachlor alters vitellogenin production and growth hormone (GH)/insulin-like growth factor-I (IGF-I) concentration in males of *Oreochromis mossambicus*. o,p'-DDE and heptachlor treatment increased plasma Vg and hepatic expression of three vitellogenin genes (Vgs A, B, and C) and estrogen receptor α (ERα), while reducing plasma levels of IGF-I and suppressing the expression of IGF-I, the GH receptor (GHR2) and the putative somatolactin receptor (GHR1) (Davi et al. 2009).

Poecilia reticulata males exposed to DDE showed impaired reproductive behavior, alterations in sexual secondary characteristics and decreased sperm count. These results suggest that DDE has a demasculinizing action on this species (Kristensen et al. 2006). *p,p'*-DDE has endocrine disrupting effects on gonadal development and gene expression of *Oryzias latipes* juvenile males. Increased hepatosomatic index (HSI) and decreased gonad-somatic index as well as intersex gonads were found in the *p,p'*-DDE-treated groups. Vitellogenins, choriogenins and estrogen αreceptor in the liver of the fish were significantly up-regulated by *p,p'*-DDE exposure (Zhaobin and Jianying 2008).

Endosulfan is a cyclodiene compound that antagonizes GABA-gated chloride channels and is used as a wood preservative and as an insecticide/acaricide on food crops. *Oreochromis mossambicus* exposed to endosulfan showed hypertrophy and hyperplasia of thyroid follicles as well as general thyroidal degeneration (Bhattacharya 1995). Exposure of *Clarias batrachus* to endosulfan increased plasma T_4 levels, but decreased those of T_3. These effects were correlated with increased thyroid iodoperoxidase activity, decreased T_4 clearance, and decreased extrathyroidal T_4 to T_3 conversion (Sinha et al. 1991).

Oryzias latipes exposed to endosulfan post-fertilization or post-hatch presented endocrine disruption in embryos and adults. Eggs exposed to endosulfan took longer to hatch, and the resulting fry was smaller at one week of age and had decreased mobility at two weeks of age. Upon reaching sexual maturity, these individuals also produced fewer eggs and these eggs took a significantly longer time to hatch (Gormley and Teather 2003). *Oreochromis niloticus* exposed through diet to endosulfan showed decreased plasma T_4 and rT_3 levels. Hepatic type I deiodinase (D1) activity was depressed and hepatic type III (D3) activity was increased by endosulfan exposure (Coimbra et al. 2005).

Lindane is used as an insecticide and fumigant on a wide range of soil-dwelling and plant-eating insects, as well as in personal care products for the control of lice and mites. Exposure of *H. fossilis* to a technical grade mixed isomers of lindane increased the plasma T_4 level, but decreased the plasma T_3 level (Yadav and Singh 1986). Toxicity experiments of lindane in embryos of *O. latipes* revealed development of testis-ova in males and induction of vitellogenesis. In the thyroid, the epithelial cells showed hypertrophy and diminished colloid content; moreover, the number of thyrotropic hormone-producing cells in the pituitary was increased (Brown et al. 2004).

Vinclozolin, a known pesticide with anti-androgenic properties, alters the embryo development in *Pimephales promelas*. Embryos showed an increased17β-estradiol serum concentration in males and a reduction in gonadal condition of female fish (Makynen et al. 2000). Mature *Carassius auratus* exposed to vinclozolin showed alternations in gonad-somatic (GSI) and hepatic-somatic indices (HSI), E_2, 11-ketotestosterone and sperm quality. Sperm volume, motility and velocity were reduced in fish exposed to vinclozolin. This was associated with the decrease in 11-KT level, suggesting direct vinclozolin effects on testicular androgenesis (Hatef et al. 2012). *Oryzias latipes* exposed to vinclozolin presented incidence of intersex gonads (i.e., testi-ova) and impaired spermatogenesis in males. Also, the vinclozolin treatments induced moderate ovarian atresia (Kiparissis et al. 2003).

Pesticides such as DDT have been well reported as endocrine disruptors, but other very commonly used pesticides such as endrin and lindane, amongst others, have also demonstrated this property. Pesticides impair mainly reproductive and thyroidal endpoints. Studies with behavioral and adrenocortical endpoints are still necessary.

Perchlorates

Perchlorates occur both naturally and through manufacturing, for use by the aerospace, weapons, and pharmaceutical industries. These compounds are extremely water-soluble, and have been detected in rain, snow, foodstuff, ground waters, and fertilizers. Perchlorates are competitive inhibitors of iodide transport into thyroid follicular cells, and are established inhibitors of TH synthesis (Jugan et al. 2010).

ClO_4^- inhibits the transport of I^- into the fish thyroid gland, thereby depressing TH formation. Patiño et al. (2003) reported the impacts of waterborne ClO_4^- on *Danio rerio* thyroid follicle histology. Exposure to ammonium perchlorate caused thyroid follicle cell (nuclear) hypertrophy and angiogenesis. The exposure to low levels of the compound produced even more pronounced thyroidal effects that included hypertrophy, angiogenesis, hyperplasia, and colloid depletion (Patiño et al. 2003).

Perchlorate had a stimulatory effect on fecundity, GSI, growth and egg/embryo mass in *Gambusia holbrooki* (Park et al. 2006). *Danio rerio* exposed to perchlorate for a period of 30 days starting at three days post-fertilization (dpf) showed that exposure to perchlorate caused changes in thyroid histology consistent with hypothyroidism. Perchlorate exposure also skewed the sex ratio toward females in a concentration-dependent manner (Mukhi et al. 2007). Larvae and adults of *Gobiocypris rarus* exposed to perchlorate showed impairment in development and deionidase activities. An up-regulation of the deionidase type 2 and the sodium iodide symporter mRNA levels were observed in the larvae and in brain of adults. Meanwhile, the expression of deionidase type 3 s was significantly down-regulated in the liver (Li et al. 2011).

Danio rerio exposed to perchlorate at early life stage showed adverse effects on the hypothalamic-pituitary-thyroid axis. Especially, higher perchlorate concentrations led to conspicuous alterations in thyroidal tissue architecture and to effects in the pituitary. In the thyroid, severe hyperplasia and an increase in follicle number could be detected. The most sensitive endpoint was the colloid, whose tinctorial properties and texture changed dramatically. The pituitary revealed significant proliferations of TSH-producing cells, resulting in alterations in the ratio of adeno to neurohypophysis (Schmidt et al. 2012).

Perchlorates are potent thyroid endocrine disruptors, blocking I⁻ uptake and consequently the thyroid hormone synthesis. Few studies have investigated the perchlorate potential to impair reproduction and steroidogenesis.

PAHs

Polycyclic aromatic hydrocarbons, or PAHs, are found in fossil fuels such as oil and coal and are released into the environment through combustion, surface runoff, oil spills, recreational boating and shipping, municipal waste effluents and atmospheric deposition, but some PAHs in the environment arise from natural combustion such as forest fires and volcanoes. There are many types of PAHs, such as benzoapyrene, phenanthrene, chrysene, anthracene among others.

Platichthys flesus exposed to phenanthrene or chrysene, during the previtellogenic phase of the annual reproductive cycle, resulted in altered plasma steroid levels. The most pronounced effect was the significant decrease in plasma 17β-estradiol and impaired ovarian growth. 17α-hydroxyprogesterone levels were also affected by both PAHs. One possible explanation is that PAH action may be mediated by a specific inhibition of steroidogenic enzymes (Monteiro et al. 2000). Histopathological alterations in the reproductive tissue of *Psammechinus miliaris* females after exposure to phenanthrene showed a severe disorganization of the acinal structure of the gonads and degeneration of previtellogenic oocytes. Growth and maturation of previtellogenic oocytes were also inhibited (Schäfer and Köhler 2009).

Benzo[a]pyrene (BaP) is a well reported AhR agonist in fish. Exposure to BaP significantly induced the gene expression of ERα, vitellogenin, AhR and CYP1A gens in *Carassius auratus* (Yan et al. 2012). Effects of BaP on the embryos of *Oryzias latipes* demonstrate that this chemical delayed the hatching time in a dose independent manner. Mortality augmented and body weight was reduced after BaP-treatment; distortion of sex ratio was noticed and GSI was decreased (Chikae et al. 2004). BaP treatment alters gonadal steroid levels in *Sebastiscus marmoratus*; the treatment significantly elevated the testosterone levels in testicle and ovaries, while it reduced the ovarian 17β-estradiol level. This change in sex hormone levels would be one of the ways to interpret the masculinization of fish by BaP (Zheng et al. 2005).

Danio rerio females exposed to BaP showed alterations on transcription of genes involved in reproduction, including gonadotropins (follicle stimulating hormone and luteinizing hormone), steroidogenic enzymes (CYP11A1, CYP17, CYP19A1, CYP19A2 and 20β-HSD), estrogen receptor and vitellogenin. A reduction in total egg output was observed in B[*a*]P exposed fish as well as an increase in CYP1A1 expression. A significant increase in 20β-HSD mRNA occurred in pre-vitellogenic oocytes from fish exposed to BaP. CYP19A2 and vitellogenin were significantly increased following the exposure (Hoffmann and Oris 2006).

Sebastiscus marmoratus embryos exposed to pyrene (Py) showed reduced concentration of T_3, but not T_4. The thyroid expression of receptor genes was down-regulated by Py. At the same time, Py exposure impaired the expression of thyroid development related genes and altered the mRNA levels of thyroid function related genes. In conclusion, the results demonstrated that Py exposure inhibited thyroid development and influenced the function of thyroid system in fish embryos (He et al. 2012).

The most studied PAH is BaP, so many studies have been conducted with the effects of this chemical on neotropical fish. The most pronounced effects are on reproductive system, but there are also reports of effects on thyroidal function. Other PAHs as well as the effects on adrenocortical cells and behavior must be studied.

Bisphenol A

Bisphenol-A (BPA) is the monomer of the plastic polycarbonate and is widely used as a component in the manufacture of phenoxy resins and corrosion-resistant unsaturated polyester-styrene resins. Furthermore, bisphenol A serves as a stabilizer for plasticizers in PVC, a thermal stabilizer for PVC resins, an antioxidant in rubber and plastics, a fungicide and other compounds used in the manufacture of flame retardants.

Pituitary gonadotropins (GTHs), follicle stimulating hormone β (FSH-β), and luteinizing hormone β (LH-β) from *Kryptolebias marmoratus* were significantly up regulated in the brain/pituitary after the exposure to BPA (Rhee et al. 2010). Exposure of *Oryzias latipes* to BPA accelerated early embryonic development, attenuated body growth, and anticipated the hatching time and reproductive maturation (Ramakrishnan and Wayne 2008). The acceleration in embryonic development and hatching time were blocked by the thyroid-hormone receptor (TH-R) antagonist amiodarone, suggesting that BPA alters global developmental timing through a thyroid-hormone pathway (Ramakrishnan and Wayne 2008). *Oryzias latipes* exposed to BPA showed decreased hatchability of fertilized eggs and increased expression of hepatic vitellogenin in males (Ishibashi et al. 2005).

Exposure of *Danio rerio* to BPA showed that this chemical affects not only the individual exposed but also the following generations. BPA increased the Vtg levels in F_1 and F_2 males and histological analyses of these generations revealed hepatocellular vacuolization, predominantly in males, after parental exposure (Keiter et al. 2012). BPA also exerts endocrine disrupting effects in growth, fecundity and fertilization rate, damages in liver, thyroid and gonads and vitellogenin (Vtg) induction in males in transgenerational protocols. BPA showed an estrogenic potential and was able to impact not only the individual but also the following generations.

Effects of tetrabromobisphenol A (TBBPA) or bisphenol A (BPA) on *Danio rerio* embryo-larvae includes alterations in the expression of genes belonging to the hypothalamic-pituitary-thyroid (HPT) axis. In larvae, both pollutants were found to have significantly induced many genes of interest, namely thyroglobulin, thyroid peroxidase, thyroid receptors, thyroid stimulating hormone, and transthyretin. In embryos, the exposure also significantly induced the sodium iodide symporter and thyroid stimulating hormone (Knoebl et al. 2004).

BPA effects are well reported as an estrogen mimic endocrine disruptor. The most pronounced effects are on reproduction and development of embryos and larvae. It is also possible to evaluate effects on thyroid function and effects of thyroid hormones. However, studies concerning other target organs for endocrine disruption are necessary.

Alkylphenols

Alkylphenols and alkylphenol polyethoxylates are used as surfactants in many applications, from soaps and detergents to pesticide formulations. Once in the environment, microbial degradation results in loss of the ethoxylates, eventually leaving the more persistent alkylpylphenol (e.g., 4-nonylphenol). These chemicals enter the aquatic environment via discharges from STPs, textile, and pulp and paper mills.

Gobiocypris rarus adults exposed to 4-nonylphenol (NP), one of the degradation products of the nonylphenol polyethoxylates, showed reduction in gonad-somatic indices (GSI) and increase in vitellogenin expression by male. At high doses, hepatic and renal tissue presented structural damages. Also at these levels of exposure, feminization of male fish could be noticed and parts of males manifested the testis-ova phenomenon. Ovaries of NP-exposed females were degenerated (Zha et al. 2007). *Cyprinodon variegates* males exposed to NP presented induction of the mRNA of vitellogenin (VTG1, VTG2) and zone radiate protein (ZRP2, ZRP3) (Knoebel et al. 2004). Mature *Danio rerio* females and males exposed to NP showed impairment in the gonad-somatic index and vitellogenin induction. The reproductive effects of NP include alteration in the embryonic cathepsin D (CAT D) activity, eggshell thickness, fecundity, hatching rate and malformation (vertebral column flexure) rate of offspring (Yang et al. 2006). None of the *Rivulus marmoratus* males exposed from hatching to high doses of NP developed testicular tissue. Oogenesis was also significantly inhibited by NP. None of the fish exposed to NP had vitellogenic oocytes. Dysplasia of the gonadal lumen also occurred in fish exposed to NP (Tanaka and Grizzle 2002).

Carassius auratus exposed to NP showed alterations in plasma levels of thyroid hormones. Nonylphenol induced a significant decrease of thyroxin levels, whereas no effect on triiodothyronine concentrations was detected. No histopathological changes were detected in thyroid or testes (Zaccaroni et al. 2009). *Clarias gariepnus* exposed to NP showed impairment in the thyroidal and gonadal functions. The levels of thyroid stimulating hormone (TSH), triiodothyronine (T_3), total thyroxine (T_4), follicle stimulating hormone, luteinizing hormone and testosterone concentrations significantly decreased, while 17β-estradiol levels significantly increased after NP exposure (Sayed et al. 2012).

Cichlasoma dimerus exposed to octylphenol (OP) showed impairment of reproductive success. Vitellogenin was detected in males exposed to OP. Morphological changes in the hepatocytes due to the accumulation of vitellogenin were also observed in OP-exposed males. Impairment of testicular structure was also observed: alteration of lobular organization with increased testicular fibrosis and progressive disruption of spermatogenesis (Vázquez et al. 2009). Exposure of *Poecilia reticulata* as adults to octylphenol (OP) alters gonads structure. Histological examinations revealed a block of spermatogonial mitosis in the testes of adult males. The post-parturition ovaries of adult females showed effects suggesting an inhibited yolk deposition (Kinnberg et al. 2003).

Alkylphenols are considered as estrogen mimics and the most common effects of the exposition of males to alkylphenols are vitellogenin induction and alterations in the testicular structure. Females are also target of these chemicals and their exposition impairs the reproductive success. Thyroidal function can also be impaired by alkylphenol.

In most cases, the evidence of a causal link between a specific physiological disruptor and a specific effect in field studies is weak, largely due to the fact that fish and, indeed, all other wildlife are exposed to a wide range of chemicals, that act at a number of different body targets, to affect a variety of physiological processes. Moreover, some investigations of the interactive effects of mixtures of estrogenic chemicals in fish have shown that combinations of steroid estrogens, alkylphenolic chemicals and a pesticide have additive effects (Thorpe et al. 2001). This highlights the fact that even chemicals that have slight effects on the endocrine system should be taken into consideration when assessing the effects of chemical mixtures in tropical fish. Exposures of fish to environmentally relevant concentrations of EDCs (and at the relevant life stages) are essential to adequately evaluate exposure/response relationships in field studies and produce credible risk assessments (Jobling and Tyler 2003).

Over the last 10 years, a number of fish species of tropical freshwater (*Catostomus commersoni, Catostomus catostomus, Coregonus clupea formis, Perca fluviatilis, Rutilus rutilus*) living downstream of pulp- and paper-mill effluents have been found to exhibit an array of altered features in their reproductive development, including reductions in gonadal growth, inhibition of spermatogenesis, depressed sex steroids, reduced pituitary hormone concentrations, and delayed sexual maturity. Moreover, other studies have suggested that the reproductive effects may (at least in part) be mediated through disruption of the process of steroidogenesis, by affecting the availability of cholesterol and pregnenolone and thus impairing steroid production by the gonads.

There is considerable (and increasing) evidence of endocrine disruption in fish populations living in river stretches downstream of treated sewage effluent discharges in America (Folmar et al. 2001). Effluents from treated sewage were estrogenic, inducing the production of vitellogenin, in male fish (Purdom et al. 1994). In addition to vitellogenin production, exposure to treated sewage effluents has also been associated with deleterious effects on gonad differentiation and development in various species of fish and with the abnormal development (feminization) of secondary sexual characteristics in male fish (Jobling and Tyler 2003).

Fish sampled from a heavily industrialized habitat in which the sediments contained more heavy metals (Hg, Pb, Cu, Zn, chromium, and Cd), organochlorines (PCBs, DDT, and DDT derivatives), and PAHs than those from a clean, non-industrialized habitat had larger thyroid follicles, a taller thyroid epithelial cell height, and higher plasma T_4 levels than fish from the clean site (Brown et al. 2004). Xu et al. (2002) measured muscle iodine and serum total and free TH in *Hypophthalmicthys molitrix* sampled along a gradient of sediment contamination. They reported that the sediments contained hexachlorobenzene, PCBs, PAHs, and various TCDD-like contaminants. Fish from the most contaminated areas exhibited elevated CYP1A and UDPGT activities along with markedly depressed concentrations of free T_3. All plasma thyroid indices negatively correlated to the elevated hepatic CYP1A activity.

An impaired adrenocortical response to stress was observed in fish living in environments polluted by polycyclic aromatic hydrocarbons (PAHs), PCBs, and heavy metals such as mercury, cadmium and zinc (Norris et al. 1999; Hontela et al. 1992). Fish from the most polluted sites had atrophied pituitary glands with a delayed rise in plasma cortisol levels after acute stress of capture; despite ACTH levels were initially elevated. It was suggested that the overall HPA-axis-dependent response to severe, short-term confinement stress was depressed in fish that were chronically exposed to heavy-metal ions.

While laboratory studies have shown effects of various EDCs on the reproduction, steroidogenesis, thyroidal and adrenocortical functions, growth and development in teleosts, much less attention has been directed at studying these disruptions in natural populations. However, the results of a number of field studies suggest that exposure to waterborne contaminants can disrupt the endocrine homeostasis and also their population health.

Acknowledgements

The author would like to thank Gabriel Adelman Cipolla (Federal University of Paraná) for his careful English translation of this chapter.

Keywords: endocrine system, tyroid dysfunction, adrenocortical cell dysfunction, reproductive dysfunction, stereoidogenesis dysfunction, liver dysfunction, mechanism of endocrine dysfunction

References

Abalos, M., E. Abad, A. Estevez, M. Sole, A. Buet, L. Quiros, B. Pina and J. Rivera. 2008. Effects on growth and biochemical responses in juvenile gilthead seabream 'Sparus aurata' after long-term dietary exposure to low levels of dioxins. Chemosphere 73: S303–S310.

Aluru, N. and M.M. Vijayan. 2006. Aryl hydrocarbon receptor activation impairs cortisol response to stress in rainbow trout by disrupting the rate-limiting steps in steroidogenesis. Endocrinology 147: 1895–1903.

Ankley, G.T., D.C. Bencic, M.S. Breen, T.W. Collette, R.B. Conolly, N.D. Denslow, S.W. Edwards, D.R. Ekman, N. Garcia-Reyero, K.M. Jensen, J.M. Lazorchak, D. Martinović, D.H. Miller, E.J. Perkins, E.F. Orlando, D.L. Villeneuve, R. Wang and K.H. Watanabe. 2009. Endocrine disrupting chemicals in fish: Developing exposure indicators and predictive models of effects based on mechanism of action. Aquat. Toxicol. 92: 168–178.

Anway, M.D. and M.K. Skinner. 2006. Epigenetic transgenerational actions of endocrine disruptors. Endocrinology 147: 43–49.

Anway, M.D. and M.K. Skinner. 2008. Transgenerational effects of the endocrine disruptor vinclozolin on the prostate transcriptome and adult onset disease. Prostate 68: 517–529.

Aoki, J., A. Hatsuyama, N. Hiramatsu and K. Soyano. 2011. Effects of ethynylestradiol on vitellogenin synthesis and sex differentiation in juvenile grey mullet (*Mugil cephalus*) persist after long-term exposure to a clean environment. Comp. Biochem. Physiol. C 154: 346–352.

Bandyopadhyay, P.S., P.P. Banjanerje and S. Bhattacharya. 1991. 17-b estradiol releases thyroxine from the thyroid follicles of a teleost fish, *Channa gachua*. Gen. Comp. Endocrinol. 81: 227–233.

Barouki, R., X. Coumoul and P.M. Fernndez-Salguero. 2007. The aryl hydrocarbon receptor, more than a xenobiotic-interacting protein. FEBS Letters. 581: 3608–3615.

Bell, A.M. 2001. Effects of an endocrine disrupter on courtship and aggressive behaviour of male three-spined stickleback, *Gasterosteus aculeatus*. Anim. Behav. 62: 775–780.

Bhattacharya, L. 1995. Histological and histochemical alterations in the thyroid activity of endosulfan treated *Oreochromis mossambicus*. J. Environ. Biol. 16: 347–351.

Bhattacharya, T., S. Bhattacharya, A.K. Ray and S. Dey. 1989. Influence of industrial pollutants on thyroid function in *Channa punctatus* (Bloch). Indian J. Exp. Biol. 27: 65–68.

Bishop, C.D., D.F. Erezyilmaz, T. Flatt, C.D. Georgiou, M.G., Hadfield, A. Heyland, J. Hodin, M.W. Jacobs, S.A. Maslakova, A. Pires, A.M. Reitzel, S. Santagata, K. Tanakay and J.H. Youson. 2006. What is metamorphosis? Integr. Comp. Biol. 46: 655–661.

Bone, Q., N.B. Marshall and J.H.S. Blaxter. 1995. Biology of fishes. Chapman and Hall, New York, USA.

Brandhof, E.J. and M. Montforts. 2010. Fish embryo toxicity of carbamazepine, diclofenac and metoprolol. Ecotoxicol. Environ. Safety. 73: 1862–1866.

Brown, A.R., A.M Riddle, I.J. Winfield and J.B. James. 2005. Predicting the effects of endocrine disrupting chemicals on healthy and disease impacted population of perch (*perch fluviatilis*). Ecol. Model. 189: 377–395.

Brown, D.D. 1997. The role of thyroid hormone in zebrafish and axolotl development. Proc. Natl. Acad. Sci. USA. 94: 13011–13016.

Brown, S.B., B.A. Adams, D.G. Cyr and L.G. Eales. 2004. Contaminants effects on the teleost fish thyroid. Environ. Toxicol. Chem. 23: 1680–1701.

Calò, M., D. Alberghina, A. Bitto, E.R. Lauriano and P.L. Cascio. Estrogenic followed by anti-estrogenic effects of PCBs exposure in juvenil fish (*Spaurus aurata*). Food Chem. Toxicol. 48: 2458–2463.

Carr, J.A. and R. Patiño. 2011. The hypothalamus-pituitary-thyroid axis in teleosts and amphibians: endocrine disruption and its consequences to natural populations. Gen. Comp. Endocrinol. 170: 299–312.

Carson, R. 1962. Silent Spring. Houghton Mifflin, Boston, USA.

Chaube, R., S. Mishra and R.K. Singh. 2010. *In vitro* effects of lead nitrate on steroid profiles in the post-vitellogenic ovary of the catfish *Heteropneustes fossilis*. Toxicology *in vitro*. 24: 1899–1904.

Chaurasia, S.S. and A. Kar. 1999. An oxidative mechanism for the inhibition of iodothyronine 59-monodeiodinase activity by lead nitrate in the fish, *Heteropneustes fossilis*. Water Air Soil Pollut. 111: 417–423.

Chikae, M., Y. Hatano, R. Ikeda, Y. Morita, Q. Hasan and E. Tamiya. 2004. Effects of bis(2-ethylhexyl) phthalate and benzo[a]pyrene on the embryos of Japanese medaka (*Oryzias latipes*). Environ. Toxicol. Pharmacol. 16: 141–145.

Clotfelter, E.D. and A.C. Rodriguez. 2006. Behavioral changes in fish exposed to phytoestrogens. Environ. Pollut. 144: 833–839.

Coe, T.S., P. Hamilton, D. Hodgson, G. Paull, J. Stevens, K. Sumner and C. Tyler. 2008. An environmental estrogen alters reproductive hierarchies, disrupting sexual selection in group-spawning fish. Environ. Sci. Technol. 42: 5020–5025.

Coimbra, A.M., M.A. Reis-Henriquesa and V.M. Darras. 2005. Circulating thyroid hormone levels and iodothyronine deiodinase activities in Nile tilapia (*Oreochromis niloticus*) following dietary exposure to Endosulfan and Aroclor 1254. Comp. Biochem. Physiol. C 141: 8–14.

Colby, H.D., H. Purcelli, S. Kominami, S. Takemori and D.C. Kossor. 1994. Adrenal activation of carbon tetrachloride: role of microsomal P450 isozymes. Toxicology 94: 31–40.

Conley, A.J. and I.M. Bird. 1997. The role of cytochrome P450 17c-hydroxylase and 3/3-hydroxylase and 3b-hydroxysteroid dehydrogenase in the integration of gonadal and adrenal steroidogenesis via the A5 and A4 pathways of steroidogenesis in mammals. Biol. Reprod. 56: 789–799.

Cook, J.C., L.S. Mullin, S.R. Frame and L.B. Biegel. 2003. Investigation of a mechanism for Leydig cell tumorigenesis by linuron in rats. Toxicol. Appl. Pharmacol. 119: 195–204.

Cooke, P.S., R.E. Peterson and R.A. Hess. 2002. Endocrine disruptors.. *In:* W.M. Haschek, C.G. Rousseaux and M.A. Wallig (eds.). Handbook of Toxicology Pathology. Elsevier, New York, USA, pp. 501–528.

Crago, J. and R.A. Klaper. 2012. A mixture of an environmentally realistic concentration of a phthalate and herbicide reduces testosterone in male fathead minnow (*Pimephales promelas*) through a novel mechanism of action. Aquat. Toxicol. 110–111: 74–83.

Crain, D., L.J. Guillette, A.A. Rooney and D. Pickford. 1997. Alterations in steroidogenesis in alligators (*Alligator mississippiensis*) exposed naturally and experimentally to environmental contaminants. Environ. Health Perspect. 105: 528–533.

Crews, D., E. Willigham and K. Skipper. 2000. Endocrine disruptors: Present issues, future directions. The Quarterly Review of Biology 75: 243–260.

Crews, D., and J.A. McLachlan. 2006. Epigenetics, evolution, endocrine disruption, health, and disease. Endocrinology 147: 4–10.

Croutch, C.R., M. Lebofsky, K.W. Schramm, P.F. Terranova and K.K. Rozman. 2005. 2,3,7,8-Tetrachlorodibenzo-p-dioxin (TCDD) and 1,2,3,4,7,8- hexachlorodibenzo-p-dioxin (HxCDD) alter body weight by decreasing insulin like growth factor I (IGF-I) signaling. Toxicol. Sci. 85: 560–571.

Cyr, D.G. and J.G. Eales. 1996. Interrelationships between thyroidal and reproductive endocrine systems in fish. Rev. Fish Biol. Fisheries. 6: 165–200.

Daouk, T., T. Larcher, F. Roupsard, L. Lyphout, C. Rigauda, M. Ledevin, V. Loizeau and X. Cousin. 2011. Long-term food-exposure of zebrafish to PCB mixtures mimicking some environmental situations induces ovary pathology and impairs reproduction ability. Aquat. Toxicol. 105: 270–278.

Davis, L.K., N. Visitacion, L.G. Riley, N. Hiramatsu, C.V. Sullivan, T. Hirano and E.G. Grau. 2009. Effects of o,p'-DDE, heptachlor, and 17β-estradiol on vitellogenin gene expression and the growth hormone/insulin-like growth factor-I axis in the tilapia, *Oreochromis mossambicus*. Comp. Biochem. Physiol. C 149: 507–514.

Diamnati-Kandarakis, E., J.P. Bourguinon, L.C. Giuduce, R. Hsuse, G.S. Prins, A.M. Soto, R.T. Zoeller and A.C. Gore. 2009. Endocrine-disrupting chemicals. An endocrine society scientific statement. The Endocrine Society. Endocrine Reviews. 30: 293–342.

Dorts, J., C.A. Richter, M.K. Wright-Osment, M.R. Ellersieck, B.J. Carter and D.E. Tillitt. 2009. The genomic transcriptional response of female fathead minnows (*Pimephales promelas*) to an acute exposure to the androgen, 17b-trenbolone. Aquat. Toxicol. 91: 44–53.

Doyle, C.J. and R.P. Lim. 2005. Sexual behavior and impregnation success of adult male mosquitofish following exposure to 17b-estradiol. Ecotoxicol. Environ. Safe. 61: 392–397.

Dzieweczynski, T. 2011. Short-term exposure to an endocrine disruptor affects behavioral consistency in male three spined stickleback. Aquat. Toxicol. 105: 681–687.

Ebrahimi, M. 2007. Effects of *in vivo* and *in vitro* zinc and cadmium treatment on sperm steroidogenesis of the African catfish Clarias gairepinus. J. Biol. Sci. 10: 2862–2867.

[EDSTAC] Endocrine Disruptors Screening and testing Advisory Committee, Final Report, 1998. EPA, Washington, USA.

Elonen, G.E., R.L. Spehar, G.W. Holcombe, R.D. Johnson, J.D. Fernandez, R.J. Erickson, J.E. Tietge and P.M. Cook. 1998. Comparative toxicity of 2,3,7,8-tetrachlorodibenzo-*p*-dioxin to seven freshwater fish species during early life-stage development. Environ. Toxicol. Chem. 17: 472–483.

Evans, J.P. and A.E. Magurran. 2000. Multiple benefits of multiple mating in guppies. Proc. Natl. Acad. Sci. USA 97: 10074–10076.

Folmar, L.C., N.D. Denslow, K. Kroll, E.F. Orlando, J. Enblom, J. Marcino, C. Metcalf and L.J. Guilette. 2001. Arch. Environ. Contam. Toxicol. 40: 392–398.

Foran, C.M., B.N. Peterson and W.H. Benson. 2002. Influence of parental and developmental cadmium exposure on endocrine and reproductive function in Japanese medaka (*Oryzias latipes*). Comp. Biochem. Physiol. C 133: 345–354.

Friedmann, A.S., M.C. Watzinb, T. Brinck-Johnsen and J.C. Leiter. 1996. Low levels of dietary methylmercury inhibit growth and gonadal development in juvenile walleye (*Stizostedion vitreum*). Aquat. Toxicol. 35: 265–278.

Gasiewicz, T.A. and S.K. Park. 2003. The Ah receptor: involvement in toxic responses. *In:* A. Schecter and T.A. Gasiewicz (eds.). Dioxins and Health. Taylor & Francis, London, UK, pp. 491–532.

Giesy, J.P., P.D. Jones, K. Kannan, J.L. Newsted, D.E. Tillitt and L.L Williams. 2002. Effects of chronic dietary exposure to environmentally relevant concentrations to 2,3,7,8-tetrachlorodibenzo-p-dioxin on survival, growth, reproduction and biochemical responses of female rainbow trout (*Oncorhynchus mykiss*). Aquat. Toxicol. 59: 35–53.

Gillesby, B.E. and T.R. Zacharewski. 1998. Exoestrogens: mechanisms of action and strategies for identification and assessment. Environ. Toxicol. Chem. 17: 3–14.

Gore, A. 2002. GnRH: the master molecule of reproduction. Norwell, MA: Kluwer Academic Publishers.

Gormley, K.M. and K.L. Teather. 2003. Developmental, behavioral, and reproductive effects experienced by Japanese medaka (*Oryzias latipes*) in response to short-term exposure to endosulfan. Ecotoxicol. Environ. Safe. 54: 330–338.

Gray, L.E., J. Ostby, V. Wilson, C. Lambright, K. Bobseine, P. Hartig, A. Hotchkiss, C. Wolf, J. Furr, M. Price, L. Parks, R.L. Cooper, T.E. Stoker, S.C. Laws, S.J. Degitz, K.M. Jensen, M.D. Kahl, J.J. Korte, E.A. Makynen and J.E. Tietge. 2002. Xenoendocrine disrupters-tiered screening and testing: Filling key data gaps. Toxicology. 181–182: 371–382.

Grün, F. and B. Blumberg. 2006. Environmental obesogens: organotins and endocrine disruption via nuclear receptor signaling. Endocrinology 147: 50–55.

Grün, F., H. Watanabe, Z. Zamanian, L. Maeda, K. Arima, R. Cubacha, D.M. Gardiner, J. Kanno, T. Iguchi and B. Blumberg. 2006. Endocrine-disrupting organotin compounds are potent inducers of adipogenesis in vertebrates. Mol. Endocrinol. 20: 2141–2155.

Gupta, P., S.S. Chaurasia, A. Kar and P.K. Maiti. 1997. Influence of cadmium on thyroid hormone concentrations and lipid peroxidation in a freshwater fish, *Clarias batrachus*. Fresenius Environ. Bull. 6: 355–358.

Hahn, M.E., R.R. Merson and S. Karchner. 2005. Xenobiotic receptors in fish: Structural and functional diversity and evolutionary insights. *In:* T. Mommsen and T.W. Moon (eds.). Biochemistry and Molecular Biology of Fishes. Elsevier, New York, USA, pp. 191–228.

Haldén, A.N., K. Arnoldsson, P. Haglund, A. Mattsson, E. Ullerås, J. Sturve and L. Norrgren. 2011. Retention and maternal transfer of brominated dioxins in zebrafish (*Danio rerio*) and effects on reproduction, aryl hydrocarbon receptor-regulated genes, and ethoxyresorufin-O-deethylase (EROD) activity. Aquat. Toxicol. 102: 150–161.

Han, S., K. Choi, J. Kim, K. Ji, S. Kim, B. Ahn, J. Yun, K. Choi, J.S. Khim, X. Zhang and J.P. Giesye. 2010. Endocrine disruption and consequences of chronic exposure to ibuprofen in Japanese medaka (*Oryzias latipes*) and freshwater cladocerans *Daphnia magna* and *Moina macrocopa*. Aquat. Toxicol. 98: 256–264.

Hanno, K., S. Oda and H. Mitani. 2010. Effects of dioxin isomers on induction of AhRs and CYP1A1 in early developmental stage embryos of medaka (*Oryzias latipes*). Chemosphere 78: 830–839.

Hatef, A., S. Mohammad, H. Alavi, Ji˘ri K˘ri˘st'an, M. Golshan, P. Fontaine and O. Linhart. 2012. Anti-androgen vinclozolin impairs sperm quality and steroidogenesis in goldfish. Aquat. Toxicol. 122–123: 181–187.

He, C., Z. Zuo, X. Shi, L. Sun and C. Wang. 2012. Pyrene exposure influences the thyroid development of *Sebastiscus marmoratus* embryos. Aquat. Toxicol. 124–125: 28–33.

Hiramatsu, N., A. Cheek, C.V. Sullivan, T. Matsubara and A. Hara. 2005. Vitellogenesis and endocrine disruption. *In:* T.P. Mommsen and T.W. Moon (eds.). Biochemstry Molecular Biology of Fishes. Elsevier Science BV, Amsterdam, Netherlands, pp. 431–471.

Hoffmann, J.L. and J.T. Oris. 2006. Altered gene expression: A mechanism for reproductive toxicity in zebrafish exposed to benzo[*a*]pyrene. Aquat. Toxicol. 78: 332–340.

Homsby, P.J. 1989. Steroid and xenobiotic effects on the adrenal cortex: mediation by oxidative and other mechanisms. Free Radic. Biol. Med. 6: 103–115.

Hong, H.N., H.N. Kim, K.S. Park, S.K. Lee and M.B. Gu. 2007. Analysis of the effects diclofenac has on Japanese medaka (*Oryzias latipes*) using real-time PCR. Chemosphere. 67: 2115–2121.

Hontela, A. 2005. Adrenal toxicology: environmental pollutants and the HPI axis. *In:* T.P. Mommsen and T.W. Moon (eds). Biochemistry and Molecular Biology of fishes. Amsterdam, Netherlands, pp. 331–363.

Hontela, A., J.B. Rasmussen, C. Audet and G. Chevalier. 1992. Impaired cortisol stress response in fish from environments polluted by PAHs, PCBs, and mercury. Arch. Environ. Con. Toxicol. 22: 278–283.

Hornung, M.W., K.M. Jensen, J.J. Korte, M.D. Kahl, E.J. Durhan, J.S. Denny, T.R. Henry and G.T. Ankley. 2004. Mechanistic basis for estrogenic effects in fathead minnow (*Pimephales promelas*) following exposure to the androgen 17α-methyltestosterone: conversion of 17α-methyltestosterone to 17α-methylestradiol. Aquat. Toxicol. 66: 15–23.

Hoshijima, K. and S. Hirose. 2007. Expression of endocrine genes in zebrafish larvae in response to environmental salinity. J. Endocrinol. 193: 481–491.

Imai, S., J. Koyama and K. Fujii. 2005. Effects of 17b-estradiol on the reproduction of Java-medaka (*Oryzias javanicus*), a new test fish species. Mar. Pollut. Bull. 51: 708–714.

Ishibashi, H.T., N. Watanabe, N. Matsumura, M. Hirano, Y. Nagao, H. Shiratsuchi, S. Kohra, S. Yoshihara and K. Arizono. 2005. Toxicity to early life stages and an estrogenic effect of a bisphenol A metabolite, 4-methyl-2,4-bis(4-hydroxyphenyl)pent-1-ene on the medaka (*Oryzias latipes*). Life Sciences 77: 2643–2655.

Jadhao, A.G., P.L. Paul and P.D. Rao. 1994. Effect of cadmium chloride on the pituitary, thyroid and gonads in the catfish, *Clarias batrachus* (Linn.). Funct. Dev. Morphol. 4: 39–44.

Jensen, K.M., M.D. Kahl, E.A. Makynen, J.J. Korte, R.L. Leino, B.C. Butterworth and G.T. Ankley. 2004. Characterization of responses to the antiandrogen flutamide in a short-term reproduction assay with the fathead minnow. Aquat. Toxicol. 70: 99–110.

Jobling, S. and C.R. Tyler. 2003. Endocrine disruption in wild freshwater fish. Pure Applied Chemistry 75: 2219–2234.

Jugan, M.L., Y. Levi and J.P. Blondeau. 2010. Endocrine disruptors and thyroid hormone physiology. Biochem. Pharmacol. 79: 938–947.

Jung, J.H., J.K. Jeon and C.H. Han. 2005. Effects of 2,2,4,4,5,5-hexachlorobiphenyl (PCB 153) on plasma sex steroids and vitellogenin in rockfish (*Sebastes schlegeli*). Comp. Biochem. Physiol. C 140: 295–299.

Kanayama, T., N. Kobayashi, S. Mamiya, T. Nakanishi and J. Nishikawa. 2005. Organotin compounds promote adipocyte differentiation as agonists of the peroxisome proliferator-activated receptor /retinoid X receptor pathway. Mol. Pharmacol. 67: 766–774.

Kato, Y., S. Ikushiro, K. Haraguchi, T. Yamazaki, Y. Ito, H. Suzuki, R. Kimura, S. Yamada, T. Inoue and M. Degawa. 2004. A possible mechanism for decrease in serum thyroxine level by polychlorinated biphenyls in Wistar and Gunn rats. Toxicol. Sci. 81: 309–315.

Katti, S.R. and A.G. Sathyanesan. 1987. Lead nitrate induced changes in the thyroid physiology of the catfish, *Clarias batrachus* (L.). Ecotoxicol. Environ. Saf. 13: 1–6.

Kavlock, R.J., G.P. Daston, C. DeRosa, P. Fenner-Crisp, L.E. Gray, S. Kaattari, G. Lucier and M. Luster. 1996. Research needs for the risk assessment of health and environmental effects of endocrine disruptors: a report of the US EPA sponsored workshop. Environ. Health. Perspect. 104: 715–740.

Keiter, S., L. Baumann, H. Färber, H. Holbech, D. Skutlarek, M. Engwall and T. Braunbeck. 2012. Long-term effects of a binary mixture of perfluorooctane sulfonate (PFOS) and bisphenol A (BPA) in zebrafish (*Danio rerio*). Aquat. Toxicol. 118–119: 116–129.

Kelce, W.R., C.R. Stone, S.C. Laws, L.G. Gray, J.A. Kemppainen and E.M. Wilson. 1995. Persistent DDT metabolite p,p'-DDE is a potent androgen receptor antagonist. Nature. 373: 581–585.

Khan, I.A. and P. Thomas. 2006. PCB congener-specific disruption of reproductive neuroendocrine function in Atlantic croaker. Mar. Environ. Res. 62: 25–28.

Kim, D.J., S.H. Seok, M.W. Baek, H.Y. Lee, Y.R. Na, S.H. Park, H.Ky. Lee, N.K. Dutta, K. Kawakami and J.H. Park. 2009. Developmental toxicity and brain aromatase induction by high genistein concentrations in zebrafish embryos. Toxicol. Mech. Methods. 19: 251–256.

Kim, E.J., J.W. Kim and S.K. Lee. 2002. Inhibition of oocyte development in Japanese medaka (*Oryzias latipes*) exposed to di-2-ethylhexyl phthalate. Environ. Internat. 28: 359–365.

Kime, D.E. 1999. Environmentally induced endocrine abnormalities in fish. Issues in environmental science and technology. Endocrine disrupting chemicals, The royal Society of Chemistry.

Kinnberg, K., B. Korsgaard and P. Bjerregaard. 2003. Effects of octylphenol and 17b-estradiol on the gonads of guppies (*Poecilia reticulata*) exposed as adults via the water or as embryos via the mother. Comp. Biochem. Physiol. C 134: 45–55.

Kiparissis, Y., G.C. Balch, T.L. Metcalfe and C.D. Metcalfe. 2003. Effects of the isoflavones genistein and equol on the gonadal development of Japanese medaka (*Oryzias latipes*). Environ. Health Perspect. 11: 1158–1163.

Kiparissis, Y., T.L. Metcalfe, G.C. Balch and C.D. Metcalfe. 2003. Effects of the antiandrogens, vinclozolin and cyproterone acetate on gonadal development in the Japanese medaka (*Oryzias latipes*). Aquat. Toxicol. 63: 391–403.

Kirubagaran, K. and K.P. Joy. 1989. Toxic effects of mercurials on thyroid function of the catfish, *Clarias batrachus* (L). Ecotox. Environ. Saf. 17: 265–271.

Kirubagaran, R. and K.P. Joy. 1991. Changes in adrenocortical-pituitary activity in the catfish, *Clarias batrachus* (L.), after mercury treatment. Ecotoxicol. Environ. Saf. 22: 36–44.

Knoebl, I., M.J. Hemmer and N.D. Denslow. 2004. Induction of zona radiata and vitellogenin genes in estradiol and nonylphenol exposed male sheepshead minnows (*Cyprinodon variegatus*). Mar. Environ. Res. 58: 547–551.

Koger, D.S., S.J. Teh and D.E. Hinton. 2000. Determining the sensitive developmental stages of intersex induction in medaka (*Oryzias latipes*) exposed to 17b-estradiol or testosterone. Mar. Environ. Res. 50: 201–206.

Köhrle, J. 2007. Thyroid hormone transporters in health and disease: advances in thyroid hormone deiodination. Best. Pract. Res. Clin. Endocrinol. Metab. 21: 173–191.

Kraugerud, M., R.W. Doughty, J.L. Lyche, V. Berg, N.H. Tremoen, P. Alestrøm, M. Aleksandersen and E. Ropstad. 2012. Natural mixtures of persistent organic pollutants (POPs) suppress ovarian follicle development, liver vitellogenin immunostaining and hepatocyte proliferation in female zebrafish (*Danio rerio*). Aquat. Toxicol. 116–117: 16–23.

Kretschmer, X.C. and W.S. Baldwin. 2005. CAR and PXR: xenosensors of endocrine disrupters? Chem. Biol. Interact. 155: 111–128.

Kristensen, T., E. Baatrup and M. Bayley. 2006. p,p-DDE fails to reduce the competitive reproductive fitness in Nigerian male guppies. Ecotoxicol. Environ. Safe. 63: 148–157.

Leaños-Castañeda, O., G.V.D. Kraak, R. Rodríguez-Canul and G. Gold. 2007. Endocrine disruption mechanism of o,p'-DDT in mature male tilapia (*Oreochromis niloticus*). Toxicol. Applied Pharmacol. 221: 158–167.

Lerner, D.T., B.T. Bjornsson and S.D. McCormick. 2007. Aqueous exposure to 4-nonylphenol and 17 beta-estradiol increases stress sensitivity and disrupts ion regulatory ability of juvenile Atlantic salmon. Environ. Toxicol. Chem. 26: 1433–1440.

LeRoy, K.D., P. Thomas and I.A. Khan. 2006. Thyroid hormone status of Atlantic croaker exposed to Aroclor 1254 and selected PCB congeners. Mar. Environ. Res. 62: 25–28.

Leusch, F.D.L. and D.L. MacLatchy. 2003. *In vivo* implants of b-sitosterol cause reductions of reactive cholesterol pools in mitochondria isolated from gonads of male goldfish (*Carassius auratus*). Gen. Comp. Endocrinol. 134: 255–263.

Li, W., J. Zha, L. Yang, Z. Li and Z. Wang. 2011. Regulation of iodothyronine deiodinases and sodium iodide symporter mRNA expression by perchlorate in larvae and adult Chinese rare minnow (*Gobiocypris rarus*). Mar. Poll. Bull. 63: 350–355.

Lintelmann, J., A. Katayama, N. Kurihara, L. Shore and A. Wenzel. 2003. Endocrine disruptors in the environment. Pure Applied Chem. 75: 631–681.

Lyche, J.L., R. Nourizadeh-Lillabadi, C. Almaas, B. Stavik, V. Berg, J.U. Skred, P. Alestrom and E. Ropstad. 2010. Natural mixtures of persistent organic pollutants (POP) increase weight gain, advance puberty, and induce changes in gene expression associated with steroid hormones and obesity in female zebrafish. J. Toxicol. Environ. Health A. 73: 1032–1057.

MacLatchy, D.L. and G.J. Van Der Kraak. 1995. The phytoestrogen beta-sitosterol alters the reproductive endocrine status of goldfish. Toxicol. Appl. Pharmacol. 134: 305–312.

Makynen, E.A., M.D. Kahl, K.M. Jensen, J.E. Tietge, K.L. Wells, G. Van Der Kraak and G.T. Ankley. 2000. Effects of the mammalian antiandrogen vinclozolin on development and reproduction of the fathead minnow (*Pimephales promelas*). Aquat. Toxicol. 48: 461–475.

Metcalfe, T.L., C.D. Metcalfe, Y. Kiparissis, A.J. Niimi, C.M. Foran and W.H. Benson. 2000. Gonadal development and endocrine responses in Japanese medaka (*Oryzias latipes*) exposed to *o,p*-ddt in water or through maternal transfer. Environ. Toxicol. Chem. 19: 1893–1900.

Migliarini, B., C.C. Piccinetti, A. Martell, F. Maradonna, G. Gioacchini and O. Carnevali. 2011. Perspectives on endocrine disruptor effects on metabolic sensors. Gen. Comp. Endocrinol. 170: 416–423.

Mimeault, C., A.J. Woodhouse, X.S. Miaob, C.D. Metcalfe, T.W. Moona and V.L. Trudeau. 2005. The human lipid regulator, gemfibrozil bioconcentrates and reduces testosterone in the goldfish, *Carassius auratus*. Aquat. Toxicol. 73: 44–54.

Miyazaki, W., T. Iwasaki, A. Takeshita, Y. Kuroda and N. Koibuchi. 2004. Polychlorinated biphenyls suppress thyroid hormone receptor-mediated transcription through a novel mechanism. J. Biol. Chem. 279: 18195–18202.

Mommsen, T.P., M.M. Vijayan and T.W. Moon. 1999. Cortisol in teleosts; dynamics, mechanisms of action, and metabolic regulation. Rev. Fish Biol. Fish. 9: 211–268.

Moncaut, N., F. Lo Nostro and M.C. Maggese. 2003. Vitellogenin detection in surface mucus of the South American cichlid fish *Cichlasoma dimerus* (Heckel 1840) induced by estradiol-17b. Effects on liver and gonads. Aquat. Toxicol. 63: 127–137.

Monteiro, P.R.R., M.A. Reis-Henriques and J. Coimbra. 2000. Plasma steroid levels in female founder (*Platichthys fesus*) after chronic dietary exposure to single polycyclic aromatic hydrocarbons. Mar. Environ. Res. 49: 453–467.

Morgado, I., M.A. Campinho, R. Costa, R. Jacinto and D.M. Power. 2009. Disruption of the thyroid system by diethylstilbestrol and ioxynil in the sea bream (*Sparus aurata*). Aquat. Toxicol. 92: 271–280.

Morthorst, J.E., H. Holbech and P. Bjerregaard. 2010. Trenbolone causes irreversible masculinization of zebrafish at environmentally relevant concentrations. Aquat. Toxicol. 98: 336–343.

Moura-Costa, D.D., F. Filipak Neto, M.D.M. Costa, R.N. Morais, J.R.E. Garcia, B.M. Esquivel and C.A. Oliveira Ribeiro. 2010. Vitellogenesis and other physiological responses induced by 17-β-estradiol in males of freshwater fish *Rhamdia quelen*. Comp. Biochem. Physiol. C 151: 248–257.

Mukhi, S., J.A. Carr, T.A. Anderson and R. Patino. 2005. Development and validation of new biomarkers of perchlorate exposure in fishes. Environ. Toxicol. Chem. 24: 1107–1115.

Mukhi, S., L. Torres and R. Patiño. 2007. Effects of larval–juvenile treatment with perchlorate and co-treatment with thyroxine on zebrafish sex ratios. Gen. Comp. Endocrinol. 150: 486–494.

Mukhi, S. and R. Patiño. 2007. Effects of prolonged exposure to perchlorate on thyroid and reproductive function in zebrafish. Toxicol. Sci. 96: 246–254.

Nakari, T. and K. Erkomaa. 2003. Effects of phytosterols on zebrafish reproduction in multigeneration test. Environ. Pollut. 123: 267–273.

Nash, J.P., D.E. Kime, V.L.T Van, P.W. Wester, F. Brion, G. Maack, A.P. Stahlschmidt and C.R. Tyler. 2004. Long-term exposure to environmental concentrations of the pharmaceutical ethynlestradiol causes reproductive failure in fish. Environ. Health. Perspect. 112: 1725–1733.

Njiwa, J.R.K., P. Muller and R. Klein. 2004. Binary mixture of DDT and Arochlor 1254: effects on sperm release by *Danio rerio*. Ecotoxicol. Environ. Safe. 58: 211–219.

Norris, D.O. 1997. Comparative aspects of vertebrate adrenals. *In:* D.O. Norris (ed.). Vertebrate Endocrinology. Academic Press, New York, USA, pp. 329–354.

Norris, D.O., S. Donahue, R.M. Dores, J.K. Lee, T.A. Maldonado, T. Ruth and J.D. Woodling. 1999. Impaired adrenocortical response to stress by brown trout, *Salmo trutta*, living in metal-contaminated waters of the Eagle River, Colorado. Gen. Comp. Endocrinol. 113: 1–8.

Odermatt, A., C. Gumy, A.G. Atanasov and A.A. Dzyakanchuk. 2006. Disruption of glucocorticoid by environmental chemicals: Potential mechanisms and relevance. J. Steroid Biochem. Mol. Biol. 102: 222–231.

Odermatt, A. and C. Gumy. 2008. Glucocorticoid and mineralocorticoid action: Why should we consider influences by environmental chemical? Biochem. Pharmacol. 76: 1184–1193.

Oshima, Y., I.K.J. Kang, M. Kobayashi, K. Nakayama, N. Imadaa and T. Honjo. 2003. Suppression of sexual behavior in male Japanese medaka (*Oryzias latipes*) exposed to 17b-estradiol. Chemosphere 50: 429–436.

Pandey, A.K., K.C. George and M.P. Mohamed. 1993. Effects of mercuric chloride on the thyroid gland of *Liza parsia* (Hamilton- Buchanan). J. Adv. Zool. 14: 15–19.

Pandey, A.K., K.C. George and M.P Mohamed. 1995. Effect of DDT on the thyroid gland of the mullet *Liza parsia* (Hamilton-Buchanan). J. Mar. Biol. Assoc. India. 37: 287–290.

Pankhurst, N.W. and G. Van Der Kraak. 2000. Evidence that acute stress inhibits ovarian steroidogenesis in rainbow trout *in vivo*, through actions of cortisol. Gen. Comp. Endocrinol. 117: 225–237.

Pait, A.S. and J.O. Nelson. 2002. Endocrine disruption in fish an assessment of recent research and results. NOAA Techinical memorandum. 63p.

Palstra, A.P., V.J. van Ginneken, A.J. Murk and G.E. van den Thillart. 2006. Are dioxin like contaminants responsible for the eel (*Anguilla anguilla*) drama? Naturwissenschaften. 93: 145–148.

Park, C.B., J. Aoki, J.S. Lee, M. Nagae, Y.D. Lee, Y. Sakakura, A. Hagiwara and K. Soyano. 2010. The effects of 17β-estradiol on various reproductive parameters in the hermaphrodite fish *Kryptolebias marmoratus*. Aquat. Toxicol. 96: 273–279.

Park, J.W., J. Rinchard, F. Liu, T.A. Anderson, R.J. Kendall and C.W. Theodorakis. 2006. The thyroid endocrine disruptor perchlorate affects reproduction, growth, and survival of mosquitofish. Ecotoxicol. Environ. Safe. 63: 343–352.

Patiño, R., M.R Wainscott, E.I Cruz-Li, S. Balakrishnan, C. McMurry, V.S. Blazer and T.A. Anderson. 2003. Effects of ammonium perchlorate on the reproductive performance and thyroid follicle histology of zebrafish. Environ. Toxicol. Chem. 22: 1115–1121.

Patyna, P.J., R.A. Davi, T.F. Parkerton, R.P. Brown and K.P. Cooper. 1999. A proposed multigeneration protocol for Japanese medaka (*Oryzias latipes*) to evaluate effects of endocrine disruptors. Sci. Total. Environ. 233: 211–220.

Patyna, P.J., R.P. Brown, R.A. Davi, D.J. Letinski, P.E. Thomas, K.R. Cooper and T.F. Parkerton. Hazard evaluation of diisononyl phthalate and diisodecyl phthalate in a Japanese medaka multigenerational assay. Ecotoxicol. Environ. Safe. 65: 36–47.

Pawlowski, S., R. van Aerle, C.R. Tyler and T. Braunbeck. 2004. Effects of 17α- ethinylestradiol in a fathead minnow (*Pimephales promelas*) gonadal recrudescence assay. Ecotoxicol. Environ. Safe. 57: 330–345.

Phillips, B. and P. Harrison. 1999. Overview of the endocrine disrupters issue. Issues in environmental science and technology. Endocrine disrupting chemicals, The royal Society of Chemistry.

Phumyu, N., S. Boonanuntanasarn, A. Jangprai, G. Yoshizaki and U. Na-Nakorn. 2012. Pubertal effects of 17a-methyltestosterone on GH–IGF-related genes of the hypothalamic–pituitary–liver–gonadal axis and other biological parameters in male, female and sex-reversed Nile tilapia. Gen. Comp. Endocrinol. 177: 278–292.

Pottinger, T.G. 2003. Interactions of endocrine-disrupting chemicals with stress responses in wildlife. Pure Appl. Chem. 75: 2321–2333.

Purdom, C.E., P.A. Hardiman, V.J. Bye, N.C. Eno, C.R. Tyler and J.P. Sumpter. 1994. Estrogenic effects of effluents from sewage treatment works. Chem. Ecol. 8: 275–285.

Power, D.M., L. Llewellyn, M. Faustino, M.A. Nowell, B.T. Bjornsson, I.E. Einarsdottir, A.V.M. Canario and G.E. Sweeney. 2001. Thyroid hormones in growth and development of fish. Comp. Biochem. Physiol. C 130: 447–459.

Quabius, E.S., P.H.M. Balm and S.E.W. Bonga. 1997. Interrenal stress responsiveness of Tilapia (*Oreochromis mossambicus*) is impaired by dietary exposure to PCB 126. Gen. Comp. Endocrinol. 108: 472–482.

Raimondo, S., B.L. Hemmer, L.R. Goodman and G.M. Cripe. 2009. Multiple generation exposure of the estuarine sheepshead minnow, *Cyprinodont variegates*, to 17β-estradiol II: population effects through two life cycles. Environ. Toxicol. Chem. 28: 2397–2408.

Ram, R.N. and A.G. Sathyanesean. 1984. Effect of mercuric chloride on thyroid function in the teleost fish, *Channa punctatus* (Bloch). Matysa. 9–10: 194–196.

Ramakrishnan, S. and N.L. Wayne. 2008. Impact of bisphenol-A on early embryonic development and reproductive maturation. Reprod. Toxicol. 25: 177–183.

Rasier, G., A.S. Parent, A. Gerard, M.C. Lebrethon and J.P. Bourguignon. 2007. Early maturation of gonadotropin-releasing hormone secretion and sexual precocity after exposure of infantile female rats to estradiol or dichlorodiphenyltrichloroethane. Biol. Reprod. 77: 734–742.

Rhee, J.S., R.O. Kim, J.S. Seo, H.S. Kang, C.B. Park, K. Soyano, J. Lee, Y.M. Lee and J.S. Lee. 2010. Bisphenol A modulates expression of gonadotropin subunit genes in the hermaphroditic fish, *Kryptolebias marmoratus*. Comp. Biochem. Physiol. C 152: 456–466.

Safe, S. and V. Krishnan. 1995. Chlorinated hydrocarbons: estrogens and antiestrogens. Toxicol. Lett. 82–83: 731–736.

Sage, M. and N.R. Bromage. 1970. Interactions of TSH and thyroid cells with gonadotropic cells in poecilid fishes. Gen. Comp. Endocrinol. 14: 137–140.

Saglio, P. and S. Trijasse. 1998. Behavioral responses to atrazine and diuron in goldfish. Arch. Environ. Contam. Toxicol. 35: 484–491.

Sayed, A.E.D., U.M. Mahmouda and I.A. Mekkawy. 2012. Reproductive biomarkers to identify endocrine disruption in *Clarias gariepinus* exposed to 4-nonylphenol. Ecotoxicol. Environ. Safe. 78: 310–319.

Schäfer, S. and A. Köhler. 2009. Gonadal lesions of female sea urchin (*Psammechinus miliaris*) after exposure to the polycyclic aromatic hydrocarbon phenanthrene. Mar. Environ. Res. 68: 128–136.

Schmidt, F., S. Schnurr, R. Wolf and T. Braunbeck. Effects of the anti-thyroidal compound potassium-perchlorate on the thyroid of the zebrafish. Aquat. Toxicol. 109: 47–58.

Scholz, S. and I. Mayer. 2008. Molecular Biomarkers of endocrine disruption in small model of fish. Mol. Cell. Endocrinol. 293: 57–70.

Schreiber, A.M. 2006. Asymmetric craniofacial remodeling and lateralized behavior in larval flatfish. J. Exp. Biol. 209: 610–621.

Schreiber, A.M. and J.L. Specker. 1998. Metamorphosis in the summer flounder (*Paralichthys dentatus*); stage-specific development response to altered thyroid status. Gen. Comp. Endocrinol. 111: 156–166.

Schussler, G.C. 2000. The thyroxine-binding proteins. Thyroid. 10: 141–149.

Shappell, N.W., K.M. Hyndman, S.E. Bartell and H.L. Schoenfuss. 2010. Comparative biological effects and potency of 17β- and 17α-estradiol in fathead minnows. Aquat. Toxicol. 100: 1–8.

Shaw, J.R., K. Gabor, E. Hand, A. Lankowski, L. Durant and R. Thibodeau. 2007. Role of glucocorticoid receptor in acclimation of killifish (*Fundulus heteroclitus*) to seawater and effects of arsenic. Am. J. Physiol. Regul. Integr. Comp. Physiol. 292: 1052–1060.

Shi, X., C. Liu, G. Wu and B. Zhou. 2009. Waterborne exposure to PFOS causes disruption of the hypothalamus-pituitary-thyroid axis in zebrafish larvae. Chemosphere. 77: 1010–1018.

Shukla, J.P. and K. Pandey. 1984. Impaired spermatogenesis in arsenic treated freshwater fish, *Colzsa fasczatus* (bl. and sch.). Toxicol. Lett. 21: 191–195.

Shukla, L. and A.K. Pandey. 1986. Restitution of thyroid activity in the DDT-exposed *Sarotherodon mossambicus*: A histological and histochemical profile. Water Air Soil Pollut. 27: 225–236.

Sinha, N., B. Lal and T.P. Singh. 1991. Pesticides induced changes in circulating thyroid hormones in the freshwater catfish, *Clarias batrachus*. Comp. Biochem. Physiol. C 100: 107–110.

Singh, P.B. and V. Singh. 2007. Exposure and recovery response of isomers of HCH, metabolites of DDT and estradiol-17b in the female catfish, *Heteropneustes fossilis*. Environ. Toxicol. Pharmacol. 24: 245–251.

Singh, T.P. 1969. Maintenance of thyroid activity with some steroids in hypophysectomized and gonadectomized catfish, *Mystis vittatus* (Bloch). Experientia. 25: 431.

Soontornchat, M.H., P.S. Cooke and L.G. Hansen. 1994. Environ. Health Perspect. 102: 568–571.

Spanò, L., C.R. Tyler, R. van Aerle, P. Devos, S.N.M. Mandiki, F. Silvestre, J.P. Thomé and P. Kestemont. 2004. Effects of atrazine on sex steroid dynamics, plasma vitellogenin concentration and gonad development in adult goldfish (*Carassius auratus*). Aquat. Toxicol. 66: 369–379.

Stocco, D.M. 2000. Intramitochondrial cholesterol transfer. Biochim. Biophys. Acta. 1486: 184–197.

Sun, L., J. Zha, P.A. Spear and Z. Wang. 2007. Tamoxifen effects on the early life stages and reproduction of Japanese medaka (*Oryzias latipes*). Environ. Toxicol. Pharmacol. 24: 23–29.

Swapna, I. and B. Senthilkumaran. 2009. Influence of ethynylestradiol and methyltestosterone on the hypothalamo–hypophyseal–gonadal axis of adult air-breathing catfish, *Clarias gariepinus*. Aquat. Toxicol. 95: 222–229.

Szczerbik, P.T., M. Mikołajczyk, Sokołowska-Mikołajczyk, M. Socha, J. Chyb and P. Epler. 2006. Influence of long-term exposure to dietary cadmium on growth, maturation and reproduction of goldfish (subspecies: Prussian carp *Carassius auratus gibelio* B.). Aquat. Toxicol. 77: 126–135.

Tanaka, J.N. and J.M. Grizzle. 2002. Effects of nonylphenol on the gonadal differentiation of the hermaphroditic fish, *Rivulus marmoratus*. Aquat. Toxicol. 57: 117–125.

Taurog, A. 2000. Hormone synthesis: thyroid iodine metabolism. *In*: L. Braverman and R. Utiger (eds.). The Thyroid: a fundamental and clinical text. Williams & Wilkins, Philadelphia, Lippincott. pp. 61–85.

Teles, M., M. Pacheco and M.A. Santos. 2006. Biotransformation, stress and genotoxic effects of 17β-estradiol in juvenile sea bass (*Dicentrarchus labrax* L.). Environ. Internat. 32: 470–477.

Thibaut, R. and C. Porte. 2004. Effects of endocrine disruptors on sex steroids synthesis and metabolismo pathways in fish. J. Steroid Biochem. Mol. Biol. 92: 485–494.

Thomas, K., B. Erik and B. Mark. 2005. 17-ethinylestradiol reduces the competitive reproductive fitness of the male guppy (*poecilia reticulata*). Biol. Reprod. 72: 150–156.

Thorpe, K., M. Hetheridge, T.H. Hutchinson, M. Scholze, J.P. Sumpter and C.R. Tyler. 2001. Environ. Sci. Technol. 35: 2476–2481.

Tillitt, D.E., D.M. Papoulias, J.J. Whyte1 and C.A. Richter. 2010. Atrazine reduces reproduction in fathead minnow (*Pimephales promelas*). Aquat. Toxicol. 99: 149–159.

Tilton, S.C., C.M. Forana and W.H. Bensond. 2003. Effects of cadmium on the reproductive axis of Japanese medaka (*Oryzias latipes*). Comp. Biochem. Physiol. C 136: 265–276.

Tindall, A.J., I.D. Morris, M.E. Pownall and H.V. Isaacs. 2007. Expression of enzymes involved in thyroid hormone metabolism during the early development of *Xenopus tropicalis*. Biol. Cell. 99: 151–163.

Toppari, J., J.C. Larsen, P. Christiansen, A. Giwercman, P. Grandjean, L.J. Guillette, B. Jegou, T.K. Jensen, P. Jouannet, N. Keiding, H. Leffers, J.A. McLachlan, 0. Meyer, J. Muller, E.R. Meyts, T. Scheike, R. Sharpe, J. Sumpter and N.E. Skakkebaek. 1996. Male reproductive health and environmental xenoestrogens. Environ. Health Perspect. 104: 741–803.

Uren-Webster, T.M., C. Lewis, A.L. Filby, G.C. Paull and E.M. Santos. 2010. Mechanisms of toxicity of di(2-ethylhexyl) phthalate on the reproductive health of male zebrafish. Aquat. Toxicol. 99: 360–369.

Van den Belt, K., R. Verheyen and H. Witters. 2003. Effects of 17a-ethynylestradiol in a partial life-cycle test with zebrafish (*Danio rerio*): effects on growth, gonads and female reproductive success. Sci. Total Environ. 309: 127–137.

Vázquez, G.R., F.J. Meijide, R.H. Da Cuña, F.L. Lo Nostro, Y.G. Piazzab, P.A. Babay, V.L. Trudeau, M.C. Maggese and G.A. Guerrero. 2009. Exposure to waterborne 4-tert-octylphenol induces vitellogenin synthesis and disrupts testis morphology in the South American freshwater fish *Cichlasoma dimerus* (Teleostei, Perciformes). Comp. Biochem. Physiol. C 150: 298–306.

Velasco-Santamaría, Y.N., B. Korsgaard, S.S. Madsenb and P. Bjerregaard. 2011. Bezafibrate, a lipid-lowering pharmaceutical, as a potential endocrine disruptor in male zebrafish (Danio rerio). Aquat. Toxicol. 105: 107–118.

Volz, D.C., D.C. Bencic, D.E. Hinton, J.M. Law and S.W. Kullman. 2005. 2,3,7,8-Tetrachlorodibenzo-p-dioxin (TCDD) induces organ-specific differential gene expression in male Japanese medaka (*Oryzias latipes*). Toxicol Sci. 85: 572–584.

Walpita, C.N., S. Van der Geyten, E. Rurangwa and V.M. Darras. 2007. The effect of 3,5,3-triiodothyronine supplementation on zebrafish (*Danio rerio*) embryonic development and expression of iodothyronine deiodinases and thyroid hormone receptors. Gen. Comp. Endocrinol. 152: 206–214.

Wang, R.L., D. Bencic, J. Lazorchak, D. Villeneuve and G.T. Ankley. 2011. Transcriptional regulatory dynamics of the hypothalamic–pituitary–gonadal axis and its peripheral pathways as impacted by the 3-beta HSD inhibitor trilostane in zebrafish (*Danio rerio*). Ecotoxicol. Environ. Safe. 74: 1461–1470.

Wannemacher, R., A. Rebstock, E. Kulzer, D. Schrenk, and K.W. Bock. 1992. Effects of 2,3,7,8-tetrachlorodibenzo-p-dioxin on reproduction and oogenesis in zebrafish . Chemosphere. 24: 1361–1368.

Wu, W.Z., W. Li, Y. Xu and J.W. Wang. 2001. Long-term toxic impact of 2,3,7,8-tetrachlorodibenzo-p-dioxin on the reproduction, sexual differentiation, and development of different life stages of *Gobiocypris rarus* and *Daphnia magna*. Ecotoxicol. Environ. Saf. 48: 293–300.

Xu, Y., J. Zhang, W. Li, K.W. Schramm and A. Kettrupp. 2002. Endocrine effects of sublethal exposure to persistent organic pollutants (POPs) on silver carp (*Hypophthalmichthys molitrix*). Environ. Pollut. 120: 683–690.

Yada, T. and T. Nakanishi. 2002. Interaction between endocrine and immune systems in fish. Int. Rev. Cytol. – a Survey of Cell Biology. 220: 35–92.

Yadav, A.K. and T.P. Singh. 1986. Effect of pesticide on circulating thyroid hormone levels in the freshwater catfish, *Heteropneustes fossilis* (Bloch). Environ. Res. 39: 136–142.

Yamaguchi, S., C. Miura, A. Ito, T. Agusa, H. Iwata, S. Tanabe, B.C. Tuyen and T. Miura. 2007. Effects of lead, molybdenum, rubidium, arsenic and organochlorines on spermatogenesis in fish: Monitoring at Mekong Delta area and *in vitro* experiment. Aquat. Toxicol. 83: 43–51.

Yan, Z., G. Lu and J. He. 2012. Reciprocal inhibiting interactive mechanism between the estrogen receptor and aryl hydrocarbon receptor signaling pathways in goldfish (*Carassius auratus*) exposed to 17β-estradiol and benzo[a]pyrene. Comp. Biochem. Physiol C 156: 17–23.

Yang, F.X., Y. Xu and Y. Hui. 2006. Reproductive effects of prenatal exposure to nonylphenol zebrafish (*Danio rerio*). Comp. Biochem. Physiol. C 142: 77–84.

Yang, L., J. Zha, W. Li, Z. Li and Z. Wang. 2010a. Atrazine affects kidney and adrenal hormones (AHs) related genes expressions of rare minnow (*Gobiocypris rarus*). Aquat. Toxicol. 97: 204–211.

Yang, L., J. Zha, X. Zhang, W. Li, Z. Li and Z. Wang. 2010b. Alterations in mRNA expression of steroid receptors and heat shock proteins in the liver of rare minnow (*Grobiocypris rarus*) exposed to atrazine and p,p_-DDE. Aquat. Toxicol. 98: 381–387.

Young, G., M. Kusakabe, I. Nakamura, M. Lokman and F.W. Goetz. 2005. Gonadal steroideogenesis in teleost fish. *In:* P. Melamed and N. Sherwood (eds.). Hormones and their receptor in Fish Reproduction. World Scientific Publishing, Singapore, pp. 155–223.

Zaccaroni, A., M. Gamberoni, L. Mandrioli, R. Sirri, O. Mordenti, D. Scaravelli, G. Sarli, and A. Parmeggiani. 2009. Thyroid hormones as a potential early biomarker of exposure to 4-nonylphenol in adult male shubunkins (*Carassius auratus*).Sci. Total Environ. 407: 3301–3306.

Zanotelli, V.R.T., S.C.F. Neuhauss and M.U. Ehrengruber. 2009. Long-term exposure to bis(2-ethylhexyl) phthalate (DEHP) inhibits growth of guppy fish (*Poecilia reticulata*). J. Allied Toxicol. 30: 29–33.

Zha, J., Z. Wang, N. Wang and C. Ingersoll. 2007. Histological alternation and vitellogenin induction in adult rare minnow (*Gobiocypris rarus*) after exposure to ethynylestradiol and nonylphenol. Chemosphere. 66: 488–495.

Zhaobin, Z. and H. Jianying. 2008. Effects of p,p-DDE exposure on gonadal development and gene expression in Japanese medaka (*Oryzias latipes*). J. Environ. Sci. 20: 347–352.

Zheng, R., C. Wang, Y. Zhao, Z. Zuo and Y. Chen. 2005. Effect of tributyltin, benzo(*a*)pyrene and their mixture exposure on the sex hormone levels in gonads of cuvier (*Sebastiscus marmoratus*). Environ. Toxicol. Pharmacol. 20: 361–367.

Zhou, J., X.S. Zhu and Z.H. Cai. 2009. Endocrine disruptors: an overview and discussion on issues surrounding their impact on marine animals. J. Mar. Anim. Ecol. 2: 7–17.

Zimmermann, M.B. 2007. The adverse effects of mild-tomoderate iodine deficiency during pregnancy and childhood: a review. Thyroid 17: 829–835.

CHAPTER 11

Nanoecotoxicology in Fish Species

Juliane Ventura-Lima,[1,2] *Alessandra Martins da Rocha,*[1]
Marlize Ferreira-Cravo,[2] *André Luís da Rosa Seixas,*[1]
Carmen Luiza de Azevedo Costa,[1] *Isabel Soares
Chaves,*[2] *Josencler Luis Ribas Ferreira,*[1,2] *Rafaela Elias
Letts,*[1,3] *Lucas Freitas Cordeiro,*[1] *Glauce R. Gouveia*[2] *and
José María Monserrat*[1,2,]*

Introduction

Nanotechnology has been developing rapidly in the last few years and, as a consequence, nanomaterials (NM) already are being used in several commercial products. As NM possess extremely low dimensions (up to 100 nm), they present a high surface/volume relationship, making possible their application in cosmetics, foods, medicine and electronics, among others (Wang et al. 2011). Owing to increased production and applications of NM, the aquatic environment is a potential sink and the exposure of these NM with aquatic biota is expected. In this context, some questions arise: (a) Do

[1]Programa de Pós-Graduação em Ciências Fisiológicas - Fisiologia Animal Comparada, Universidade Federal de Rio Grande – FURG, Rio Grande, Brasil.
[2]Instituto de Ciências Biológicas (ICB), Universidade Federal do Rio Grande–FURG, Rio Grande, Brasil.
[3]Universidad Peruana Cayetano Heredia, Lima, Perú.
*Corresponding author: josemmonserrat@pesquisador.cnpq.br

NM have the potential to induce toxicological effects in aquatic organisms? and (b) what are the toxicity mechanisms elicited by these NM?

Some studies have evaluated the effects of several NM in different aquatic organisms; however, the toxicity effects of materials at the nano-scale dimension has sometimes shown differences to those induced by bulk materials of the same composition (Li et al. 2008). Thanks to their unique characteristics, NM have been attracted great interest in application of many biological fields (Kim et al. 2011). However, the physical-chemical properties, large surface area, chemical activity and high penetration power of NM can be also regarded as potentially harmful to both environment and living organisms (Matranga and Corsi 2012).

The development of nanotechnology is ahead of the evaluation of the impact of NM on the environment, plants and animals. As a matter of fact, data about potential effects of NM are still limited, especially concerning aquatic organisms. Therefore, there is a need to raise investments (both economically and intellectually) in toxicological studies that assess the effects of NM on aquatic organisms from an ecotoxicological perspective. For this, it is necessary to consider some aspects related to the characteristics of NM, route and condition of exposure and mostly the species analyzed.

NM are grouped according to their chemical composition such as carbon (fullerene, graphene and nanotubes), metallic oxides (for example: titanium dioxide, zinc oxide, cerium dioxide), metals (silver, gold) and semiconductors (for example: quantum dots) (Krysanov et al. 2010). Characteristics such as shape, mass, surface area, aggregation, agglomeration, solubility, and surface chemical can influence the NM's biological fate; for example, if the particle will be dispersed, deposited or eliminated (Moore 2006). Consequently, these characteristics will influence the biological effects such as inflammatory responses, cytotoxicity, immunomodulation, genotoxicity, cell transformation and oxidative stress (Bergamaschi 2012).

Although nanotoxicological studies do not accompany the development of nanotechnology at the same scale, some studies have analyzed the effect of several NM in different species including fish. In this chapter, we will discuss some important topics such as sources and environmental fate, routes of exposure, toxicity of NM considering biotic and abiotic factors, interaction of NM with other pollutants, mechanisms of action and biological effects. This body of information certainly will contribute towards clarifying some aspects that are extremely important for the development of nanotoxicology.

In this context, the use of fish as animal models can be compared with other organisms. The review of Kahru and Dubourguier (2010), analyzed the toxicological knowledge of several biological groups, shown in Table 11.1.

Table 11.1. Sensitivity of different organisms to inorganic (nano TiO_2, nano ZnO, nano CuO, nano Ag) and organic (SWCNT, MWCNT, C_{60}) nanomaterials. Up arrows indicate the most sensitive organism group. Gray boxes stand for existing toxicological data. White boxes indicate absence of toxicological information. Data based on the study of Kahru and Dubourguier (2010). SWCNT: single-wall carbon nanotubes. MWCNT: multi-wall carbon nanotubes. C_{60}: fullerene.

Organism	NanoTiO$_2$	Nano ZnO	Nano CuO	Nano Ag	SWCNT	MWCNT	C$_{60}$
Crustaceans				↑		↑	
Bacteria							
Algae	↑	↑	↑	↑			
Fish							
Ciliates							↑
Nematodes							
Yeasts							

As can be seen, toxicological data for both organic and inorganic NM exist for fish organisms, although it seems less sensitive than other biological groups such as algae. It is also important to consider quantitative aspects of the published studies in econanotoxicology and the impact on the scientific community. Using SCOPUS (www.scopus.com) as a database of scientific publications, surveys were carried out on a number of papers and the *h* index registered (Hirsch 2005), using key-words like "nanoparticles", "toxicity" and the name of each organism group shown in Table 11.1.

It is important to note that, according to data from Fig. 11.1, fish is the second organism group in terms of number of published articles and

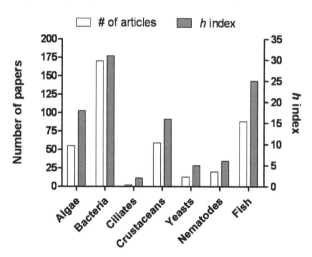

Figure 11.1. Number of published papers (white bars) and *h* index (grey bars) registered on 06/28/2012 in SCOPUS database (www.scopus.com).

h index, surpassed only by the bacteria group. However, it is important to state that several papers on bacteria deal with antibiotic responses induced by NM, more focused on biomedical than on ecotoxicological or environmental aspects. It is evident that nanotoxicology is still "crawling" and more studies should be performed with different species and NM to extend our knowledge about biological effects induced by NM in fish and other aquatic organisms and their interaction with the environment.

Variety and quality of toxicological information are thus needed in order to implement new policies considering the release of these NM into the environment and the associated risks. The current state of knowledge and scientific needs will also be discussed in this chapter.

Source and Environmental Fate of Nanomaterials

NM are abundant in nature, produced in processes as volcanic eruption, photochemical reactions, forest fires and simple erosion, for example. However, many NM are of anthropogenic origin, as they are by-products of cooking, ore refining and smelting, simple combustion, chemical manufacturing, welding, combustion in vehicles, airplane engines and others. In this context, Hoyt and Mason (2008) suggest that the term "nanoparticles" (NP) or "nanomaterials" (NM) should only be applied to those from engineered technology, such as fullerenes, metal oxides, carbon nanotubes, etc. (Buzea et al. 2007).

Engineered nanomaterials present many unique properties when compared to conventional formulations of the same material (Ferrari 2005; Vasir et al. 2005; Qin et al. 1999; Webster et al. 1999, 2000). These singularities led to their use in a wide range of fields, from medical applications to environmental sciences. However, there is no complete and comprehensive review of nanotechnology applications, likely due to the rapid development of this field (Arora et al. 2012).

NM are increasingly being used for commercial purposes and are commonly used in many products on the market as in sporting goods, cosmetics (Farkas et al. 2011), stain-resistant clothing, surface disinfectants, food additives, medical applications (Ferrari 2005; Vasir et al. 2005) besides their general use as microelectronics, synthetic rubber, catalytic compounds, photographic supplies, inks and pigments, coatings and adhesives, ultrafine polishing compounds, UV absorbers for sun screens, synthetic bone, hazardous chemical neutralizers, diesel and fuel additives, fluids iron, optical fiber cladding and other related products (Buzea et al. 2007).

A recent report from a project on Emerging Nanotechnologies shows more than 1,000 nanotechnology based consumer products were on the market in 2009 (Woodrow Wilson International Center for Scholars 2009), where silver nanoparticles (AgNP) were the most frequently used NM due to their antibacterial/antifungal effects (Pal et al. 2007; Martínez-Castanõñ et al. 2008). Following this study, major areas of use of AgNP in consumer products include fitness and health care products such as cosmetics, cloth and sporting devices, personal care products and others like air sanitizer sprays, socks, pillows, slippers, detergent, air filters, coatings of refrigerators, vacuum cleaners, washing machines, food storage containers, cellular phones, and even in liquid condoms (Buzea et al. 2007). Other engineered NM that have been in focus are titanium dioxide (nano-TiO$_2$), zinc oxide (nano-ZnO) and aluminum oxide (nano-Al$_2$O$_3$), which are widely used in many products such as white pigment, food colorant, sunscreens and cosmetic creams (Donaldson et al. 2004) and sterilizing equipment of environmental microorganisms in health care facilities (Shimtani et al. 2006).

With their large production and widespread application, quantities of NM are growing and it is expected that significant levels of such materials will inevitably enter the environment and end up in water either from industrial and domestic products, and wastes containing NM, both directly into rivers and lakes via outdoor use of sunscreens, or indirectly via surface run-off, domestic or industrial wastewater (Zhang 2003; Aitken et al. 2006; Moore 2006; Vaseashta et al. 2007; Baun et al. 2008; Wiesner et al. 2009; Gottschalk et al. 2011; Bhawana and Fulekar 2012). Besides, the use of NM foreseen for applications in environmental remediation techniques is another expected way of its direct release into the aquatic environment. However, little is known about the environmental fate and effects of NM leached from consumer products. Also, unfortunately, the quantitative detection of NM and the distinction between naturally NM occurring in the environment are still extremely limited (Limbach et al. 2008; Tiede et al. 2008; Hasellöv and Kaegi 2009; Gottschalk et al. 2011).

In this regard, scientific efforts have been particularly focused on aquatic pollution (Battin et al. 2009; Farre et al. 2009; Perez et al. 2009). Although it has shown that NM can promote some level of protection against oxidative stress (Ahlbom et al. 2009), many studies have observed that NM can accumulate and are toxic (Hussain et al. 2005; Vinardell 2005; Adams et al. 2006; Dhawan et al. 2006; Griffitt et al. 2007; Heinlaa et al. 2008). Thus, there are serious concerns over NM risks once released into the environment; understanding the transfer, transport and fate of NM in the environment is significant for evaluating their environmental and health impact (Helland et al. 2007; Nowack and Bucheli 2007).

Given its complexity, the mineral and organic compositions and structural heterogeneity of natural media must be taken into account to understand the transport and fate of NM under natural conditions. Almost nothing is known about how NM interact with soils and sediments (Oberdörster et al. 2006; Wiesner et al. 2006). Solid and dissolved matter can be powerful geosorbents that affect NM mobility. Recently, it has been shown that organic matter can interact with NM in a way that may influence their transport and dispersion (Hyung et al. 2007; Yang et al. 2009), and some environmental factors, such as pH and ionic strength, may determine if NM are attached within or transported out of soils and sediments towards the water (Luthy et al. 1997; Brant et al. 2005). Furthermore, behavior of NM in the environment can be influenced by their structure, concentration, physical-chemical properties and technically intended modifications in functionalization or coating characteristics (Huuskonen 2002; Schwarzenbach et al. 2003; Mathivanan et al. 2012).

NM employed in laboratory assays have been so far characterized by using a multiplicity of techniques, but accurate analysis for detection in environmental samples still requires the development of new and effective methods, and the use of complex detection techniques of variable resolution power (nanometric, micrometric and metric). Clearly, available analytical methods for characterizing NM in the environment must be improved in terms of sensitivity and selectivity (Tiede et al. 2009; Simonet and Valcárcel 2009).

As addressed above, consumer products are a likely source of NM entering into the environment. According to Gottschalk et al. (2009), it must be assumed that different release pathways of these materials to the environment can occur during the product's life-cycle. Indeed, some studies have already shown the release of NM from commercially available products as nanosilver washing machines and nano-textiles products (Benn and Westerhoff 2008; Kaegi et al. 2008; Geranio et al. 2009; Kaegi et al. 2010; Farkas et al. 2011). At the same time, some other attempts have been made to predict future environmental concentrations of some of the most frequently used NM through computational models using probabilistic/stochastic methods of environmental exposure analysis (McKone and Bogen 1991; van der Voet and Slob 2007; Mueller and Nowack 2008). These approaches intend to consider all possible inputs in the models, also covering extreme events. They also provide an insight into the frequency probability of each simulated outcome. These probabilistic modeling approaches have been used to describe flows of environmental NM in Switzerland (Gottschalk et al. 2009; Gottschalk et al. 2010).

According to a few models also used to assess aquatic exposure to NM, probabilistic methods of environmental exposure analysis allow for explaining some inconsistency and variability of input parameters by using probability (or density) distributions. Those input distributions may be constructed based on empirical data, on expert judgment or on a combination of these sources (Scheringer et al. 1999; Blaser et al. 2008; Arvidsson et al. 2011; Tuoriniemi et al. 2012).

For the future, there is a need to check and amend the methodological variability of the data applied to existing models that assess the fate of NM in the environment. It is important that standardization of the methodology be developed and implemented so that variability may be maintained at a minimum, with results widely accepted and replicated.

Besides, design and development of models to assess the release of NM and their fate in the environment are essential to estimate environmental exposure. Such estimations have to cover diffuse emissions from a large number of relevant NM containing products and life-cycle stages. These include, but are not limited to, nanoparticles release into the environment from NM production, nanoparticles incorporation into products and storage, use, waste generation and disposal of such products (Gottschalk et al. 2011). These models need to be challenged with high quality data and modified as necessary to take into account any effect that is "nano-specific". This will be an iterative and continuous process, that should lead to improvements in models for faster evaluations.

Routes of Exposure

The exposure routes to xenobiotics, including NM, in juvenile and adult fish are the gills surface and food intake. In fish embryos, the egg surface is the target, although the chorion represents a protection, with pores of 0.5–0.7 µm in diameter (Rawson et al. 2000).

Exposure through Water

Three fish species have been the most studied using NM exposure via water: the medaka fish (*Oryzias latipes*), the rainbow trout (*Oncorhynchus mykiss*) and the zebrafish (*Danio rerio*). In the adult medaka fish exposed to latex nanoparticles, accumulation mainly occurred in the gills and intestine, but they were also detected in brain, testis, liver and blood (Kashiwada 2006). In another study, morphological and histopathological changes were observed in the gills and intestine of medaka fish exposed to nano-iron, besides a disturbance in the activity of the antioxidant enzyme superoxide dismutase (SOD) and levels of reduced glutathione (GSH) in liver and brain (Li et al. 2009). Adult rainbow trouts exposed to single-wall carbon nanotubes (SWCNT) also presented gill histopathologies and augmented ventilation rate, increased activity of Na^+K^+-ATPase in the gills and intestine, and pathologies in the brain and liver (Smith et al. 2007). TiO_2 nanoparticles injured rainbow trout gills (Federici et al. 2007) and Ag nanoparticles were also concentrated in the gills, inducing the expression of *cyp1a2* (Scown et al.

2010). In zebrafish gills, Cu nanoparticles exposure enlarged gill filaments and induced a distinctive gene expression profile (Griffitt et al. 2009). These studies demonstrate that exposure to NP via water affects mainly gills and intestines at both morphological, biochemical and genetic level.

Exposure through Food Intake

Few studies have considered NM incorporation through food. Juvenile rainbow trouts that consumed ration containing SWCNT for six weeks had only one modification on the several parameters analyzed, an increase in brain TBARS (lipid peroxidation) at week four (Fraser et al. 2011). The crustacean *Daphnia magna*, treated with TiO_2 nanoparticles, transferred these NP via diet to zebrafish, although no biomagnification was observed (Zhu et al. 2010).

Nanomaterials Exposure to Embryos and Larvae

Although there are few studies in the area, they make it evident that exposure to nanoparticles may influence the development of embryos and larvae of fish. Zebrafish embryos exposed to fullerene showed an increase in body malformations, pericardial edema, necrotic and apoptotic death and mortality (Usenko et al. 2008). Furthermore, at 36 hours and 48 hours after fertilization, modifications occurred in the expression of several genes, like ferritin, α-tocopherol transport protein, heatshock protein 70, glutathione-S-transferase and glutamate cysteine ligase (Usenko et al. 2008). In addition, studies with TiO_2 nanoparticles were implicated in impairment of zebrafish reproduction (Wang et al. 2011), with a reduction in the number of zebrafish eggs (Wang et al. 2011) and affected larval swimming parameters (Chen et al. 2011a). The medaka embryos exposed to TiO_2 nanoparticles also presented pericardial edema and other indicators of toxicity such as premature hatching and, after hatching, moribund swimming behavior and mortality (Paterson et al. 2011).

Toxicological Effects of Nanomaterials on Fish Species

Understanding the effects of NM on fish is therefore an important aspect when considering the effects of NM on the aquatic environment as a whole. Potential routes of uptake for NM in fish include absorption via the gill epithelia, via the gut epithelia as a result of dietary exposure and drinking or via the skin (Handy et al. 2008), as explained in the previous section.

Carbon-based Nanomaterials

One of the first *in vivo* exposure studies of fish to NM to emerge was one that analyzed fullerene toxicity responses in juvenile largemouth bass *Micropterus salmoides* (Oberdorster 2004). The finding that oxidative stress was detected in the brain alerted the scientific community to the potential hazards associated with the release of NM into the aquatic environment. Henry et al. (2007) conducted a similar study using larval zebrafish and found changes in gene expression and reduced survival in zebrafish exposed to fullerenes dispersed using the organic solvent tetrahydrofuran (THF), as well as in water to which THF had been added and evaporated off. Analysis of the THF-fullerene media and THF-water by gas chromatography-mass spectrometry did not detect THF but detected an oxidation product, butyrolactone, which has an LC_{50} of 47 mg/L for zebrafish, suggesting that the effects observed in fish exposed to fullerene prepared in this way may be attributable to this degradation product. As a result of the issues surrounding the preparation of stable suspensions of fullerene, the present literature concerning *in vivo* exposure of fish species to fullerene does not yet provide clear information regarding the effects, as so many studies have employed the use of solvents to aid dispersion. However, some studies have in fact showed that aqueous suspensions of fullerene can in fact induce oxidative stress in fish or even in bacteria associated with mucous fish surface (Letts et al. 2011; Britto et al. 2012; Ferreira et al. 2012).

A number of studies have assessed the effects of carbon nanotubes in zebrafish embryos. Embryos treated with simple-walled carbon nanotubes (SWCNT) and double-walled carbon nanotubes (DWCNT) caused hatching delay at concentrations over 120 mg/L and 240 mg/L, although 99% of embryos hatched by 72 hours post fertilization (hpf). Embryonic development was not found to be affected and the authors suggested that as chorion pores are nano-scaled, they may provide a protective barrier against penetration by the micro-scaled carbon nanotubes (CNT) aggregates. Hatching delay was therefore attributed to trace levels of residual cobalt and nickel catalysts in the CNT. Carbon black nanoparticles were found to have no effect on hatching at similar concentrations (Cheng et al. 2007). Embryos treated with multi-walled carbon nanotubes (MWCNT) exhibited dose dependent increases in mortality and decreased hatching as well as bradycardia, slowed blood flow and apoptosis. Mortality of 100% was reached at a concentration of 200 mg/L and at concentrations of 60 mg/L and above, embryos showed deformation of the notochord and increased mucus production in the intrachorion region (Asharani et al. 2008a). The authors did not provide details of trace metal concentrations in nanotubes preparations; however, the concentrations where hatching delay was observed were similar to that seen in the study by Cheng et al. (2007).

A further study by Cheng et al. (2009) found that fluorescent-labeled MWCNT microinjected into zebrafish embryos at the 1-cell stage was distributed to all blastoderm cells and excluded from the yolk cell; after being introduced into the circulation system, it was removed from the body after 96 hours. In juvenile rainbow trout *Oncorhynchus mykiss* exposed to SWCNT, a dose-dependent rise in ventilation rate was observed. Accumulation of SWCNT aggregates was associated with gill mucus (Smith et al. 2007).

The results of these studies suggest that carbon-based NM, such as fullerenes and nanotubes, have the capacity to induce toxicity in aquatic vertebrates both as a function of their chemistry by inducing oxidative stress and as a result of their aggregation, probably causing physical damages.

Titanium Dioxide Nanoparticles

Exposure of zebrafish embryos to TiO_2 nanoparticles have so far been shown to be relatively non-toxic when compared with effects seen in exposures to carbon-based nanoparticles. No toxicity was observed in zebrafish embryos exposed to 30 nm TiO_2 comprised of 20% rutile and 80% anatase crystals at concentrations of up to 10 mg/L (Griffitt et al. 2008); a further study by the same research group found exposure of adult zebrafish to 1.0 mg/L of the same TiO_2 nanopartices caused no changes in molecular or histological parameters in zebrafish gills (Griffitt et al. 2009). In contrast, however, a 14-day semi-static exposure of rainbow trout resulted in gill oedema and thickening of the gill lamellae as well as decreases in Na^+, K^+-ATPase activity in the gills and intestine (Federici et al. 2007).

A subsequent 8-week oral exposure to rainbow trout by the same research group found that concentrations of up to 100 mg kg^{-1} in the food had no impact on growth. An increase in TiO_2 content in the liver and spleen was observed in the early stages of the exposure but no effect was seen on haematological parameters or TBARS (lipid peroxidation), suggesting that adverse effects of TiO_2 nanoparticles on rainbow trout are more severe as a result of water-borne exposure than by exposure via the diet (Handy et al. 2008). In a study using a rainbow trout gonadal cell line (RTG-2), no adverse effects were observed in cells at concentrations of up to 50 µg TiO_2 nanoparticles/mL, although increased levels of DNA strand breaks were reported after exposure to TiO_2 under UVA radiation (3 kJ/m^2) (Vevers and Jha 2008).

Very few studies in fish have examined the uptake and partitioning of TiO_2 nanoparticles within the body as result of exposure, probably due in part to the difficulties involved in measuring low levels of TiO_2 and limitations in analytical equipment. A study by Moger et al. (2008), however, used coherent anti-Stokes Raman Scattering (CARS) to examine the gills of rainbow trout exposed to 5,000 µg/L TiO_2 nanoparticles and confirmed the presence of small numbers of particle aggregates within the gill tissue. Two other studies have demonstrated the enhanced accumulation of the heavy metals arsenic and cadmium in the viscera and gills of carp when exposed to the presence of TiO_2 nanoparticles (Sun et al. 2007; Zhang et al. 2007).

Silver Nanoparticles

To date, the majority of *in vivo* exposures of silver nanoparticles to fish have been in zebrafish models. Effects in embryo exposures have shown similarities between effects seen with carbon-based nanoparticles although the exact mechanisms of toxicity have not yet been elucidated.

Zebrafish embryos exposed to 5–20 nm silver nanoparticles, capped with starch or bovine serum albumin (BSA) to aid dispersion, exhibited a dose-dependent increase in mortality and hatching delay and dose-dependent toxicity which was typified by larvae with deformations of the notochord, slow blood flow, pericardial oedema and cardiac arrhythmia. Distribution of silver nanoparticles in the brain, heart, yolk and blood was demonstrated by transmission electron microscopy (TEM) and apoptosis was also seen in 50% of the embryos treated with 50 µg/mL and above (Asharani et al. 2008b). Another study also found decreased hatch rate, notochord abnormalities and weak heart beat in embryos treated with 10–20 nm silver nanoparticles. Catalase activity was found to be increased in exposed embryos and the expression of Sel N1, a gene associated with notochord development and heart disease in zebrafish, was significantly lower in exposed fish. The study also confirmed the presence of silver ions (Ag^+) in the exposure media to which the authors attributed the detrimental effects seen (Yeo and Kang 2008).

Unlike the aggregates of carbon nanotubes that were unable to pass through chorion pores, silver nanoparticles of 5–46 nm have been shown to be transported in and out of chorion pore channels by Brownian diffusion (Lee et al. 2007). Another study that evaluated the effects of silver nanoparticles in zebrafish fry found that although 26 nm silver particles exhibited toxicity, silver ions were found to be over 300 times more toxic to zebrafish fry on a mass basis (Griffitt et al. 2008). This led to a further study where zebrafish were exposed to 26 nm silver nanoparticles at their previously ascertained no observable effect concentration (NOEC) of 1,000 µg/L. Analysis of global gene expression in the gills found differences in response between nanoparticle exposed fish and fish exposed to soluble silver ions, suggesting that the biological effects of exposure to silver nanoparticles do not appear to be driven solely by the release of silver ions (Griffitt et al. 2009).

Effects of Nanomaterials on Nervous System

In general, most molecules cannot cross the blood-brain barrier (BBB) but NM made of certain materials and with varying sizes can cross this barrier and enter into the brain or enter by olfactory bulb and these NM can be administered to the animal body via several routes including inhalation, oral administration and injection (Hu and Gao 2010).

Experiments performed with rats and fish have arisen the idea that carbon NM can be taken up by olfactory neurons and are translocated to the brain (Oberdorster 2004; Simkó et al. 2008; Belyanskaya et al. 2009); it was also shown that some NM can cross the BBB, affecting brain signaling linked to Alzheimer's and Parkinson's diseases and decrease of cognitive function (Wu et al. 2012).

The blood-brain-barrier (BBB) protects the central nervous (CNS) system from harmful xenobiotics and endogenous molecules but it has been shown that NM from the blood circulation may influence endothelial cell membrane and/or disrupt the BBB, inducing vesicular transport to gain access into CNS (Chen et al. 2008). Moreover, NM can induce oxidative stress generating free radicals that could disrupt the BBB and cause dysfunctions (Simkó and Mattsson 2010). According to Smith and colleagues (2007), SWCNT may damage the cardiovascular system of fish or alter the permeability of the BBB and when vascular lesions are associated with a vital organ such brain, these responses will increase the risk of stroke and mortality. Altered neurotransmitter levels (dopamine and serotonin) were observed in zebrafish larvae exposed to Ag^+ (released by silver nanoparticles), neurotransmitters that play important roles in reward, anxiety and sensorimotor integration such the deficits in swimming behavior (Powers et al. 2011). Also, inhibition of Na^+, K^+-ATPase activity in the rainbow trout brain was observed after exposure to TiO_2 nanoparticles (Ramsden et al. 2009) .

Abiotic Factors Influencing Nanomaterials Toxicity

The aggregation state, morphology and surface charge of NP in the aquatic environment determines its reactivity, transportation, fate and distribution, and consequently, its bioavailability. From an environmental point of view, the inter-relationship between NM and abiotic variables is a key issue because abiotic environmental factors influence and control the stability, aggregation dynamics and surface chemistry of NM and, consequently, its effects on fish and other organisms (Handy et al. 2008; Kusk et al. 2008; Peralta-Videa et al. 2011). Some representative abiotic factors are discussed below.

Ionic Strength and pH

The persistence of NM in the water column strongly depends on the formation of stable colloidal particles. The diffuse electrostatic double layer surrounding the particles which prevents aggregation is maintained by a balance between the van der Waals attraction forces and the electrostatic repulsion forces, as stated by the classic DLVO theory (Derjaguin and Landau 1941; Verwey and Overbeek 1948). The presence of ions in the medium compress the diffuse layer, increasing the attraction forces, decreasing the zeta potential and causing aggregation and sedimentation (Zha et al. 2002; Cosgrove 2005). The valence of the surrounding ions also determines the rate and size of the aggregation. Depending on the surface charge density of the NM, bivalent cations can cause faster and more pronounced aggregation than monovalent cations in the same concentrations. Likewise, ionic strength and pH of the medium interrelate with the particle surface charge in a way that changes in the pH decrease the electrolytes concentration necessary to cause aggregation (Chen and Elimelech 2006; Lowry and Wiesner 2007; French et al. 2009; El Badawy et al. 2010). These factors must be taken into account when predicting the impacts of NM in aquatic environments that present particularities as variations in pH, salinity and water hardness as seen in estuarine, coastal and some surface waters.

Organic Matter

Natural dissolved organic matter (DOM) is ubiquitous in aquatic environments and can reach up to 100 mg/L in fresh water (Paul et al. 2006). Depending on their sources, such as degradation of lignins, tanins and algae, DOM is composed by a complex mixture of humic substances (humic and fulvic acids) and non-humic compounds (as polysaccharides) that can react with ions forming colloids in water (Uyguner-Demirel and Bekbolet 2011). Surface coating of NM by DOM regulates the behavior of NM by enhancing the colloidal stability, mobility and deposition in water (Keller et al. 2010; Jiang et al. 2012; Kim et al. 2012). For organic NM, whose main mechanism of reaction is through π-π interactions, DOM coating increases steric repulsion and avoids particle aggregation. This is due, in part, to the content of phenolic substances as phenolate groups that enhance the colloidal stability. Other properties of DOM include the ability to promote redox reactions, which can alter NM chemistry in the environment (Hyung et al. 2007; Lin and Xing 2008; Chappell et al. 2009).

Effects of DOM on NM toxicity in aquatic organisms apparently depend on the species analyzed and are linked to the chemical nature of both humic and non-humic compounds. For example, harmful effects of multi-walled carbon nanotubes on zebrafish embryos, and quantum dots on daphnids have been reported to be ameliorated only by typically hydrophobic DOM (Lee et al. 2011; Kim et al. 2012). For microorganisms, humic acids also mitigated the toxicity of zero-valent iron and silver nanoparticles (Fabrega et al. 2009; Chen et al. 2011b). In all of these cases, the protection performed by DOM was attributed to NM coating that should prevent the direct contact of NM with organisms. However, it was also found that deleterious effects of CuO nNP to algae *Microcystis aeruginosa* were enhanced by pond water with high organic matter content, and this finding was ascribed to the intake of CuO NP bound to nutrients present in DOM (Wang et al. 2011).

Ultraviolet Radiation Incidence

Ultraviolet (UV) radiation is among the factors which may contribute to the increased toxicity of NM. These radiations comprise a range of electromagnetic wavelengths ranging between 200 and 400 nm. The organic NM fullerene (C_{60}) has strong light absorption within the solar spectrum, especially in the UV range and it was demonstrated that C_{60} undergoes surface modification or decomposition under sunlight or UV irradiation (Hwang and Li 2010; Hou and Jafvert 2009; Lee et al. 2009), suggesting that surface-oxidized C_{60} may be an important form of C_{60} in the aquatic environment. Authors like Kamat et al. (2000) have already reported that C_{60} generates reactive oxygen species in the presence of both UV and visible light.

The manufacture of NP such as zinc oxide, aluminum and titanium dioxide (TiO_2) finds wide use in cosmetics and sunscreens because of its ability to reflect, scatter and absorb UV radiation, preventing sunlight related skin disorders. Although TiO_2 has been considered to be biologically inert, recent reports have demonstrated that when TiO_2 is under UVA or UVB radiation, oxidative, genotoxic and cytotoxic effects are observed in cells. UV radiation by itself is known to cause several deleterious effects, ranging from molecular and tissue damage to population level effects and include alterations of relevant biological molecules such as proteins, lipids and DNA (Sinha and Häder 2002; Gouveia et al. 2005).

In this sense, some authors have been concerned with the understanding of ecotoxicological interactions between radiations and TiO_2 NP. Hund-Rinke and Simon (2006) did not registered algae and daphnid mortality using pre-illuminated TiO_2 to examine potential phototoxicity. More recently, Miller et al. (2012) showed that relatively low levels of ambient UV radiation can induce toxicity of TiO_2 to marine phytoplankton, increase overall oxidative stress and cause decreased resiliency of marine ecosystems.

Studying cell lines, Vevers and Jha (2008) showed that TiO_2 engineered nanoparticles plus UVA radiation increases the cytotoxic potential of established fish cell line derived from rainbow trout (*Oncorhyncus mykiss*).

In an *ex vivo* study with gills extracted from carps, Britto et al. (2012) also attributed generation of oxidative stress by singlet oxygen when the organs were exposed to fullerene under UVA incidence. Park et al. (2011) saw that TiO_2 NP in the absence of photoactivation are cytotoxic to HaCaT cells (human keratinocyte) and this effect becomes more pronounced in the simultaneous irradiation of UVA dependent on photocatalytic potential of TiO_2 NP, also causing oxidative stress. Yin et al. (2012) studying the same cell line showed that phototoxicity is mediated by reactive oxygen species (ROS) generated during UVA irradiation. The same phototoxic damage is true in retinal pigment epithelial cells (ARPE-19) (Sanders et al. 2012). Tu et al. (2012) showed that nitrite increased the photo-toxicity of TiO_2 NP in a dose dependent manner, and generated protein tyrosine nitration in keratinocyte cells. Murrat et al. (2012) saw that co-exposure to UVB and superparamagnetic iron oxide (SPIONs) was associated with induction of oxidative stress and release of inflammatory mediators, showing the need to evaluate dermal toxicity of engineered NP on human skin.

Therefore, these results illustrate that phototoxicity must be considered when evaluating impacts the level of organism or environment of NM, many of which are photoactive. Moreover, some scientists also have been concerned about the use of TiO_2 NP in photodynamic therapy for the treatment of cancer cells that has been proposed following studies of cultured cancer cells (Yu et al. 2011).

In aquatic environments, abiotic factors are obviously not isolated, and the interactions among them result in a buffered medium that constantly modifies and is modified by the organisms in a dynamic interchange. Such interactions rule NP fate and effects on aquatic life. For example, it was demonstrated that sunlight incidence in the presence of DOM can induce reduction of $AuCl_4^-$ and Ag^+ forming nano-Au and nano-Ag (Yin et al. 2012). This is important because nanoscaled forms of these metals sometimes are more toxic than their ionic counterparts. DOM can also increase the colloidal stability of CeO_2 NP, but the increase in water pH and ionic strength thwarts DOM adsorption to the NP, causing particle aggregation, and this modulates the NP toxicity to algae (Van Hoecke et al. 2011). These examples reinforce the urgency of understanding the actual consequences to fish, and aquatic life in general, of the entrance of NM in aquatic environments.

Effects of Co-exposure of Nanomaterials with other Pollutants on Fish

NP are diverse substances that, when released in the environment, can interact with each other and other kinds of pollutants (Baun et al. 2008; Handy et al. 2008; Sánchez et al. 2011; Costa et al. 2012), so it is of special importance to take into consideration not only the impact a NP can cause in the environment, but the interactions it can have with other pollutants possibly present in that environment as well.

Several studies have been performed aiming to analyze the interactions of NM with other kinds of compounds, taking into account the adsorptive interactions between them, generally aiming for their employment as decontaminants of polluted waters (Filho et al. 2007; Pourata et al. 2009; Chen et al. 2011c; Bikshapathi et al. 2012). Some of these works focus on the use of NM for other objectives, like increasing the efficiency of antifoulings, mainly trying to reduce their release into the water column from the surface on which they would be applied (Shtykova et al. 2009; Karmali et al. 2012).

It is interesting to note, however, that the interactions between NM and other kinds of substances may cause especially harmful effects, even if at first glance these may appear positively promising. Although, for example, the adsorption between a NM like fullerene (C_{60}) and a pollutant, like arsenic (As), may mean that nC_{60} can be used in the remediation of an environment polluted with this semi-metal, it is known that C_{60} has the propriety of carrying As inside living organisms, raising its incorporation under a "Trojan horse" effect. (Costa et al. 2012). A similar effect seems to happen with C_{60} and phenanthrene (Baun et al. 2008). It is of great importance to first investigate if the cleaning product is toxic by itself or if it interacts negatively with the pollutant it is intended to remove or degrade or even with other substance present in the effluent treated, because depending on these factors, the benefits of using a NM under this context can be very limited (Sánchez et al. 2011).

Taking into account the "Trojan horse" concept of, where contaminants can have their input in organisms enhanced by association with a NM (Limbach et al. 2007), there is a real need to evaluate the consequences of these associations and if they can cause bioaccumulation (Costa et al. 2012). The study of Costa and co-workers (2012), carried out with zebra fish hepatocytes, analyzed the interaction of $C_{60'}$ and arsenic. The results showed higher arsenic (as As^{III}) accumulation in cells co-exposed to C_{60} and As^{III}, although no alterations in cell viability were observed. C_{60} (1 mg/L) when co-exposed to arsenic (2.5 µM and 100 µM) decreased the levels of lipid peroxidation and a peak of reduced glutathione (GSH) registered in cells exposed to the lowest As^{III} concentration (2.5 µM). Thus, although C_{60} seems to act as a "Trojan Horse", increasing the concentration of intracellular arsenic, no evidence of higher toxicity was observed, suggesting that the dynamics of absorption and further releases of toxic molecules once inside the cells should be a critical point for the onset of toxicity under the "Trojan Horse" paradigm. Also, Sun et al. (2009) reported that carp (*Cyprinus carpio*) co-exposed to As^{III} and TiO_2 NP showed significant increase (44%) in the accumulation of arsenic when compared with carps only exposed to As^{III}.

Other studies conducted with carp showed that animals exposed TiO$_2$ NP and cadmium presented higher intracellular concentration of cadmium, confirming that NP have the ability to deliver compounds to the intracellular environment. In fact, the significant increase in cadmium caused biochemical changes which can lead to cell damage (Zhang et al. 2007). The study of Reeves et al. (2008) in goldfish skin cells exposed to TiO$_2$ NP and UVA showed that cells exposed only to TiO$_2$ NP did not show a significant reduction in its viability, even at the highest concentration (1,000 µg/ml). However, cells exposed to TiO$_2$ NP and UVA presented a dose dependent reduction in cell viability, with further increases in DNA damage.

Although there is a good deal of study on bioaccumulation of pollutants in organisms like fish, after co-exposure with different kinds of NM, it seems that studies on the biochemical toxicological interactions between these compounds in fish is being at least comparatively overlooked. This is an important issue, as the uses of NM can be limited in some situations, depending on the toxic interactions these can present with other kind of compounds (Sánchez et al. 2011).

Nanomaterials Contamination Policies

Nanotechnology development has been growing exponentially in the lasts years and consequently, the use of NM in commercial products has accompanied this development. However, little is known about risk assessment in biological systems with these NM (Moore 2006; Kahru and Dubourguier 2010). Inevitably, different kinds of NM should reach the environment and depending on their chemical properties, volume and mode of use, they can affect human and environmental health in the long term (Guillén et al. 2012). Once into the environment, these NM can be incorporated by living organisms and bioaccumulated, with the potential to induce several deleterious effects through many biological pathways.

Thus, it is clearly necessary to increment the knowledge about the risks and benefits of several NM that are being launched in the global market at present. As commented in Section 1, Kahru and Dubourguier (2010) revised the toxicity of several NM in different species considering LC(E)$_{50}$. These authors classified different NM according to the degree of toxicity: extremely toxic (< 0.01 mg/L); very toxic (0.1–1 mg/L); toxic (1–10 mg/L) and harmful (10–100 mg/L). It is important to underline that some of compounds considered very toxic, for example, carbon nanomaterials, are widely used in several products (Krysanov et al. 2010). Although some existing studies have evaluated the toxicological effects of NM, they are still are behind the rapid development of nanotechnology. As a consequence, the potential risk of these NM being released into the environment, without the knowledge about their effects on living organisms, cannot be disregarded. Perhaps many of the effects that might be caused by NM exposure will be known only in the future. So, there is a need for new legislation that stipulates a mandatory compliance with standards of risk assessment considering several types of NM and organisms (Lapresta-Fernández 2011).

From this, the importance is evident of biological studies using biomarkers, which are early signs, mostly reversible, that indicate an actual or potential condition of exposure, effect and susceptibility that can trigger damages in several organisms (Bergamaschi 2012). Data of exposure biomarkers can help to understand the behavior and the risks that NM offer to the environment and organisms being, in this sense, useful in evaluating risk assessments of these compounds. Once the early effects triggered by NM can be established, security procedures can be defined before these NM may reach the environment, affecting several ecologically relevant parameters such as social behavior and reproduction.

In this context, there is a need to establish regulatory norms, considering the toxicity of NM. Some efforts have been undertaken to ensure the safety of living organisms; however, much remains to be done. Obviously, for the implementation of new legislation to regulate the development of nanotechnologies, breakthroughs in toxicological studies are necessary. .

Acknowledgements

J.M. Monserrat receives a productivity research fellowship from CNPq (process number PQ 306027/2009-7). The logistic and material support from the Instituto Nacional de Ciência e Tecnologia de Nanomateriais de Carbono (CNPq) and the Electronic Microscopy Center, at the Universidade Federal do Rio Grande do Sul (UFRGS) were essentials for the execution of studies commented upon in this chapter. The support of DECIT/SCTIE-MS through Conselho Nacional de Desenvolvimento Científico e Tecnológico (CNPq) and Fundação de Amparo à Pesquisa do Estado do Rio Grande do Sul (FAPERGS, Proc. 10/0036-5–PRONEX/Conv. 700545/2008) is also acknowledged. J.M. Monserrat acknowledges the support of Nanotoxicology Network (MCTI/CNPq process number 552131/2011-3).

Keywords: nanotoxicology, nanotechnology, fish, biochemical responses, fullerene, carbon nanotubes, inorganic nanomaterials

References

Adams, L.K., D.Y. Lyon and P.J.J. Alvarez. 2006. Comparative eco-toxicity of nanoscale TiO_2, SiO_2, and ZnO water suspensions. Water Res. 40: 3527–3532.

Ahlbom, A., J. Bridges, W. Jong, P. Hartemann, T. Jung, M. Mattsson, J. Pagès, K. Rydzynski, D. Stahl and M. Thomsen. 2009. Risk assessment of products of nanotechnologies. Report of the Scientific Committee on Emerging and Newly Identified Health Risks–SCENIHR. European Commission Health & Consumers DG. Brussels.

Aitken, R.J., M.Q. Chaudhry, A.B.A. Boxall and M. Hull. 2006. Manufacture and use of nanomaterials: current status in the UK and global trends. Occup. Med. 56: 300–306.

Arora, S., J.M. Rajwade and K.M. Paknikar. 2012. Nanotoxicology and *in vitro* studies: The need of the hour Toxicol. Appl. Pharmacol. 258: 151–165.

Arvidsson, R., S. Molander, B.A. Sanden and M. Hassellöv. 2011. Challenges in exposure modeling of nanoparticles in aquatic environments. Hum. Ecol. Risk Assess. 17: 245–262.

Asharani, P.V., N.G.B. Serina, M.H. Nurmawati, Y.L. Wu and Z.S.V. Gong. 2008a. Impact of multiwalled carbon nanotubes on aquatic species. J. Nanosci. Nanotechnol. 8: 3603–3609.

Asharani, P.V., Y.L. Wu, Z.Gong and S. Valiyaveettil. 2008b. Toxicity of silver nanoparticles in zebrafish models. Nanotechnology 19: 2255102–2255107.

Battin, T.J., F.V.D. Kammer, A. Weilhartner, S. Ottofuelling and T. Hofmann. 2009. Nanostructured TiO_2: Transport behavior and effects on aquatic microbial communities under environmental conditions. Environ. Sci. Technol. 43: 8098–8104.

Baun, A., N.B. Hartmann, K. Grieger and K.O. Kusk. 2008. Ecotoxicity of engineered nanoparticles to aquatic invertebrates: A brief review and recommendations for future toxicity testing. Ecotoxicology. 17: 387–395.

Baun, A., S.N. Sørensen, R.F. Rasmussen, N.B. Hartmann and C.B. Koch. 2008. Toxicity and bioaccumulation of xenobiotic organic compounds in the presence of aqueous suspensions of aggregates of nanoC_{60}. Aquat. Toxicol. 86: 379–397.

Belyanskaya, L., S. Weigel, C. Hirsch, U. Tobler, H.F. Krug and P. Wick. 2009. Effects of carbon nanotubes on primary nervous and glial cells. Neurotoxicol. 30: 702–711.

Benn, T.M. and P. Westerhoff. 2008. Nanoparticle silver released into water from commercially available sock fabrics. Environ. Sci. Technol. 42: 4133–4139.

Bergamaschi, E. 2012. Human biomonitoring of engineering nanoparticles: An appraisal of critical issues and potential biomarkers. J. Nanomaterials 2012: 1–12.

Bhawana, P. and M.H. Fulekar. 2012. Nanotechnology: Remediation technologies to clean up the environmental pollutants. Res. J. Chem. Sci. 2: 90–96.

Bikshapathi, M., S. Singh, B. Bhaduri, G.N. Mathura, A. Sharma and N. Verma. 2012. Fe-nanoparticles dispersed carbon micro and nanofibers: Surfactant-mediated preparation and application to the removal of gaseous VOCs. Colloids and Surfaces A: Physicochem. Eng. Aspects 399: 46– 55.

Blaser, S.A., M. Scheringer, M. MacLeod and K. Hungerbuehler. 2008. Estimation of cumulative aquatic exposure and risk due to silver: Contribution of nano-functionalized plastics and textiles. Sci. Total Environ. 390: 396–409.

Brant, J., H. Lecoanet and M.R. Wiesner. 2005. Aggregation and deposition characteristics of fullerene nanoparticles in aqueous systems. J. Nanopart. Res. 7: 545–553.

Britto, R.S., M.L. Garcia, A. Martins da Rocha, J.A. Flores, M.V.B. Pinheiro, J.M. Monserrat and J.L.R. Ferreira. 2012. Effects of carbon nanomaterials fullerene C_{60} and fullerol $C_{60}(OH)_{18-22}$ on gills of fish *Cyprinus carpio* (Cyprinidae) exposed to ultraviolet radiation. Aquat. Toxicol. 114–115: 80–87.

Buzea, C., I.I.I. Pacheco Blandino and K. Robbie. 2007. Nanomaterials and nanoparticles: Sources and toxicity. Biointerphases. 2: 17–172.

Chappell, M.A., A.J. George, K.M. Dontsova, B.E. Porter, C.L. Price, P. Zhou, P.E. Morikawa, A.J. Kennedy and J.A. Steevens. 2009. Surfactive stabilization of multi-walled carbon nanotube dispersions with dissolved humic substances. Environ. Pollut. 15: 1081–7.

Chen, K.L. and M. Elimelech. 2006. Aggregation and deposition kinetics of fullerene (C_{60}) nanoparticles. Langmuir 22: 10994–1001.

Chen, L., R.A. Yokel, B. Hennig and M. Toborek. 2008. Manufactured aluminum oxide nanoparticles decrease expression of tight junction proteins in brain vasculature. J. Neuroimm. Pharmacol. 3: 286–295.

Chen, T.-H., C.-Y. Lin and M.-C. Tseng. 2011a. Behavioral effects of titanium dioxide nanoparticles on larval zebrafish (*Danio rerio*). Mar. Poll. Bull. 63: 303–308.

Chen, J., Z. Xiu, G.V. Lowry and P.J.J. Alvarez. 2011b. Effect of natural organic matter on toxicity and reactivity of nano-scale zero-valent iron. Water Res. 45: 1995–2001.

Chen, G.C., X.Q. Shan, Z.G. Pei, H. Wang, L.R. Zheng, J. Zhang and Y.N. Xie. 2011c. Adsorption of diuron and dichlobenil on multiwalled carbon nanotubes as affected by lead. J. Haz. Mat. 188: 156–163.

Cheng, J.P., E. Flahaut and S.H. Cheng. 2007. Effect of carbon nanotubes on developing zebrafish (*Danio rerio*) embryos. Environ. Toxicol. Chem. 26: 708–716.

Cheng, P., C.M. Chan, L.M. Veca, W.L. Poon, P.K. Chan, L. Qu, Y.-P. Sun and S.H. Cheng. 2009. Acute and long-term effects after single loading of functionalized multi-walled carbon nanotubes into zebrafish (*Danio rerio*). Toxicol. Appl. Pharmacol. 235: 216–227.

Cosgrove, T. 2005. Colloid science: Principles, methods and applications. Blackwell Publishing, UK.

Costa, C.L.A., I.S. Chaves, J. Ventura-Lima, J.L. Ribas Ferreira, L. Ferraz, L.M. Carvalho and J.M. Monserrat. 2011. *In vitro* evaluation of co-exposure of arsenium and an organic nanomaterial (fullerene, C_{60}) in zebra fish hepatocytes. Comp. Biochem. Physiol. C 155: 206–212.

Derjaguin, B.V. and L.D. Landau. 1941. Theory of the stability of strongly charged lyophobic sols and of the adhesion of strongly charged particles in solutions of electrolytes. Acta Physicochim. URSS 14: 733–62.

Dhawan, A., J.S. Taurozzi, A.K. Pandey, W. Shan, S.M. Miller, S.A. Hashsham and V.V. Tarabara. 2006. Stable colloidal dispersions of C_{60} fullerenes in water: Evidence for genotoxicity. Environ. Sci. Technol. 40: 7394–7401.

Donaldson, K., V. Stone, C. Tran, W. Kreyling and P.J.A. Borm. 2004. Nanotoxicology. Occup. Environ. Med. 61: 727–728.

El Badawy, A.M., T.P. Luxton, R.G. Silva, K.G. Scheckel, M.T. Suidan and T.M. Tolaymat. 2010. Impact of environmental conditions (pH, ionic strength, and electrolyte type) on the surface charge and aggregation of silver nanoparticles suspensions. Environ. Sci. Technol. 44: 1260–1266.

Fabrega, J., S.R. Fawcett, J.C. Renshaw and J.R. Lead. 2009. Silver nanoparticle impact on bacterial growth: effect of pH, concentration, and organic matter. Environ. Sci. Technol. 43: 7285–7290.

Fabrega, J., S.N. Luoma, C.R. Tyler, T.S. Galloway and J.R. Lead. 2011. Silver nanoparticles: Behaviour and effects in the aquatic environment. Environ. Int. 37: 517–531.

Farkas, J., H. Peter, P. Christian, J.A. Gallego-Urrea, M. Hassellöv, J. Tuoriniemi, S. Gustafsson, E. Olsson, K. Hylland and K.V. Thomas. 2011. Characterization of the effluent from a nanosilver producing washing machine. Environ. Int. 37: 1057–1062.

Farre, M., K. Gajda-Schrantz, L. Kantiani and D. Barcelo. 2009. Ecotoxicity and analysis of nanomaterials in the aquatic environment. Anal. Bioanal. Chem. 393: 81–95.

Federici, G., B.J. Shaw and R.D. Handy. 2007. Toxicity of titanium dioxide nanoparticles to rainbow trout (*Oncorhynchus mykiss*): Gill injury, oxidative stress, and other physiological effects. Aquat. Toxicol. 84: 415–430.

Ferrari, M. 2005. Cancer nanotechnology: Opportunities and challenges. Nat. Rev. Cancer 5: 161–171.

Ferreira, J.R.F., D.M. Barros, L.A. Geracitano, G. Fillmann, C.E. Fossa, E.A. De Almeida, M.C. Prado, B.R.A. Neves, M.V.B. Pinheiro and J.M. Monserrat. 2012. Influence of *in vitro* exposure to fullerene C_{60} in redox state and lipid peroxidation of brain and gills of carp *Cyprinus carpio* (Cyprinidae). Environ. Toxicol. Chem. 31: 961–967.

Filho, N.L.D., R.M. Costa, F. Marangoni and D.S. Pereira. 2007. Nanoparticles of octakis[3-(3-amino-1,2,4-triazole)propyl]octasilsesquioxane as ligands for Cu(II), Ni(II), Cd(II), Zn(II), and Fe(III) in aqueous solution. J. Colloid Interf. Sci. 316: 250–259.

Fraser, T.W.K., H.C. Reinardy, B.J. Shaw, T.B. Henry and R.D. Handy. 2011. Dietary toxicity of single-walled carbon nanotubes and fullerenes (C_{60}) in rainbow trout (*Oncorhynchus mykis*). Nanotoxicol. 5: 98–108.

French, R.A., A.R. Jacobson, B. Kim, S.L. Isley, R.L. Penn and P.C. Baveye. 2009. Influence of ionic strength, pH, and cation valence on aggregation kinetics of titanium dioxide nanoparticles. Environ. Sci. Technol. 43: 1354–1359.

Geranio, L., M. Heuberger and B. Nowack. 2009. The behavior of silver nanotextiles during washing. Environ. Sci. Technol. 43: 8113–8118.

Gottschalk, F., T. Sonderer, R.W. Scholz and B. Nowack. 2009. Modeled environmental concentrations of engineered nanomaterials (TiO_2, ZnO, Ag, CNT, Fullerenes) for different regions Environ. Sci. Technol. 43: 9216–9222.

Gottschalk, F., C. Ort, R.W. Scholz and B. Nowack. 2011. Engineered nanomaterials in rivers —Exposure scenarios for Switzerland at high spatial and temporal resolution. Environ. Pollut. 159: 3439–3445.

Gottschalk, F., R.W. Scholz and B. Nowack. 2010. Probabilistic material flow modeling for assessing the environmental exposure to compounds: Methodology and an application to engineered nano-TiO_2 particles. Environ. Modell. Softw. 25: 320–332.

Gouveia, G.R., D.S. Marques, B.P. Cruz, L.A. Geracitano, L.E.M. Nery and G.S. Trindade. 2005. Antioxidant defenses and DNA damage induced by UVA and UVB radiation in the crab *Chasmagnathus granulata* (Decapoda, Brachyura). Photochem. Photobiol. 81: 398–403.

Griffitt, R.J., J. Luo, J. Gao, J.-C. Bonzongo and D.S. Barber. 2008. Effects of particle composition and species on toxicity of metallic nanomaterials in aquatic organisms. Environ. Toxicol. Chem. 27: 1972–1978.

Griffitt, R.J., K. Hyndman, N.D. Denslow and D.S. Barber. 2009. Comparison of molecular and histological changes in zebrafish gills exposed to metallic nanoparticles. Toxicol. Sci. 107: 404–415.

Griffitt, R.J., R. Weil, K.A. Hyndman, N.D. Denslow, K. Powers, D. Taylor and D.S. Barber. 2007. Exposure to copper nanoparticles causes gill injury and acute lethality in zebrafish (*Danio rerio*). Environ. Sci. Technol. 41: 8178–8186.

Guillén, D., A.Ginebreda, M. Farré, R.M. Darbra, M. Petrovic, M. Gros and D. Barceló. 2012. Prioritization of chemical in the aquatic environment based on risk assessment: analytical, modeling, and regulatory perspective. Environ. Int. *in press*.

Handy, R., T. Henry, T. Scown, B. Johnston and C. Tyler. 2008. Manufactured nanoparticles: Their uptake and effects on fish. A mechanistic analyses. Ecotoxicol. 17: 396–409.

Handy, R.D., R. Owen and E. Valsami-Jones. 2008. The ecotoxicology of nanoparticles and nanomaterials: current status, knowledge gaps, challenges, and future needs. Ecotoxicol. 17: 315–25.

Hassellöv, R. and R. Kaegi. 2009. Analysis and characterization of manufactured nanoparticles in aquatic environments. *In*: J.R. Lead and E. Smith (eds.). Nanoscience and Nanotechnology: Environmental and Human Health Implications. Wiley, New Jersey, USA, pp. 211–266.

Heinlaan, M., A. Ivask, I. Blinova, H.-C. Dubourguier and A. Kahru. 2008. Toxicity of nanosized and bulk ZnO, CuO and TiO_2 to bacteria *Vibrio fischeri* and crustaceans *Daphnia magna* and *Thamnocephalus platyurus*. Chemosphere. 71: 1308–1316.

Helland, A., P. Wick, A. Koehler, K. Schmid and C. Som. 2007. Reviewing the environmental and human health knowledge base of carbon nanotubes. Environ. Health. Perspect. 115: 1125–1131.

Henry, T.B., F.M. Menn, J.T. Fleming, J. Wilgus, R.N. Compton and G.S. Sayler. 2007. Attributing effects of aqueous C_{60} nanoaggregates to tetrahydrofuran decomposition products in larval zebrafish by assessment of gene expression. Environ. Health Perspect. 115: 1059–1065.

Hirsch, J.E. 2005. An index to quantify an individual's scientific research output. Proc. Natl. Acad. Sci. USA 102: 16569–16572.

Hou, W.C. and C.T. Jafvert. 2009. Photochemical transformation of aqueous C_{60} clusters in sunlight. Environ. Sci. Technol. 43: 362–367.

Hoyt, V.W. and E. Mason. 2008. Nanotechnology: emerging health issues. J. Chem. Health Saf. 15: 10–15.

Hu, Y.-L. and J.-Q. Gao. 2010. Potencial toxicity of nanoparticles. Int. J. Pharm. 394: 115–121.

Hund-Rinke, K. and M. Simon. 2006. Ecotoxic effect of photocatalytic active nanoparticles TiO_2 on algae and daphnids. Environ. Sci. Pollut. Res. 13: 225–232.

Hussain, S.M., K.L. Hess, J.M. Gearhart, K.T. Geiss and J.J. Schlager. 2005. In vitro toxicity of nanoparticles in BRL 3A rat liver cells. Toxicol. Vitro. 19: 975–983.

Huuskonen, J. 2002. Prediction of soil sorption of a diverse set of organic chemicals from molecular structure. J. Chem. Inf. Model. 43: 1457–1462.

Hwang, Y.S. and Q.L. Li. 2010. Characterizing photochemical transformation of aqueous nC_{60} under environmentally relevant conditions. Environ. Sci. Technol. 44: 3008–3013.

Hyung, H., J.D. Fortner, J.B. Hughes and J.H. Kim. 2007. Natural organic matter stabilizes carbon nanotubes in the aqueous phase. Environ. Sci. Technol. 41: 179–184.

Jiang, X., M. Tong and H. Kim. 2012. Influence of natural organic matter on the transport and deposition of zinc oxide nanoparticles in saturated porous media. J. Colloid Interf. Sci. 386: 34–43.

Kaegi, R., A. Ulrich, B. Sinnet, R. Vonbank, A. Wichser, S. Zuleeg, H. Simmler, S. Brunner, H. Vonmont, M. Burkhardt and M. Boller. 2008. Synthetic TiO_2 nanoparticle emission from exterior facades into the aquatic environment. Environ. Pollut. 156: 233–239.

Kaegi, R., B. Sinnet, S. Zuleeg, H. Hagendorfer, E. Mueller, R. Vonbank, M. Boller and M. Burkhardt. 2010. Release of silver nanoparticles from outdoor facades. Environ. Pollut. 158: 2900–2905.

Kahru, A. and H.-C. Dubourguier. 2010. From ecotoxicology to nanoecotoxicology. Toxicology 269: 105–119.

Kamat, J.P., T.P.A. Devasagayam, K.I. Priyadarsini and H. Mohan. 2000. Reactive oxygen species mediated membrane damage induced by fullerene derivatives and its possible biological implications. Toxicol. 155: 55–61.

Karmali, P.P., Y. Chao, J.H. Park, M.J. Sailor, E. Ruoslahti, S.C. Esener and D. Simberg. 2012. Different effect of hydrogelation on antifouling and circulation properties of dextran–iron oxide nanoparticles. Mol. Pharm. 9: 539–545.

Kashiwada, S. 2006. Distribution of nanoparticles in the see-through medaka (*Oryzias latipes*). Environ. Health Perspect. 114: 1697–1702.

Keller, A.A., H. Wang, D. Zhou, H.S. Lenihan, G. Cherr, B.J. Cardinale, R. Miller and Z. Ji. 2010. Stability and aggregation of metal oxide nanoparticles in natural aqueous matrices. Environ. Sci. Technol. 44: 1962–1967.

Kim, S., W.-K. Oh, Y.-S. Jeong, J.-Y. Hong, B.-R. Cho, J.-S. Hahn and J. Jang. 2011. Citotoxicity of, and innate immune response to size-controlled polypyrrole nanoparticles in mammalian cells. Biomaterials 32: 2342–2350.

Kim, K., M. Jang, J. Kim, B. Xing, R.L. Tanguay, B. Lee and S.D. Kim. 2012. Embryonic toxicity changes of organic nanomaterials in the presence of natural organic matter. Sci. Total Environ. 426: 423–429.

Krysanov, E.Y., D.S. Pavlov, T.B. Demidova and Y.Y. Dgebuadze. 2010. Effect of nanoparticles in aquatic organisms. Biol. Bull. 37: 406–412.

Lapresta-Fernadéz, A. 2011. Public concern over ecotoxicology risks from nanomaterials: pressing need for research-based information. Environ. Int. 39: 148–149.

Lee, K.J., P.D. Nallathamby, L.M. Browning, C.J. Osgood and X.-H.N. Xu. 2007. *In vivo* imaging of transport and biocompatibility of single silver nanoparticles in early development of zebrafish embryos. ACS Nano 1: 133–143.

Lee, J., M. Cho, J.D. Fortner, J.B. Hughes and J.H. Kim. 2009. Transformation of aggregate C_{60} in the aqueous phase by UV irradiation. Environ. Sci. Technol. 43: 4878–4883.

Lee, S., K. Kim, H.K. Shon, S.D. Kim and J. Cho. 2011. Biotoxicity of nanoparticles: effect of natural organic matter. J. Nanoparticle Res. 13: 3051–3061.

Letts, R.E., T.C.B. Pereira, M. Reis Bogo and J.M. Monserrat. Biological responses of bacteria communities livin at the mucus secretion of common carp (*Cyprinus carpio*) after exposure to the carbon nanomaterial fullerene (C_{60}). Arch. Environ. Contam. Toxicol. 61: 311–317.

Li, H., J. Zhang, T. Wang, W. Luo, Q. Zhou and G. Jiang. 2008. Elemental selenium particles at nano-size (Nano-Se) as a consequence of hyper-accumulation of selenium: A comparison with sodium selenite. Aquat. Toxicol. 89: 251–256.

Li, H., Q. Zhou, Y. Wu, J. Fu, T. Wang and G. Jiang. 2009. Effects of waterborne nano-iron on medaka (*Oryzias latipes*): Antioxidant enzymatic activity, lipid peroxidation and histopathology. Ecotox. Environ. Safety 72: 684–692.

Limbach, L.K., R. Bereiter, E. Müller, R. Krebs, R. Gälli and W.J. Stark. 2008. Removal of oxide nanoparticles in a model wastewater treatment plant: Influence of agglomeration and surfactants on clearing efficiency. Environ. Sci. Technol. 42: 5828–5833.

Lin, D. and B. Xing. 2008. Tannic acid adsorption and its role for stabilizing carbon nanotube suspensions. Environ. Sci. Technol. 42: 5917–23.

Lowry, V.G. and M.R. Wiesner. 2007. Environmental considerations: occurrences, fate, and characterization of nanoparticles in the environment. *In:* N.A. Monteiro-Riviere and C.L. Tran (eds.). Nanotoxicology: Characterization, Dosing and Health Effects. Informa Healthcare USA, Inc. New York, pp. 369–389.

Luthy, R.G., G.R. Aiken, M.L. Brusseau, S.D. Cunningham, P.M. Gschwend, J.J. Pignatello, M. Reinhard, S.J. Train, W.J. Weber and J.C. Westall. 1997. Sequestration of hydrophobic organic contaminants by geosorbents. Environ. Sci. Technol. 31: 3341–3347.

Martínez-Castanôñ, G.A., N. Niño-Martínez, F. Martínez-Gutierrez, J.R. Martínez-Mendoza and F. Ruiz. 2008. Synthesis and antibacterial activity of silver nanoparticles with different sizes. J. Nanopart. Res. 10: 1343–8.

Mathivanan, V., S. Ananth, P.G. Prabu and Selvisabhanayakam. 2012. Role of silver nanoparticles: Behaviour and effects in the aquatic environment—a review. Int. J. Dev. Biol. 2: 77–82.

Matranga, V. and I. Corsi. 2012. Toxic effect of engineered nanoparticles in the marine environment: Model organisms and molecular approach. Mar. Environ. Res. 76: 32–40.

McKone, T.E. and K.T. Bogen. 1991. Predicting the uncertainties in risk assessments: A California groundwater case-study. Environ. Sci. Technol. 25: 1674–1681.

Miller, R.J., S. Bennett, A.A. Keller, S. Pease and H.S. Lenihan. 2012. TiO_2 nanoparticles are phototoxic to marine phytoplankton. PLoSONE 7: e30321.

Moger, J., B.D. Johnston and C.R. Tyler. 2008. Imaging metal oxide nanoparticles in biological structures with CARS microscopy. Optics Express 16: 3408–3419.

Moore, M.N. 2006. Do nanoparticles present ecotoxicological risks for the health of the aquatic environment? Environ. Int. 32: 967–976.

Mueller, N.C. and B. Nowack. 2008. Exposure modeling of engineered nanoparticles in the environment. Environ. Sci. Technol. 42: 4447–4453.

Nowack, B. and T.D. Bucheli. 2007. Occurrence, behavior and effects of nanoparticles in the environment. Environ. Pollut. 150: 5–22.

Oberdorster, E. 2004. Manufactured nanomaterials (fullerenes, C_{60}) induce oxidative stress in the brain of juvenile largemouth bass. Environ. Health Perspect. 112: 1058–1062.

Oberdörster, E., S.Q. Zhu, T.M. Blickley, P. McClellan-Green and M.L. Haasch. 2006. Ecotoxicology of carbon-based engineered nanoparticles: Effects of fullerene (C_{60}) on aquatic organisms. Carbon. 44: 1112–1120.

Pal, S., Y.K. Tak and J.M. Song. 2007. Does the antibacterial activity of silver nanoparticles depend on the shape of the nanoparticle? A study of the Gram-negative bacterium *Escherichia coli*. Appl. Environ. Microbiol. 73: 1712–20.

Park, Y.H., S.H. Jeong, S.M. Yi, B.H. Choi, Y.R. Kim, I.K. Kim, M.K. Kim and S.W. Son. 2011. Analysis for the potential of polystyrene and TiO_2 nanoparticles to induce skin irritation, phototoxicity, and sensitization. Toxicol. Vitro 25: 1863–1869.

Paterson, G., J.M. Ataria, M.E. Hoque, D.C. Burns and C.D. Metcalfe. 2011. The toxicity of titanium dioxide nano powder to early life stages of the Japanese medaka (*Oryzias latipes*). Chemosphere 82: 1002–1009.

Paul, A., R. Stösser, A. Zehl, E. Zwirnmann, R.D. Vogt and C.E.W. Steinberg. 2006. Nature and abundance of organic radicals in natural organic matter: Effect of pH and irradiation. Environ. Sci. Technol. 40: 5897–5903.

Peralta-Videa, J.R., L. Zhao, M.L. Lopez-Moreno, G. de la Rosa, J. Hong and J.L. Gardea-Torresdey. 2011. Nanomaterials and the environment: A review for the biennium 2008–2010. J. Hazardous Mat. 186: 1–15.

Perez, S., M. Farre and D. Barcelo. 2009. Analysis, behavior and ecotoxicity of carbon-based nanomaterials in the aquatic environment. TrAC-Trends. Anal. Chem. 28: 820–832.

Pourata, R., A.R. Khataee, S. Aber and N. Daneshvar. 2009. Removal of the herbicide bentazon from contaminated water in the presence of synthesized nanocrystalline TiO_2 powders under irradiation of UV-C light. Desalination 249: 301–307.

Powers, C.M., E.D. Levin, F.J. Seidler and C.A. Slotkin. 2011. Silver exposure in development zebrafish produces persistent synaptic and behavioral changes. Neurotoxicol. Teratol. 33: 329–332.

Qin, X.Y., J.G. Kim and J.S. Lee. 1999. Synthesis and magnetic properties of nanostructured γ-Ni–Fe alloys. Nanostruct. Mater. 11: 259–270.

Ramsden, C., T. Smith, B. Shaw and R. Handy. 2009. Dietary exposure to titanium dioxide nanoparticles in rainbow trout, *Onchorynchus mykiss*: no effect on growth, but subtle biochemical disturbances in the brain. Ecotoxicology 18: 939–951.

Rawson, D.M., T. Zhang, D. Kalicharan and W.L. Jongebleod. 2000. Field emission scanning electron microscopy and transmission electron microscopy studies of the chorion, plasma membrane and syncytial layers of the gastrula-stage embryo of the zebrafish *Brachydanio rerio*: A consideration of structural and functional relationships withrespect to cytoprotectant penetration. Aquac. Res. 3: 325–336.

Reeves, J.F., S.J. Davies, N.J.F. Dodd and A.N. Jha. 2008. Hydroxyl radicals (•OH) are associated with titanium dioxide (TiO_2) nanoparticle-induced cytotoxicity and oxidative DNA damage in fish cells. Mut. Res. 640: 113–122.

Sánchez, A., S. Recillas, X. Font, E. Casals, E. González and V. Puntes. 2011. Ecotoxicity of, and remediation with, engineered inorganic nanoparticles in the environment. Trends Anal. Chem. 30: 507–516.

Sanders, K., L.L. Degn, W.R. Mundy, R.M. Zucker, K. Dreher, B. Zhao, J.E. Roberts and W.K. Boyes. 2012. *In vitro* phototoxicity and hazard identification of nano-scale titanium dioxide. Toxicol. Appl. Pharmacol. 258: 226–236.

Scheringer, M., D. Halder and K. Hungerbuhler. 1999. Comparing the environmental performance of fluorescent whitening agents with peroxide bleaching of mechanical pulp. J. Ind. Ecol. 3: 77–95.

Schwarzenbach, R.P., P.M. Gschwend and D.M. Imboden. 2003. Environmental Chemistry. John Wiley and Sons, New Jersey. USA.

Scown, T.M., E.M. Santos, B.D. Johnston, B.Gaiser, M. Baalousha, S. Mitov, J.R. Lead, V. Stone, T.F. Fernandes, M. Jepson, R. van Aerleand and C.R. Tyler. 2010. Effects of aqueous exposure to silver nanoparticles of different sizes in rainbow trout. Toxicol. Sci. 115: 521–534.

Shaw, B.J. and R.D. Handy. 2011. Physiological effects of nanoparticles on fish: A comparison of nanometals versus metal ions. Environ. Int. 37: 1083–1097.

Shinohara, N., T. Matsumoto, M. Gamo, A. Miyauchi, S. Endo, Y. Yonezawa and J. Nakanishi. 2009. Is lipid peroxidation induced by the aqueous suspension of fullerene C_{60} nanoparticles in the brains of *Cyprinus carpio*? Environ. Sci. Technol. 43: 948–953.

Shintani, H., S. Kurosu, A. Miki, F. Hayashi and S. Kato. 2006. Sterilization efficiency of the photocatalyst against environmental microorganisms in a health care facility. Biocontrol Sci. 1: 17–26.

Shtykova, L., C. Fant, P. Handa, A. Larsson, K. Berntsson, H. Blanck, R. Simonsson, M. Nydén and H.I. Härelind. 2009. Adsorption of antifouling booster biocides on metal oxide nanoparticles: Effect of different metal oxides and solvents. Prog. Org. Coatings 64: 20–26.

Simkó, M. and M.O. Mattsson. 2010. Risks from accidental exposures to engineered nanoparticles and neurological health effects: A critical review. Particle and Fibre Toxicol. 7: 42.

Simonet, B.M. and M. Valcárcel. 2009. Monitoring nanoparticles in the environment. Anal. Bioanal. Chem. 393: 17–21.

Smith, C.J., B.J. Shaw and R.D. Handy. 2007. Toxicity of single walled carbon nanotubes on rainbow trout (*Oncorhynchus mykiss*): Respiratory toxicity, organ pathologies, and other physiological effects. Aquat. Toxicol. 82: 94–109.

Sinha, R.P. and D.P. Häder. 2002. Life under solar UV radiation in aquatic organisms. Adv. Space Res. 30: 1547–1556.

Sun, H., X. Zhang, Z. Zhang, Y. Chen and J.C. Crittenden. 2009. Influence of titanium dioxide nanoparticles on speciation and bioavailability of arsenite. Environ. Pollut. 157: 1165–1170.

Tiede, K., A.B.A. Boxall, S.P. Tear, J. Lewis, H. David and M. Hassellöv. 2008. Detection and characterization of engineered nanoparticles in food and the environment. Food Addit. Contam. 25: 795–821.

Tiede, K., M. Hasselöv, E. Breitbarth, Q. Chaundhry and A.B.A. Boxall. 2009. Considerations for environmental fate and ecotoxicity testing to support environmental risk assessments for engineered nanoparticles. J. Chromatogr. 1216: 503–509.

Tu, M., Y. Huanga, H.L. Li and Z.H. Gao. 2012. The stress caused by nitrite with titanium dioxide nanoparticles under UVA irradiation in human keratinocyte cell. Toxicol. 299: 60– 68.

Tuoriniemi, J., G. Cornelis and M. Hasselöv. 2012. Size discrimination and detection capabilities of single-particle ICPMS for environmental analysis of silver nanoparticles. Anal. Chem. 84: 3965–3972.

Usenko, C.Y., S.L. Harper and R.L. Tanguay. 2008. Fullerene C_{60} exposure elicits an oxidative stressresponse in embryonic zebrafish. Toxicol. Appl. Pharmacol. 229: 44–55.

Uyguner-Demirel, C.S. and M. Bekbolet. 2011. Significance of analytical parameters for the understanding of natural organic matter in relation to photocatalytic oxidation. Chemosphere 84: 1009–31.

Van Hoecke, K., K.A.C. De Schamphelaere, P. Van der Meeren, G. Smagghe and C.R. Janssen. 2011. Aggregation and ecotoxicity of CeO_2 nanoparticles in synthetic and natural waters with variable pH, organic matter concentration and ionic strength. Environ. Pollut. 159: 970–976.

van der Voet, H. and W. Slob. 2007. Integration of probabilistic exposure assessment and probabilistic hazard characterization. Risk Anal. 27: 351–365.

Vaseashta, A., M. Vaclavikova, S. Vaseashta, G. Gallios, P. Roy and O. Pummakarnchana. 2007. Nanostructures in environmental pollution detection, monitoring, and remediation. Sci. Technol. Adv. Mater. 8: 47–59.

Vasir, J.K., M.K. Reddy and V.D. Labhasetwar. 2005. Nanosystems in drug targeting: Oportunities and challenges. Curr. Nanosci. 1: 47–64.

Verwey, E.J.W. and J.T.G. Overbeek. 1948. Theory of the stability of lyophobic colloids. Amsterdam: Elsevier.

Vinardell, M.P. 2005. *In vitro* cytotoxicity of nanoparticles in mammalian germ-line stem cell. Toxicol. Sci. 88: 285–286.

Vevers, W.F. and A.N. Jha. 2008. Genotoxic and cytotoxic potential of titanium dioxide (TiO_2) nanoparticles on fish cells *in vitro*. Ecotoxicol. 17: 410–420.

Wang, J., X. Zhu, X. Zhang, Z. Zhao, H. Liu, R. George, J. Wilson-Rawls, Y. Chang and Y. Chen. 2011. Disruption of zebrafish (*Danio rerio*) reproduction upon chronic exposureto TiO_2 nanoparticles. Chemosphere 83: 461–467.

Wang, Y., W.G. Aker, H.-M. Hwang, C.G. Yedjou, H. Yu and P.B. Tchounwou. 2011. A study of the mechanism of *in vitro* cytotoxicity of metal oxide nanoparticles using catfish primary hepatocytes and human HepG2 cells. Sci. Total Environ. 409: 4753–4762.

Wang, Z., J. Li, J. Zhao and B. Xing. 2011. Toxicity and internalization of CuO nanoparticles to prokaryotic alga *Microcystis aeruginosa* as affected by dissolved organic matter. Environ. Sci. Technol. 45: 6032–6040.

Webster, T.J., C. Ergun and R.H. Doremus. 2000. Enhanced functions of osteoblasts on nanophase ceramics. Biomaterials. 21: 1803–1810.

Webster, T.J., R.W. Siegel and R. Bizios. 1999. Osteoblast adhesion on nanophase ceramics. Biomaterials. 20: 1221–1227.

Wiesner, M.R., G.V. Lowry, K.L. Jones, M.F. Hochella, R.T. Di Giulio, E. Casman and E.S. Bernhardt. 2009. Decreasing uncertainties in assessing environmental exposure, risk, and ecological implications of nanomaterials. Environ. Sci. Technol. 43: 6458–6462.

Wiesner, M.R., G.V. Lowry, P. Alvarez, D. Dionysiou and P. Biswas. 2006. Assessing the risks of manufactured nanomaterials. Environ. Sci. Technol. 40: 4336–4345.

Woodrow Wilson International Center for Scholars. Nanotechnology Consumer Product Inventory. 2009. [Internet]. Available from: http://www. nanotechproject.org.

Wu, D., E.S. Pak, C.J. Wingard and A.K. Murashov. 2012. Multi-walled carbon nanotubes inhibit regenerative axon growth of dorsal root ganglia neurons of mice. Neurosci. Lett.: 507: 72–77.

Yang, K., D. Lin and B. Xing. 2009. Interactions of humic acid with nanosized inorganic oxides. Langmuir. 25: 3571–3576.

Yeo, M.K. and M. Kang. 2008. Effects of nanometer sized silver materials on biological toxicity during zebrafish embryogenesis. Bull. Korean Chem. Soc. 29: 1179–1184.

Yin, J.J., J. Liu, M. Ehrenshaft, J.E. Roberts, P.P. Fu, R.P. Mason and B. Zhao. 2012. Phototoxicity of nano titanium dioxides in HaCaT keratinocytes—Generation of reactive oxygen species and cell damage. Toxicol. Appl. Pharmacol. 263: 81–88.

Yin, Y., J. Liu and G. Jiang. 2012. Sunlight-induced reduction of ionic Ag and Au to metallic nanoparticles by dissolved organic matter. ACS Nano 6: 7910–7919.

Yu, C.K., K.H. Hu, S.H. Wang, T. Hsu, H.T. Tsai, C.C. Chen, S.M. Liu, T.Y. Lin and C.H. Chen. 2011. Photocatalytic effect of anodic titanium oxide nanotubes on various cell culture media. Appl. Phys. A 102: 271–274.

Zha, L., J. Hu, C. Wang, S. Fu and M. Luo. 2002. The effect of electrolyte on the colloidal properties of poly(N-isopropylacrylamideco-dimethylaminoethylmethacrylate) microgel latexes. Colloid Polym. Sci. 280: 1116–1121.

Zhang, W.X. 2003. Nanoscale iron particles for environmental remediation: An overview. J. Nanopart. Res. 5: 323–332.

Zhang, X., H. Sun, Z. Zhang, Q. Niu, Y. Chen and J.C. Crittenden. 2007. Enhanced bioaccumulation of cadmium in carp in the presence of titanium dioxide nanoparticles. Chemosphere 67: 160–166.

Zhu, X., J. Wang, X. Zhang, Y. Chang and Y. Chen. 2010. Trophic transfer of TiO_2 nanoparticles from daphnia to zebrafish in a simplified freshwater food chain. Chemosphere 79: 928–933.

CHAPTER 12

Effect of Pollutants on Condition Index

Larissa Paola Rodrigues Venancio* and *Claudia Regina Bonini Domingos*

Introduction

Organisms can be understood as simple input-output systems. Processes such as photosynthesis or foraging represent the input of materials and energy, and production is represented by their offspring (Pianka 2000). Therefore, it is possible to understand the dynamics of a population and to evaluate how both abiotic and biotic interactions affect a given specimen's fitness (Camara et al. 2011).

Environmental stressors can alter a species' physiology and biochemistry, including antioxidant defenses, metabolic parameters and morphometric indexes (Li et al. 2010). In most contaminated environments, organisms are exposed to mixtures of pollutants, the synergistic or antagonistic effects of which are difficult to interpret. These effects are predicted exclusively in chemical analyses (Regoli et al. 2004). In recent years, different biological tests have been developed to evaluate biological responses to environmental pollution at molecular, cellular and organism levels (Shugart et al. 1992; Viarengo et al. 1997). However, it is still complex and difficult to predict

[1]Department of Biology, Center for Chelonia Studies (CEQ) and the Hemoglobin and Genetics of Hematologic Diseases Laboratory (LHGDH), IBILCE, UNESP–Sao Paulo State University, Sao Jose do Rio Preto, SP 15054-000, Brazil.
*Corresponding author: larissa_biorp@yahoo.com.br

the biological effects caused by different classes of chemicals during co-exposures when reciprocal interactions and indirect mechanisms can either enhance or suppress the expected responses (Benedetti et al. 2007).

According to Green (2001), an important determinant of fitness is the "condition" of the body. "Condition" is defined by the magnitude of organic energy reserves, which affects the probability of survival at both the individual and the population level (Ardia 2006; Jennings et al. 2006). Body condition indexes are used to assist in describing the health and welfare of several wild species, and to define the influence of factors such as environmental degradation, life history parameters and ecological interactions on animal health (Stevenson and Woods 2006).

Changes in condition resulting from changes in biochemical composition and tissue mobilization of energetic reserves may be related to the seasons, periods of life, reproduction, health and/or exposure to stressors (Barton et al. 2002). The reproductive success and, therefore, the population dynamics depend on this condition (Stevenson and Woods 2006; Labrada-Martagón et al. 2010). Therefore, understanding the causes of condition variation allows scientists to assess and to predict demographic changes, as well as the population's ability to respond to environmental stressors (Sandeman et al. 2008).

Usually there is interest in evaluating the conditions of animals using non-invasive techniques and submitting the individuals to the lowest stress levels possible. This can be accomplished by the use of morphometric data, which, when combined in an index, provides indirect energy storage estimates. This is a common practice in studies with fish, which estimate their conditions using weight and length (Camara et al. 2011). Weight-length relationships are used to estimate the weight corresponding to a given length, and condition factors are used to compare the "condition", "fatness", or "well-being" of fish, based on the assumption that heavier fish of a given length are in better condition (Le Cren 1951; Tesch 1968; Froese 2006).

This index, therefore, has been accepted as an integrative indicator of a fish's overall condition. It can provide information on an animal's ability to tolerate environmental stressors. The condition factor can be estimated for comparative purposes to assess the impact of environmental alterations to fish performance (Barton et al. 2002), but it does not give information on specific responses to toxic substances in the environment (Linde-Arias et al. 2008). This is a low-cost index and it can be assessed with simple morphological data, making it a valuable tool to indicate the general effects of pollution on fish (Van der Oost et al. 2003). For example, in a multi-biomarker approach, which consists of the combined use of different biomarkers can both indicate the exposure to contaminants as the quantification of their effects on the health of organisms (Broeg and Lehtonen 2006; Humphrey et al. 2007).

In this chapter, we present a review about the use and evaluation of length and weight data as factors that are indicative of the condition, with a description of mathematical analyses, experimental design and application in ecotoxicological studies.

The Condition Factor

In the beginning, with the need to provide metric relationships for the dimensions and development of the body, Galileo Galilei (1564–1642) proposed that the volume increases as the cube of linear dimensions, whereas strength increases only as the square. In *The Principles of Biology* (1864–1867), Herbert Spencer revisited the first part of Galileo's law and suggested the "cube law": "In similarly shaped bodies, the masses, and therefore the weights, vary as the cubes of the dimensions."

In 1904, Thomas Wemyss Fulton evaluated approximately 6,000 fish specimens and stated that the cubic relationship does not apply to fishes with precision. In his analysis, Fulton observed that most species increase in weight more than they increase in length, and he concluded that the weight for a given length could change in different species. These changes were associated at different places and at certain times of year, particularly in periods associated with reproduction. Fulton noted that in early stages, fishes grow more in length than they grow in other dimensions, and that the variation in weight at a given size in the same species increases greatly as the fish grows in length (Froese 2006). Thus, Fulton was the first scientist who linked weight and length measurements to condition (Nash et al. 2006).

Fulton's Condition Factor

Fulton did not abandon the cube law. He presented tables for calculating weight from length based on a fixed weight-length ratio, and he elaborated the equation which is today known as Fulton's condition factor:

$K = 100 \ W/L^3$

where K is a Fulton's condition factor that is calculated with W (whole body wet weight in grams) and L (length in cm); the factor 100 is used to bring K close to unity.

Despite the shortcomings already pointed out by Fulton in 1904 and confirmed by subsequent researchers, the cube law remained in use in fisheries for estimating weight from length for two more decades (Froese 2006). Currently, this index is heavily criticized due to its cubic (isometric) variation (Vazzoler 1996), which makes it dependent on length. For comparative purposes, it is limited to fish samples of similar size unless the scaling coefficient of population or species is exactly 3.0 (Pope and

Kruse 2001; Couture and Rajotte 2003). Baigún et al. (2009), conducted tests to assess the applicability of different indexes to *Odontesthes bonariensis* populations. They found that Fulton's condition factor must be disregarded in cases in which the population presents allometric growth. Most research that uses Fulton's condition factor considered intervals that were shorter in length, which did not require a comparison between length classes of species (Tanck et al. 2001; Chellappa et al. 2003; Van den Heuvel et al. 2008).

Weight-length Relationship and Condition

The analysis of length-weight data has usually been directed towards two rather different objectives: 1) towards mathematically describing the relationship between length and weight, primarily so that one may be converted into the other; and 2) to measure the variation from the expected weight for length of individual fish or relevant groups of individuals as indications of fatness, general "well-being", gonad development, etc. More specifically, the length-weight relationship is associated with the first objective, while the term *condition* is applied to the length-weight analysis of the second objective (Le Cren 1951).

In 1928, Keys formally established the form of the weight-length relationship (WLR) that is used today, as well as its logarithmic equivalent. These equations are presented below:

WRL: $W = aL^b$

Logarithmic equivalent: $log\ W = log\ a + b\ log\ L$

where W and L are the previously defined variables, and a and b are parameters to be calculated. Hile (1936) evaluated the publications and found that within a species, "the values of the coefficient a depended primarily not on the heaviness of the fish, but rather, on the value of the exponents. A large value of b is associated with a small value of the coefficient a and the reverse." This author presented a first interpretation of the exponent b, namely that the difference from 3.0 indicates the direction and "rate of change of form or condition." The exponent b when < 3.0 indicates a decrease in condition or elongation in form with increase in length, whereas b > 3.0 indicates an increase in condition or in height or width with increase in length. This exponent usually lies between 2.5 and 4.0 (Hile 1936; Martin 1949); however, for an ideal fish, it maintains the same shape ($b = 3.0$) (Allen 1938).

Le Cren (1951) gave an excellent review of WLRs and the condition factor. This author evaluated the principal factors that can affect the value of K and grouped them into three different groups. First, the fish does not in fact obey the cube law in its weight-length relationship, as noted previously, because the length itself and any correlated factor will affect the values of K. Further, factors such as age, sex or maturity, which may affect the value of *b*, may in turn affect the values of K. Secondly, the values of K may be affected by selection in sampling. Several authors described the effect of gill-nets on the computation of WRL, which may also be selective for the condition factor (Farran 1936; Deason and Hile 1947). Thirdly, there are certain features that are usually associated with K—namely environment, food supply and degree of parasitism (Brown 1946). These three factors can affect the condition factor, and, because of this, K interpretation is difficult and often leads to erroneous results (Le Cren 1951).

Therefore, the conclusion is that, when using the condition factor, great care must be exercised to decide exactly what purpose the information on condition is required for, and whether the condition factor will yield this information and give results that are unaffected by the effect of variables other than those being studied (Le Cren 1951).

How to use the Condition Factor

Despite the problems presented above, there are alternative methods for analyzing condition that may be more suitable. It is relatively easy to eliminate the effect of length and correlated factors on K by calculating a "condition factor" based not on the "ideal" WLR, where the equation exponent is 3.0, but on empirical data based on WRL, where the equation exponent is *b*. In 1951, Le Cren proposed the idea of the *relative condition factor*, designated by K_n, to distinguish it from the condition factor K based on the cube law, and then proposed the following equation:

$$K_n = W / aL^b$$

where the parameter *a* is the coefficient of the arithmetic form of the WLR and the intercept of the logarithmic form, and the parameter *b* is the exponent of the arithmetic form of WLR and the regression line slope of the logarithmic form (Froese 2006).

In general, condition factor studies have presented length and weight in centimeters and grams, but other units can be used. This did not affect the exponent *b*, but the intercept *a* needed to be converted into the examples presented below ($a^* =$ parameter *a* with length in cm and weight in g):

1) Length was given in mm and weight in g: $a^* = a10^b$

2) Length was given in cm and weight in kg: $a^* = a1000$

3) Length was given in mm and weight in mg: $a^* = 10^b/1000$

4) Length was given in mm and weight in kg: $a^* = a10^b1000$

The Weight-length Relationship Evaluation

Determining the WLR is important for plotting the data (weight and length) in a log-log plot with regression line. The logarithmic presentation not only linearizes the relationship but also corrects for the increase in variation according to length (Froese 2006).

The Parameter *b*

As mentioned previously, the parameter *b* is the exponent of the arithmetic form of the WLR and the slope of the regression line in logarithmic form (Fig. 12.1).

When *b* = 3, the small specimens in the sample have the same form and condition as large specimens. On the other hand, when the value of *b* is > 3.0, the large specimens have increased in height or width more than in length, either as the result of a notable ontogenetic change in body shape with a change in size, which is rare, or because most large specimens in the sample were thicker than the small specimens. When *b* < 3.0, it likely means that large specimens have changed their body shape in that they have become more elongated, or that small specimens were in better nutritional condition at the time of sampling (Froese 2006).

It is important to evaluate specific aspects of parameter *b*. When evaluating the exponent *b* of single WLR, one should consider differences in condition between small and large individuals in the respective area at that point in time. An important observation is that allometric growth

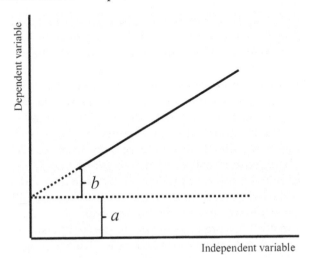

Figure 12.1. Graphical representation of the regression line. The x-axis plots the independent variable data (size), while the y-axis is the variable dependent data (weight).

patterns are rare and should be accompanied by an examination of growth and an evaluation of any potential benefits associated with ontogenetic changes in the body.

The Parameter *a*

The parameter *a* is the intercept of the logarithmic form of the WLR equation (Fig. 12.1). Based on the logarithmic form of WLR and from the corresponding plot of weight versus length, it is evident that every decrease in the slope of the regression line will lead to an increase in the intercept, and vice-versa. Froese (2000) points out that if several WLR are available for a species, then a plot of log *a* over *b* will form a straight line and can be used to detect outliers. The strong relationship between parameters *a* and *b* is linearized in a plot of log *a* over *b* and helps in detecting questionable WLRs (Froese 2006).

Making Decisions for the Calculation of the Condition Factor

In an excellent revision about the use and evaluation of the condition factor, Froese (2006) suggested some important observations to be made during the sampling and analysis of data:

1) One must pay attention to the specimen sampling equipment, because there is the possibility of introducing a bias with respect to length or weight! The gill nets tend to select fat fish among the shorter ones and thin fish among the longer ones, which introduces a bias in *b* (Kipling 1962). In exposure experiments, it is always important to pay attention to the morphometric parameters, and to organize the specimens into homogeneous experimental groups with respect to the morphometric data.

2) Pay attention to the size range when selecting specimens for measuring weight-length data. Do not evaluate fish that have not yet obtained an adult body shape (Le Cren 1951; Carlander 1969; Safran 1992). The best way to evaluate animals that are at different stages of development and/or age is to estimate the WLRs separately (Froese 2006).

3) When planning data collection, try to sample as many months as possible. This is important for detecting seasonal variation.

4) Evaluate the differences between the sexes. When the differences are important, estimate separate relationships for males and females.

5) The way that WLR is presented is very important. Make sure to indicate the number of specimens processed, as well as the range and type of length measurements, units, parameters *a* and *b* with their respective 95% confidence limits, and the coefficient of correlation.

Ecotoxicological Analysis: The Effect of Pollutants on the Condition Factor

In ecotoxicological evaluations, the association of data generated by the assessment of detoxification enzymes and antioxidant capacity with body condition is very common. The association of different biomarkers is important, because single biomarkers cannot reflect the impairment of an organism's health. However, when the use of several biomarkers can reflect an exposure to contaminants and assist in quantifying their effects on the health of organisms, a more comprehensive and integrative assessment of environmental quality is possible (Fonseca et al. 2011).

The contamination of aquatic ecosystems by heavy metals has seriously garnered worldwide attention (Wagner and Boman 2003; Yilmaz et al. 2007). Authman (2008) evaluated the heavy metal concentrations in water and in the muscles of *Oreochromis niloticus* in order to study the potential human risk of consumption and the relationship between the heavy metal load of fish and some of their biological aspects (age, size, sex and condition factor). The results showed that there were positive relationships between fish length and weight, age, and metal concentration. The increases in copper, iron, lead, manganese and zinc values were observed with increasing fish length and weight, and these results were enhanced by a positive correlation between the metals cited and the body parameters. The author suggested that the positive correlation between metal concentration and fish length, weight and age may be due to the loss of homeostasis capacity of *O. niloticus* under chronic metal exposure leading to bioaccumulation.

In the same study, the negative relationship between the heavy metal concentration and the condition factor reveals the relative dilution effect of the lipid content of tissues, which changes according to the reproductive period and the age of the fish (Weatherly and Gill 1987). Another factor that affects the concentration of metals is the sex of the fish, since the differences in nature of hormones and the available number of active sites in the protein and cytochrome P450 between female and male fish may account for this behavior (Al-Yousuf et al. 2000).

Initially, a reduction in K would be expected in cases of contaminated sites, changes in the environment and animal stress, but an increase in the stability and the K values in impacted areas has been previously reported (Wolf and Wolfe 2005). Knapen et al. (2009) showed that a specific population of *Gobio gobio* located in a pollution gradient of cadmium and zinc presented a higher-than-average condition factor than the reference populations, and the population from the most contaminated site exhibited the highest condition factor. This observation indicates that the condition factor is an important variable to consider in studies that evaluate the contamination effects and the implication in adaptation of populations inhabiting polluted environments.

The condition factor is an important parameter to consider in metal stress evaluation, since it can allow for a clearer distinction between yellow perch (*Perca flavescens*) of clean and polluted areas. Couture and Rajotte (2003) identified that the condition factor can show evidence of stress caused by metals, and that this morphometric indicator should be coupled with a measurement of the concentration of metals in animal tissue.

Conclusion

The condition factor is a relevant parameter to consider in morphometric and physiological studies, including ecotoxicology analyses. The index reflects the body condition and how it is associated with environmental changes. However, the analysis of this parameter must follow certain data collection routines in order to minimize possible biases, and situations such as sex and reproduction periods should be considered in the evaluation in order to make consistent inferences in relation to the biology of the organism being studied.

Acknowledgements

We are grateful to Luis Dino Vizotto, Ph.D, for his important suggestions and the CEQ students Tiago Silva, Maria Isabel Silva, Nathalia Rossigali, Vanessa Cardoso e Jéssica Bacchi.

Keywords: condition factor, weight-length relationship, morphometric and physiological analysis.

References

Allen, K.R. 1938. Some observation on the biology of the trout (*Salmo trutta*) in Windermere. J. Anim. Ecol. 7: 333–349.

Al-Yousuf, M.H., M.S. El-Shahawi and S.M. Al-Ghais. 2000. Trace metals in liver, skin and muscle of *Lethrinus lentjan* fish species in relation to body length and sex. Sci. Total Environ. 256: 87–94.

Ardia, D. 2006. Geographic variation in the trade-off between nestling growth rate and body condition in the tree swallow. Condor. 108: 601–611.

Authman, M.M.N. 2008. *Oreochromis niloticus* as a biomonitor of heavy metal pollution with emphasis on potential risk and relation to some biological aspects. Global Vet. 2(3): 104–109.

Baigún, R.M., D.C. Colautti and F. Grosman. 2009. Assessment of condition in pejerry *Odontesthes bonariensis* (Atheriniformes: Atherinopsidae) populations: which index works best? Neotrop. Ichthyol. 7: 439–446.

Barton, B.A., C.B. Schreck and L.D. Barton. 1987. Effects of chronic cortisol administration and daily acute stress on growth, physiological conditions, and stress responses in juvenile rainbow trout. Dis. Aquat. Organ. 2: 173–185.

Barton, B.A., J.D. Morgan and M.M. Vijayan. 2002. Physiological and condition-related indicators of environmental stress in fish. *In:* S.M. Adam (ed.). Biological Indicators of Aquatic Ecosystem Stress. American Fisheries Society, Bethesda, USA, pp. 111–148.

Benedetti, M., G. Martuccio, D. Fattorini, A. Canapa, M. Barucca, M. Nigro and F. Regoli. 2007. Oxidative and modulatory effects of trace metals on metabolism of polycyclic aromatic hydrocarbons in the Antartic fish *Trematomus bernacchii*. Aquat. Toxicol. 85: 167–175.

Broeg, K. and K.K. Lehtonen. 2006. Indices for the assessment of environment pollution of the Baltic Sea coasts: integrated assessment of a multi-biomarker approach. Mar. Pollut. Bull. 53: 508–522.

Brown, M.E. 1946. The growth of Brown trout (*Salmo trutta* Linn.). II. The growth of two-year-old trout at a constant temperature of 11.5°C. J. Exp. Biol. 22: 130–144.

Camara, E.M., E.P. Caramaschi and A.C. Petry. 2011. Fator de condição: bases conceituais, aplicações e perspectivas de uso em pesquisas ecológicas com peixes. Oecol. Aust. 15(2): 249–274.

Carlander, K.D. 1969. Handbook of freshwater fishery biology, Vol.2. The Iowa State University Press, USA.

Chellappa, S., M.R. Câmara and N.T. Chellappa. 2003. Ecology of *Cichla monoculus* (Osteichthyes: Cichlidae) from a reservoir in the semi-arid region of Brazil. Hydrobiologia. 504: 267–273.

Couture, P. and J.W. Rajotte. 2003. Morphometric and metabolic indicators of metal stress in wild yellow perch (*Perca flavescens*) from Sudbury, Ontario: a review. J. Environ. Monitor. 5: 216–221.

Deason, H.J. and R. Hile. 1947. Age and growth of the kiyi, *Leucichthys kiyi* Koelz, in Lake Michigan. T. Am. Fish. Soc. 74: 88–142.

Farran, G.P. 1936. On the mesh of herring drift-nets in relation to the condition factor of the fish. J. cons. – Cons. Int. explor. mer. 11: 43–52.

Fonseca, V.F., S. Franc, A. Seragim, R. Company, B. Lopes, M.J. Bebianno and H.N. Cabral. 2011. Multi-biomaker reponses to estuarine habitat contamination in three fish species: *Dicentrarchus labrax*, *Solea ssenegalensis* and *Pomatoschistus microps*. Aquat. Toxicol. 102: 216–227.

Froese, R. 2000. Evaluation length-weight relationships. *In:* R. Froese and D. Pauly (eds.). FishBase 2000: Concepts, Design and Data Sources.. ICLARM, Los Baños, Laguna, Philippines. p. 133.

Froese, R. 2000. Cube law, condition factor and weight-length relationships: history, meta-analysis and recommendations. J. Appl. Ichthyol. 22: 241–253.

Green, A.J. 2001. Mass/length residuals: measures of body condition or generators of spurious results? Ecology 82: 1473–1483.

Hile, R. 1936. Age and growth of the cisco *Leucichthys artedi* (Le Sueur), in the lakes of the north-eastern highlands, Wisconsin. Bull. U.S. Bureau Fish. 48: 211–317.

Humphrey, C.A., S.C. King and D.W. Klumpp. 2007. A multibiomarker approach in barramundi (*Lates calcarifer*) to measure exposure to contaminants in estuaries of tropical North Queensland. Mar. Pollut. Bull. 54: 1569–1581.

Jennings, N., R.K. Smith, K. Hackländer, S. Harris and P.C.L. White. 2006. Variation in demography, condition and dietary quality of hares, *Lepus europeaus*, from high-density and low-density populations. Wildlife Biol. 12: 179–189.

Keys, A.B. 1928. The weight-length relationship in fishes. Proceedings of the National Academy of Science, Vol. XIV, n.12, USA, pp. 922–925.

Kipling, C. 1962. The use of scales of the brown trout (*Salmo trutta* L.) for the back-calculation of growth. J. cons. – Cons. Int. explor. mer. 27: 304–315.

Knapen, D., H. de Wolf, G. Knaepkens, L. Bervoets, M. Eens, R. Blust and E. Verheyen. 2009. Historical metal pollution in natural gudgeon populations: inferences from allozyme, microsatellite and condition factor analysis. Aquat. Toxicol. 95: 17–26.

Labrada-Martagón, V., L.C. Méndez-Rodriguez, S.C. Gardner, V.H. Cruz-Escalona and T. Zenteno-Savín. 2010. Health indices of the green turtle (*Chelonia mydas*) along the Pacific Coast of Baja California Sur, Mexico. II. Body Condition Index. Chelonian Conserv. Bi. 9: 173–183.

Le Cren, E.D. 1951. The length-weight relationship and seasonal cycle in gonad weight and condition in the perch (*Perca fluviatilis*). J. Anim. Ecol. 20(2): 201–219.

Li, Z.H., V. Zlabek, J. Velisek, R. Grabic, J. Machova and T. Randak. 2010. Physiological condition status and muscle-based biomarkers in rainbow trout (*Oncorhynchus mykiss*), after long-term exposure to carbamazepine. J. Appl. Toxicol. 30: 197–203.

Linde-Arias, R.A., F.A. Inácio, C. de Albuquerque, M.M. Freire and C.J. Moreira. 2008. Biomarkers in an invasive fish species, *Oreochromis niloticus*, to assess the effects of pollution in a highly degraded Brazilian River. Sci. Total Envir. 399: 186–192.

Martin, W.R. 1949. The mechanics of environmental control of body form in fishes. University Toronto Studies Biology 58 (Publ. Ont. Fish. Res. Lab. 70): 1–91.

Nash, R.D.M., A.H. Valencia and J.G. Audrey. 2006. The origin of Fulton's condition factor—setting the record straight. Fisheries. 31(5): 236–238.

Pianka, E.R. 2000. Evolutionary ecology. Sixth Edition. Addison Wesley Longman, San Francisco, USA.

Pope, K.L. and C.G. Kruse. 2001. Assessment of fish condition data. *In:* C. Guy and M. Brown (eds.). Statistical Analyses of Freshwater Fisheries Data. American Fisheries Society Publication, North Bethesda, USA, pp. 51–56.

Regoli, F., G. Frenzilli, R. Bocchetti, F. Annarumma, V. Scarcelli, D. Fattorini and M. Nigro. 2004. Time-course variations of oxyradical metabolism, DNA integrity and lysosomal stability in mussels, *Mytilus galloprovincialis*, during a field translocation experiment. Aquat. Toxicol. 68(2): 167–178.

Safran, P. 1992. Theoretical analysis of the weight-length relationship in fish juveniles. Mar. Biol. 112: 545–551.

Sandeman, L.R., N.A. Yaragina and C.T. Marshall. 2008. Factors contributing to inter- and intra-annual variation in condition of cod *Gadus morhua* in the Barents Sea. J. Anim. Ecol. 77: 725–734.

Shugart, L.R., J.F. McCarthy and S.H. Halbrook. 1992. Biological marker of environmental and ecological contamination: an overview. Risk Anal. 12: 345–360.

Stevenson, R.D. and W.A. Woods. 2006. Condition indices for conservation: new uses for evolving tools. Integr. Comp. Biol. 46: 1169–1190.

Tanck, M.W.T., K.J. Vermeulen, H. Bovenhuis and H. Komen. 2001. Heredity of stress-related cortisol reponse in androgenic common carp (*Cyrpinus carpio* L.). Aquaculture. 199: 283–294.

Tesch, F.W. 1968. Age and growth. *In:* W.E. Rickers (ed.). Methods for Assessment of Fish Production in Fresh Water. Blackwell Scientific Publications, Oxford, UK, pp. 93–123.

Van den Heuvel, M.R., M.J. Landman, M.A. Finely and D.W. West. 2008. Altered physiology of rainbow trout in response to modified energy intake combined with pulp and paper effluent exposure. Ecotox. Environ. Safe. 69: 187–198.

Van der Oost, R., J. Beyer and N.P.E. Vermeulen. 2003. Fish bioaccumulation and biomarkers in environmental risk assessment: a review. Environ. Toxicol. Phar. 13: 57–149.

Vazzoler, A.E.A.M. 1996. Biologia de peixes teleósteos: teoria e prática. Editora da Universidade Estadual de Maringá, BR.

Viarengo, A., E. Bettell, R. Fabbri and B. Burlando. 1997. Heavy metal inhibition of EROD activity in liver microsomes from the bass *Dicentrarchus labrax* exposed to origanic xenobiotics: role of GSH in the reduction of heavy metal effects. Mar. Environ. Res. 44(1): 1–11.

Wagner, A. and J. Boman. 2003. Biomonitoring of trace elements in muscle and liver tissue of freshwater fish. Spectrochim. Acta B 58: 2215–2226.

Weatherly, A.H. and H.S. Gill. 1987. The biology of fish growth. Academic Press, USA.

Wolf, J.C. and M.J. Wolfe. 2005. A brief overview of nonneoplastic hepatic toxicity in fish. Toxicol. Pathol. 33: 75–85.

Yilmaz, F., N. Özdemir, A. Demirak and A. Levent Tuna. 2007. Heavy metal levels in two fish species *Leuciscus cephalus* and *Lepomis gibbosus*. Food Chem. 100: 830–835.

CHAPTER 13

Behavioral Biomarkers and Pollution Risks to Fish Health and Biodiversity

Paulo Sérgio Martins de Carvalho

Role of Behavioral Endpoints in Ecological Risk Assessment and Ecotoxicology

Ecological risk assessment is a tool used to support the risk-based, decision making process in environmental issues (Clark et al. 1999). Both ecological risk assessment and ecotoxicology can use endpoints of contaminant effect from all levels of biological organization, ranging from genes to ecosystems. These endpoints provide different and complementary perspectives, though. Effects of contaminants at lower levels of biological organization might provide information on specific mechanisms of action and typical effects of certain groups of chemicals, such as acetylcholinesterase (AChE) inhibition by organophosphates (see Chapter 4 for more information). AChE inhibition may be critical to the identification of an important source of pollution if it is detected in the brain of an important fish species whose population is in danger, for example. However, sublethal AChE inhibition per se does not tell much about the overall ecological well being of the individuals in this population. Endpoints of effect related to the ability of

Av. Prof. Moraes Rego, 1235, Centro de Ciências Biológicas–Depto. de Zoologia–Cidade Universitária, Recife–PE–CEP: 50670-901.
Email: pcarvalho@ufpe.br

these fish to swim, catch their prey, avoid predation and find partners to reproduce are considered more ecologically relevant, although these types of effect alone usually do not allow the identification of a specific group of contaminants as the causal factors. Ecological relevance is a rather subjective term, which depends on what risk managers (decision-makers) and risk assessors (ecotoxicologists) agree upon. However, it is generally accepted that contaminant effects at the population level should be avoided, and that the evaluation of evidence of contaminant related effects at this level of biological organization should be the target for fish and wildlife contaminant risk assessments (Clark et al. 1999). A less conservative application of the term defines as ecologically relevant any effect that would probably decrease the chances of survival or the reproductive fitness of individuals, a definition which would encompass a broader array of endpoints, including those at suborganismic levels.

Fish have been used to provide an early warning of the presence of toxic materials in aquatic environments for a long time, and behavioral parameters such as swimming movements and breathing rates have been monitored in selected individuals either in the laboratory or confined near effluent disposal sites, so that alterations in these endpoints would signal alarm of toxic releases to aquatic systems (Cairns et al. 1973; Cairns and van der Schalie 1980). This approach, focused on real-time detection of toxicity, has evolved significantly in recent years with the advance of computer, video and software automation (Kane 2005; Gerhardt 2007). These advancements have allowed the utilization of automated systems for laboratory, on-site or field monitoring of behavioral effects related to locomotion and ventilation in fish, such as the non-optical multispecies freshwater biomonitor system, based on monitoring of changes in an electrical field provoked by movements of the organism inside a chamber that can be applied to fresh and marine water species (Mohti et al. 2012). Other technologies involve video-tracking systems to evaluate behaviors in experimental animals, allowing the detailed recording of swimming speeds, trajectories, distances traveled, complex social behavior interactions together with a wide range of analysis parameters, such as the software packages Ethovision (Noldus Information Technology, Wageningen, The Netherlands) or Smart (PanLab Harvard Apparatus, Holliston, Mass., USA).

The use of the term "early warning for detection of real-time toxicity events" mentioned above differs from the equally common term "early warning systems of potential impending population level effects". The latter statement frequently used in fish ecotoxicology articles expresses the idea that we can have early warning suborganismal biological metrics responding before and therefore possibly anticipating that measurable effects on individuals and populations would happen after longer exposures (Forbes et al. 2006). However, as part of the homeostatic responses at the organismal and suborganismal levels, suborganismal effects of contaminants may not be obviously related to higher level effects (Calow 1994; Forbes et al. 2011). The importance of the evaluation of "ecologically relevant" endpoints in aquatic ecotoxicology, in conjunction with the potential causal mechanisms associated with them has been stressed since the 1980s (Cairns 1989).

Mechanisms of toxic action are characterized by the primary interaction of a contaminant with biomolecules at its site of action, whereas modes of toxic action are defined as secondary effects based on physiological and behavioral endpoints that characterize some kind of adverse biological response, including effects on reproduction and growth (Vogl et al. 1999). Recently, the concepts of mechanism and modes of action have been framed into the Adverse Outcome Pathway (AOP) concept, which is based on a better understanding of mechanistic pathways of toxicity and linkages between levels of effects. Furthermore, an AOP requires an anchor to both a molecular initiating event and an adverse outcome relevant to risk assessment (Ankley et al. 2010). It is commonplace in ecotoxicological literature to find information about the effects of chemicals at one level of biological organization, together with speculations about potential ecologically relevant consequences at the next higher level. The challenge in fish ecotoxicology is to gain a proper understanding of the effects of chemicals across different levels of biological organization, including mechanisms at the molecular, cellular and physiological levels that lead to individual level performance deficits. Fish are undoubtedly relevant to society, and are an ideal group for behavioral tests based on ecologically relevant behaviors that are readily observed and quantified in a controlled setting (Kane 2005). Additionally, fish physiology is well developed, and the establishment of the desired links between physiologic effects and corresponding behavioral changes is attainable (Scott and Sloman 2004; Schlenk et al. 2008). Furthermore, it is essential to take into account ecological mechanisms where individual level behavioral endpoints can provide the basis for the prediction of potential population or higher level effects.

Concepts about Behavioral Biomarkers in Ecotoxicology

In ecotoxicology, behavioral endpoints provide an important perspective between organisms and how they interpret and respond to their environment, between ecophysiology and population ecology (Little and Brewer 2005). These essential links need to be better understood (Amiard-Triquet 2009) in order to assign proper risk factors to environmental contaminants in biodiversity protection issues. Behavioral parameters have been included in the biomarker concept in Ecotoxicology by Depledge et al. in 1995. Behavioral biomarkers of contaminants in fish are defined as quantitative measurements of adaptive abilities that contribute to their Darwinian fitness, and so they should be important to survival, growth and/or reproduction, have relatively low variability and indicate a dose-response relationship (Beitinger 1990). Furthermore, behavioral endpoints measured in individuals integrate important physiologic processes essential for life, allowing adjustment to internal homeostatic processes that enable efficient resource exploration in a particular habitat (Little and Brewer 2005).

Several studies have shown that altered behavioral functions are often the first responses exhibited by contaminant exposed fishes (Little et al. 1990). Additionally, mal-adaptive behavioral effects frequently occur at concentrations that are significantly below the LC50 (concentration lethal to 50 percent of the test population). Contaminant concentrations that induce changes in swimming, feeding behavior or predator avoidance by exposed fish can be lower than 2 percent of the LC50 (Beitinger 1990; Little and Finger 1990; Little et al. 1993).

Population declines of fishery resources can occur due to acute mortality of individuals exposed to lethal concentrations of contaminants in a situation of major accidents. However, sublethal chronic behavioral effects can lead to so called ecological death (Scott and Sloman 2004), as the individuals can no longer feed, grow, reproduce and avoid predators satisfactorily. Improvements in environmental laws and their enforcement regarding the control of effluents tend to make acute mortalities of fish due to major chemical spills less frequent than sublethal effects of contaminants in aquatic systems. If this scenario does not change, evaluation of behavioral effects will become even more important, as they are considered more sensitive indicators of potential for impacts on ecological survival in the field than are measures of lethality (Robinson 2009). As an example, bluegill *Lepomis macrochirus* exposed to fluorine exhibited a reduction in feeding in laboratory studies that accurately predicted a reduction in growth and survival in more environmentally realistic studies, where the species was exposed to the same chemical within an aquatic mesocosm community (Finger et al. 1985).

Recruitment of early life stages of fish to the adult population is a fundamental process in the maintenance of any fish population. It is therefore important to evaluate if chemical contamination is a significant causal agent to recruitment issues, separately or in combination with other ecological factors in the life cycle of fish. Information derived from ecological and physiological bioenergetics theory have been used to generate the so called individual-based models (IBMs), which use attributes of individual fish through time as input, aggregate them, and generate insights into population function (Van Winkle and Rose 1993). Measuring how chemical contaminants affect these individual level performance parameters has been proposed as an efficient way of translating individual level effects into growth, survival and reproduction in population dynamics models (Rose et al. 1999). Figure 13.1 provides a conceptual flow diagram illustrating the sequence of processes a model individual fish goes through during growth towards reproductive size, including prey search and encounter, prey capture, energy assimilation and growth, predator evasion, and eventually reproduction, when appropriate size is reached. Many of these processes important to survival of larval and juvenile fish are size dependent, and therefore all functions in the flow diagram are size dependent (Letcher et al. 1996).

The potential of the use of the AOP concept together with population models in predictive ecotoxicology has been recently emphasized, and behavioral endpoints are an important component of this approach (Kramer et al. 2011). An example is the study based on laboratory and field data coupled with a modeling approach used to demonstrate that sublethal exposure to organophosphate and carbamate insecticides can

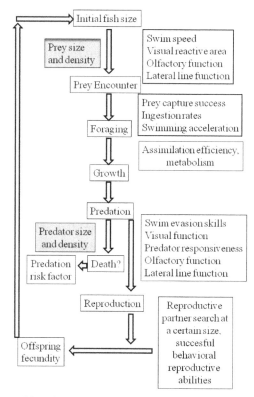

Figure 13.1. Conceptual flow diagram of a theoretical individual based model for fish involving submodels for prey encounter, foraging, growth, predation risk and reproduction. Major behavioral aspects affecting each submodel are inside boxes on the right. Characteristics of the larvae's environment (food and predators) are inside gray boxes.

lead to acetylcholinesterase inhibition, reductions in feeding activity, reduced growth of juvenile chinook salmon (*Oncorhynchus tshawytscha*), and a reduction of productivity of wild salmon populations (Baldwin et al. 2009).

Behavioral Effects of Contaminants in Sensorial and Neuromotor Systems

During early life stages of fishes, specific behaviors are established when sense organs and the respective motor and neuronal circuitry associated with them become functional (Blaxter 1986; Noakes and Godin 1988; Browman 1989). In fact, during initial and later stages of development, practically all fish depend on the proper functioning of all of their sense organs, including the visual, olfactory and lateral line system (Fig. 13.2) for detection and capture of prey, as well as for predator evasion, selection of adequate habitats and reproduction. Theoretically, all submodel components illustrated in Fig. 13.1 can be influenced by the effects of contaminants in sense organs and associated neuromotor systems. Inorganic and organic pollutants such as metals and organometals, organophosphate and carbamate insecticides,

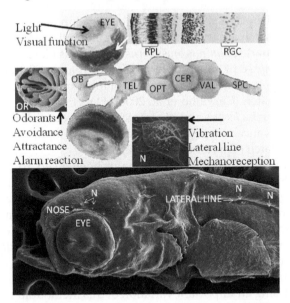

Figure 13.2. Overview of major sensory organs in fish that can be affected by chemical contaminants. Eye and retina with retinal photoreceptor layer (RPL) and retinal ganglion cell layer (RGC). Olfactory system with detail of olfactory rosette (OR) and olfactory bulb (OB). Typical brain of a teleost with telencephalon TEL; (optic tectum OPT); CER: cerebellum; VAL: vagal lobe; SPC: spinal cord. Lateral line with detail of a neuromast (N) in head and trunk of a larval tambaqui *Colossoma macropomum*.

aromatic hydrocarbons, dioxin, among other chemicals, can cause partial or complete failure of sense organs' function, therefore influencing fish behavior by reducing or changing the flow of sensory information channeled to the brain (Blaxter and Hallers-Tjabbes 1992).

Visual System Ecotoxicology

Proper vision is fundamental for survival especially during early developmental phases, as most fish are essentially sight feeders during the initial exogenous feeding stage (Gerking 1994). Additionally, visual cues are also important for predator avoidance at early and later stages of development (Fuiman and Magurran 1994), whereas color vision can be important for mate selection during reproduction (Sabbah et al. 2010). The duplicity theory of vision was established in the 19th century, and its fundamental concept is based on the presence of two types of retinal neurons or photoreceptors located in the photoreceptor layer in the retina (Fig. 13.2), cones that are involved in photopic or daylight visual function, and rods involved in night or scotopic visual function in teleost fishes (Fernald 1990). Phototoreceptor neurogenesis in teleost fish has been classified in three basic schemes: indirect, intermediate and direct development (Evans and Fernald 1990). Tropical species with indirect development are common, and they typically hatch as larvae where the eye and retina are not formed yet, such as the dourado *Salminus brasiliensis*. Salmonids like the rainbow trout *Oncorhynchus mykiss* fit into the intermediate developing fish, where the eye is already formed at hatch, but vision is not functional until approximately 10 days after hatch (Carvalho et al. 2004). Direct developing species like the medaka *Oryzias latipes* and the tropical viviparous fish *Poecilia vivipara* hatch at an advanced stage of visual system development, and vision is functional at hatch (Carvalho et al. 2002).

Different behavioral methods have been used to test whether contaminants affect fish vision. One of them is based on instinctive behaviors presented by the fish where they follow the movement direction of black and white stripes within their field of vision, either by swimming in the same direction of the rotating stripes (optomotor response), or by making eye movements that follow the stripes movement (optokinetic responses). The optomotor response has been used to demonstrate the effects of the organophosphate insecticide diazinon in bluegill *Lepomis macrochirus* (Dutta et al. 1992). Using an important variation of this methodology Carvalho et al. (2008) exposed the dourado *Salminus brasiliensis* to phenanthrene and quantified the visual acuity angle of the tested fish also based on optomotor and optokinetic responses. A visual acuity angle can be calculated based on the smallest stripe width to which the fish responds positively (Carvalho et al. 2002), if the testing system allows the manipulation of the width of the stripes, not available to Dutta et al. (1992). The actual measurement of a visual acuity angle in ecotoxicological studies has important ecological implications because the visual acuity angle can be expressed in terms of a reaction distance to a prey of certain size, an important parameter used in fish foraging models (Blaxter 1986; Letcher et al. 1996). Fish with poor vision will have a larger acuity angle and detect prey of a certain size at a small reactive distance compared with fish with good vision (Fig. 13.3). Reactive distances to prey allow the calculation of two-dimensional reactive areas similar to the area of a tow net that the predator fish would be using to detect prey. If swimming speeds are known, it is possible to calculate volumes of habitat water searched per unit time for a prey of certain size, the essence of the volume searched feeding model illustrated in Fig. 13.3 (Blaxter and Staines 1971). Deleterious effects on visual acuity of a fish involve a decrease in its reaction distance to prey, that can lead to decreased energy intake and growth deficits. Pessoa et al. (2011) exposed early life stages of Nile tilapia *Oreochromis niloticus* to carbofuran and found a dose dependent effect on visual acuity. The lowest observed effect concentration (LOEC) for visual acuity was 40 µg.L^{-1}, and a change in the visual acuity angle from 0.4 degree in control fish to 2.1 degrees at the LOEC was quantified. This small increase in visual acuity angle translated into a decrease of 80% in the reactive

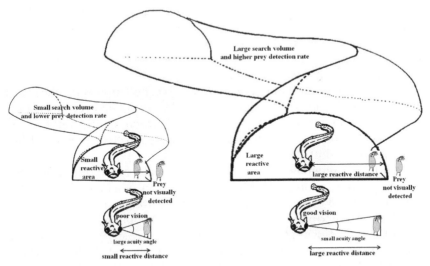

Figure 13.3. Visual abilities of fish based on the visual acuity angle and the reactive distance to a prey of fixed size (maximum distance a prey can be away from the predator and still be visually detected). Fish with poor vision will detect prey of fixed size at a small reactive distance compared with fish with good vision. Reactive distances to prey allow calculation of reactive areas to prey in two dimensions and of search volume if fish swimming speed is quantified. All parameters can fit into the prey encounter submodel of individual based models generalized in Fig. 13.1.

distance to *Daphnia magna* of 2 mm total length, from 28 cm in control fish to 5.6 cm in treated fish at the LOEC. Even without a significant decrease in swimming speeds of exposed tilapias, the volume searched for prey decreased by 97%, from 1.1 $L.s^{-1}$ to 0.04 $L.s^{-1}$. This predicted decrease in prey encounter rates of exposed tilapia at environmentally realistic carbofuran concentrations was confirmed by data showing reduced attacks on Daphnia and by a significantly decreased growth rate at the same concentrations.

The use of optokinetic or optomotor responses allows the testing of other ecologically important aspects of visual function in fishes. The quantification of flicker fusion thresholds using the optomotor method is considered an efficient way to evaluate the motion detection ability of animals (Muntz 1974), a skill important for detection of prey and predators and of evolutionary significance. Flicker fusion thresholds are quantified by determination of the maximum speed of the striped cylinder within the fish field of vision that can still elicit a positive optomotor response. The other visual function parameter that can also be evaluated using the optomotor response is the night vision ability of fish, or scotopic threshold. Individuals are kept in total darkness for an acclimation period, and illumination is gradually increased until a positive optomotor response is detected. This illumination, where a positive response is detected, is the scotopic threshold. Optomotor response based evaluation of flicker fusion thresholds (Carvalho et al. 2002) and night vision skills of fishes (Fuiman and Delbos 1998) have been measured in a few species to characterize the improvement of these skills along development. The flicker fusion threshold is related to a general motion detection ability that can be correlated with other more general behaviors, including detection of potential predators as well as prey.

Although these endpoints are important to better understand the visual abilities of fish, to our knowledge they have been applied in ecotoxicology only with rainbow trout exposed to dioxin, when acuity, night vision and flicker fusion based motion detection were affected in exposed fish (Carvalho and Tillitt 2004).

Another different method to evaluate a visually based response of fish in ecotoxicology involves simulations of the attack of a potential predator, by the movement of a black ellipse on a white card towards larval fish as a visual cue. The endpoint was named Apparent Looming Threshold (ALT) (Dill 1974), which would be the rate of the image enlargement on the fish retina at the moment where an escape response is elicited. Faulk et al. (1999) detected a decrease in mean burst speeds of atlantic croaker *Micropogonias ungulatus* larvae parentally exposed to DDT during evaluations of Apparent Looming Thresholds. ALT can also be considered a measurement of the fishes' motion detection abilities, and theoretically ALT would apply to the specific situation of an ongoing frontal predatory strike. Successful escape from this extreme situation is unlikely in early life stages of marine fish larvae (Folkvord and Hunter 1986), whereas escape is more likely when the predator is sighted before its attack, and therefore can be evaded.

Olfactory System Ecotoxicology

Chemoreception is important to basic behavioral functions that involve intra and interspecific communication like foraging, reproduction, homing behavior, recognition and avoidance of predators. Additionally, fish chemoreception is fairly well developed and allows the detection of chemicals in water through three primary systems: olfactory (smell), gustatory (taste) and difused chemosensory (Little and Brewer 2005; Daghfous et al. 2012). The olfactory system (Fig. 13.2) is by far the best studied and where most information on behavioral effects of contaminants is available. Sensory neurons in the olfactory epithelium of the olfactory rosette are in direct contact with the aquatic environment, and dissolved contaminants can easily interact with them, disrupting signals eliciting important motor behaviors (Sloman and Wilson 2005). Several reviews have been made where evidence is provided that environmentally realistic concentrations of metals and pesticides can interfere with fish olfaction and disrupt life history processes that determine individual survival and reproductive success (Beitinger 1990; Tierney et al. 2010). Behavioral effects can involve a fish losing partially or completely, its ability to smell a certain molecule, depending on the exposure concentrations; alternatively, fish can incorrectly interpret olfactory information, when they might be attracted to these contaminants, increasing exposure even more (Little and Brewer 2005).

Avoidance-attractance Behavioral Responses

Olfactory based behavioral effects of contaminants in fish can be adaptive when avoidance of an effluent plume or a region with some deleterious chemical is detected and the fish swims away from this region, decreasing exposure. However, reduction of fishery resources or indirect deleterious effects to individuals can still happen if the fish avoids contaminants and moves away from a suitable habitat, selecting another region where resources are not as readily available or abiotic conditions are not optimal. Laboratory experiments on the behavioral avoidance responses of rainbow trout *Oncorhynchus mykiss* and brown trout *Salmo trutta* showed that both species will avoid typical metal concentrations observed on the Clark Fork River, where metal mining was intense, possibly explaining significant decreases of trout populations in the area (Woodward et al. 1995; Hansen et al. 1999). Beitinger (1990) reviewed avoidance-attractance studies and found that fish avoided one-third of 75 tested contaminants, whereas two-thirds did not trigger an avoidance response or induced inconsistent responses. These numbers point out that a significant number of xenobiotics might not be detected by fish, which therefore will not minimize exposure by avoiding that region. Such an example was described for cutthroat trout *Oncorhynchus clarki*, which did not avoid an estuary where carbaryl was applied, and exposure to the insecticide led to disruption of swimming abilities and increased predation of the individuals (Labenia et al. 2007).

Some anthropogenic compounds are identical to substances used in natural communication systems in fish. The infochemical effect describes that anthropogenic substances can influence an organism so that it perceives its chemical environment differently (Klaschka 2008). The relevant concentrations of natural infochemicals are usually in the nano- to micromolar range, and therefore contaminants at minor concentrations can interfere with the complex chemical communication of fishes. Some compounds may be confounded with natural chemical cues, and in fact attract fish at low concentrations, such as xylene or ammonia (Beitinger 1990).

An interesting new approach in behavioral ecotoxicology is the development of models to predict metal toxicity based on metal speciation and binding dynamics of waterborne copper (Cu) in the olfactory epithelium (OE), linking Cu-OE binding to changes in olfactory acuity based on avoidance responses (Green et al. 2010). A recent avoidance study with tropical marine fishes demonstrated that ocean acidification and predicted increases in dissolved CO_2 to 700 ppm disrupts coral reef fish larval olfactory senses used to detect chemical cues to locate suitable adult habitat and avoid predators during the settlement process. These effects lead to 5 to 9 times higher mortality from predation in fish exposed to higher dissolved CO_2 than control fish, potentially impairing recruitment and the sustainability of fish populations (Munday et al. 2010).

Alarm Reactions to Imminent Predatory Risk

Since Karl von Frisch accidentally found in 1938 that injured fish released a substance within their skin ("schreckstoff", which is German for "scary stuff") that would induce other conspecific fish in the tank to dart back and forth and freeze in place as evading a predator, this substance was recently identified as a mixture that includes the sugar glycosaminoglycan (GAG) chondroitin (Mathuru et al. 2012). These alarm substances elicit innate anti-predator motor behaviors through the olfactory system in fish (Døving and Lastein 2009), and the alarm reaction behavior that is triggered in fish that are exposed to conspecific skin extract has been used as a tool in behavioral ecotoxicology to evaluate the functional status of the olfactory system. The typical experimental design involves exposure of fish to contaminants followed by an evaluation of the behavioral alarm response in the different concentrations compared to the alarm reaction in control fish. Partial or complete loss of the alarm reaction indicates different levels of disruption of this important anti-predator behavior. The tropical estuarine guppy *Poecilia vivipara* has been recently shown to present this response, which had not been tested previously in the species. Additionally, the alarm response, assessed by frequency of occurrence of fast "C-start" swimming events in individuals of *P. vivipara* decreased with exposure to copper with a lowest observed effect concentration (LOEC), compared with controls of 5 µgCu.L^{-1} at salinity 25 (Ribeiro 2012). A reduction of alarm responses was also detected in the amazonic matrinxã *Brycon amazonicus* following exposure to 9 µgCd.L^{-1} in freshwater (Honda et al. 2008). Brazilian laws regarding maximum acceptable concentrations in class 1 waters (most restrictive criteria) for copper in saline waters and cadmium in freshwater are 5 µgCu.L^{-1} and 1 µgCd.L^{-1}, respectively (Brasil 2005). This copper criteria of 5 µgCu.L^{-1} is exactly the LOEC for *P. vivipara* and the cadmium criteria is below the concentration where *Brycon amazonicus* was affected, although Honda, Fernandes-de-Castilho, and Val (2008) did not test concentrations below 9 µgCd.L^{-1}, where effects were detected. Cadmium and copper also led to disruption of the alarm response in salmonids (Scott et al. 2003; Sandahl et al. 2007). In addition to metals, the herbicides diuron, atrazine, the organophosphate diazinon and a carbamate have also disrupted the alarm response in fish, as discussed in the good review recently published by Tierney et al. 2010. In summary, complete or partial loss of the alarm response by contaminants may increase the risk of a predator attack, and therefore lead to higher mortality.

Lateral Line System Ecotoxicology

Fish are unique vertebrates because they have evolved a lateral line system which functions exclusively in an aquatic medium, detecting local water flow and low frequency vibration, and mediating behaviors such as schooling, feeding and predator avoidance (Blaxter and Hallers-Tjabbes 1992). The neuromast (Fig. 13.2) is the functional unit of the lateral line, and its hair cells that extend into the water column act as a mechanosensory organ detecting changes in water pressure. Several contaminants can disrupt lateral line function, and it has been suggested as a promising biological endpoint (Froehlicher et al. 2009). The lateral line of brook trout exposed to DDT at 0.1–0.3 ppm for 24 h became hypersensitive to low frequency shock waves (Blaxter and Hallers-Tjabbes 1992). Fish can detect hydrodynamic stimuli elicited by swimming predators with the lateral line, resulting in an escape response characterized by the fish body bent into a shape that resembles a "C", named C-start or startle response, followed by swimming acceleration through the caudal fin (Faucher et al. 2006). These authors developed a system to evaluate the functionality of the lateral line system, consisting of a pipette used to inject a water jet between the water surface and the base of the tank when fish swam in the vicinity. The lateral line system of the fish was considered functional when the water jet stimulation provoked a sudden escape reaction or startle response, characterized by the bending of the fish's body into a C-like shape. Immobility or a constant swimming velocity was noted as a negative response. Escape behavior in response to the water jet was disrupted in the sea bass *Dicentrarchus labrax* after exposure to cadmium ions in laboratory-controlled conditions during acute exposure to 5 µgCd.L^{-1} (Faucher et al. 2008), or chronic exposure to 9 µgCd.L^{-1} (Faucher et al. 2008).

Copper also elicits significant effects on the function of the lateral line system, and Johnson et al. (2007) found a concentration dependent decrease on functional neuromasts and on behavioral rheotaxis evaluated in individual larval zebrafish. Rheotaxis was quantified based on the proportion of time within a three-minute period that zebrafish larvae in a 50mL beaker could remain actively swimming against the water flow created by a magnetic flea and a magnetic stirrer set to a constant speed. Given the importance of the lateral line system to fish, the effects of contaminants on mechanosensory hair cell disruption can provide powerful tools for future risk assessment (Froehlicher et al. 2009).

Effects on Swimming Behavior

Contaminants can disrupt spontaneous swimming as well as swimming resistance of fish, behaviors that are central to important life processes such as migration, territory maintenance, resource searching behavior, predation and escape responses (Baatrup and Bayley 1998). Methods for the evaluation of spontaneous swimming in fishes range from simple direct counts of squares crossed by fish in a grid system, to software analysis of recorded video in three dimensions. Quantitative measurements include frequency, duration, pattern of movements, distance, linear and angular velocity, position in the water column, and swimming posture. Little and Finger (1990) reviewed the subject, and found that alterations in swimming behavior were reported at concentrations as low as 0.7% of the LC50 for some contaminants. Drummond and Russom (1990) introduced the idea of behavioral toxicity syndromes, where fish locomotion was a key parameter. After monitoring behavioral and morphological endpoints in the fathead minnows (*Pimephales promelas*), three general responses were correlated with three classes of contaminants. The hypoactivity syndrome, or Type I responses are characterized by depressed locomotor activity, loss of startle responses, and rapid shallow opercular rates, and are indicative of narcosis-producing chemicals such as ethers and alcohols, depressors of central and peripheral nervous system activities. The hyperactivity syndrome or Type II responses involved heightened locomotory activity, hyperactivity to stimulation, increased rate and amplitude of opercular activity, slight darkening and edema, and were indicative of chemicals such as rotenone, benzene, and phenol, which disrupt metabolic activity. The physical deformity syndrome or Type III responses usually exhibit depressed locomotory activity with hyperactivity to stimulation, typical of neurotoxicants such as carbamates and organophosphates. These behavioral syndromes are useful to characterize specific, underlying mechanisms for different classes of compounds. Another aspect related to contaminant induced alterations in swimming speeds involve alterations of the volume searched for prey discussed previously, where increases in swimming speed lead to a higher probability of prey detection if sense organs related to prey detection are not affected.

Critical swimming speed is a parameter used to evaluate the swimming resistance of fishes, where a fish is placed in a tunnel of water where flow or velocity are known, and forced to swim against this water flow which is slow in the beginning and incremented at fixed intervals until the fish can no longer resist the flow, and is carried downstream into a net or reservoir. The time and velocity at which the fish fatigue are used to calculate an index of swimming resistance, a method that has been frequently used in ecotoxicology (Plaut 2001). Several classes of contaminants have been shown to affect swimming resistance in fish. Swimming resistance is affected by the organophosphates malathion and fenitrothion in brook trout at concentrations of about 33% of the LC50 (Little and Finger 1990). Critical swimming speed (Ucrit) was significantly reduced in Pacific herring *Clupea pallasi* exposed to the highest concentration of water soluble fraction (WSF) of crude oil for 96 hours (Kennedy and Farrell 2006). A reduction in swimming resistance of rainbow trout exposed to pulp and paper effluents was also found (Landman et al. 2006). Coho salmon exposed to ammonia showed a significant linear decrease in critical swimming speeds (Wicks et al. 2002). Copper exposure leads to significant reductions in swimming resistance of several species, including brown trout *Salmo trutta* (Beaumont et al. 1995), the estuarine fish *Pomatoschistus microps* (Vieira et al. 2009), rainbow trout *Oncorhynchus mykiss*, common carp *Cyprinus carpio*, gibel carp *Carassius auratus gibelio* (De Boeck et al. 2006; Waser et al. 2009), and the tropical guppy *Poecilia vivipara* (Ribeiro 2012). Copper induced reductions in swimming resistance have been attributed to an impairment of gas exchange across the gill following increases in the blood-water diffusion distance in gills (Heerden et al. 2004). On the other hand, some authors have shown that the reduction in swimming resistance are related to plasma ammonia accumulation in salmonids (McKenzie et al. 2003).

Behavioral Effects of Contaminants on Predator-prey Interactions

It is important to emphasize that effects of contaminants on sensory and neuromotor systems discussed above can have critical consequences to individuals of different species that interact during their life cycle in their typical habitats. Among these more complex adaptive behaviors that can be affected, ecological theory has shown that predator-prey interactions can be important to organize communities and influence the abundance of interacting species. However, studies that analyze the effects of contaminants on predator-prey interactions are not common in Ecotoxicology, although it is quite clear that various aspects of a predator–prey interaction can be affected. Available laboratory studies have typically focused on combinations of the interaction between exposed predators and unexposed prey, and exposed prey escaping unexposed predators. It is important to consider that realistic field situations would also include exposure of both predators and prey, when the outcomes of this interaction would be dependent on the sensitivity of each interacting species to the contaminants in question (Newman and Clements 2008). Quantitative behavioral endpoints can measure the ability of an exposed fish (acting as a predator) to feed. These endpoints will evaluate changes in the ability of a fish to detect, pursue, capture, and eventually ingest prey, which will then be converted to energy that can additionally affect growth and survival. Therefore, these behavioral parameters may provide an ecologically relevant measure of contaminant effects. Typical experiments use a model predator fish exposed to contaminants which is later allowed to feed on unexposed prey for a fixed time period, followed by quantification of the total number of attacks on prey, total number of prey captured, and prey capture efficiency (total attacks/total captures). These quantitative endpoints of feeding behavior can be quite sensitive, as significant effects have been detected at concentrations of less than 1% of the LC50 (Little et al. 1990).

Kasumyan (2001) reviewed data on the effects of different contaminants on the foraging behavior and sensorial ability of fish, and discussed that cessation of feeding was a nonspecific response to high concentrations of toxicants, whereas effects in abilities to detect, pursue and capture prey occur during exposure to low concentrations of pollutants, which can also be explained by structural and functional changes in sensory systems. Kasumyan (2001) also pointed out that information on functional disturbances in sensory systems by behavioral methods was scarce, with more studies focused on chemosensory systems of olfaction and taste, and less focused on other senses, including vision.

Larval stages of the carnivorous tropical freshwater fish dourado *Salminus brasiliensis* were exposed to phenanthrene and later allowed to feed individually on 5 larvae of unexposed prey fish *Prochilodus lineatus* for 12 min, while monitored by video cameras. Analysis of the interactions revealed that the endpoints number of prey captured and prey capture efficiency were equally sensitive, with a lowest observed effect concentration (LOEC) of 50 µg.L^{-1}. Swimming speeds were not affected by phenanthrene exposure, but effects on visual acuity of exposed predators were detected, as well as clear predator attack errors towards prey were seen in analyzed videos, suggesting that visual impairment of PHE exposed fish could be the major mechanism leading to reductions in prey capture (Carvalho et al. 2008). Altered prey capture ability was detected in rainbow trout *Oncorhynchus mykiss* larvae treated with 2,3,7,8-tetrachlorodibenzo-p-dioxin (TCDD), at a dose of 300 pg TCDDg^{-1} egg, when larvae also presented some level of edema and craniofacial deformity (Carvalho and Tillitt 2004). In another study, embryonic exposure of *Fundulus heteroclitus* larvae to 3,3'4,4',5 pentachlorobiphenyl (PCB126) affected prey capture ability at environmentally relevant concentrations of 5 pg PCB126 egg^{-1}, where no mortality or obvious deformity was seen in exposed larvae (Couillard et al. 2011). Carvalho and Tillitt (2004) provided evidence that TCDD effects on visual acuity and retinal ganglion cells are important factors explaining prey capture deficiencies, and Couillard et al. (2011) also noted that visual function seemed important to explain prey capture deficits in PCB-126 exposed *F. heteroclitus* larvae. Exposure of zebrafish *Danio rerio* to lead at 100µg.L^{-1} decreased the reaction distance to living *Daphnia* and increased the duration of keeping of a food item after grasping (Nyman 1981). *Anabas scandens* exposed to mercuric nitrate at 3 mg.L^{-1} for 4 weeks caused 71% of fish to become blind (Panigrahi and Misra 1978). Atlantic salmon exposed to organophosphate fenotrothion had a significant reduction in reaction distance and feeding efficiency at 0.3% of the LC50 (Morgan and Kiceniuk 1990). Visual acuity of nile tilapia *Oreochromis niloticus* exposed to carbofuran at 18% of the LC50 was significantly affected and correctly predicted deficits in prey capture and growth (Pessoa et al. 2011).

It is evident from the references above that many different classes of contaminants, aromatic hydrocarbons, planar halogenated hydrocarbons, metals, organophosphates and carbamates can affect both visual function and prey capture behavior. Sandheinrich and Atchison (1990) stressed the need to develop mechanistic approaches to feeding studies using optimal foraging theory, and focused on understanding effects of contaminants on critical steps of the predation sequence, including searching for and detecting prey, pursuit, capture and handling of prey after capture. The mechanistic approach taken by (Pessoa et al. 2011) with *O. niloticus* exposed to carbofuran involved the application of the volume searched feeding model (Blaxter and Staines 1971), taking into account sublethal alterations in reaction distance to prey and in average swimming speed depicted in Fig. 13.3. A concentration of 18% of the LC50 led to a decrease in search volumes for prey to less than 5% of control volumes, predicting effects in prey capture and in growth rates. This approach can be applied to different chemicals or even mixtures of chemicals, allowing behavioral data to be incorporated into IBM foraging models, as well as AOP models mentioned in the beginning of the chapter. These mechanistic approaches using behavioral information may allow the development of testable hypotheses of contaminant-altered feeding and growth in the field (Sandheinrich and Atchison 1990).

The ability of a fish to avoid predation in any stage of its life cycle involves the integration of environmental awareness, sensorial predator detection based on visual, olfactory, pressure senses, swimming skills, habitat choice, any of which can be impacted by contaminants (Beitinger 1990). From the perspective of a contaminant exposed prey, behaviors related to predator detection and evasion skills have been evaluated in ecotoxicology. Contaminant exposed prey can fail to detect predators, take more risks in search of food, have deficient swimming acceleration capabilities, be unable to form schools, be hyperactive and more visible to the predator, all of these situations leading to a higher predation vulnerability (Newman and Clements 2008). Quantitative behavioral endpoints applied to predator-prey interactions can measure the risk of exposed prey fish being eaten by unexposed predators. Parameters can include death risk factors, or other endpoints expressing predation risk such as the feeding efficiency of the predator at different prey exposure situations, or simply the number of attacks a predator needs to perform in order to capture prey exposed to different doses. Such approach was used by (Pessoa et al. 2011), where early juvenile tilapia exposed to carbofuran at 70µg.L^{-1} (one third of the LC50) were captured by a predator fish on its first attack, whereas the predator needed 5 attacks to capture unexposed control tilapia. The method above involved video analysis of interactions between one exposed prey and one unexposed predator. A different and more challenging method to evaluate predation risk is to analyze the interaction between a single unexposed predator and groups of different prey fish exposed to different contaminant concentrations. The challenge here is that development of marking methods that allow the identification of exposed prey from each concentration tested is necessary. Counts of surviving prey from different contaminant exposure treatments and control are made when approximately 50 percent of the prey population have been captured, and predation risk factors can be calculated for each concentration (Little et al. 1985). Early juvenile rainbow trout exposed to the carbamate carbaryl, the organophosphate methylparathion and the organochloride chlordane were more susceptible to predation at concentrations of 0.5%, 2.7% and 5% of the LC50 using this approach (Little et al. 1990). Juvenile guppies *Poecilia reticulata* exposed to pentachlorophenol at 500 and 700 pg.L^{-1} were more easily captured by largemouth bass, which performed fewer strikes and chases and spent less time chasing guppies that had been exposed than control guppies (Brown et al. 1985). Cadmium exposed fathead minnows were more vulnerable to largemouth bass *Micropterus salmoides* predation, and loss of proper schooling behavior might have been a causal mechanism (Sullivan et al. 1978).

Although most studies have focused on the effects of contaminants on predator–prey interactions restricted to the laboratory, an interesting approach involving realistic field exposure to contaminants has been used, where fish from field contaminated sites and from clean sites were brought to the laboratory and evaluated for behavioral performance both in terms of their prey capture and predator avoidance skills (Weis and Candelmo 2012). *Fundulus heteroclitus* from a mercury polluted tidal creek captured its typical prey shrimp at a slower rate than fish from an uncontaminated environment, where lower levels of mercury were found in their brain tissues. Additionally, *F. heteroclitus* from the polluted environment were more vulnerable to predation by the blue crab *Callinectes sapidus*, than fish from the unpolluted environment (Smith and Weis 1997).

Although laboratory studies are simplifications of real world scenarios, prey capture data were the only laboratory endpoint (Finger et al. 1985) that accurately predicted an adverse impact on growth and survival of bluegills *Lepomis macrochirus* exposed to the hydrocarbon fluorene at environmentally realistic parallel mesocosm studies (Boyle et al. 1985).

Behavioral Effects of Contaminants in Reproduction

Endocrine disrupting chemicals (EDCs) have become a hot topic as evidence is available that a wide variety of chemical contaminants can disrupt hormonal systems in wildlife (Colborn et al. 1993), leading to adverse consequences ranging from the molecular to the population level of biological organization in fish (Kidd et al. 2007). A significant portion of the evidence that EDCs affect sexual and reproductive behaviors in wildlife comes from studies with fish (Zala and Penn 2004). Effects detected include decreases in nest construction activities and offspring defense, and changes in parental care division between sexes. The density of *Tilapia rendalli* nests in areas sprayed with endosulfan in Botswana decreased to 25% of the density in unsprayed areas, indicating a probable effect on the population.

Furthermore, juveniles of the cichlid *Sarotherodon mossambicus* exposed to endosulfan presented delayed breeding displays which could explain *T. rendalli* nest reduction (Jones and Reynolds 1997; Scott and Sloman 2004). Sand goby (*Pomatoschistus minutus*) males exposed to 4 ng.L^{-1} of 17alfa-ethinyl estradiol (EE2) were not able to acquire or defend a nest, and females preferred to mate with control males (Saaristo et al. 2009). Another experiment showed that male *P. minutus* exposed to 11 ng.L^{-1} of EE2 delayed nest building, spent less time in courtship and were less aggressive against rivals than control males (Saaristo et al. 2010). These studies suggest that environmentally relevant concentrations of EE2 are likely to have negative implications for *P. minutus* male reproductive success. Male three-spined stickleback *Gasterosteus aculeatus* exposed to environmentally relevant levels of EE2 decreased their aggressive response to a conspecific male over time (Bell 2001). Male fathead minnows *Pimephales promelas* practice paternal nest care, and males have to conquer and defend a nest site, attract one or more females, and successfully guard the developing larvae until hatching. Inability to acquire and defend a nest until larvae hatch will prevent males from reproducing. Fathead minnow larvae were exposed for 64 days to a mixture of nonylphenolethoxylate/octylphenolethoxylate (NPE/OPE) compounds found in wastewater treatment plants, and a significant decrease in reproductive behavior was evident, as exposed males were unable to acquire and hold a nest site required for reproduction (Bistodeau et al. 2006). Exposure of zebrafish *Danio rerio* larvae to EE2 led to a significant alteration in juvenile zebrafish behavior patterns, with a decrease in swimming activity even at the lowest nominal concentration of 0.19 ng.L^{-1}, showing that behavioral endpoints could be more sensitive than typical endpoints measured at adulthood such as vitellogenin expression (Sárria et al. 2011). EE2 at all exposure concentrations reduced male aggression during male–male encounters and caused a social dominance reversal in 50% of the fish at the highest exposure dose (50 ng.L^{-1} EE2). Additionally, the frequency of courtship-specific behaviors decreased in dominant males of *Danio rerio* exposed to EE2 at environmentally relevant levels (Colman et al. 2009).

Sexually mature male poecilids display courtship behavior during the whole day, characterized by the male positioning himself in front of the female's visual field. From this position he displays to the female with his body assuming a "S or C shape", and this behavior is named sigmoid display. From this display, the male swims along the side of the female trying to position him to fertilize the female internally with his modified anal fin, the gonopodium. These behaviors are highly stereotyped and frequent, and even automated software analysis of them are possible, making them amenable to become standard behavioral biomarkers for the evaluation of endocrine disruptor contaminants (Baatrup 2009). Effects on courtship include decreases or increases in frequency of displays, increased courtship duration, or performance of male-like behavior by masculinized females. A seminal paper on the subject detected females of the viviparous mosquitofish, *Gambusia affinis*, sampled downstream a paper mill in Florida, exhibiting male physical secondary sexual characteristics with development of male-like gonopodium and displaying male reproductive behavior towards other females (Howell et al. 1980). The hypothesis that chemicals released from the paper mills were causing this masculinization of females was proven by laboratory experiments where degradation products of plant sterols present in paper mill effluents could reproduce these observed effects. Females of *G. affinis* exposed in the laboratory to the microbially degraded stigmastanol or beta-sitosterol developed male-like gonopodial structures within 6 to 11 days after exposure (Denton et al. 1985). Furthermore, these laboratory masculinized females behaved like males in that they followed, swung, and thrust their gonopodium at non-treated females and larger masculinized females (Krotzer 1990). In a field survey in Texas, masculinized female mosquitofish were smaller, less fecund and did not carry fertilized embryos, indicating that female masculinization can significantly influence life history and population dynamics (Deaton and Cureton 2011). Male *G. affinis* exposed to EE2 at concentrations of 20 ng.L^{-1} were less capable of impregnating females, displaying decreased sexual activity in terms of number of approaches and copulation attempts toward nonexposed females (Doyle and Lim 2005). Male mosquitofish *Gambusia holbrooki* exposed to water from Lake Apopka presented decreased sexual behavior and sperm count after exposure for one month (Toft and Guillette Jr. 2005).

Courtship behaviors have been frequently used to address contaminant effects on reproduction, especially those related to viviparous fish like poecilids (Baatrup 2009). Both estrogenic and anti-androgenic compounds significantly suppress the number and duration of displays that male poecilids perform towards females. Freshwater guppies *Poecilia reticulata* were exposed to the fungicide vinclozolin, the DDT metabolite p,p'-DDE, or the known antiandrogen flutamide from birth to adulthood. All chemicals skewed the sex ratio towards males, which had reduced orange display coloration, inhibited gonopodium development, lower sperm counts and reduced courtship behavior, indicating that these contaminants may seriously compromise male reproductive fitness (Bayley et al. 2002). In another study, male guppies *P. reticulata* exposed to ethinylestradiol EE2 from birth to adulthood produced significant reductions in male sperm count, testis weight, body coloration and courtship behavior. Only one of the 17 EE2 exposed males sired offspring in competition with unexposed control males, suggesting EE2-treated males have reduced reproductive fitness compared with untreated males (Kristensen et al. 2005).

In summary, research has shown that critical reproductive fish behaviors can be affected by EDCs with potentially wide implications for individual fitness and population level outcomes. Furthermore, EDC-induced changes in behavior can be even more sensitive than widely accepted physiological biomarkers such as vitellogenin induction in male fish (Söffker and Tyler 2012).

Conclusion

The reviewed information on the effects of contaminants in different sensorial system specific behaviors, as well as in more complex social interactions clearly indicates that the major advantages of behavioral effects is that they are more sensitive indicators of the potential for impacts on survival in the field than are measures of lethality. Behavioral tests provide definable, interpretive endpoints that could be used for regulatory purposes in product registration, damage assessments and in the formulation of water quality criteria. The avoidance response discussed previously is legally accepted as evidence of injury for Natural Resource Damage Assessments from discharges of oil or hazardous substances in the US, as it has been verified as a response to contaminants in the field. Other behavioral endpoints are still not accepted because there are no well established field validations where causal links between behavioral impairment and population or community effects are shown, although their biological significance is very clear (NRDA 1986; Little and Brewer 2005). New technologies that can be used for field validation of behavioral toxicity include wireless submersible video cameras, telemetry, infrared light beams, and "on-line" biomonitors in various rivers around the globe (Gerhardt 2007). Although behavioral testing in Ecotoxicology is relatively old, there has been an increase in the use of behavioral tests in recent years (Sloman and McNeil 2012). One possible explanation is the simple fact that the incredible development in knowledge about mechanisms and modes of action based on molecular, biochemical and physiological biomarkers need to be linked coherently to population level effects, as advocated by the adverse outcome pathway concept and by societal environmental concerns. As stated earlier by different authors (Little, Flerov, and Ruzhinskaya 1985; Beitinger 1990; Little 1990), behavioral endpoints provide the sensitive, ecologically relevant information, and obvious link between suborganismal and population level effects of contaminants that society cares about. Studies featuring several different methods demonstrate that there is no single best behavioral bioassay method. However, depending upon the expected effects and possible mode of action, there probably is a "best" behavioral bioassay to illustrate the effects of each potential stressor (Beitinger 1990). Carefully selected behavioral endpoints will be necessary to understand ecological mechanisms of contaminant effect that will eventually allow the prediction of population level effects from suborganismal biomarkers in conjunction with behavioral biomarkers.

Acknowledgements

I am grateful to Dr. Donald E. Tillitt and Dr. Edward E. Little from Columbia Environmental Research Center, USA, who provided me initial guidance with behavioral studies in ecotoxicology. I am also grateful to all students of our laboratory who were essential to some our research reviewed here.

Keywords: biomarkers, sensorial systems, vision, olfaction, lateral line, swimming, predator-prey, reproduction, ecological risk

References

Amiard-Triquet, Claude. 2009. Behavioral disturbances: The missing link between suborganismal and supra-organismal responses to stress? Prospects based on aquatic research. Hum. Ecol. Risk Assess. 15(1): 87–110.

Ankley, G.T., R.S. Bennett, R.J. Erickson, D.J. Hoff, M.W. Hornung, R.D. Johnson, D.R. Mount, J.W. Nichols, C.L. Russom, P.K. Schmieder, J.A. Serrrano, J.E. Tietge and D.L. Villeneuve. 2010. Adverse outcome pathways: A conceptual framework to support ecotoxicology research and risk assessment. Environ. Toxicol. Chem. 29(3): 730–741.

Baatrup, E. and M. Bayley. 1998. Animal locomotor behaviour as a health biomarker of chemical stress. Arch. Toxicol. Supplement. = Archiv fur Toxikologie. Supplement 20: 163–178.

Baatrup, Erik. 2009. Measuring complex behavior patterns in fish—effects of endocrine disruptors on the Guppy reproductive behavior. Hum. Ecol. Risk Assess.: An International Journal 15(1): 53–62.

Baldwin, D.H., J.A. Spromberg, T.K. Collier and N.L. Scholz. 2009. A fish of many scales: extrapolating sublethal pesticide exposures to the productivity of wild salmon populations. Ecol. Appl. 19(8): 2004–2015.

Bayley, Mark, Mette Junge and Erik Baatrup. 2002. Exposure of juvenile guppies to three antiandrogens causes demasculinization and a reduced sperm count in adult males. Aquat. Toxicol. 56(4): 227–239.

Beaumont, M.W., P.J. Butler and E.W. Taylor. 1995. Exposure of brown trout, Salmo trutta, to sub-lethal copper concentrations in soft acidic water and its effect upon sustained swimming performance. Aquat. Toxicol. 33(1): 45–63.

Beitinger, Thomas L. 1990. Behavioral reactions for the assessment of stress in fishes. J. Great Lakes Res. 16(4): 495–528.

Bell, Alison M. 2001. Effects of an endocrine disrupter on courtship and aggressive behaviour of male three-spined stickleback, Gasterosteus aculeatus. Anim. Behav. 62(4): 775–780.

Bistodeau, Travis J., Larry B. Barber, Stephen E. Bartell, Roberto A. Cediel, Kent J. Grove, Jacob Klaustermeier, Janet C. Woodard, Kathy E. Lee and Heiko L. Schoenfuss. 2006. Larval exposure to environmentally relevant mixtures of alkylphenolethoxylates reduces reproductive competence in male fathead minnows. Aquat. Toxicol. 79(3): 268–277.

Blaxter, J.H.S. and C.C. Ten Hallers-Tjabbes. 1992. The effect of pollutants on sensory systems and behaviour of aquatic animals. Netherlands J. Aquat. Ecol. 26(1): 43–58.

Blaxter, J.H.S. 1986. Development of sense organs and behaviour of teleost larvae with special reference to feeding and predator avoidance. Trans. Am. Fish. Soc. 115: 98–114.

Blaxter, J.H.S. and M. Staines. 1971. Food searching potential in marine fish larvae. *In*: D.J. Crisp (eds.). Fourth European Marine Biology Symposium. Cambridge University Press, Cambridge, UK, pp. 467–485.

Boyle, T.P., S.E. Finger, R.L. Paulson and C.F. Rabeni. 1985. Comparison of laboratory and field assessment of fluorene. Part II. Effects on the ecological structure and function of experimental pond ecosystems. Validation and Predictability of Laboratory Methods for Assessing the Fate and Effects of Contaminants in Aquatic Ecosystems, ASTM STP 865. American Society for Testing and Materials, Philadelphia, PA.

Brasil. 2005. Resolução no. 357, 17 de março de 2005. edited by CONAMA. Brasília: Ministério do Meio Ambiente.

Browman, H.I. 1989. Functional development of sensory systems and aquisition of behavior in fish larvae. Brain Behav. Evol. 34(1): 1–72.

Brown, J.A., P.H. Johansen, P.W. Colgan and R.A. Mathers. 1985. Changes in the predator-avoidance behaviour of juvenile guppies (*Poecilia reticulata*) exposed to pentachlorophenol. Can. J. Zool. 63(2001–2005).

Cairns I., J.W. Hall, E.L. Morgan, R.E. Sparks, W.T. Waller and G.F. Westlake. 1973. The development of an automated biological monitoring system for water quality. Vol. 59, *Virginia Water Resources Research Center Bulletin 59*. Springfield: Virginia Water Resources Research Center, Virginia Polytechnic Institute and State University.

Cairns, J.J., J.W. Hall, E.L. Morgan, R.E. Sparks, W.T. Waller and G.F. Westlake. 1973. The development of an automated biological monitoring system for water quality. Virginia Water Resources Research Center, Virginia Polytechnic Institute and State University, Springfield.

Cairns, J. Jr. and W.H. van der Schalie. 1980. Biological monitoring part I—Early warning systems. Water Res. 14(9): 1179–1196.

Cairns, J. Jr. 1989. Will the real ecotoxicologist please stand up? Environ. Toxicol. Chem. 8: 843–844.

Calow, P. 1994. Ecotoxicology: what are we trying to protect? Environ. Toxicol. and Chem. 13(10): 1549.

Carvalho, P.S., D.D. Kalil, G.A. Novelli, A.C. Bainy and A.P. Fraga. 2008. Effects of naphthalene and phenanthrene on visual and prey capture endpoints during early stages of the dourado Salminus Brasiliensis. Mar. Environ. Res. 66: 205–207.

Carvalho, P.S. and D.E. Tillitt. 2004. 2,3,7,8-TCDD effects on visual structure and function in swim-up rainbow trout. Environ. Sci. Technol. 38(23): 6300–6306.

Carvalho, P.S.M., D.B. Noltie and D.E. Tillitt. 2002. Ontogenetic improvement of visual function in the medaka Oryzias latipes based on an optomotor testing system for larval and adult fishes. Anim. Behav. 64: 1–10.

Carvalho, P.S.M., D.B. Noltie and D.E. Tillitt. 2004. Biochemical, histological and behavioural aspects of visual function during early development of rainbow trout. J. Fish Biol. 64: 833–850.

Clark, J., K. Dickson, J. Giesy, R. Lackey, E. Mihaich, R. Stahl and M. Zeeman. 1999. Using reproductive and developmental effects data in ecological risk assessments for oviparous vertebrates exposed to contaminants. *In*: R.T. Di Giulio and D.E. Tillitt (eds.). Reproductive and Developmental Effects of Contaminants in Oviparous Vertebrates. SETAC Pellston Workshop on Reproductive and Developmental Effects of Contaminants in Oviparous Vertebrates. SETAC, Pensacola, Florida, USA, pp. 366–401.

Colborn, T., F.S. vom Saal and A.M. Soto. 1993. Developmental effects of endocrine-disrupting chemicals in wildlife and humans. Environ. Health Perspect. 101(5): 378–384.

Colman, Jamie R., David Baldwin, Lyndal L. Johnson and Nathaniel L. Scholz. 2009. Effects of the synthetic estrogen, 17α-ethinylestradiol, on aggression and courtship behavior in male zebrafish (Danio rerio). Aquat. Toxicol. 91(4): 346–354.

Couillard, Catherine M., Benoît Légaré, Andréane Bernier and Zara Dionne. 2011. Embryonic exposure to environmentally relevant concentrations of PCB126 affect prey capture ability of Fundulus heteroclitus larvae. Mar. Environ. Res. 71(4): 257–265.

Daghfous, Gheylen, Warren W. Green, Barbara S. Zielinski and Réjean Dubuc. 2012. Chemosensory-induced motor behaviors in fish. Curr. Opin. Neurobiol. 22(2): 223–230.

De Boeck, G., K. van der Ven, J. Hattink and R. Blust. 2006. Swimming performance and energy metabolism of rainbow trout, common carp and gibel carp respond differently to sublethal copper exposure. Aquat. Toxicol. 80(1): 92–100.

Deaton, Raelynn and James C. Cureton II. 2011. Female masculinization and reproductive life history in the western mosquitofish (Gambusia affinis). Environ. Biol. Fishes 92(4): 551–558.

Denton, Thomas E., W. Mike Howell, J.J. Allison, J. McCollum and B. Marks. 1985. Masculinization of female mosquitofish by exposure to plant sterols and Mycobacterium smegmatis. Bull. Environ. Contam. Toxicol. 35(1): 627–632.

Depledge, M.H., A. Aagaard and P. Györkös. 1995. Assessment of trace metal toxicity using molecular, physiological and behavioural biomarkers. Mar. Pollut. Bull. 31(1–3): 19–27.

Dill, L.M. 1974. The Escape Response of the Zebra Danio (*Brachydanio rerio*) I. The Stimulus for Escape. Anim. Behav. 22: 711–722.

Døving, Kjell B. and Stine Lastein. 2009. The alarm reaction in fishes—odorants, modulations of responses, neural pathways. Ann. N. Y. Acad. Sci. 1170(1): 413–423.

Doyle, Christopher J. and Richard P. Lim. 2005. Sexual behavior and impregnation success of adult male mosquitofish following exposure to 17[beta]-estradiol. Ecotoxicol. Environ. Saf. 61(3): 392–397.

Drummond, R.A. and C.L. Russom. 1990. Behavioral toxicity syndromes: A promising tool for assessing toxicity mechanisms in juvenile fathead minnows. Environ. Toxicol. Chem. 9(1): 37–46.

Dutta, H., J. Marcelino and Ch. Richmonds. 1992. Brain acetylcholinesterase activity and optomotor behavior in bluegills, Lepomis macrochirus exposed to different concentrations of diazinon. Arch. Physiol. Biochem. 100(5): 331–334.

Evans, B.I. and R.D. Fernald. 1990. Metamorphosis and fish vision. J. Neurobiol. 21(7): 1037–1052.

Faucher, K., D. Fichet, P. Miramand and J.P. Lagardère. 2008. Impact of chronic cadmium exposure at environmental dose on escape behaviour in sea bass (Dicentrarchus labrax L.; Teleostei, Moronidae). Environ. Pollut. 151(1): 148–157.

Faucher, Karine, Denis Fichet, Pierre Miramand and Jean Paul Lagardère. 2006. Impact of acute cadmium exposure on the trunk lateral line neuromasts and consequences on the "C-start" response behaviour of the sea bass (Dicentrarchus labrax L.; Teleostei, Moronidae). Aquat. Toxicol. 76(3–4): 278–294.

Faulk, C.K., L.A. Fuiman and P. Thomas. 1999. Parental exposure to ortho, para-dichlorodiphenyltrichloroethane impairs survival skills of atlantic croaker (*Micropogonias undulatus*) larvae. Environ. Toxicol. Chem. 18(2): 254–262.

Fernald, R.D. 1990. The optical system of fishes. In: R.H. Douglas and M. Djamgoz (eds.). The Visual System of Fish. Chapman and Hall, New York, USA, pp. 45–59.

Finger, S.E., E.F. Little, M.G. Henry, J.F. Fairchild and T.P. Boyle. 1985. Comparison of laboratory and field assessment of fluorene - Part I: effects of fluorene on the survival, growth, reproduction, and behavior of aquatic organisms in laboratory tests. In: T.P. Boyle (ed.). Validation and Predictability of Laboratory Methods for Assessing the Fate and Effects of Contaminants in Aquatic Ecosystems. STP 865. American Society for Testing and Materials, Philadelphia, PA. pp.134–151.

Folkvord, A. and J.R. Hunter. 1986. Size-specific vulnerability of northern anchovy, *Engraulis mordax*, larvae to predation by fishes. Fishery Bulletin 84: 859–869.

Forbes, V.E., P. Calow, V. Grimm, T.I. Hayashi, T. Jager, A. Katholm, A. Palmqvist, R. Pastorok, D. Salvito, R. Sibly, J. Spromberg, J. Stark and R.A. Stillman. 2011. Adding value to ecological risk assessment with population modeling. Hum. Ecol. Risk Assess. 17(2): 287–299.

Forbes, Valery E., Annemette Palmqvist and Lis Bach. 2006. The use and misuse of biomarkers in ecotoxicology. Environ. Toxicol. Chem. 25(1): 272–280.

Froehlicher, Mirjam, Anja Liedtke, Ksenia J. Groh, Stephan C.F. Neuhauss, Helmut Segner and Rik I.L. Eggen. 2009. Zebrafish (Danio rerio) neuromast: Promising biological endpoint linking developmental and toxicological studies. Aquat. Toxicol. 95(4): 307–319.

Fuiman, L.A. and B.C. Delbos. 1998. Developmental changes in visual sensitivity of Red Drum, *Sciaenops ocellatus*. Copeia 1998(4): 936–943.

Fuiman, L.A. and A.E. Magurran. 1994. Development of predator defenses in fishes. Rev. Fish Biol. Fish. 4: 145–183.

Gerhardt, A. 2007. Aquatic behavioral ecotoxicology—Prospects and limitations. Hum. Ecol. Risk Assess. 13(3): 481–491.

Gerking, S.D. ed. 1994. Feeding ecology of fish. 1st Ed. ed. San Diego: Academic Press.

Green, Warren W., Reehan S. Mirza, Chris M. Wood and Greg G. Pyle. 2010. Copper binding dynamics and olfactory impairment in fathead minnows (Pimephales promelas). Environ. Sci. Technol. 44(4): 1431–1437.

Hansen, James A., Daniel F. Woodward, Edward E. Little, Aaron J. DeLonay and Harold L. Bergman. 1999. Behavioral avoidance: Possible mechanism for explaining abundanc and distribution of trout species in a metal-impacted river. Environ. Toxicol. Chem. 18(2): 313–317.

Heerden, Daléne van, André Vosloo and Mikko Nikinmaa. 2004. Effects of short-term copper exposure on gill structure, metallothionein and hypoxia-inducible factor-1α (HIF-1α) levels in rainbow trout (Oncorhynchus mykiss). Aquat. Toxicol. 69(3): 271–280.

Honda, R.T., M. Fernandes-de-Castilho and A.L. Val. 2008. Cadmium-induced disruption of environmental exploration and chemical communication in matrinxã, Brycon amazonicus. Aquat. Toxicol. 89(3): 204–206.

Howell, W. Mike, D. Ann Black and Stephen A. Bortone. 1980. Abnormal expression of secondary sex characters in a population of Mosquitofish, Gambusia affinis holbrooki: Evidence for environmentally-induced masculinization. Copeia 1980(4): 676–681.

Johnson, A., E. Carew and K.A. Sloman. 2007. The effects of copper on the morphological and functional development of zebrafish embryos. Aquat. Toxicol. 84(4): 431–438.

Jones, J.C. and J.D. Reynolds. 1997. Effects of pollution on reproductive behaviour of fishes. Rev. Fish Biol. Fish. 7(4): 463–491.

Kane, A.S., J.D. Salierno and S.K. Brewer. 2005. Fish models in behavioral toxicology: Automated techniques, updates and perspectives. *In*: G. Ostrander (ed.). Methods in Aquatic Toxicology. Lewis Publishers, Boca Raton, FL, USA, pp. 32: 559–590.

Kasumyan, A.O. 2001. Effects of chemical pollutants on foraging behavior and sensitivity of fish to food stimuli. J. Ichthyol. 41(1): 76–87.

Kennedy, Christopher J. and Anthony P. Farrell. 2006. Effects of exposure to the water-soluble fraction of crude oil on the swimming performance and the metabolic and ionic recovery postexercise in Pacific herring (Clupea pallasi). Environ. Toxicol. Chem. 25(10): 2715–2724.

Kidd, Karen A., Paul J. Blanchfield, Kenneth H. Mills, Vince P. Palace, Robert E. Evans, James M. Lazorchak and Robert W. Flick. 2007. Collapse of a fish population after exposure to a synthetic estrogen. Proc. Natl. Acad. Sci. USA Biol. Sci. 104(21): 8897–8901.

Klaschka, Ursula. 2008. The infochemical effect—a new chapter in ecotoxicology. Environ. Sci. Pollut. Res. 15(6): 452–462.

Kramer, Vincent J., Matthew A. Etterson, Markus Hecker, Cheryl A. Murphy, Guritno Roesijadi, Daniel J. Spade, Julann A. Spromberg, Magnus Wang and Gerald T. Ankley. 2011. Adverse outcome pathways and ecological risk assessment: Bridging to population-level effects. Environ. Toxicol. Chem. 30(1): 64–76.

Kristensen, Thomas, Erik Baatrup and Mark Bayley. 2005. 17α-Ethinylestradiol reduces the competitive reproductive fitness of the Male Guppy (Poecilia reticulata)1. Biol. Reprod. 72(1): 150–156.

Krotzer, MaryJane. 1990. The effects of induced masculinization on reproductive and aggressive behaviors of the female mosquitofish, Gambusia affinis affinis. Environ. Biol. Fishes 29(2): 127–134.

Labenia, Jana S., David H. Baldwin, Barbara L. French, Jay W. Davis and Nathaniel L. Scholz. 2007. Behavioral impairment and increased predation mortality in cutthroat trout exposed to carbaryl. Mar. Ecol. Prog. Ser. 329: 1–11.

Landman, Michael J., Michael R. van den Heuvel, Megan Finley, Henry J. Bannon and Nicholas Ling. 2006. Combined effects of pulp and paper effluent, dehydroabietic acid, and hypoxia on swimming performance, metabolism, and hematology of rainbow trout. Ecotoxicol. Environ. Saf. 65(3): 314–322.

Letcher, B.H., J.A. Rice, L.B. Crowder and K.A. Rose. 1996. Variability in survival of larval fish: disentangling components with a generalized individual-based model. Can. J. Fish. Aquat. Sci. 53: 787–801.

Little, E.E., R.D. Archeski, B.A. Flerov and V.I. Kozlovskaya. 1990. Behavioral indicators of sublethal toxicity in rainbow trout. Arch. Environ. Contam. Toxicol. 19(3): 380–385.

Little, E., J.F. Fairchild and A.J. DeLonay. 1993. Behavioral methods for assessing impacts of contaminants on early life stage fishes. Am. Fish. Soc. Symp. 14: 67–76.

Little, E.E. and S.K. Brewer. 2005. Neurobehavioral toxicity in fish. In: D. Schlenk and W.H. Benson (eds.). Target Organ Toxicity in Marine and Freshwater Teleosts New Perspectives: Toxicology and the Environment. Taylor and Francis, London, UK, pp. 141–176.

Little, E.E. and S.E. Finger. 1990. Swimming behavior as an indicator of sublethal toxicity in fish. Environ. Toxicol. Chem. 9: 13–19.

Little, E.E., B.A. Flerov and N.N. Ruzhinskaya. 1985. Behavioral approaches in aquatic toxicity investigations: a review. In: P.M.J. Mehrle, R.H. Gray and R.L. Kendall (eds.). Toxic Substances in the Aquatic Environment. American Fisheries Society. pp. 72–78.

Little, Edward E. 1990. Behavioral toxicology: stimulating challenges for a growing discipline. Environ. Toxicol. Chem. 9(1): 1–2.

Mathuru, Ajay S., Caroline Kibat, Wei Fun Cheong, Guanghou Shui, Markus R. Wenk, Rainer W. Friedrich and Suresh Jesuthasan. 2012. Chondroitin fragments are odorants that trigger fear behavior in fish. Curr. Biol.: CB 22(6): 538–544.

McKenzie, D.J., A. Shingles and E.W. Taylor. 2003. Sub-lethal plasma ammonia accumulation and the exercise performance of salmonids. Comp. Biochem. Physiol. Part A: Molecular & Integrative Physiology 135(4): 515–526.

Mohti, A., M. Shuhaimi-Othman and A. Gerhardt. 2012. Use of the multispecies freshwater biomonitor to assess behavioral changes of Poecilia reticulata (Cyprinodontiformes: Poeciliidae) and macrobrachium lanchesteri (Decapoda: Palaemonidae) in response to acid mine drainage: Laboratory exposure. J. Environ. Monit. 14(9): 2505–2511.

Morgan, M.J. and J.W. Kiceniuk. 1990. Effect of fenitrothion on the foraging behavior of juvenile atlantic salmon. Environ. Toxicol. Chem. 9(4): 489–495.

Munday, Philip L., Danielle L. Dixson, Mark I. McCormick, Mark Meekan, Maud C.O. Ferrari and Douglas P. Chivers. 2010. Replenishment of fish populations is threatened by ocean acidification. Proc. Natl. Acad. Sci. USA Biol. Sci.

Muntz, W.R.A. 1974. Comparative aspects in behavioural studies of vertebrate vision. In: H. Davson and L.T. Grahan (eds.). The Eye. Comparative Physiology. Academic Press, New York, USA, pp. 155–227.

Newman, Michael C. and William H. Clements. 2008. Biotic and abiotic factors that regulate communities. In: M.C. Newman and W.H. Clements (eds.). Ecotoxicology: A Comprehensive Treatment. CRC Press, Boca Raton, FL, USA, pp. 379–403.

Noakes, D.L.G. and J.D.G. Godin. 1988. Ontogeny of behavior and concurrent developmental changes in sensory systems in teleost fishes. In: W.S. Hoar and D.J. Randall (eds.). Fish

Physiology. The Physiology of Developing Fish. Academic Press, San Diego, CA, USA, pp. 345–384.

NRDA. 1986. (Natural Resource Damage Assessments) Final rule. Federal Register 51: 27674–27753.

Nyman, H.G. 1981. Sublethal effects of lead (Pb) on size selective predation by fish: Applications on the ecosystem level. Verh. Intern. Verein. Limnol. 21: 1126–1130.

Panigrahi, A.K. and B.N. Misra. 1978. Toxicological effects of mercury on a freshwater fish, Anabas scandens, Cuv. & amp; Val. and their ecological implications. Environ. Pollution (1970) 16(1): 31–39.

Pessoa, P.C., K.H. Luchmann, A.B. Ribeiro, M.M. Veras, J.R.M.B. Correa, A.J. Nogueira, A.C.D. Bainy and P.S.M. Carvalho. 2011. Cholinesterase inhibition and behavioral toxicity of carbofuran on Oreochromis niloticus early life stages. Aquat. Toxicol. 105(3–4): 312–320.

Plaut, Itai. 2001. Critical swimming speed: its ecological relevance. Comp. Biochem. Physiol. A—Part A: Molecular & Integrative Physiology 131(1): 41–50.

Ribeiro, Anderson Brito. 2012. Efeitos do cobre no comportamento de *Poecilia vivipara*. M.S. Thesis, Universidade Federal de Pernambuco, Recife.

Robinson, Peter D. 2009. Behavioural toxicity of organic chemical contaminants in fish: application to ecological risk assessments (ERAs). Can. J. Fish. Aquat. Sci. 66(7): 1179–1188.

Rose, K.A., L.W. Brewer, L.W. Barnthouse, G.A. Fox, N.W. Gard, M. Mendonca, K.R. Munkittrick and L.J. Vitt. 1999. Ecological responses of oviparous vertebrates to contaminant effects on reproduction and development. *In*: R.T. Di Giulio and D.E. Tillitt (eds.). Reproductive and Developmental Effects of Contaminants in Oviparous Vertebrates. SETAC Pellston Workshop on Reproductive and Developmental Effects of Contaminants in Oviparous Vertebrates. SETAC, Pensacola, Florida, USA, pp. 225–281.

Saaristo, Minna, John A. Craft, Kari K. Lehtonen and Kai Lindström. 2009. Sand goby (Pomatoschistus minutus) males exposed to an endocrine disrupting chemical fail in nest and mate competition. Horm. Behav. 56(3): 315–321.

Saaristo, Minna, John A. Craft, Kari K. Lehtonen and Kai Lindström. 2010. Exposure to 17α-ethinyl estradiol impairs courtship and aggressive behaviour of male sand gobies (Pomatoschistus minutus). Chemosphere 79(5): 541–546.

Sabbah, S., R.L. Laria, S.M. Gray and C.W. Hawryshyn. 2010. Functional diversity in the color vision of cichlid fishes. BMC biology 8: 133.

Sandahl, Jason F., David H. Baldwin, Jeffrey J. Jenkins and Nathaniel L. Scholz. 2007. A sensory system at the interface between Urban Stormwater Runoff and Salmon survival. Environ. Sci. Technol. 41(8): 2998–3004.

Sandheinrich, M.B. and G.J. Atchison. 1990. Sublethal toxicant effects on fish foraging behavior: Empirical vs. mechanistic approaches. Environ. Toxicol. Chem. 9: 107–119.

Sárria, M.P., J. Soares, M.N. Vieira, L. Filipe C. Castro, M.M. Santos and N.M. Monteiro. 2011. Rapid-behaviour responses as a reliable indicator of estrogenic chemical toxicity in zebrafish juveniles. Chemosphere 85(10): 1543–1547.

Schlenk, Daniel, Richard Handy, Scott Steinert, Michael H. Depledge and William Benson. 2008. Biomarkers. *In*: R.T. Di Giulio and D.E. Hinton (eds.). The Toxicology of Fishes. CRC Press, Boca Raton, FL, USA, pp. 684–713.

Scott, G.R. , K.A. Sloman, C. Rouleau and C.M. Wood. 2003. Cadmium disrupts behavioural and physiological responses to alarm substance in juvenile rainbow trout (Oncorhynchus mykiss). J. Exp. Biol. 206(Pt 11): 1779–1790.

Scott, Graham R. and Katherine A. Sloman. 2004. The effects of environmental pollutants on complex fish behaviour: integrating behavioural and physiological indicators of toxicity. Aquat. Toxicol. 68(4): 369–392.

Sloman, K.A. and P.L. McNeil. 2012. Using physiology and behaviour to understand the responses of fish early life stages to toxicants. J. Fish Biol.

Sloman, Katherine A. and Rod W. Wilson. 2005. Anthropogenic impacts upon behaviour and physiology. *In*: R.W.W. Katherine, A. Sloman and B. Sigal (eds.). Fish Physiology. Academic Press, New York, pp. 413–468.

Smith, Graeme M. and Judith S. Weis. 1997. Predator-prey relationships in mummichogs (Fundulus heteroclitus (L.)): Effects of living in a polluted environment. J. Exp. Mar. Biol. Ecol. 209(1–2): 75–87.

Söffker, Marta and Charles R. Tyler. 2012. Endocrine disrupting chemicals and sexual behaviors in fish—a critical review on effects and possible consequences. Crit. Rev.Toxicol. 42(8): 653–668.

Sullivan, J.F., G.J. Atchison, D.J. Kolar and A.W. McIntosh. 1978. Changes in the predator-prey behavior of fathead minnows (Pimephales promelas) and largemouth bass (Micropterus salmoides) caused by cadmium. J. Fish. Res. Board Can. 35: 446–451.

Tierney, Keith B., David H. Baldwin, Toshiaki J. Hara, Peter S. Ross, Nathaniel L. Scholz and Christopher J. Kennedy. 2010. Olfactory toxicity in fishes. Aquat. Toxicol. 96(1): 2–26.

Toft, Gunnar and Louis J. Guillette Jr. 2005. Decreased sperm count and sexual behavior in mosquitofish exposed to water from a pesticide-contaminated lake. Ecotoxicol. Environ. Saf. 60(1): 15–20.

Van Winkle, W. and K.A. Rose. 1993. Individual-based approach to fish population dynamics: an overview. Trans. Am. Fish. Soc. 122: 397–403.

Vieira, L.R., C. Gravato, A.M.V.M. Soares, F. Morgado and L. Guilhermino. 2009. Acute effects of copper and mercury on the estuarine fish Pomatoschistus microps: Linking biomarkers to behaviour. Chemosphere 76(10): 1416–1427.

Vogl, C., B. Grillitsch, R. Wytek, O.H. Spieser and W. Scholz. 1999. Qualification of spontaneous undirected locomotor behavior of fish for sublethal toxicity testing. Part I. Variability of measurement parameters under general test conditions. Environ. Toxicol. Chem. 18(12): 2736–2742.

Waser, Wolfgang, Olga Bausheva and Mikko Nikinmaa. 2009. The copper-induced reduction of critical swimming speed in rainbow trout (Oncorhynchus mykiss) is not caused by changes in gill structure. Aquat. Toxicol. 94(1): 77–79.

Weis, Judith and Allison Candelmo. 2012. Pollutants and fish predator/prey behavior: {A} review of laboratory and field approaches. Curr. Zool. 58(1).

Wicks, B.J., R. Joensen, Q. Tang and D.J. Randall. 2002. Swimming and ammonia toxicity in salmonids: the effect of sub lethal ammonia exposure on the swimming performance of coho salmon and the acute toxicity of ammonia in swimming and resting rainbow trout. Aquat. Toxicol. 59(1–2): 55–69.

Woodward, Daniel F., James A. Hansen, Harold L. Bergman, Aaron J. DeLonay and Edward E. Little. 1995. Brown trout avoidance of metals in water characteristic of the Clark Fork River, Montana. Can. J. Fish. Aquat. Sci. 52(9): 2031–2037.

Zala, Sarah M. and Dustin J. Penn. 2004. Abnormal behaviours induced by chemical pollution: a review of the evidence and new challenges. Anim. Behav. 68(4): 649–664.

Index

17α–methyltestosterone 281
17α,20β-dihydroxy-4-pregnen-3-one 256, 260
17α–ethynilestradiol 256, 280
17β-estradiol 6, 38, 39, 245, 256–258, 267, 278, 284, 291, 292, 294–296, 298
17β-estradiol (E2) 278
17-β-stradiol 224
17β-trenbolone 281, 282
2,3,7,8-Tetrachlorodibenzo-p-dioxin (TCDD) 22, 23, 40
3'-phosphoadenosine 5'-phosphosulfate (PAPS) 20
3-methylcholanthrene (3-MC) 27, 40
454 GS Junior (Roche) 29
5alpha-cyprinol 27-sulfate 45
8-hydroxyguanosine 91

A

A. brasiliensis 221, 225
Aberrations 134–140, 142, 144, 148, 156, 158
Acará-Açú 25, 28
acetylcholinesterase 55, 57, 68–70
AChE activity 170
Acid-base regulation 231
acidophilic foci 224, 225
Acridine 145, 149
acrocentric 135
acute 213, 219, 228, 235–237
acyl-CoA oxidase 42, 43
Acyl-CoA oxidase 1 42
Acyl-CoA synthetase 42
adenoma 224, 226
adipose tissue 271, 273, 285
adrenocortical cells 246, 247, 249, 250, 258, 269–272, 279, 282, 291, 296
adverse effects 206–208
Adverse Outcome Pathway 352, 370
agonist 250–252, 254, 272, 286, 292
agricultural 218, 220–222
AHR nuclear translocator (ARNT) 22
ALA-D 194

Alarm Reactions 360
aldehyde dehydrogenases (ALDHs) 19
aldo-keto reductases (AKR) 16, 17, 20
alkylphenol polyethoxylates 297
Alkylphenols 297, 298
alteration 133, 145, 210–212, 216, 218, 220–222, 228, 237
Amazon 218, 219
Amazonian fish species 28
Ames 133
ammonia 233
Anablepidae 24
androgen 36, 38
Androgen Receptor 38
anemia 184, 194, 196
Anesthetized 188
aneugenic events 141
Aneuploidy 135, 136, 140, 141
antagonist 250, 252, 297
apical surface area 231, 234, 235
Apoptosis 154, 155, 158, 188, 196, 223, 228
apoptotic 235
aquatic pollution 207
armored suckermouth catfish 9–11
Aryl hydrocarbon receptor 22, 23, 40, 46, 267, 268, 272
Arylamine N-acetyltransferases (NATs) 20
arylamines 20, 23
Assay 134, 138, 139, 144, 147–158
Astronotus ocellatus 24, 25, 28
Astyanax bimaculatus 25
Astyanax sp. 214, 223, 228–230
Atherinela brasiliensis 220–222, 224–226
Atlantic cutlassfish 25
ATP-binding cassette (ABC) transporters 21
ATP synthase 24
Avoidance-attractance 359

B

β-naphthoflavone (BNF) 27, 40
Bagre 25
Baixada Santista 186, 187

BaP 218, 219, 227, 228
Barrigudinho 25
basophilic 217, 222
basophils 183, 185
behavioral testing 27
Benign 210, 222, 224, 233
benign tumor 222, 233
benzo[*a*]pyrene 3
Benzocaine 188
bile 210, 212, 213, 217, 221, 222
bile duct 210, 213, 217
biliary duct 224, 227
biliary epithelial 224
bioaccumulation 187, 327, 328
bioativation 218
bioavailability 35, 165, 168, 175
biochemical 187
biological factors 213
biomarker 22, 27, 94, 98, 99, 101, 103–105,
        108, 110–120, 122, 164–182, 185, 188,
        197, 206–208, 212, 213, 215, 218, 225,
        228, 229, 234, 237, 238, 352
biomonitoring 188, 212, 229, 238
biomonitoring programs 212, 229
Biotransformation 15–34, 208, 213, 218, 224,
        237
biotransformation enzymes 169, 173, 176
bisfenol A 4, 5
Bisphenol-A (BPA) 296, 297
Blebbed 142, 145, 146
bleeding 184, 194
Blood 182–205
blood-brain barrier 323
blood vessels 193
Boca-de-rato 25
Body condition indexes 339
Brazil 15, 24, 25

C

C. parallelus 187–193, 196
Cadmium 60, 62, 63, 67, 73
calyptospora 229, 230
Cananeia-Iguape 186
cancer 222
Cará 25
carbamates 55, 70
carbon nanomaterials 328
carbon nanotubes 314, 315, 318, 320, 324,
        329
Carcinogenic 190, 222
carcinogens 222, 224
carcinoma 222, 224, 226
carnivorous 212

carp *Cyprinus carpio* 16
CAT 95–97, 99–105, 107–115, 117–122
catalase 94–96, 104, 119, 123
catfish 25, 186, 194–197
Cathorops spixii 25
cell proliferation 232, 233, 235
cell rupture 231
cellular deformation 228
cellular pathology 228
Centropomidae 191
Centropomus parallelus 25, 186, 188, 189, 191
Characiformes 18, 25
Characins 25
chemical pollutant 165
chlordanes 207
chloride cells 231
cholangio-carcinoma 224
cholangioma 222, 224
cholestasis 220–223
chromatid 138, 139, 148, 151
chromosomal 133–142, 144, 146–148,
        156–158
chromosome 133–135, 137, 138, 140, 141,
        144, 146, 157
chronic 207, 208, 213, 219, 222, 229, 236, 237
chronic exposure 213, 229, 236, 237
circulatory system 228
citogenotoxicity 188
clastogenic 138, 141, 143
climate changes 3, 6, 12
clofibric acid 44
clotrimazole 44
Cnesterodon 27
Colossoma macropomum 28
Comet 134, 138, 139, 144, 147–158
common carp 7
community 218
complex mixture 3, 4, 206, 209
Condition Factor 339–342, 344–346
Constitutive Androstane Receptor 38, 44
contaminants 208, 219, 224, 231, 233, 235,
        237
contamination 183, 186, 187, 190, 191, 196,
        198
control region (COI) 24
Copper 60, 63, 64, 189, 190, 193
corticoisteroid hormones 269
Corticosteroid 269, 271, 273, 277
cortisol 185, 193, 231, 234, 235, 245, 247,
        249, 252, 258, 269–273, 279, 286, 288,
        290–292, 300
COUP transcription factor 37
Courtship 254, 274, 275,
courtship behavior 27, 368, 369

crude oil 23
Curimbata 25
cyanobacteria 52, 67–69
Cyanomethemoglobin 188
cyanotoxin 67, 69, 218, 219, 224, 226, 228
CYP1A 4, 6
CYP1s 15, 21–24, 29
CYP3A 43, 44, 45
Cypriniformes 16
Cyprinodontiformes 16, 18, 21, 24, 25, 28, 29
cytochrome b 24
cytochrome c oxidase (COX) 24
cytochrome P450 (CYP) 16–18, 22, 29, 42, 45
cytochrome P450 17-hydroxylase (P450C17) 260
Cytochrome P450 1A (CYP1A) 4, 170–173, 176
Cytochrome P450 enzymes 266
cytochrome P450 side-chain cleavage (P450scc) 260
cytochrome P4503A (CYP3A) 43, 44, 45
cytochromes 193
cytoskeleton 222

D

damage 209, 211–213, 218, 229, 231, 235
DAX-1 38
DBT 224, 226
DDT 40, 45, 207, 216, 222–225, 227–229
debris 185
decondensations 138
defense responses 231
degenerate primers 28
Degree of Tissue Change 209
deiodinase 262, 264–267, 286, 290, 293
deletions 135, 140
detergents 187
detoxification 189, 218, 237
diagnostic 218
dibenzofuran (PCDF) 22
dibenzo-p-dioxin (PCDD) 22
dieldrin 207
differential counting 185
Diffusion 134, 154–156, 158
dioxin-like toxicity 23
dioxin responsive elements (DRE) 22
Dioxins 245, 253, 257, 287, 288
disease 206, 220, 222
DNA 132–134, 140, 141, 146–158
DNA-binding domain 37
DNA damage 188, 198
double-strand breaks 132, 140, 149, 150

double-walled carbon nanotubes 320
drugs 21, 23
duplications 135

E

early warning systems 351
ecological death 353
Ecological relevance 351
ecotoxicology 8, 98, 106, 112, 113, 115, 123, 346, 350–352, 356, 358–361, 363, 364, 366, 370, 371
edema 210, 231, 233, 237
emerging contaminants 1, 3–6, 12
Endocrine disrupting chemicals 367
endocrine disruption 6, 40, 244, 245, 250–252, 254, 256, 258, 265, 268, 272, 275, 277, 293, 297, 299
endocrine disruptor compounds (EDCs) 39, 244, 248, 250–252, 255–258, 260, 264, 265, 268, 272, 274–277, 299, 300
endocrine disruptors 27
endocrine system 5, 243–311
endocrine tissue 215
endothelial cells 213, 222
environmental changes 165, 169
environmental health 165, 166, 173, 176
environmental mixtures 22
Environmental Monitoring 21, 22, 27
eosinophilic 217, 222, 224, 225
eosinophilic focus 224, 225
eosinophils 183, 185
epigenetic disruption 35
epithelial lifting 231–233, 237
epithelium 210, 229, 233–237
epoxide hydrolase (EPHX) 16, 17, 20
ER 38–42, 46
ERE 39, 40, 42
EROD 22, 23, 170, 172, 176
ERRγ 37
erythrocytes 134, 139, 142–147, 152, 153, 155, 156, 158, 183, 184, 187–189, 191, 192, 194–197
erythropoiesis 184, 191, 194
ERα 39, 41
ERβ1 39
ERβ2 39
Estradiol Receptors 38
estrogen 36, 38–41, 43, 46
estrogen responsive elements (ERE) 39, 40, 42, 252
estrogens 23
Estuarine 182, 186, 187, 189, 191, 193, 196–198

Estuary 186, 187, 191
ethanol 19
European sea bass 7
Eutrophication 191
Exchange 151
excretion 208, 218, 229, 236
experimental studies 209, 212
Exposure to Pollutants 207, 208, 212, 215,
    229, 230

F

fat snook 25, 186, 189, 196, 197
fatty liver 218–220
females 212, 224
fibrosis 233
Filament 231
filament epithelium 233–235
fish 51–83
Fish bile 171
Fish Endocrine System 246
Fish health 1–377
fish histology 208
fish livers 218
fitness 338, 339
flame retardants 5
flavin-containing monooxygenases (FMO)
    16, 17, 19
flurbiprofen 45
foraging behavior 364
freshwater fish 231, 234
Fugu rubripes 37, 38, 42
fullerene 313, 314, 319, 320, 325–327, 329

G

G protein-coupled receptors 39
gall bladder 213
Gasterosteus aculeatus 21
Gene 15–34, 133, 135, 157
gene/protein annotations 16
gene/protein nomenclature 18
genetic damage 35
Genidens genidens 186, 187, 194–197
genomes 29
genotoxic 224
Genotoxicity 132–163
Geophagus brasiliensis 25
ghosts 154, 156
gill 99, 104, 106–113, 117, 118, 121, 193, 207,
    208, 229, 231, 232, 234–238
gill epithelial 229
gill functions 231, 236
gill lamellae 193

gill morphology 231
Glucocorticoid Receptor 38
Glucocorticoids 196
glutamate cysteine ligase 319
glutathione 90, 92, 94–97, 104, 110, 116, 119,
    121, 123
glutathione peroxidase 90, 95–97, 104, 123
glutathione S-transferases (GSTs) 17, 19, 20,
    43, 90, 94, 97, 123
glycogen 215
GnRH 247, 248, 255, 259, 280
goldfish 6, 7
gonadotrophin 248
gonadotrophin releasing hormone (GnRH)
    247, 248, 255, 259, 280
gonads 246–250, 255–258, 271, 274, 278, 281,
    290, 293–295, 297–299
gonopodial thrust 27
GPx 90, 95–97, 99–112, 115, 117–121
graphene 313
growth hormone 245, 249, 261, 292
GSH 92, 94–97, 99–107, 109–115, 118, 121,
    122
GST 90, 94, 97, 99–103, 105, 107, 110–115,
    117–121
Guarú 25
Guppy 25, 29

H

Harbor 219–221, 224–226
head kidney 244, 246, 249, 269, 291
health 338–349
health status 211
heat-shock proteins 37
Heavy metals 60, 70–72, 233, 245, 255, 256,
    288, 289, 291, 299, 300
helix–loop–helix–PAS 40
Hematocrit (Ht) 188, 189, 194
hematological 183, 185, 186, 188–192,
    194–196, 198
hematopoietic 184, 190
Hemodilution 192
hemoglobin (Hb) 183, 184, 188, 189, 191,
    192, 194, 195, 196
hemolysis 184, 189, 192, 194
hemorrhage 193, 232
Hemorrhagic congestion 228
hemosiderin 226
hepatocellular 222, 224, 226
hepatocyte 213, 216, 233
Hepatocyte Nuclear Factor 4 38
Herbicide 189
herbivorous 212

high-throughput sequencing 28
histopathological 206–242
Histopathological findings 206–212, 215, 237
histopathology 206, 207, 209, 218, 237, 238
Histopatological Index 209
homeostasis 182
Hoplias malabaricus 25, 133–136, 138–140, 145, 156, 214, 215, 218–220, 222–224, 226–228, 234
hormonal 185
Hormone Receptors 244, 249, 251–253, 264
Hydrocarbons 187, 189
hydrogen peroxide 85–87, 94–97
hydropic 223
hydroxyl radical 85, 87, 93, 95, 96, 112
hypercapnia 231
hyperplasia 209, 210, 222, 231–236
hypertrophy 209, 210, 216, 223, 228, 231–235
hypothalamus 245–248, 250, 254, 255, 257, 261, 263, 269, 272, 280, 286
hypoxia 7, 10, 183, 189, 191, 231, 232

I

ibuprofen 44
Idiopathic 224
immune 185, 190, 196
immune response 228, 229
immune system 229
immunity 190
immunological 190, 197
Immunosuppression 194
immunotoxicity 190
importance factor 209–211
Index lesion 212
Index of Lesion 212
individual-based models 353
infection 193, 218
inflammation 210, 212
Inflammatory 223, 228–230
Inflammatory response 223, 228, 229
Inorganic 138, 156
interlamellar cell mass 232, 233
interlamellar space 233
interrenal 269, 272, 286
interstitial tissue 210
intraperitoneal exposure 224, 226, 227
inversions 135, 138, 139, 157
Ion Torrent PGM (Life Technologies) 29
irreversible 207, 210
islets of Langerhans 215

isoforms 16, 18–21
Itanhaem River Estuary 186, 187

J

Japanese killfish 7
Jenynsia multidentata 24–26
Jundiá 25

K

karyograms 136
karyotype 136, 137
ketoconazole 6
kidney 85, 102, 103, 107–110, 114, 116, 118–120, 193
Killifishes 25
Kupfer Cells 213

L

lamellar epithelial 231, 233, 237
lamellar epithelium 229, 233, 234, 237
Lateral Line System 355, 361, 362
Lead 52, 55, 56, 60–65, 67, 69, 73, 74, 138, 140, 156
lesion 207, 209–212, 215, 216, 218, 221, 222, 224, 226, 233, 237
Lesion Index 210
leucopenia 185, 194, 196
leukocyte infiltration 231, 233
leukocytes 183, 185, 187, 190, 192, 193, 196
leukocytosis 190, 192, 196
leukopenia 185, 194, 196
leukopoietic centers 196
ligand binding domain 36
light microscopy 213, 228
Lipid peroxidation 89, 90, 91, 92, 96, 97, 99–105, 109, 111–113, 116–121, 123
Lipodystrophy 219
lipofuscin 226
liposoluble 226
live bearers 24, 25, 28, 29
liver 193, 207, 208, 210, 212–216, 218–222, 224, 226, 229, 237, 238, 247, 248, 253, 256, 257, 264–268, 271, 278, 281, 284, 287, 290, 292, 293, 295, 297, 300
Liver lesions 215, 216, 224
liver tissue 208, 210
livers of fish 215, 218, 219, 222–225, 227–230
Lobed 142, 145
long-term 218, 222, 237
lymphocytes 183, 185
lymphocytopenia 185

M

M. furnieri 187, 191–193, 196, 197
macrocytic anemia 194
macrophage centers 224
macrophages 213, 225, 226, 228
Malic Enzyme 42
malignancy 224
malignant 210, 222, 224
marine fishes 184, 186
maturation inducing hormone 248, 256, 259, 260
MDR 21
mean corpuscular hemoglobin concentration (MCHC) 188, 191, 192, 194–196
Mean corpuscular volume (MCV) 188, 189, 192, 194, 195, 196
Mechanisms of toxic action 352
Medaka 21
medications 183
melanin 226
Melanomacrophage 225–227, 229
membrane associated proteins involved in eicosanoid and glutathione metabolism (MAPEG) 16, 19
mercury 52, 60–63, 65, 66, 71–73, 218, 219, 222, 223
metabolism 208, 218, 222, 233
metacentric 135
metallothioneins 169, 172
Metals 187, 193, 194
methyl mercury 189, 228
microarray 22
Microhematocrit 188
micronuclei 141–144, 146, 147, 188, 191, 194, 197, 198
Micronucleus 134, 138, 141–144, 148, 152, 156, 158
Micropogonia furnieri 220, 221
Micropogonias furnieri 186, 191, 192, 194
microsatellites 24
Mineralocorticoid Receptor 38
MiSeq (Illumina) 29
mitochondria-rich cells 231, 232, 236
mixture 206, 227, 229
modes of toxic action 352
monitoring 183, 198
monocytes 183, 185
morphological 206–208, 231, 234, 235, 237
Morphological parameters 170, 171
MRC density 233–236
MRC fractional surface area 234
mucous cells 231, 232

Multidrug resistance 2 42
multidrug resistance-associated protein 2 44
multidrug resistance protein 1 44
Multidrug Resistance Proteins MRP1 21
Multidrug Resistance transporter 21
Multidrug Transporter (MXR) 21
multi-walled 320, 324
mutagenic 190
Mutagenicity 132–163
mutations 133, 135
MXR proteins 169

N

Na+/K+-ATPase 234–236
NAD(P)H-quinone oxidoreductase (NOQ1) 16, 17, 20
NADPH 224
nanoecotoxicology 312–337
nanoengineered materials 4
nanomaterials 5, 12, 312, 314, 315, 319, 320, 323, 326, 328, 329
nanoparticles 314, 315, 318–323, 325
nanotechnology 312, 313, 315, 328, 329
Necrosis 196, 209, 210, 217–219, 221, 231–233
neoplasms 207, 224
neoplastic changes 222
Netuma barba 25
neurotoxicity 51–54, 57–60, 63, 64, 67, 68, 70, 73, 154, 170
neutralized 210
neutrophils 183, 185
next-generation sequencing platforms 29
nifurtimox 19
Nile tilapia 8–10
non-destructive 182, 185
non-model fish species 29
normoxia 232
Notched 143, 145, 146
nuclear abnormalities 184, 188, 191, 194, 197, 198
Nuclear alterations 210, 217, 228
Nuclear Receptors 35–50

O

Olfactory System 355, 359, 360
Oligosarcus hepsetus 218, 219, 224, 225, 227
Oncorhynchus mykiss 21, 44
Onesided livebearer 25
oocytes 247, 248, 256–258, 278, 280, 283–285, 287, 289, 292, 295, 296, 298
oral exposure 213, 217, 229

Orange 145, 149
Oreochromis niloticus 44
Organ Index 211
organic anion transporter polypeptide 2 44
organic pollutants 213, 233
organochlorine 229, 235
organochlorine compounds 53–55
organophosphate pesticide 3, 9
Organophosphates 53, 55, 67
Oryzias latipes 21
Oscar 25, 28
ovoviviparous 27
oxidants 222
oxidative damage 84, 91, 93, 101, 109, 122
oxidative stress 28, 84–131, 313, 316, 320, 321, 323, 325, 326
oxyradical 224

P

P450 aromatase 253, 257, 258, 260
pancreas 2113
pancreatic 135, 138, 215, 224, 227
parasite infestation 184
parasites 229, 230
pathogens 190
pathological importance 209, 210
pavement cells 232
PCB congeners 22, 23
PCB126 23
Pearl eartheater 25
Peixe-espada 25
Perch-likes 25
Perchlorates 294, 295
Perciformes 16, 18, 24, 25, 28
peri 135
peritoneal cavity 229
Peroxisome Proliferator-Activated Receptor 38, 41
personal care products 4, 5, 12
Pesticides 51–53, 55, 56, 75, 243, 245, 255, 291, 294
petroleum 23
P-glycoprotein 21
PGP 21
pH 231, 233, 235
phagocytes 190
Phalloceros 26–28
Phallotorynus 27
pharmaceuticals 1, 4, 5, 36, 40, 43, 282, 283, 288
Phtalates 5
Phthalates 4, 245, 284, 285
physiological biomarker 228

Phytoestrogens 279, 282
Pimelodus maculatus 218, 219, 222, 223, 227, 229, 230
pituitary 244–250, 254–257, 261, 263, 269, 272, 279–281, 284, 286, 291, 293, 295–297, 299, 300
pituitary gland 246–248, 254, 255, 263
plaice 42
planar molecules 22
plasmatic membrane 213
pleomorphic 217, 223, 228
pleomorphic nuclei 217, 223, 228
pleomorphism 224
Poecilia reticulate 27–29
Poecilia vivipara 21, 24–26
Poecilidade 24
polarized cells 213
pollutant mixtures 4
pollutants 183, 185, 188–190, 207, 208, 212, 213, 215, 216, 218, 219, 222, 229–231, 233, 237
polluted 229, 230
pollution 206–242
pollution biomarkers 176, 182
poly-aromatic hydrocarbons 187
polychlorinated biphenyl (PCB) 22, 23, 40, 43, 187, 207, 244, 253, 255, 257, 264, 266, 271, 285, 286, 299, 300
polycyclic aromatic hydrocarbons (PAHs) 22, 28, 40, 43, 207, 220, 255, 257, 266, 295, 296, 299, 300
population 207, 213, 218, 235, 237
portal triads 213
PPAR/RXR 42
PPARα 41–43
PPARβ 41, 42
PPARβ/δ 42
PPARγ 273, 274
Predator-prey Interactions 364, 366, 368
Pregnane X Receptor 38, 43, 46
Pregnenolone 259, 260, 270, 271, 299
pregnenolone 16α-carboninitrile 44
prenecrotic 219, 220
preneoplastic 207, 222
primary culture 224
Prochilodus lineatus 25
progesterone 259, 260, 270, 271, 280, 289, 290
Progesterone Receptor 38
progressive changes 210, 212
proliferation 222, 224, 231–233, 235
propranolol 44
Prostaglandins 23

Protein carbonylation 102, 103, 109, 113,
    116, 117, 120, 122, 123
Proteins 15–34
Pyrethroids 53, 58–60

R

R. quelen 223, 225, 229
rainbow trout 7, 21
RAR 37, 38
Reaction Index 210–212
reactive oxygen species 325, 326
reactive species 84, 85, 87, 93, 99, 102
reactivity 35
real time PCR (RT-qPCR) 22, 23, 27
recombinant expressed proteins 23
red blood 226
red blood cell count (RBC) 188, 189, 192,
    194–196
redox 85, 87, 88, 92–94, 96, 99, 100, 103, 105,
    108, 109
redox cycling 87, 88, 100, 105, 109
reduced glutathione 318, 327
regressive 210, 212
regulators 36
repair 132, 140, 149–152
reproduction behavior 27
reproductive hormones 245
reproductive system 244, 278, 296
reservoir 218, 219, 222–225, 227, 229, 230
resistance/susceptibility to PAHs 28
respiratory organs 229
retina 65–67, 72, 73
Retinoic Acid Receptors 38
reversible 207, 210, 218, 228, 232, 234
Rhamdia quelen 25, 133–135, 137, 139, 153,
    158, 218, 219, 223–226, 228, 229
Ribosomal 24
rifampicin 44
risk 22, 23
risk assessment 207
river 223, 229, 230
Rivulines 25
Robalo 25
ROS 84–94, 96–102, 105, 106, 109, 111, 113,
    114, 119, 120, 123
rough endoplasmic reticule 213
rough endoplasmic reticulum 214, 222
RXR 38, 42, 43, 46
RXRβ 37

S

Salmo salar 21
Santos Estuarine System 186, 187, 189

sasonal 187, 190, 194, 197
Sciades herzbergii 25
Seasonality 192, 197
secondary sexual characteristics 260, 274,
    278
sensitive 208, 218, 226
sequencing platforms 29
Sewage 187, 189, 193
shapes 135, 142
SHP 38
sigmoid display 27
signal transduction 35, 39
Siluriformes 16, 18, 25
silver nanoparticles 229, 315, 322, 323, 325
single-strand breaks 140, 149, 150
single-wall carbon nanotubes 314, 318
Sinusoid 213–216, 220, 222
sister 138, 148, 151
skin 229
SOD 95–97, 99–115, 117–122
sodium/iodide symporter (NIS) 262, 263
space of Disse 213, 214
sperm 248, 255–257, 260, 274, 275, 278, 280,
    282, 289, 292–294
Splenomegaly 192
Starvation 226
steatorrhea 222
Steatosis 216, 218, 220–222
Stellifer brasiliensis 25
steroid hormones 248, 253–256, 259, 260,
    266, 271, 272, 278, 282
steroidogenic acute regulatory protein
    (StAR) 258–260, 270–272, 282
Stickleback 21
stress 182, 185, 190, 191, 193
subchronic 224, 227, 235
submetacentric 135
subtelocentric 135
Subtractive Suppressive Hybridization
    (SSH) 28
sulfotransferases (SULT) 16, 17, 20, 23, 29,
    43
superoxide 85–88, 92, 94–96, 123
superoxide dismutase 94, 95, 123
surrogate models 27
susceptibility 35
Swimming 232, 351, 353, 356–358, 360–366,
    368, 371

T

Takifugu rubripes 44
tambaqui 28

Target Organs 212, 237
target tissue 207
TBT 223, 224, 226, 227
telangiectasis 232, 233, 237
teleost liver 212
Temperature 189, 193, 231, 232, 235
teratogenic 190
Test 133, 134, 136, 137, 141–145, 148, 150, 151, 154, 156, 157
testis 247, 248, 255–258, 260, 278, 280, 285, 287–289, 291, 293, 297
Tetra 22, 25
Tetraeodon nigroviridis 38
tetrahydrofuran 320
thrombocyte (TRB) count 188
thrombocytes 184, 185, 189, 190, 192, 193, 195, 196
Thrombocytopenia 194, 196
thrombocytosis 196
thyroid 244–253, 261–267, 269, 275, 277–279, 282, 284–286, 289–298
Thyroid Hormones 38
thyroid hormones (THs) 245, 277, 261, 263–267, 288–290
thyroperoxidase (TPO) 262, 263
thyroxine (T4) 262
tissue differentiation 224, 226
Total Index 211
total leukocyte (WBC) count 188
Total Reaction Index 211, 212
toxic equivalent factor (TEF) 23
Toxicology 15, 21, 24, 25, 28, 29
Trahira 25
Traira 25
transcription factors 36, 39
transcriptional activation 22
transcriptomes 29
translocations 135, 144, 157
transmission electron microscopy 322
tributyltin 138, 156
Trichiurus lepturus 25
triiodothyronine (T3) 262
"Trojan horse" effect 327
Tropical Ecosystems 1–14
Tropical Fish 1–377
Tropical Fish Species 7
Tumor 210, 212, 222, 224, 231–233

U

UDP-glucuronic acid 20
UDP-glucuronosyl transferases (UGT) 16, 17, 20, 43
UDP-glucuronosyl-transferase 1A9 42
Ultrastructure 214, 220
undifferentiated cells 234
urban sludge 218, 220–223, 228–230

V

vacuolated 142, 222, 224, 226
vessel changes 209, 231, 237
vessels 213, 227, 233
visual system 65, 72, 356
Vitamin D Receptor 38, 44
vitellogenesis 257, 260
vitellogenin 39, 46, 247, 248, 252, 254, 257, 266–268, 278, 280–287, 289, 291–293, 295–299

W

water acidification 231
Water pollution 207
Weight-length Relationship 341, 343, 346
welfare 339
whitemouth croaker 186, 192, 196, 197

X

xenobiotic 87, 95, 106, 166, 170, 173, 208, 213, 218, 228
xenobiotic response element 40
xenobiotic responsive elements (XRE) 22, 23
xenobiotics 35, 36, 43, 45

Y

yolk protein 248, 254, 257, 266, 267

Z

zebrafish 15, 19
zebrafish (Danio rerio) 15, 39, 44
zona radiata proteins 39

Color Plate Section

Chapter 6

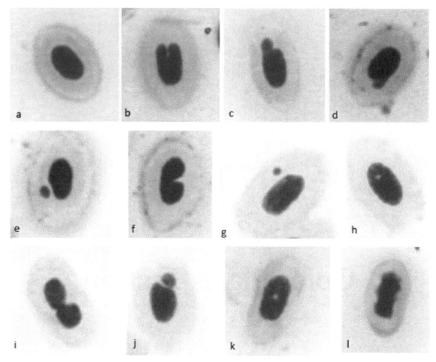

Figure 6.5. Several morphological alterations colored with Giemsa: a) normal; b and f) notched; c, d and j) blebbed; e and g) micronucleus; h and k) vacuolated; i) binucleous; l) lobed. Species *Astyanax fasciatus*. Source: Antonio E.L.M. Marques (Master's student at PPGGEN-UFPR).

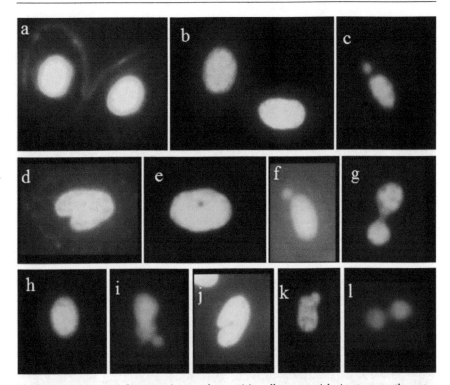

Figure 6.6. Mature erythrocytes (normochromatic), yellow greenish; imature erythrocytes (polychromatic) with reddish cytoplasm: a) erythrocytes normochromatics normals; b and h) erythrocytes polychromatics normals; c, f, i and k) blebbed; d) notched—erythrocyte polychromatic; e) vacuolated; g and l) binucleous; j) notched. Species *Astyanax fasciatus.* Source: Emanuele Cristina Pesente (Master's student at PPGGEN-UFPR).

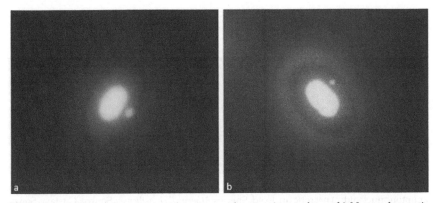

Figure 6.7. a) Normochromatic erythrocytes with one micronucleous; b) Normochromatic erythrocytes with one micronucleous. Species *Astyanax altiparanae.* Source: Galvan 2011 (Master in PPGGEN-UFPR).

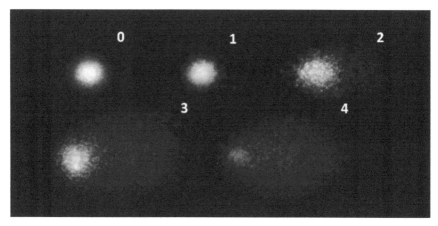

Figure 6.8. Nucleoids with damage levels (or classes) 1, 2, 3 and 4. Species *Astyanax altiparanae*. Source: Galvan 2011 (Master in PPGECO-UFPR).

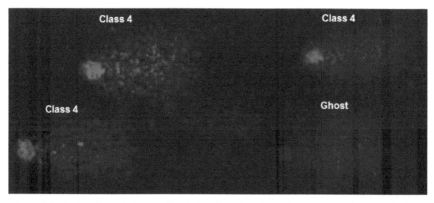

Figure 6.9. Nucleoides wth damage levels (or class) 4 and one ghost. Species *Astyanax fasciatus*. (Ramsdorf 2011—Thesis PPGGEN-UFPR).

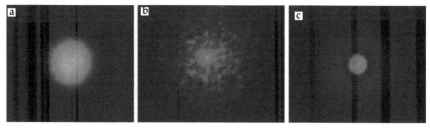

Figure 6.10. Apoptotic and necrotic cells in the DNA diffusion test: a) normal nucleoid; b) apoptotic nucleoid; c) necrotic nucleoid. Species *Atherinella brasiliensis*. Source: Gustavo Souza Santos (PhD student at PPGECO-UFPR).